Solid State Dosimetry

Solid State Dosimetry

Author:

Klaus Becker

Head, Applied and Solid-State
Dosimetry Research
Oak Ridge National Laboratory
Oak Ridge, Tennessee

published by:

18901 Cranwood Parkway, Cleveland, Ohio 44128

To my parents, wife, and sons

THE AUTHOR

Klaus Becker is Head of the Applied Dosimetry Research Group, Health Physics Division, Oak Ridge National Laboratory, and Professor of Physics at the University of Tennessee in Knoxville.

Dr. Becker received his B.S. and M.S. degrees at the Free University in Berlin and his Ph.D. degree in 1961 at the Technical University of Munich.

Prior to his present appointment, he was in charge of personnel radiation protection and dosimetry research groups at the Nuclear Research Establishment in Jülich, Germany. He has been Scientist-in-Residence at the USNRDL in San Francisco, and served as visiting professor and government advisor for the International Atomic Energy Agency, the World Health Organization, and the Organization of American States in various Asian and Latin American countries. He has been President of the Central European Section of the Health Physics Society and of the German-Swiss Radiation Protection Association and was a member and Vice-Chairman of task groups of the International Commission for Radiation Protection and the International Commission for Radiation Units and Measurements. He was also Principal Investigator for the NASA moon rock program, organized and chaired several international conferences, and has been the editor of a series of bibliographies for the German Nuclear Documentation Center.

Dr. Becker has authored or co-authored more than 100 papers and holds several patents. A book that he wrote on film dosimetry has been published in German, English, and Russian editions. His current research interests are biophysics, the development of new techniques in solid-state dosimetry, and their applications in radiation research and protection.

PREFACE

Ever since the discovery of ionizing radiation, the development of radiation detectors has played a key role in promoting our understanding of the nature of radiation and its interactions with matter. It is essential for most industrial, medical, or research uses of ionizing radiation, and it enhances our awareness of the benefits as well as the hazards implicit in the use of radiation.

The inhabitants of certain areas in Colorado, India, and Brazil, uranium miners, astronauts, and the crews and passengers of high-flying planes are among those people who may be exposed to high *natural* radiation levels. More important, an increasingly large fraction of the population is exposed to *artificial* sources of radiation in medical diagnostics and therapy, in or around nuclear power plants and research installations, or during industrial uses of radiation sources — in short, wherever radiation is used or occurs as a regular or potential by-product of technological processes. Furthermore, the vast arsenal of nuclear weapons that are produced and stored in various parts of the world makes it necessary to consider exposure of large fractions of armed forces and/or civil populations to high radiation levels in cases of accidents or nuclear warfare.

Personnel radiation dosimeters are now prescribed by law in all advanced countries for the continuous monitoring of persons who are engaged in radiation work. Millions of such dosimeters are in use, and this number will probably rise with the increasing use of radiation, as well as the increasing concern about the potential hazards even of very low radiation doses. However, personnel dosimetry is only one of the many applications of dosimeters. Another area of increasing importance is medical dosimetry, such as the measurement of depth-dose distributions, the calibration of radiation sources, and in vivo dosimetry in "hospital physics" (dose deviations of as little as $\pm10\%$ can, for example, dramatically influence the success rate in the therapy of certain types of cancer).

Integrating dosimeters are also widely used in biomedical and materials research, for space radiation studies, in food processing, radiation sterilization of medical instruments and pharmaceuticals, in radiation genetics (plant breeding), for the mapping of radiation fields in or around reactors,

accelerators, and in nuclear weapons research, to name only a few other applications. The precise long-term measurement of low doses in the environment of nuclear facilities may be related to economic, ecological, and energy policy problems, such as the siting of nuclear power stations. In other words, reliable, simple measurements of accumulated radiation exposures are essential for technical, medical, administrative, legal, and psychological reasons.

For the first half of this century, radiation dosimetry has been almost synonymous with types of detectors which are based either on the ionization of gases or on the darkening of photographic emulsions. With the increasing need for more sensitive, more accurate, and less expensive dosimeters, this situation is presently undergoing a dramatic change. Modern integrating solid-state dosimeters that are based on the measurement of subtle changes occurring in certain solids as a result of irradiation are in an increasing number of areas rapidly replacing the "classical" detector systems.

A large quantity of information on solid-state dosimetry is available, reflecting, in approximately 5,000 publications, the accumulated results of almost 10^4 man-years of research and development work (including 12 by the author and another 40 by his associates and graduate students in Jülich and Oak Ridge). Unfortunately, no reasonably complete and up-to-date compilation exists. A new student to this field, or a scientist considering the application of solid-state dosimetry to his particular problem, may find himself confronted with the choice of risking the chance of drowning as a victim of the information explosion, or remaining in a blissful state of semi-ignorance based on reading some outdated papers or textbooks. This book attempts to fill this gap. In compiling it, it was obviously not advisable to give a truly complete review of all the articles, reports, lectures, patents, etc. This would imply the description of many erroneous results, obsolete techniques, and redundant studies which are not even of historical interest. It also would result in a large volume which would be difficult to read and to comprehend. Instead, it is the goal of this book to give an up-to-date introduction into the field of solid-state dosimetry which also can serve as a laboratory manual. Only relevant publications that had been available to the author by mid-1973 could be considered.

It has been pointed out (Loevinger, 1972) that almost all known radiation effects on matter have at least once been suggested as a "simple" method of radiation dosimetry, with the claim of simplicity based largely on the fact that the investigators had not taken the time and effort to discover its particular problems. Other materials such as photographic emulsions and lithium fluoride are among the best-studied materials on earth, and the fact is indisputable that some detectors are much more suitable for certain types of measurements than others. It is hoped that this book will help select the best detector for a given purpose and to indicate areas where further research is needed.

An effort has been made not only to present the most up-to-date information (which is crucial in such a rapidly developing field), but to look ahead in certain areas and anticipate some future trends and developments. In selecting references, usually the more recent, more comprehensive, and most easily accessible ones have been chosen. The content of some chapters is related to previous reviews by the same author on film dosimetry (Becker, 1966a),* radiophotoluminescence dosimetry (Becker, 1967 and 1969a), exoelectron dosimetry (Becker, 1970a and 1972) and nuclear track etching in solids (Becker 1972), but does not duplicate their content and/or purpose.

Some use has also been made of several other fairly recent reviews on related areas which are scattered in the literature, in particular those by Dudley (1966), Becker (1966a), Herz (1969), and McLaughlin (1970) on film dosimetry; Spurny (1965), Fowler and Attix (1966), Cameron et al. (1968), and Schwarz et al. (1968) on thermoluminescence dosimetry; Becker (1967), IAEA (1970a) and Piesch (1972) on radiophotoluminescence; and by Frank and Stolz (1969), Amelinckx et al. (1969); and Becker (1970b) on solid-state dosimetry in general, as well as of a series of bibliographies on these subjects (Angino et al. 1965; IAEA 1966; Becker 1966b and 1969b, and 1964 to 1971; Lin and Cameron, 1968; Spurny, 1967 and 1969; and Spurny and Sulcova, 1973). Recommended sources for additional references are the proceedings of a number of international meetings, e.g., the First, Second, and Third International Conference on Luminescence Dosimetry (Lumi-

nescence Dosimetry, 1967, 1968, and 1972); the Track Etching Conference in Clermont-Ferrand (Isabelle and Monnin, 1969); several IAEA Symposia (IAEA, 1965, 1967, 1969, 1971, and 1972), and technical reports (IAEA, 1970a and b).

Pulse-type ("active") solid-state radiation detectors and spectrometers (scintillators, semiconductor detectors, etc.) are not covered. However, a brief description of photographic emulsions (which have to be regarded as solid-state dosimeters) is given even if this field will not be treated exhaustively — partly because this has been done previously and partly because its importance is declining. Neutron activation is also omitted because the frequently rapid decay of the induced activity excludes such detectors from the category of truly integrating techniques. In general, emphasis is placed not on reviewing older, well-known techniques, but on promising modern methods, especially those which have not yet been treated in book form.

The book is intended for the rather large audience of all those who are sufficiently interested in radiation effects and radiation dosimetry and desire more than the rather superficial, frequently outdated information on solid-state dosimetry found in conventional textbooks, handbooks, and encyclopedias. Radiologists, public health officials, practical and research health physicists, medical physicists, radiation biologists, and perhaps even some solid-state physicists and radiation chemists may find it helpful, although the treatment of the more fundamental aspects of the various detector systems has been kept to a minimum in order to cover the practical aspects more adequately.

The support and comments of many friends and colleagues helped to improve the manuscript: Miss A. Lawrence typed part of it; F. H. Attix, J. A. Auxier, M. Ehrlich, R. B. Gammage, F. F. Haywood, J. E. McLaughlin, E. Piesch, R. L. Shoup, W. S. Snyder, and J. E. Turner are among those who suggested improvements in one or more chapters. The author is particularly grateful to his wife, Gisela, for her valuable assistance in typing, proofreading, and the preparation of the bibliographies and index. Comments and suggestions from the readers concerning improvements for possible future editions of this book are appreciated by the author.

*The references in the Preface are given in the bibliography at the end of Chapter 1.

TABLE OF CONTENTS

1. INTRODUCTION

1-1. History

Even medieval alchemists knew that certain minerals, such as fluorite, when heated in darkness, exhibited a transient glow. On October 28, 1663, Robert Boyle reported to the Royal Society in London observations of a strange "glimmering light" when he warmed a diamond in the dark of his bedroom. As early as 1895, the underlying physical process of thermal release of stored, radiation-induced luminescence (thermoluminescence) was used for the detection of ionizing radiation (Wiedemann and Schmidt, 1895). Solid-state dosimetry, i.e., the integrating measurement of directly or indirectly ionizing radiation by means of radiation-induced changes in inorganic or organic crystals, glasses, and polymers, is, therefore, by no means new.

Early observations of photographic effects of ionizing radiation had been reported in 1842 by Moser and 1867 by de St. Victor, long before Röentgen used photographic plates for quantitative radiation measurements beginning in 1895. The principle of radiophotoluminescence, e.g., the permanent alternation of the ultraviolet-excited luminescence spectrum of numerous inorganic compounds after exposure to beta and gamma radiation, was discovered in 1912 (Goldstein). Exoelectron emission had already been studied in 1902 (McLennan), 55 years before it was first suggested as a possible means of dosimetry (Kramer, 1957 and Gourgé and Hanle, 1957). Only nuclear track registration in solids by etching, first described in 1958 (Young), is a relatively late development.

During the first few decades after the discovery of x-rays and radioactivity, research on and the practical use of radiation (mostly in medicine) were severely restricted by the lack of sufficiently simple, sensitive, and precise techniques for quantitative measurements. This was the period of the "pastille dose" (defined as the amount of radiation required to change the color of a pressed barium platinocyanide tablet from apple-green to red-brown) or the "skin erytherma dose" (that dose which reddens the skin in 80% of the cases within three weeks). From the 1930's into the 1960's, ionization chambers and photographic films dominated the field of integrating radiation dosimetry. Many investigators, beginning between 1914 and 1926, studied the effect of x-rays on photographic emulsions and used them for such purposes as depth-dose studies, determination of isodose patterns, and scattered radiation fields (for a brief historical review, see Becker, 1966a). In 1929, Eggert and Luft designed the first film badge with metal filters of different thicknesses. In the following year, film dosimeters with filters made from different metals were first used (Bouwers and van der Tuuk, 1930).

The basic design, although much varied in minor detail, remains unchanged to the present day, when badges of this type are regularly worn by about a million radiation workers in more than 50 countries. The only significant improvements were the addition of a pair of suitable metal filters to permit crude thermal neutron measurements (Dessauer and Lennox, 1944), and the use of fine-grain "nuclear track" emulsions for fast-neutron dosimetry (Cheka, 1944).

With the rapidly increasing use of radiation sources and reactors in the civilian as well as the military fields after World War II, it was realized that the capabilities of film as a large-scale, long-term dosimeter are rather limited. The main reasons are its inherent problems of a strong energy dependence; pronounced fading at higher temperatures and humidities; poor reproducibility; high sensitivity to disturbing agents such as light, pressure, and certain chemicals; limited shelf life, dose-range, and sensitivity; and the need for a rather complex darkroom processing procedure involving many potential sources of error.

An intense search for alternative, sufficiently stable, and sensitive physical effects in solids produced first results in the late 1940's and early 1950's. Radiophotoluminescence (RPL) dosimeters based on silver-activated phosphate glasses were developed (Weyl et al., 1949 and Schulmann et al., 1951), which later became the first mass-produced solid-state dosimetry system ($>10^6$ units for use in the U.S. Navy, Schulman et al., 1953). Various investigators in Europe and the U.S. studied the dosimetric potentials of thermoluminescence (TL) in such materials as LiF (Daniels, 1950), $CaSO_4$ (Kossel et al., 1954), and CaF_2 (Ginther and Kirk, 1956).

Tremendous efforts have since been made in all relevant areas, namely, the search for new detection principles and new or improved detector materials; the better understanding of the mechanism and kinetics of the radiation effects in various detector systems; detailed studies of their radiation response characteristics; development of more reliable, sensitive, and simplified readout instrumentation; and application in many different fields of dosimetry, including such seemingly unrelated areas as geological dating and the detection of art forgeries.

A second generation of materials including greatly improved RPL glasses and TL phosphors and a wide variety of commercial systems became available during the 1960's. The new techniques of exoelectron dosimetry and track etching were also explored during this period. The amount of published papers in all these areas is growing rapidly and doubles every few years (Figure 1-1A), making it increasingly difficult for individual scientists, particularly in small institutions and/or developing countries, to keep their knowledge up-to-date. It is estimated that in personnel dosimetry alone, about half of all monitored persons will be provided with solid-state dosimeters before 1975 (Attix, 1972b).

Table 1-1 summarizes some of the radiation effects in solids which have been considered or are presently used for dosimetric purposes. The particularly important or widely used ones that are discussed in this book in more detail are italicized; others that have received relatively little attention thus far (although perhaps of future interest) are given in parentheses. Naturally, there are various other ways of classifying solid-state dosimeters. One would be according to their sensitivity. Another would be to distinguish between reversible detectors, based on trapping phenomena, and semireversible or irreversible effects. One could also classify them based on the evaluation principle (for example, single-event detection vs. "amplification" of the radiation effect by built-in mechanism or "development" processes). Table 1-2 lists some leading research groups in solid-state dosimetry and their main areas of interest. In Table 1-3, the main applications are compiled.

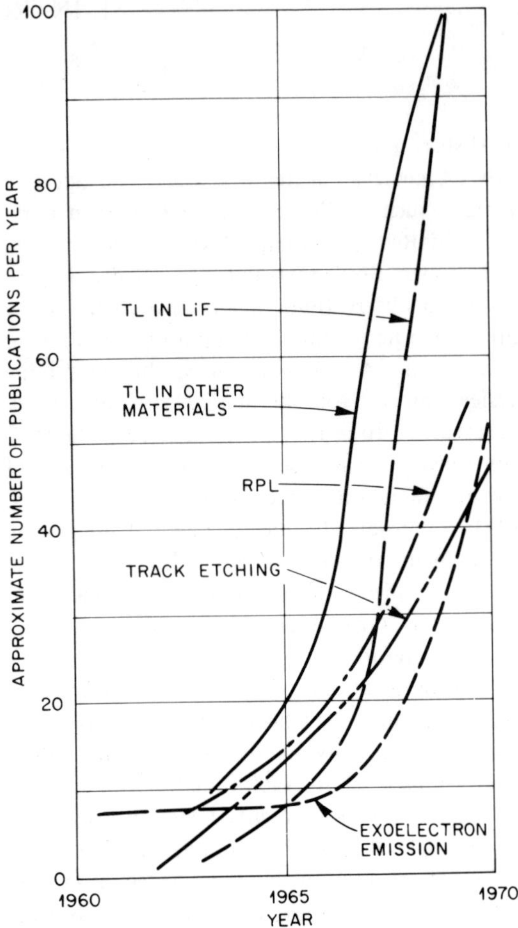

FIGURE 1-1A. Approximate number of publications per year on thermoluminescence (TL) in LiF and in other materials, on radiophotoluminescence (RPL), track etching, and exoelectron emission. (Based on data by Attix, Fleischer, and Becker.)

1-2. Personnel and Accident Dosimetry

It cannot be the purpose of a book on dosimetry to cover adequately the complex subject of justifications for radiation protection programs, of which personnel monitoring is an important part. There are legal, administrative, psychological, and educational reasons for equipping persons with dosimeters, but there should be no doubt that the basic reason is to assess the wearer's external radiation exposure and to help minimize such exposures.*

Many questions treated in Table 1-3 regarding

*According to recent studies (Sagan, 1972), 1 rem of whole-body exposure may result in one day of life shortening. Depending on the average monetary value of a human life (current estimates vary around $400,000), this would imply a "cost" of $30 to 250 per man-rem. In the U.S. less than 500 man-rem/year are presently due to power reactors, but $\sim 2 \times 10^7$ are due to diagnostic x-rays.

TABLE 1-1

Some Radiation Effects in Solids of Dosimetric Interest

Type of radiation effect

Type of solid	Optical	Electrical	Chemical	Localized structural damage
Inorganic crystals, minerals	Coloration or discoloration; changes in optical activity; *thermoluminescence;* IR-Stimulated luminescence; radiophotoluminescence	*Optically or thermally stimulated exoelectron emission* (thermally stimulated current or conductivity)	*Photographic effects in silver halides*	Preferential etchability along heavy charged particle tracks; *nuclear track photography;* decoration by Ag depostion along particle tracks
Semiconductors		*Neutron-induced permanent resistance changes*		
Inorganic glasses	*Coloration or discoloration, thermoluminescence; radiophotoluminescence*	Optically or thermally stimulated exoelectron emission		Preferential etchability along heavy charged particle tracks
Organic crystals and polymers	*Coloration or discoloration,* fluorescence degradation; (thermoluminescence, triplet exciton annihilation fluorescence changes)	Electron spin resonance; (optically stimulated exoelectron emission, thermostimulated current)	(Increased bulk etching rate)	*Preferential etchability along nuclear tracks;* (optical amplification of radiation damage along nuclear tracks)

dose-range, desirable accuracy, and especially the choice of the best detector for a particular application will be discussed in some detail in later chapters. Nevertheless, there are a number of factors related to the use of all integrating radiation dosimeters, for example, in personnel dosimetry, which should be mentioned here. However, this is not the place to attempt another summary of the theoretical fundamentals of radiation physics, medical physics, radiation biology, and dosimetry, or to define units and such commonly used terms as photon energy dependence, neutron flux, self-shielding, RBE, oxygen effect, LET, etc. The unfamiliar reader is referred to the more recent reports of the International Commissions for Radiation Units and Measurements (ICRU) and for Radiation Protection (ICRP) and books on the subject (Fitzgerald et al., 1967; Young, 1967; Morgan and Turner, 1967; Attix and Roesch, 1968; Johns and Cunningham, 1969; Cember, 1969; Fitzgerald, 1969–70; Hendee, 1970; Hurst and Turner, 1970; Nachtigall, 1971; Meredith and Massey, 1972;

Attix, 1972a; Selman, 1972; United Nations, 1972; Willis and Handloser, 1973; etc.).

If any dosimeter is attached to the surface (or introduced into the interior) of an irradiated sample, it will at best provide information on the dose at the location of the dosimeter, and not throughout the sample. Whenever the spatial dose distribution in the object of interest is more complex, more than one dosimeter and/or the application of correction factors are required. For an example of particular importance, consider the case of a personnel dosimeter attached to the front surface of the human trunk. Only in rare cases, such as isotropic exposure of a person to 1 to 2 MeV gamma radiation, may it be assumed that the indication of the dosimeter represents the actual critical organ or average (mid-line) whole-body dose within less than ±50%. (For some recent depth dose calculations in phantoms, see Jones et al., 1973; ICRU, 1973; Frigerio et al., 1972 and 1973; and Frigerio and Coley, 1973.) In the majority of personnel exposures, the situation is more complicated. The two main factors affecting

3

TABLE 1-2

Some Solid-state Dosimetry Research Groups

Country	Institute	Principal investigators	Major fields of activity*	Remarks
Austria	IAEA, Vienna	Y. Nishiwaki W. Köhler	TL, RPL, Track etching	
Belgium	MBLE, Brussels	R. Schayes M. Lheureux	TL	∿1962–1967
Brazil	Institute of Atomic Energy, Sao Paulo	S. Watanabe M. Mayhugh	TL	Since ∿1971
Canada	Atomic Energy of Canada Chalk River	W.G. Cross A.R. Jones	Track etching TL	
Denmark	Danish AEC, Risö Roskilde	V. Mejdahl L. Bötter-Jensen P. Christensen	TL	
England	Hammersmith Hospital, London Res. Lab. for Archeology Oxford University, Oxford Natl. Radiolog. Protect. Board, Glasgow/Scotland	J.F. Fowler D.K. Bewley M. Aitken S.J. Fleming E.W. Mason, G.S. Linsley	TL TL TL	
Finland	Instit. Radiat. Physics, Helsinki	M. Toivonen	RPL	
France	Sect. Dosimétrie, CEA, Fonteney-aux-Roses University of Toulouse	H. François G. Portal D. Blanc	RPL, TL RPL	
Germany, West	Kernforschungszentrum, Karlsruhe	E. Piesch H. Kiefer R. Maushart	RPL Track etching	Since ∿1961
	Institute of Physics, University of Giessen Institute für Strahlenschutz Neuherberg b. München Physikal.-Techn. Bundesanstalt Braunschweig and Berlin	W. Hanle A. Scharmann G. Burger D.F. Regulla J. Kramer G. Holzapfel	TL TSEE TL, RPL, TSEE Track etching TSEE, OSEE	Since ∿1956 Since ∿1967 Since ∿1956
	KFA Jülich	K. Becker M. Heinzelmann	RPL, Track etching	
Germany, East	Inst. Anwend. Radioakt. Isotope, T. U. Dresden	M. Frank W. Stolz	TL, Track etching Foil discoloration	
Hungary	Institute Nucl. Res., Hungary Academy of Sciences, Debrecen	L. Medveczky G. Somogyi	Track etching	
India	Bhabha Atomic Research Centre Trombay, Bombay	C.M. Sunta	TL	
Israel	Soreq Nuclear Research Centre Yavne	Y. Feige	TL, RPL	

*RPL = Radiophotoluminescence
TL   = Thermoluminescence
TSEE = Thermally stimulated exoelectron emission
OSEE = Optically SEE

TABLE 1-2 (Continued)

Some Solid-state Dosimetry Research Groups

Country	Institute	Principal investigators	Major fields of activity*	Remarks
Italy	Dosimetry Laboratory, CNEN Rome	G. Scarpa E. Rotondi	TL, TSEE	
Japan	Toshiba Research Laboratory Kowasaki	R. Yokota	RPL	Since ∿1960
	Dai Nippon Toryo Co; Kanagawa Ken	H. Sakamoto	TL	
	Matsushita Electric Ind., Osaka	T. Yamashita	TL	Since ∿1966
	Division of Physics, National Institute Rad. Health Chiba	T. Nakajima	TL	
Netherlands	Euratom-Ital Wageningen	K.J. Puite	TL, TSEE	
Poland	Institute Nuclear Physics, Cracow	T. Niewiadomski	TL, TSEE	
Soviet Union	State Com. Use Nucl. Energy, Moscow	V.V. Kuzmin A.I. Beskorskii	TL TSEE	
Sweden	Department Rad. Physics, University of Lund	C. Carlsson	TL, TSEE	
Switzerland	CERN, Geneva	J. Dutrannois J. W. N. Tuyn	Track etching	
U.S.	Natl. Bureau of Standards Washington, D.C.	M. Ehrlich	Film, TL	
	U.S. Naval Research Laboratory Washington, D.C.	J.H. Schulman F.H. Attix R.J. Ginther V.H. Ritz	RPL, TL TSEE	Since 1948
	Health Physics Division Oak Ridge National Laboratory Oak Ridge, Tenn.	J.S. Cheka K. Becker R.B. Gammage D.R. Johnson	Track etching, Films, RPL, TSEE, TL	Since 1944
	Department of Radiology and Physics, University of Wisconsin Madison, Wisc.	J.R. Cameron N. Suntharalingam P.R. Moran D.W. Zimmerman	TL	Since 1960
	U.S. Naval Radiolog. Defense Laboratory San Francisco, Calif.	E. Tochilin N. Goldstein E. Benton	TL, Track etching	∿1964–1968
	Teledyne/Isotopes, Westwood, N.J.		TL	

*RPL = Radiophotoluminescence
TL = Thermoluminescence
TSEE = Thermally stimulated exoelectron emission
OSEE = Optically SEE

5

Main Areas of Application for Solid-state Dosimeters

Application	Approximate dose range (rad)	Main types of radiation	Minimum desirable accuracy (±%)
Personnel dosimetry			
a. Routine monitoring	10^{-2} to 10^{2}	X, γ, β, n	10-20
b. Radiation accident	10 to 3×10^{3}	γ, n	$\sim$10
c. Military, civil defense	10 to 3×10^{3}	γ, n (β)	$\sim$20
Clinical dosimetry	10 to 10^{4}	X, γ, e^{-}, n	$\sim$ 5
Environmental monitoring	10^{-4} to 1	γ	5-10
Radiobiology	10 to 10^{6}	X, γ, n	5-10
Radiation chemistry and technology (food processing, material testing, radio-sterilization, etc.)	10^{4} to $>10^{7}$	X, γ, e^{-}	10
Reactor dosimetry	10^{3} to 10^{9} (10^{10} to 10^{19} n/cm^{2})	γ, n, (β)	3-5

the relation between observed dose and the biologically significant dose to the bone marrow, gonads, eye lenses, etc. are absorption and build-up and backscatter from the body.

As can be seen in Figure 1-1B, for 1 rad absorbed dose at 10 cm depth in a tank of water, an absorbed dose of $\sim$1,000 rad would be found at the surface for 20 keV x-rays, 30 rad for 30 keV, 2 rad for 100 keV, and 170 mrad for 50 MeV photons. If broad-beam geometry is replaced by narrow-beam and/or a reduced source-skin distance is used, the depth-dose distribution will also change (for detailed high energy depth-dose information, see Webster and Tsien, 1965). At photon energies above several MeV, the maximum dose is observed at an increasing depth in the body (Figure 1-2). Around 35 MeV corresponding to an approximately 70 MV x-ray beam (source-skin distance 1 m), maximum ionization occurs 10 cm from the body surface. At low x-ray energies on the other hand, a considerable dose reduction is observed in thin tissue layers (Figure 1-3).

Depending on photon energy, the dose at the body surface is up to 50% (at 70 keV) higher than at the same location in the absence of the body (Figure 1-4). This difference decreases with increasing distance between the detector and the body, but may still amount to 20% for a distance of 5 cm (Figure 1-5).

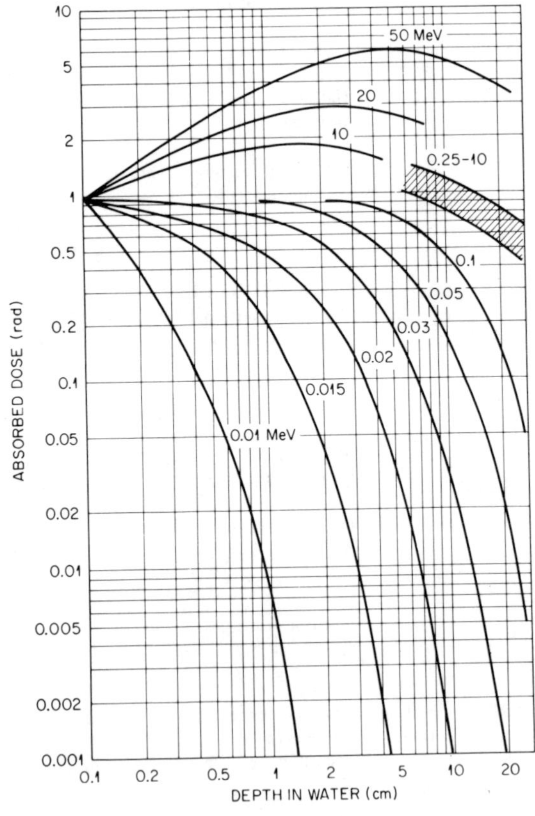

FIGURE 1-1B. Absorbed dose per unit surface dose (rad) as a function of depth in water, calculated for different photon energies. (After Joffre, 1963.)

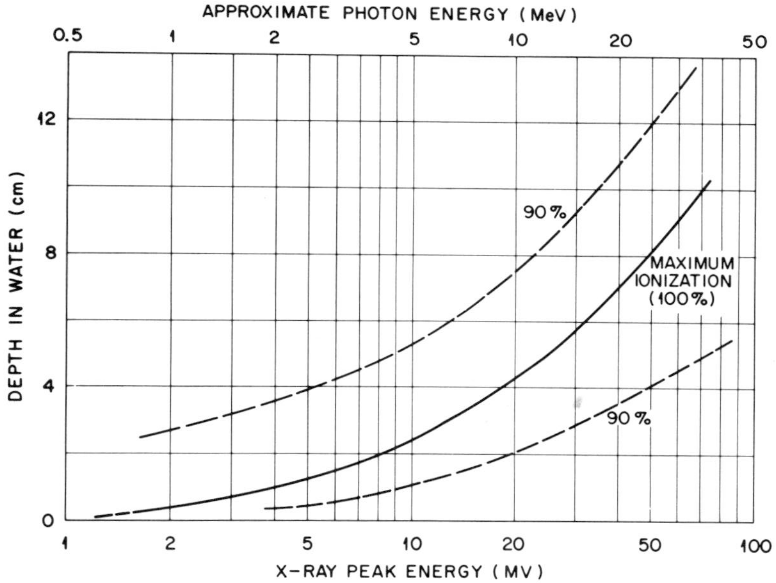

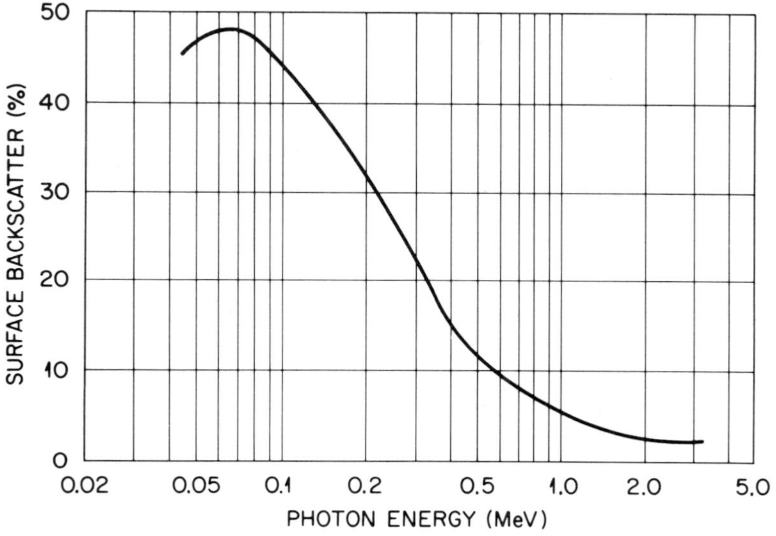

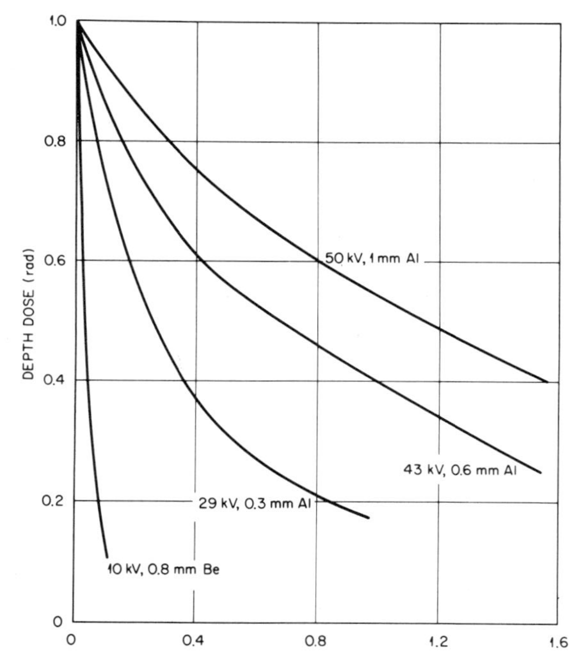

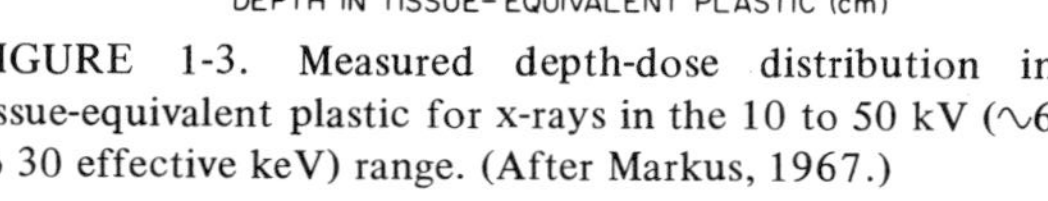

FIGURE 1-3. Measured depth-dose distribution in tissue-equivalent plastic for x-rays in the 10 to 50 kV ($\sim$6 to 30 effective keV) range. (After Markus, 1967.)

FIGURE 1-2 (left top). Estimated position of maximum ionization (100% dose) and 90% ionization in a water phantom as a function of x-ray peak voltage (lower scale), and approximate photon energy. (After Webster and Tsien, 1965.)

FIGURE 1-4 (left bottom). Measured backscatter of radiation at the surface of a phantom in percent of the dose measured in the absence of a phantom, as a function of photon energy. (After Delafield, 1963.)

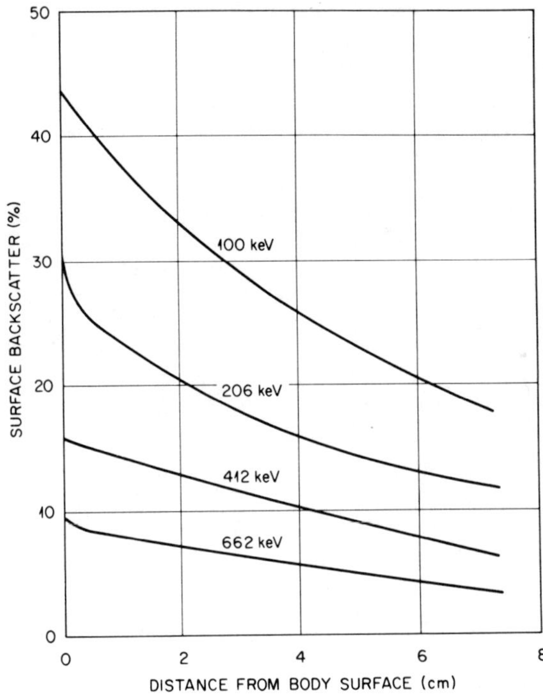

FIGURE 1-5.   Surface backscatter as a function of the distance from a phantom at various photon energies. (After Delafield, 1963.)

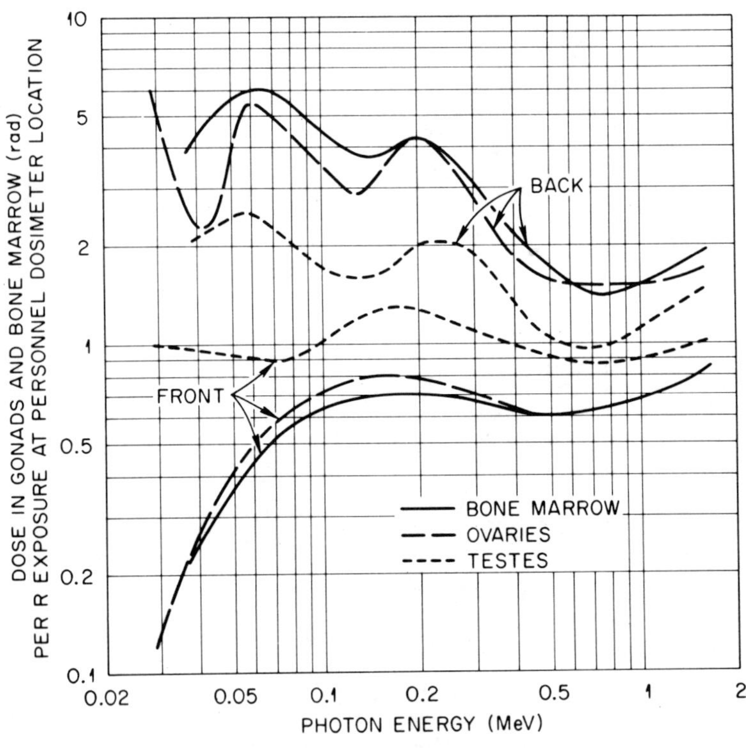

FIGURE 1-6.   Dose in male and female gonads and in bone marrow (rad) for 1 R exposure at the location of a personnel dosimeter, located at the front of the trunk, as a function of photon energy. (After Jones, 1966.)

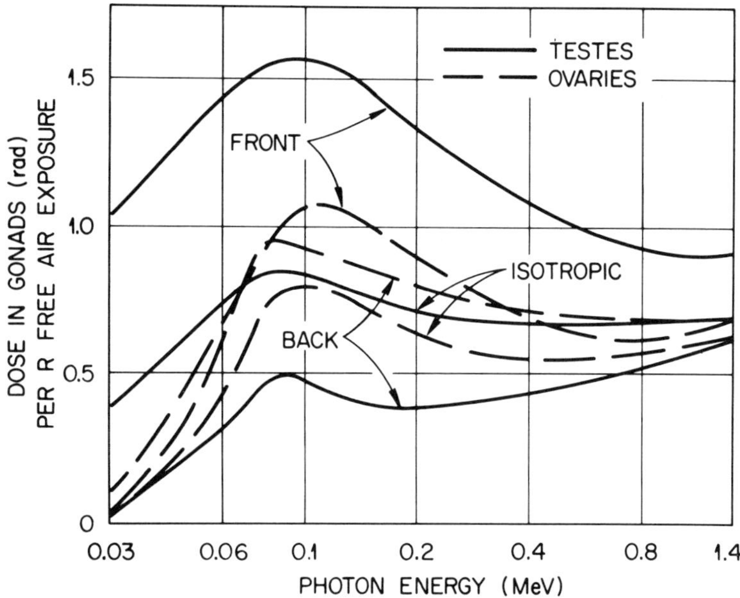

FIGURE 1-7.   Dose in human testes and ovaries in rad per R free-air exposure as a function of photon energy for frontal, isotropic, and irradiation from the back. (After Jones, 1966.)

These effects result in considerable differences between surface (or free air) dose and organ dose. If the body is exposed to broad-beam frontal radiation from a distant 50 keV x-ray source (corresponding to a tube voltage of approximately 80 kV and medium filtration) and the recorded surface dose is 1 rad, the actual average dose to the bone marrow and to the ovaries would be less than 0.4 rad. If irradiated from the back, the actual bone marrow dose would be 5.5 rad, the ovaries dose approximately 3 rad, and the testes dose 2.5 rad (Figure 1-6). If the exposure measurement is performed in free air in the absence of the body, 1 R of 50 keV radiation would correspond to 1.3 rad in the testes for frontal, 0.7 rad for isotropic, and 0.25 for irradiation from the back (Figure 1-7). The results of calculations for monoenergetic external photon sources of different energies (Bennett, 1970 and Jones et al., 1973) agree reasonably well with the experimental data. Figure 1-8 presents typical depth-dose distributions for photons incident unilaterally on one (small) side of.the chest at different heights.

To compensate for some of these errors, correction factors for different instrument types have been proposed (Figure 1-9A). For very high accidental radiation levels (blood-forming organs and gut mucosa taken as critical organs), the correction factors are different from those required for low dose levels of possible genetic interest. Within limitations, it is possible to design

personnel dosimeters which have an energy dependence paralleling that of a certain critical organ (Piesch, 1972a and b) and/or use two dosimeters (one to be worn on the front, the other one on the back of the trunk) for a better estimation of the effective organ (mostly gonad) dose (Figure 1-9B).

For neutrons, the situation is also complex. In Figure 1-10, the calculated dose equivalent in rem for fast neutrons (based on the ICRP values for the quality factor) is given as a function of depth along a line parallel to the direction of a broad beam of incident neutrons in the central part of a cylindrical phantom of 30 cm diameter and 60 cm height. The dose contribution due to (n,γ) reactions is considered. Such values may be used for an estimation of the ratio between free-field or surface neutron dose determination and actual organ, or total body dose (Auxier et al. in Attix and Roesch, 1968).

Deviations of those values may occur due to a variety of reasons, for example, if the distance between neutron source and body becomes small, or the broad neutron beam is replaced by a narrow beam, or if the neutron incidence is not normal to the surface. As an example, Figure 1-11 presents the reduction of the multicollision absorbed dose of a neutron beam through slabs of polyethylene or water for different angles of neutron incidence. (For further, more detailed data and references, see NCRP Report No. 38 (1971) and the recent

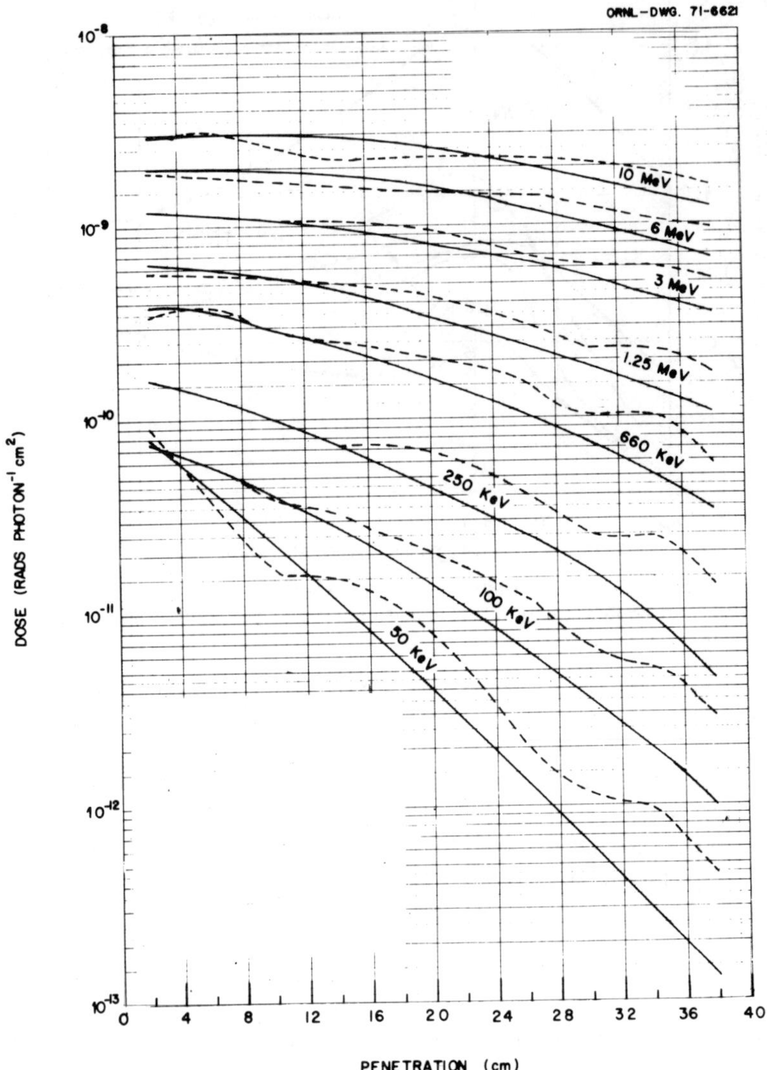

FIGURE 1-8. Dose from a broad, parallel beam of monoenergetic photons incident unilaterally on one side of the trunk of a standard 70 kg human, calculated for the pelvis (solid lines) and the middle of the lung (dashed lines). (After Jones, 1973.)

literature on neutron shielding.) Making some simplifying assumptions, the maximum absorbed dose in the human body as a function of neutron energy can, for the cases of a normally incident broad neutron beam, for isotropic incidence, and for situations with partially normal and partially isotropic incidence be represented by a "strip," as indicated in Figure 1-12.

Only very little general advice can be given regarding the organization of a personnel monitoring service. A classification of the personnel in the installations to be monitored appears to be desirable:

1. Only the rather small group of "high-risk" persons likely to work with radiation sources or radioactive materials (usually less than 5 to 10% of the total employees in a nuclear or medical installation) should be equipped with the best (which usually means most complex and expensive) dosimeters adjusted to the types of radiation they are encountering, with relatively frequent (biweekly or monthly) as well as accumulative annual readings.

2. The larger fraction of "low-risk" personnel (persons unlikely to be routinely exposed to radiation, but potentially exposed in an

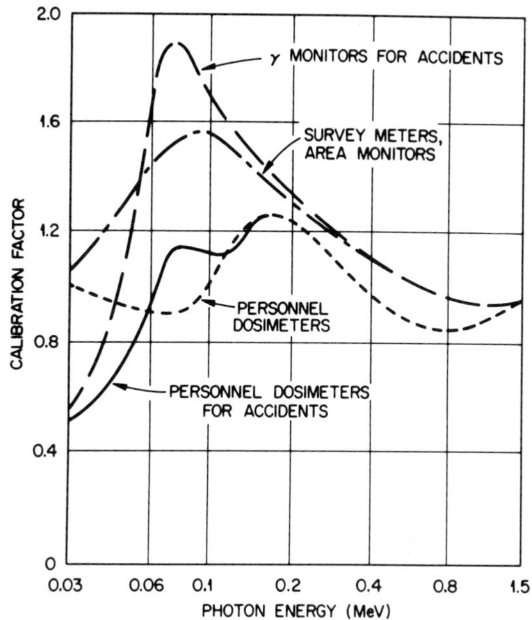

FIGURE 1-9A. Proposed calibration factors for different types of dosimeters and "routine" as well as "accident" dose levels (with the gonads and the bone marrow being the critical organ, respectively) as a function of photon energy. (After Jones, 1966.)

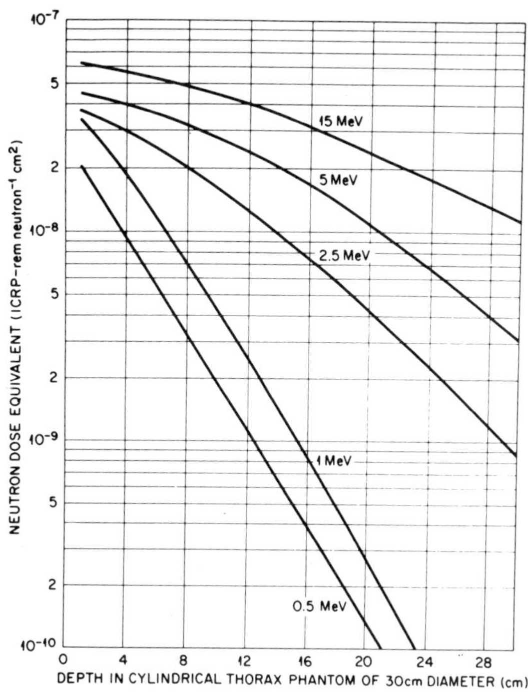

FIGURE 1-10. Calculated dose equivalent (rem) as a function of depth in a cylindrical thorax phantom for different neutron energies. (After Auxier et al., in Attix and Roesch, 1968.)

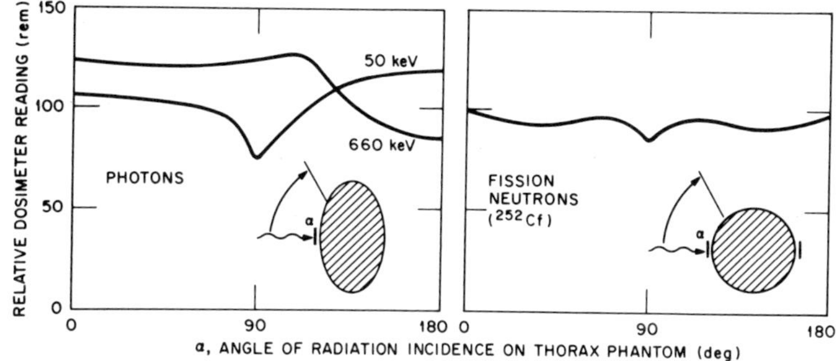

FIGURE 1-9B. Directional response of asymmetrically shielded phosphate glass dosimeters (see Figure 4-28) for photon measurements (left), and of ^{6}LiF/^{7}LiF albedo dosimeter pairs in boron shields to be worn at front and back of the chest for fission neutron measurements, right (after Picseh, 1972b.)

accident) may be equipped with a quite simple, inexpensive one- or two-detector unit to be read once or twice a year. Even in such simple dosimeters it is, however, desirable to distinguish between penetrating and nonpenetrating radiation. The actual natural radiation background should not be considered a part of the occupational exposure, but artificially increased "background" should, of course, be taken into account.

Much of the data that have been accumulated during the last 30 years on large-scale personnel dosimetry are highly questionable for reasons of insufficient accuracy of the dosimeters and over-simplified interpretation of the results. Now that more sensitive and accurate detectors are available at least for X- and gamma radiation, some guidelines on the desirable accuracy of personnel dosimeters are needed. Unfortunately, the state-

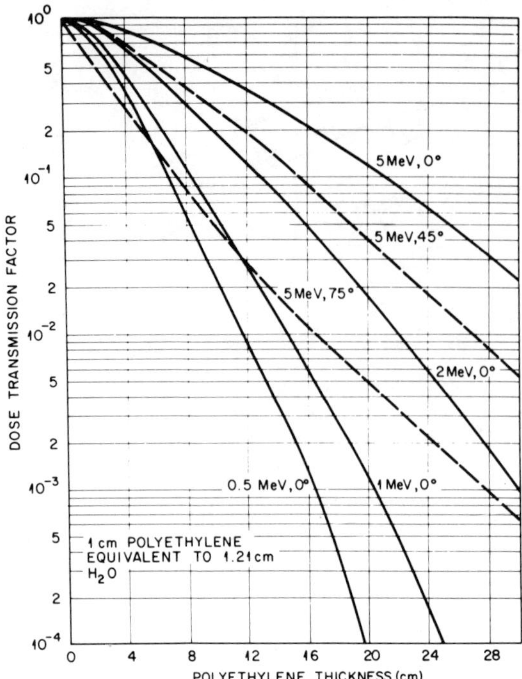

FIGURE 1-11.   Attenuation of absorbed neutron dose in polyethylene or water slabs for different energies and angles of incidence. (After Allen and Futtener, 1963.)

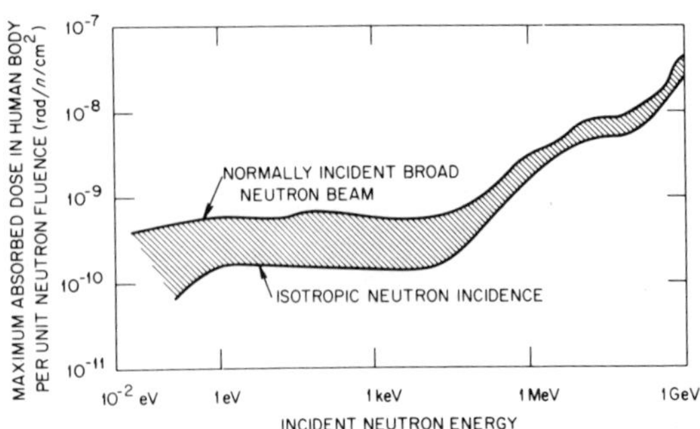

FIGURE 1-12.   Maximum absorbed dose in the human body per unit neutron fluence in rad per n/cm² for a normally incident broad neutron beam and isotropic neutron incidence, as a function of neutron energy. (After Prêtre, 1972.)

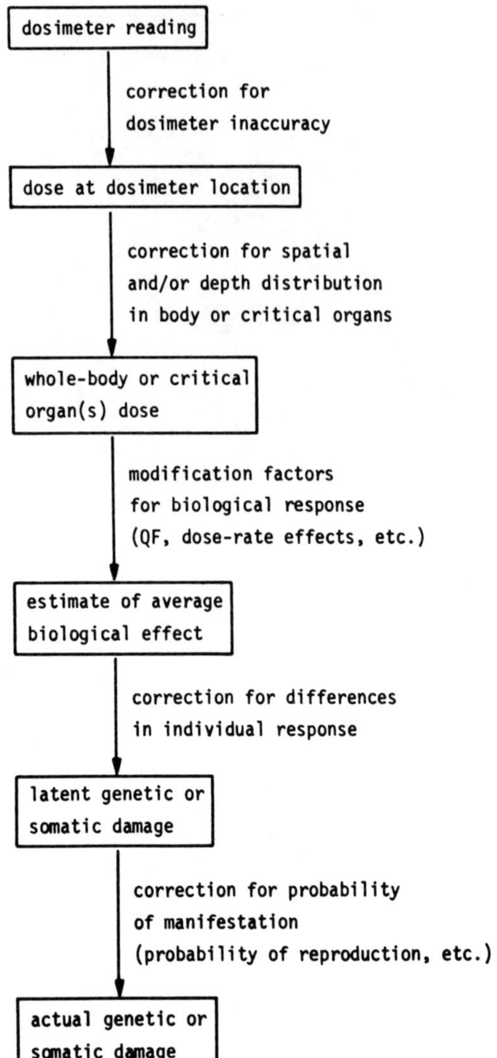

FIGURE 1-13.   Schematical presentation of the relation between personnel dosimeter reading and the actual somatic or genetic damage in an individual. (After Becker, 1972c.)

ments that can be found in more or less official recommendations (ICRP, 1969 and IAEA, 1971) are rather vague, suggesting tolerable uncertainties in the ±25 to 50% range. ICRP Report No. 12 states that "the uncertainty in assessing the annual dose . . . should not exceed 50% of the recorded dose, or 1 rem, whichever is larger."

The relation between the dose reading, which is usually derived from a single dosimeter worn on the front of the trunk, and what one would actually like to know, namely, the degree of somatic or genetic radiation damage in a particular monitored person, is quite complicated (Figure 1-13 presents a schematic summary of this relationship). With the large correction factors involved in the later steps, one might infer that the initial accuracy requirements for the dosimeters could be rather low. Indeed, substantial total errors in dose estimates by a factor of 3 or even 10 in the very low dose range can be tolerated if they are nonsystematic on the assumption that it is likely that such errors will compensate each other during many monitoring periods. Maximum possible accuracy is, however, required in the accident range ($\sim$50 to 1,000 rad) and the accuracy requirements can be relaxed again in the range of absolutely lethal overexposure (Becker, 1972c). The higher accuracy can only be obtained by a careful investigation of the individual case. A reenactment of the actual exposure conditions with a phantom may be the only way to obtain the desired precision in the dose estimate.

There is no universal ideal personnel dosimeter, and there probably will never be one. Usually not all desirable features for a given monitoring situation can be met and priorities have to be set. For example, the difference between high-risk and low-risk persons is reflected not only in the complexity and versatility of the dosimeter and the frequency of change, but also in the goal of the measurement; in the low-risk group, a single number could reflect the dose in the organ of interest as precisely as possible. Such a dosimeter should, therefore, not necessarily attempt an "energy-independent" measurement in the sense of attempting a free-air equivalent dose indication at the location of the dosimeter, but may try to simulate the effective energy response of the shielded critical organs such as the gonads in the routine dose range, and the bone marrow in the accident range. This has, for instance, been attempted with specially designed RPL glass dosimeter encapsulations (Piesch, 1968 and 1972).

On the other hand, the dosimeter for the high-risk category should yield as detailed information as possible on the radiation field. Every piece of information that can help to interpret LET and depth-dose distribution in the body from an accidental exposure, such as information on the types of radiation, their energy and directional distribution, and perhaps even the time distribution (dose rate, fractionation) of the exposure, should be indicated by such a dosimeter.

The extremes of the scale of possible personnel dosimeters are represented by something like a single or double detector which is able to distinguish between penetrating and nonpenetrating (skin) dose, and by an ultrasophisticated dosimeter with a dozen or more components. Where exactly the optimum compromise for a given situation can be found between these extremes depends on a careful consideration of all the parameters involved. Some of the more important ones are listed here.

1. The types of radiation to be measured — In the simplest case, only gamma radiation above a few hundred keV is present. Even in the environment of gamma radiation sources such as ^{60}Co and ^{137}Cs, however, scattered radiation may have a substantially lower energy, and most of the personal dose may be due to radiation with an effective energy in the 50 to 150 keV range. In more complex cases, intermediate neutrons, soft X-radiation, and medium-energy electrons also have to be measured. Thermal neutrons and electrons below a few hundred keV represent an external radiation hazard only in very rare cases and can usually be ignored.

2. Dose range(s) to be covered — Most experts seem to agree that for monthly dosimeter readings, the minimum dose that can be detected with high reliability should not exceed 10 mrad and be preferably in the 1 to 5 mrad range; this number would have to be reduced if the recommendations for permissible external radiation exposures should become more restrictive. For yearly readings, a minimum detectable dose around 50 mrad is sufficient. The upper dose limit should not be less than 1,000 rad, but does not have to exceed 3,000 rad.

3. Reproducibility (accuracy) of readings — In discussions about minimum accuracy requirements for personnel dosimeters, the opinion is sometimes expressed that the main value of such dosimeters is of a psychological and educational nature; they make people feel supervised. The fact of an established radiation exposure of a person can lead to an investigation of the working conditions and the accuracy of the measurement is consequently of minor importance. Most health physicists do not, however, agree with such a lenient point of view and, despite all known limitations, would prefer to consider a personnel dosimeter as a means of reasonably reliable quantitative measurements.

In the past, standards have been tailored to fit the performance which is obtainable with a film badge under idealized laboratory conditions (Barber, 1966). It appears desirable that performance standards should be based instead on the actual needs. Unlike the total discrepancy between dose reading and the dose in the critical organ(s) which may be substantial, the inherent inaccuracy of a personnel dosimeter reading as measured with a free-air gamma radiation exposure should perhaps not exceed ±10% at least for gamma radiation from 1 to 1,000 rad in the energy range ∿0.1 to 2 MeV. This limit will have to be somewhat relaxed in case of long storage times at extreme climatic conditions, in the presence of soft X-radiation, but should stay within approximately ±20% to establish an overall accuracy of about ±30%, as required in accident dosimetry. In performance tests, one should make sure that the tested dosimeters receive no special treatment, but exactly as much (or as little) attention by the evaluation service as the actual dosimeters used by its customers.

4. Speed and convenience of readout — Especially after accidents, it is important for decision-making concerning the medical treatment planning and for psychological reasons that the dosimeters are readable within a few hours. Tamperproof badge design, which is generally desirable, should not interfere with fast readout.

5. Automation — Automation may prove to be more economical and provide more accurate results in very large monitoring services, but should not be overemphasized. It tends to reach an equilibrium point where the costs of the maintenance of the automatic system equal those of a manual system. For example, most electro-optical automatic track counters for photographic nuclear track emulsions never even came close to economic feasibility. Some automatic TLD systems also do not seem to offer great economic or other advantages over well-organized manual or semi-automatic systems.

6. Cost factors — Wages and training level of the staff, available space, amount of mailing and bookkeeping required, restrictions on the purchase of materials and equipment, and many other factors contribute in a complex way to the economics of a personnel monitoring service. There are reports that TLD or RPL glass services can be operated at a substantially lower cost than a comparable film badge service, while others claim that solid-state dosimeters are more costly (they usually require a larger initial investment). As a guideline, the simple version of the badge (low-risk group, visitors) should, including reuseable detector(s), cost not more than $1 to 2, while for the sophisticated (high-risk group) badge $5 to 10 may be a reasonable limit. Only in exceptional cases of very large evaluation centers (more than several thousand readings/month) will the invest-

ment for the evaluation equipment exceed $10,000. In a semiautomatic service including evaluation, data-handling, mailing of the results and new dosimeters, and intercalibration (but not including research and development), three to five qualified technicians can handle about 10^4 readings per month. The price for monitoring low-risk persons can be kept below $5 per year per person, while high-risk personnel may require as much as $25 per year.

7. Location of the dosimeter — For special purposes such as monitoring of the exposure of the fingertips, the hands, gonads, etc., obviously special dosimeters have to be designed, some of which, such as finger-ring detectors, are to be discussed later. In the vast majority of situations in personnel dosimetry, however, a single badge-type dosimeter is worn. The precise location where the dosimeter should be worn has been the subject of some controversy, in particular if a lead-rubber apron is used, but the majority of experts agree that the dosimeter should in this case be worn underneath the apron, preferably on the front side at waist level (Langmead and Farmer, 1971). A constant location is important if a precise interpretation of the reading is to be attempted, as illustrated in Figure 1-14, which shows the

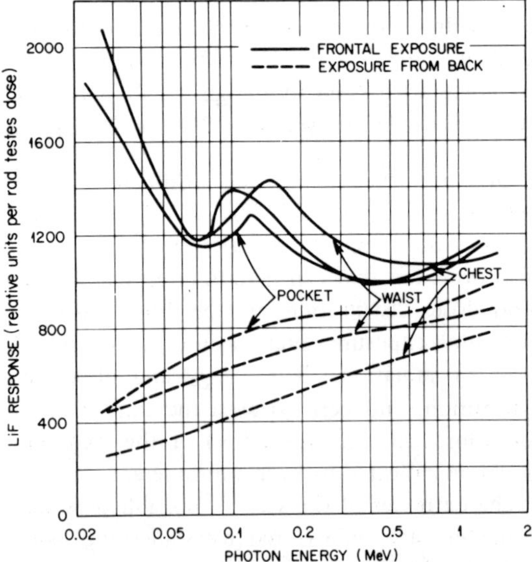

FIGURE 1-14. Response of an energy-independent LiF TLD badge (in relative units per rad of absorbed dose in the tests) as a function of photon energy and direction of radiation incidence, if the badge is worn on the waist, the middle of the chest, and on the side of the chest (shirt pocket); measured with a Rando phantom. (After Ambiger, 1969.)

differences in the relative response of an "energy-independent" LiF dosimeter located on the front of the waist, the middle of the chest, and the side of the chest (shirt pocket) for different photon energies and directions of radiation incidence.

8. **Miscellaneous factors** – Obviously, a personnel dosimeter should not exceed certain size and weight limits (for example, albedo dosimeters for intermediate energy neutrons tend to become rather bulky). About 50 g seem to be the limit. On the other hand, if the dosimeter is also used as a security badge, there is usually a minimum size requirement for picture, name, and number. Credit card-type dosimeters (Nash and Attix, 1971) to be worn in the wallet or a pocket can be considered for the low-risk group, but the front of the trunk between belt and collar should be specified for the high-risk group. More thought should be given to less conventional design possibilities, such as button dosimeters on protective clothing. The design and large-scale use of new dosimeters may be complicated by legal considerations: One may have to prove in court that the methods received "general scientific recognition" for reliability and accuracy (Hart, 1973). It is now generally accepted that proper record keeping does not imply the saving of exposed dosimeters as legal evidence, but that a punch card or recorded glow curve is sufficient (still, the admissibility of computer printouts in court appears to be somewhat questionable). Other factors related to the choice of personnel dosimeters may be of a more psychological nature (monitored persons seem to be more inclined to follow protection rules). A personnel dosimeter should, however, never become exclusively a security badge, a status symbol, an educational tool, or a good luck charm.

The speed with which modern solid-state dosimeters would replace the photographic films has been overestimated by some and underestimated by others. There has been a surprising resistance by many applied health physicists and administrators against any sort of change, often caused by lack of adequate information. For example, nuclear track film for fast neutron measurements may give reasonable results when used for short-term monitoring around a 14 MeV source in a dry climate. If used at a high relative humidity, 50% fading occurs in such a film in one or two days, and after a week no tracks are left. The dramatic effect of humidity on the fading in nuclear track emulsions has been known as long as the NTA film

has been used (see Chapter 6.2), and it requires only a simple test to find out that the common additional sealing of the film only delays the penetration of the humidity by a short time. Yet, NTA films are still widely used even under the most adverse conditions, without adequate protection, and with dangerously misleading results.

Undoubtedly, the photographic film – recently called "the senior citizen of personnel dosimeters" – has some properties which are difficult to match for other types of detectors. From a large quantity of developed films, the few darkened ones can easily be detected by visual scanning. Contamination of the badge and direction of radiation incidence can be determined by visual inspection. The cost of each film is low, in particular if the cutting and packaging of sheets of x-ray film is done by the dosimetry service. The processed films can be kept as further evidence of the radiation exposure, or to impress the careless wearer with his darkened film.

However, film dosimetry has in some monitoring services degenerated into a ritual, performed according to ancient traditions. A steady flow of data from the monitoring service to the customer, which is well suited for producing impressive statistics and reports, was gladly accepted by the customers without questioning the actual meaning and the reliability of the data. Occasional disturbing reports on the results of test exposures and fading experiments were dismissed as nontypical. Despite frequent claims that a new film badge had been "invented" or an old one "improved" (which is a rather easy task if one keeps in mind that the average badge has 4 filters, and about 20 different filter materials can be chosen in an almost unlimited variety of thicknesses and combinations with one or more different films), most radiation protection experts now realize that the potentials of the film dosimeter are essentially exhausted and further research in this area is unlikely to lead to substantial improvements.

In recent years, an increasing number of test reports showed that both thermoluminescence and glass dosimeters clearly were performing more reliably and accurately than the photographic film (see Chapters 2, 4, and 6). Solid-state dosimetry became a serious alternative even in the eyes of the more conservative personnel dosimetrists. The switch to solid-state systems is contemplated by an increasing number of installations, and has been accelerated by such recent developments as the discontinuation of the production of the popular DuPont dosimeter film. A recent survey (Attix, 1972b) was conducted of personnel dosimetry services in different countries with a total of 200,000 monitored persons. It was found that 70,000 of these are already equipped with TLD and it will be used for 40,000 more in the near future (to these numbers, about 10,000 present users of radiophotoluminescent glass dosimeters should be added).

Facing the conflicting statements by the proponents of various systems, in particular the confusing claims of some commercial manufacturers, the potential user's reaction often is a simple postponement of any decision in the hope that newer, simpler, and less costly systems will become available in the near future. In general, there is no doubt that personnel dosimetry has entered a period of transition and the time has come to take a fresh look at some old problems, redefine needs, and review critically the methods available as well as the trends of research and development. It is one of the purposes of this book to provide some assistance in doing this.

The relative merits and handicaps of various nonphotographic methods in personnel dosimetry have been evaluated in many publications (Becker, 1966). The present situation can be summarized as follows: In thermoluminescence dosimetry (TLD), LiF:Mg,Ti (Harshaw TLD-100) with its somewhat complicated response characteristics and the time-consuming annealing procedure got strong competition by equally sensitive and energy-independent systems, such as $Li_2B_4O_7$:Mn and BeO:Na and new types of LiF which are easier to anneal. Highly sensitive and stable phosphors such as Mg_2SiO_4:Tb, $CaSO_4$:Dy, and $CaSO_4$:Tm permit precise, if energy-dependent measurements at dose levels as low as 0.5 mR. A wide choice of sophisticated dosimeters and automatic readers has been designed.

For a number of mostly personal and historic reasons, glass dosimetry has in recent years been overshadowed by TLD, and it still has to suffer, especially in the U.S., from the poor reputation of some early RPL glasses. It is somewhat difficult to predict the future of glass dosimetry, but it should be stressed that impressive progress has been made. New RPL glasses with low inherent energy dependence and higher sensitivity have been developed. Using a new reading principle, cumbersome cleaning may not be required anymore, and a fully automatic reader with data printout is not commercially available. The potential role of advanced glass systems in future dosimetry programs should not be underestimated.

Dosimeters based on thermally stimulated exo-electron emission (TSEE) have not yet, but may become a rather serious competitor for TLD in the future. Some detectors permit dose measurements in the microrad range. Instead of the rather limited temperature range for TLD peaks of $\sim$170 to 250°C, considerably more stable TSEE peaks up to $\sim$550°C can be used for sensitive dose measurements; the fact that the sensitive layer is extremely thin permits some interesting applications, for example, in fast-neutron dosimetry. One sealed TSEE dosimeter is already in mass production in the U.S.S.R., while other interesting types are being developed in the U.S. and Germany (Chapter 3.4).

In the field of track etching, it appears that the presently available choice of systems has been narrowed to only a few alternatives. Either fission-fragment tracks from a well-protected, thin ^{237}Np or (less desirably) ^{232}Th layer are counted by sparking, or use is made of recoil nuclei production in sensitive polymers, preferably with etch pit amplification to simplify evaluation. The latter method avoids the somewhat problematic use of fissile materials in personnel dosimeters.

No general advice can be given with regard to the actual badge design. Which detector or detector combination is selected will depend on the needs, economic factors, and currently available instruments, as well as on the knowledge, ability, ingenuity, experience, and personal philosophy of those who have to make the decision in a particular case. However, it simplifies performance checks and comparisons of data if the number of designs which are in use is kept as small as feasible — perhaps not more than two to three types per country.

There are several special problems of accident dosimetry which deserve brief mentioning. One is personnel dosimetry for critical accidents, involving as a main problem the measurement of a more or less moderated fission neutron spectrum in a mixed neutron and gamma radiation field (IAEA, 1970b). Rather similar are the problems in the design of a "tactical" military dosimeter for personnel considered likely to be exposed to small nuclear weapons with a relatively high contribution of fission neutrons to the total dose. As the

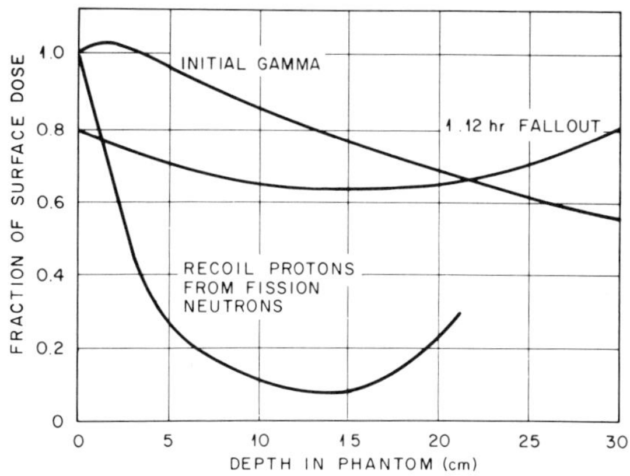

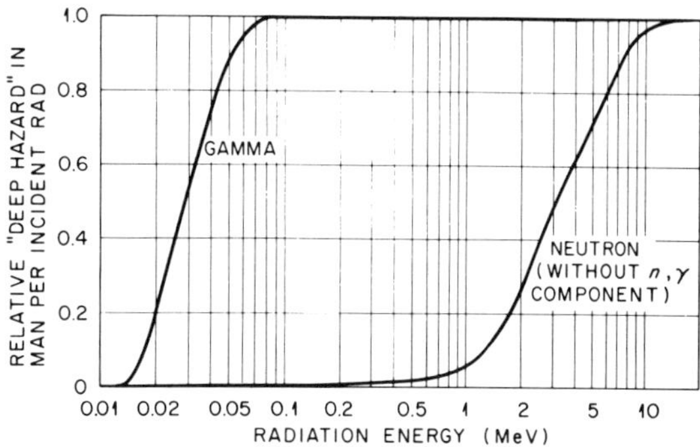

FIGURE 1-15. Approximate depth-dose distribution of the initial gamma and fission neutron radiation (unilateral incidence) from a nuclear explosive of the fission type; and of (isotropically incident) gamma radiation from fresh (1.12 hr) fallout.

FIGURE 1-16. Correction for the depth-dose distribution of photons and neutrons in man as a function of their energy for the crude estimation of the bone marrow (radiation accident) damage.

usual goal of this type of measurement is only to assist the formation of groups among the survivors (for example (1) no immediate biological consequences to be expected, (2) medical treatment advisable, and (3) lethal dose range), accuracy requirements are not very high, but emphasis has to be placed on low price, long shelf life and the possibilities for fast, simple evaluation even under very adverse conditions such as personnel with little training (Becker, 1968).

The same is true for dosimeters which are designed for the civil population in cases of the explosion of large-yield nuclear bombs. However, in this case the problem is simplified by the fact that the weapon's initial radiation can be neglected. As can be seen in Figure 1-15, initial radiation is unilateral and there is a substantial drop, in particular of neutron dose, in the body. On the other hand, in a fairly homogeneous fallout field, the gamma radiation dose is distributed rather uniformly through the body. In a somewhat oversimplified way, the relevant bone marrow dose in radiation accidents can be estimated as a function of photon or neutron energy according to Figure 1-16.

With regard to further research needs in personnel dosimetry, efforts should not be focused on detectors for gamma, x-rays, and high energy beta radiation, and particularly not on detector materials with a high Z. With the wide choice of excellent detectors already available, there is no need for more unless they are really much simpler, less expensive, more stable, and/or more sensitive than the existing one.

On the other hand, there is still no sufficiently reliable dosimeter for intermediate and fast neutrons. The frequently rediscovered "albedo" techniques based on the moderating and backscattering effects of the human body tend to disappear quietly from the scene after awhile, when it is fully realized how much their response depends on such factors as the orientation to the neutron source and the distance from the body surface. A complex hemispherical device to be worn on a belt by workers who are exposed to a moderated fission spectrum appears to be no exception to to this rule (Hoy, 1972).

More promising than differential techniques, where the neutron dose in a mixed field is found as the small difference between two large numbers, and dosimeters based on (n,f) reactions, appear to be solid-state detectors with a high inherent fast neutron sensitivity and the ability to discriminate against low LET radiation. This could be an inorganic detector which is highly sensitive to high LET recoil protons and imbedded in a hydrogeneous matrix, or perhaps organic TLD, RPL, or exoelectron detectors with a high hydrogen content and a favorable LET response. Other possible solutions would be solid-state recoil proton track detectors which are sufficiently fading-resistant, not subject to interference from other types of low LET radiation, and whose evaluation could easily be automated, or more sensitive and stable semiconductors similar to the silicon diodes in which the forward resistance depends on the integrated fast-neutron flux.

Also, some educational efforts are needed. For example, the use of photographic film in tropical countries should be strongly discouraged. In particular, developing countries that are just starting personnel dosimetry programs (many of them have a tropical climate) should be encouraged to begin with a simple, inexpensive nonphotographic solid-state dosimetry system.

During the last 20 or 30 years of large-scale personnel dosimetry, the need for external, physical radiation detectors has frequently been questioned on the basis of great hopes for "biological dosimetry." Intense studies have been carried out on radiation-induced changes in the human body from the point of view of their potential use as a reliable indicator of radiation damage. A good biological dosimeter could indeed have a number of attractive advantages over physical dosimeters: It could not be lost, forgotten or falsified, provide more accurate information when the dose distribution is very uneven, reflect individual differences in radiation response, correctly add the effects of external and internal irradiation, and realistically indicate the degree of repair which has taken place.

Among the many techniques which have been considered are

a. Biochemical indicators. Most of them are based on changes in the chemistry of body fluids which are connected with the degradation and/or inhibition of the biosynthesis of proteins. Some examples are the excretion.in urine of creatine, pseudouridine, deoxycystidine, taurine, etc. (see Proc. IAEA/WHO Panel on Biochemical Indicators of Radiation Injury in Man, Vienna, 1971) and the quenching effect of pre-irradiated blood plasma on liquid scintillators.

b. Changes occurring in the number, size distribution, age, mechanical stability, electrophoretical mobility, etc. of peripherical blood cells, using more or less sophisticated "scoring" matrices as an indicator of the type and degree of radiation damage.

c. Easily detectable physical and chemical changes in body constituents, such as neutron-activation of the whole-body or blood sodium, the sulfur in the hair, or the gold in teeth fillings; EPR signals due to free radicals in tissue and fingernails; thermoluminescence of teeth enamel, etc.

d. Characteristic chromosome aberrations, especially the formation of ring, fragments, and dicentrics in peripherical blood lymphocytes (Figure 1-17; for a recent discussion see Evans, 1971 and 1972).

Each of these indicators can be of some importance for the interpretation of a radiation accident within hours or days after it occurred, but only the chromosome analysis has so far shown potentials as a long-term integrating indicator of biological radiation damage. Unfortunately, it is unlikely that this technique can find any widespread application in routine monitoring, partly because of its poor sensitivity (doses below 10 to 50 rad are difficult or impossible to detect) and the effect of repair mechanisms, but mostly because the evaluation of chromosome aberrations is very complex and time-consuming even in computer-aided systems. In the foreseeable future, it is unlikely that a biological indicator will replace physical dosimeters.

1-3. Medical Physics, Environmental Monitoring, and Other Applications

Many of the more general aspects of the proper choice, use, and interpretation of results of integrating dosimeters mentioned in Section 1-2 also

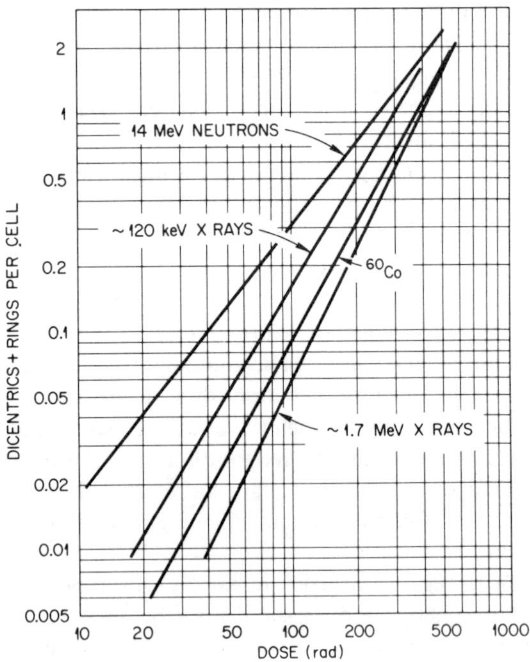

FIGURE 1-17. The yield of two types of chromosome aberrations (dicentrics and rings per cell) in human peripherical blood lymphocytes as a function of dose for different types of radiation. (After Sasaki, 1971.)

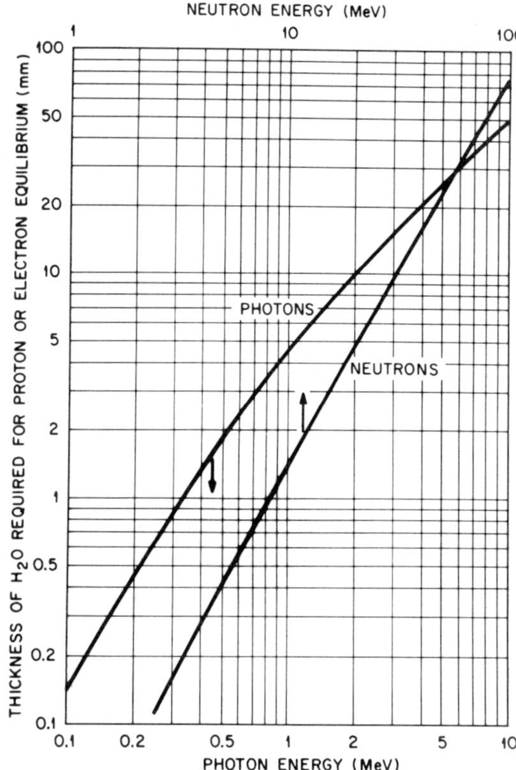

FIGURE 1-18A. Approximate thickness of unit density, hydrogeneous material (water, polyethylene) in front of a thin detector required to establish electron or proton equilibrium for monoenergetic photons and neutrons of various energies. (After ICRU, 1962.)

apply to uses other than in personnel dosimetry. In addition, some other factors may also have to be considered.

With increasing photon energy, an increasing thickness of preferably low Z, tissue or air-equivalent material between radiation source and detector is required to establish electron equilibrium. For photon energies up to a few hundred keV, this is usually accomplished by the encapsulation (or wrapping) of the detector, but for ^{60}Co gamma radiation, ∿4 mm of plastic is required in front of a thin detector. As can be seen in Figure 1-18A, the approximate thickness of unit density material that is necessary to establish complete electron equilibrium for monoenergetic photons rapidly increases with energy. Also given in Figure 1-18A are the corresponding data for fast neutrons.

The basic calibration of solid-state detectors is usually carried out with gamma radiation. The immediate environment of source and detectors should be free from heavy scattering materials. A

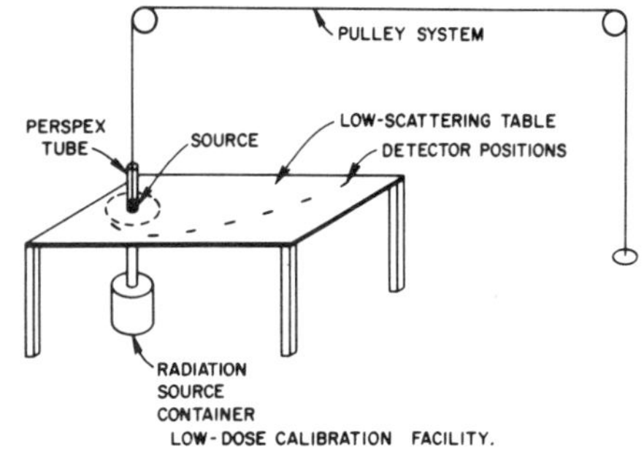

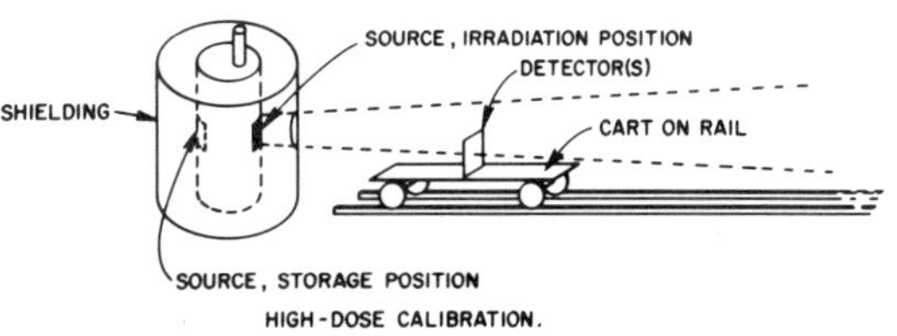

FIGURE 1-18B. Schematical diagram of a simple gamma radiation calibration facility for solid-state and photographic film dosimeters.

simple, inexpensive set-up consists of a wooden or plastic table in the middle of a room (the detectors should always be closer to the radiation source than to the next floor or walls), where a small radionuclide source (a few mCi of ^{137}Cs, ^{60}Co, or ^{226}Ra in a steel capsule, for example) can be placed in a fixed location, and the detectors are arranged either in a circle around the source or in a series of varying distances from it (Figure 1-18B). If the largest distance is 2 m, the following distances from the source will increase the dose-rate by a factor of two for each step according to the inverse square law: 141.4 cm, 100 cm, 70.7 cm, 50 cm, and 35.35 cm.

If the samples are too far from the source or too close, scattering or geometric factors may increase the errors. The same device can be used for neutron calibrations, using ^{252}Cf, Po/Be, or another isotopic neutron source of known strength. If detectors are to be calibrated with high gamma radiation doses, it has advantages to keep the source in the shield and use a narrow, vertical or horizontal beam. Care has to be taken that both detectors and the reference ionization chamber(s) are fully within the beam and that scattering from heavy shielding materials is minimized.

A photon energy response that differs from that of air, water, and tissue, in particular at energies below 100 to 200 keV and above several

TABLE 1-4

Characteristics of K-Fluorescence Radiation Sources

Radiator	Density of radiator (g/cm^2)	Effective energy (keV)	Additional Al filter (g/cm^2)
Ag	0.074	22.6	0.1
Ba	0.13	32.8	0.1
Gd_2O_3	0.23	43.8	0.1
Dy_2O_3	0.26	46.8	0.22
Er_2O_3	0.30	50.0	0.22
Yb_2O_2	0.35	53.3	0.44
HfO_2	0.40	56.7	0.44
Ir	0.55	65.9	0.44
Pb	0.72	76.1	0.44
U	1.50	99.6	1.45

(After Storm and Shlaer, 1965.)

MeV, is a general characteristic of detector systems whose effective atomic number Z* differs substantially from that of the reference materials (see, for example, the compilation of energy dependence measurements by Burgkhardt and Piesch, 1971). In calculating or measuring this "energy

$$*Z = 3\sqrt{\frac{\sum_i n_i Z_i^4}{\sum_i n_i Z_i}}$$

(n_i = molar fraction of element i in percent, Z_i = atomic number of element i)

dependence" of detectors, geometrical factors affecting the escape and reabsorption probabilities of electrons and fluorescence X-radiation have to be considered. This results, for example, in a different energy dependence for thermoluminescent phosphors as a function of grain size (Chan and Burlin, 1970). If a calculated photon energy response is to be compared with experimental data, one should use either monoenergetic (K_a fluorescence) radiation (Table 1-4) and/or radionuclides which provide "monoenergetic" radiation, or correct for the energy distribution of the more or less well-filtered X-radiation. Table 1-5 presents a compilation of effective photon energies which can be obtained by "hard" filtration of bremsstrahlung.

Another important factor to be considered in particular in long-term experiments is the fading stability of a detector. In photographic emulsions it is the relative humidity and the grain size more than the temperature that determines the total fading rate, which consequently can be reduced by extremely careful protection of the film, especially nuclear track emulsions, from humidity (sealing in

TABLE 1-5

Filtration of X-Radiation for Achieving a Relatively Narrow Energy Distribution

Tube voltage (kV)	Additional filtration (mm)*	Effective energy measured (keV)
15	0.2 Al + 0.1 Cu	10
25	2 Al + 0.15 Cu	20
35	2 Al + 0.5 Cu	30
45	2 Al + 1.35 Cu	40
56	2 Al + 1.2 Cu	50
78	2 Al + 3.5 Cu	65
100	2 Al + 2 Cu + 1.5 Pb	80
130	2 Al + 7.5 Cu + 2 Sn	108
200	2 Al + 5 Cu + 4 Sn + 2 pb	185
300	2 Al + 5 Cu + 10.5 Pb	270
^{137}Cs		660
^{60}Co		1250

The sequence of the filters is important. Usually, the filters with higher Z should face the x-ray source.

(After Drexler and van Goss, 1968.)

a thin polymer foil is not sufficient; see Chapter 6.2). In most solid-state detectors, however, it is almost exclusively the storage temperature that determines fading rates. This temperature can be substantially higher than "room temperature" in many cases, for instance, when detectors are implanted in animals or during measurements of background radiation in hot climates.

As the fading rates of various solid-state detectors are quite different and fading experiments at the actual temperature at which detectors are used are time-consuming and/or require high instrument stability, it is tempting to accelerate the fading by measuring it at various increased temperatures and extrapolate to the fading rate of a lower temperature. This technique, although useful in some cases, should be used with caution because of the occurrence of abnormally high fading rates, in particular in some TL materials (see Wintle et al., 1971).

As in personnel dosimetry, the proper choice of detectors is crucial for the special applications. Only a few more common cases can be mentioned here. For measurements at the hands of workers handling beta emitters in glove boxes or implanting radioactive needles, where a soft and flexible fingertip detector is desirable, the use of TL phosphors which are imbedded in thin Teflon foils probably represents the best method.

Energy independence, which can be accomplished either by the use of an energy-independent detector (LiF, BeO, $Li_2B_4O_7$) or by a proper energy compensation filter, is an important factor not only for x-ray dosimetry, but also whenever the spectrum of a high energy radiation source has been "softened" by multiple scattering, for example, inside large irradiated volumina (after approximately one mean free path length of gamma radiation in water, a low energy equilibrium peak at $\sim$50 keV, Figure 1-19, is observed). Depending on the angle and the initial photon energy, even single scattering may result in a substantial reduction of the effective energy of the scattered radiation (Figure 1-20). X-ray fluorescence, which is induced in the scattering material (filters, walls, shielding etc.), further contributes to the degradation of high photon energies.

In clinical dosimetry (treatment planning), the highest possible accuracy is desirable — at least ±5%, preferably less. Radiobiological experiments have demonstrated that a 10% dose more or less can, under certain conditions, result in differences of cell survival rates by a factor of almost 100, and in clinical radiotherapy it has been reported that a 20% change in dose resulted in 60% increase in the

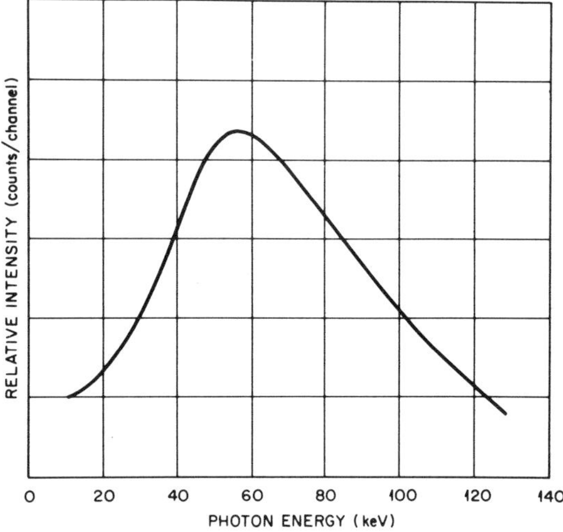

FIGURE 1-19. Low energy equilibrium spectrum of ^{60}Co gamma radiation in water. (After Weiss and Bernstein, 1953.)

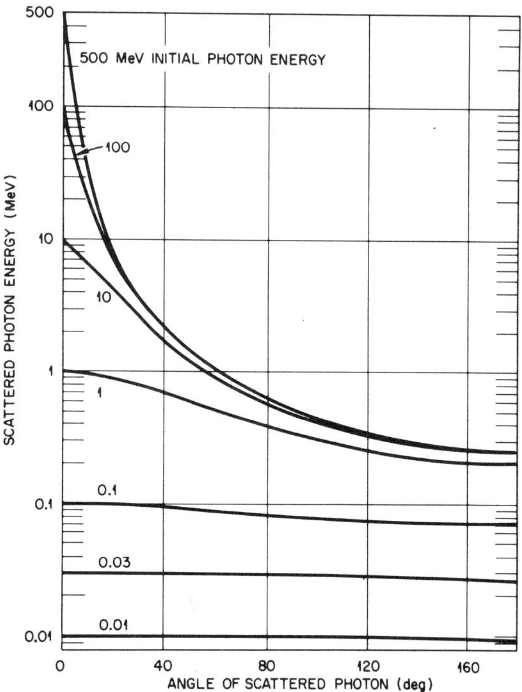

FIGURE 1-20. Scattered photon energy as a function of the angle of the scattered photons for different initial photon energies, calculated with the Klein-Nishina formula. (After NBS Circular 542, 1953.)

cure rate of squamous cell sarcoma (Shukovsky, 1970).

The present situation in clinical dosimetry needs much improvement. In a recent survey of 71 hospitals in the U.S., 80% of the therapy units delivered doses within ±5% of the prescribed dose, but some were more than 20% off (Grant et al., 1972). According to the IAEA Directory of High Energy Radiotherapy Center, about 2,000 ^{60}Co and ^{137}Cs units and about 300 medical accelerators were in use throughout the world in 1970, many of them in developing countries without any qualified medical physicist. A recent review on medical dosimetry technology in developing countries (LeVan, 1972) states that only 11% of the radiotherapy centers in one large region have water phantoms, and 23% have no operating dosimeters of any sort. Obviously, this is not the type of hospital in which one would like to undergo radiation therapy.

High sensitivity and thermal stability are required for measurements of the natural radiation background. Photographic film does not meet these requirements. Modern radiophotoluminescent glasses are sufficiently stable, but limited precision in the measurement of doses below approximately 50 mrad makes them only suitable for monitoring periods of more than one year. So far, only little information on the long-term stability of TSEE detectors is available. The choice may be, therefore, essentially limited to advanced TLD detectors. Of those, LiF:Mg,Ti is less sensitive (measurements below ∿10 mrad are difficult to perform with sufficient accuracy) than some newer TLD materials such as Mg_2SiO_4:Tb and $CaSO_4$:Dy, which permit dose integration with a precision of about ±0.2 mrad or better than ±3%, whichever is larger, even under adverse tropical climate condition and can be used for reliable measurements involving storage times of as little as one week.

It is not completely clear how important photon energy independence is in environmental monitoring. It is known that there is a contribution of low energy photons to the external environmental background radiation exposure (Figure 1-21), but in measurements involving energy-independent as well as strongly energy-

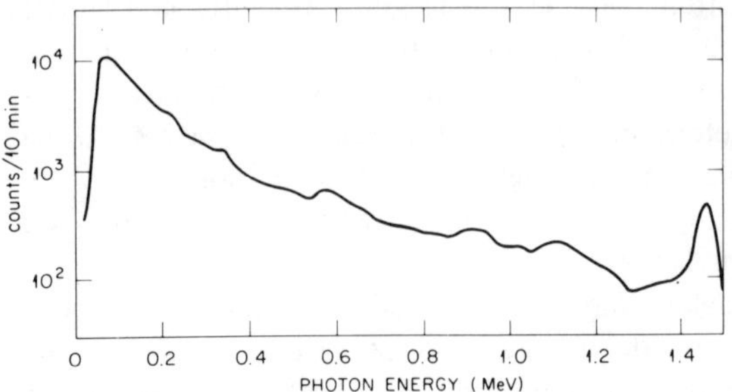

FIGURE 1-21. Natural background radiation spectrum as measured with a NaJ scintillator in the Livermore Valley, Calif. (After Jones et al., 1971.)

dependent detectors (Becker et al., 1971 and unpublished results) no difference in the results was observed, indicating that this contribution may be negligible in many, if not all situations. Care should, however, be taken that the exposure (time) during which the detectors are not "on duty," namely, between annealing and distribution and again between collection and reading, is small compared to the storage time at the location in question (for details, see Chapter 2).

An entirely different problem is the measurement of very high dose-levels in the 10^4 to 10^7 rad range, as encountered in radiation chemistry and technology, food processing, radiation sterilization, material testing, etc.* Some of the better techniques used in this field such as calorimetry, chemical dosimetry with the Fricke (Fe^{2+}/Fe^{3+}) dosimeter, ceric sulfate, the ferric-cupric system and oxalic acid, or with organic dyes in the form of liquid solutions are subjects not to be covered in this book. Reference is made to recent reviews of those techniques (Fricke and Hart, 1966; Holm and Berry, 1970; Stolz, 1972; IAEA, 1972). According to a recent comparison of TLD systems for the 10^3 to 10^6 R gamma radiation dose range (Gorbics et al., 1973), a CaF_2:Mn/Teflon compound is more suitable than other phosphors. For high dose-level studies with LiF:Mg,Ti, RPL glasses, and TSEE materials, see Chapters 2 to 4.

Dose-rate dependence of response rarely poses a problem in solid-state dosimetry unless extremely low or extremely high dose-rates are involved. In the first case, some fading may already take place during exposure, and in photographic emulsions

---

*If sterility is defined by the reduction of the number of the most radiation-resistant spores in dust (*Streptococcus faecium*) by factors of 10^7 and 10^8, doses of 2.5 and 4.5 Mrad, respectively, are required. On the other hand, sprout inhibition in potatoes occurs after 7 to 15 krad and grain disinfestation after 20 to 50 krad.

the well-known "reciprocity failure" (Schwarzschild effect) is observed. In the other extreme of dose-rates exceeding $\sim 10^{10}$ rad/sec (field-emission x-ray and electron sources, nuclear explosions), a drop in sensitivity also occurs in some detector systems, but few reliable data are available.

With the exception of nuclear track detectors that have a distinct registration threshold, and a few little-studied TLD phosphors, all solid-state dosimeters exhibit a marked drop in sensitivity with increasing ionization density of the radiation. The sensitivity usually begins to decrease around 10 to 100 keV/μm and reaches 10 to 20% of its gamma radiation sensitivity at about 10^3 keV/μm. This LET dependence unfortunately complicates the measurement of charged particles (neutron-induced recoil protons, alpha particles) in mixed radiation fields.

High spatial resolution may be required in case of dosimetric studies in radiation fields involving steep dose gradients. The resolution of dosimeter films is limited to about 5 to 20 μm. With fine-grained TLD phosphors that are imbedded in very thin heat-resistant polymers (Teflon, silicon rubber), track etch detectors, and TSEE systems, a similar resolution can be obtained. TSEE, with, in principle, monoatomic "sensitive layers," has the highest potentials for extremely high spatial resolution.

Most solid-state detectors respond to high energy electrons as they would to gamma radiation on a basis of rad-equivalence. With decreasing electron energy, however, some secondary effects may occur. If the detector is voluminous and/or protected by an absorbing layer, a substantial attenuation or build-up of the electron dose in the sensitive volume may occur. Exposure of an insulating glass or crystal to very low energy electrons, for example, by submersion into a solution containing tritium, can lead to a local charge build-up which eventually prevents electrons from entering the sensitive volume. Self-shielding is also an important factor in the use of voluminous thermal neutron detectors with a high thermal neutron cross section (detectors containing Li, B, Dy, Cd, etc.).

Other factors to be considered in the choice of detectors for a particular application are the hygroscopic nature or solubility which limits the use of some unprotected detectors such as phosphate and borate glasses and lithium borate under humid conditions; another is toxicity, which can be an important factor in the use, for example, of LiF for in vivo dosimetry and which restricts the use of unfired BeO powders. Others are related to the cost and availability of materials and readout instrumentation (it may be advisable for a developing country to use cheap locally made products even if they are somewhat inferior to the best ones on the market). These parameters will be discussed in more detail in the "Applications" section of the following chapters.

There are several areas of application of solid-state detectors which are not strictly dosimetric and in which interest is likely to increase in coming years. One example is the "dosimetry" of unconventional types of radiation, for example, microwaves or ultraviolet radiation. Ultraviolet has adverse effects such as eye damage, erythrema, and skin cancer, as well as beneficial (germicidal) effects on human health, for workers who are professionally exposed to UV in the pharmaceutical industry, as well as the total population. It has, for example, been estimated (Setlow, R. B., unpublished) that a 1% decrease in atmospheric ozone due to high-flying aircraft would increase the intensity of solar UV in the biologically relevant wavelength region (Figure 1-22) to such a degree that the incidence of skin cancer in the white population of the U.S. would increase by 6,000 cases each year. There are potentials for long-term integrating UV detectors among exo-electron and thermoluminescence materials.

The imaging of the spatial distribution of radiation fields (photon and neutron radiography,

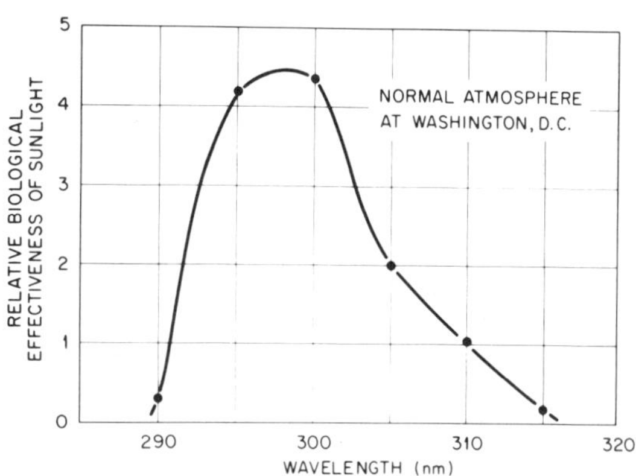

FIGURE 1-22. Relative biological effectiveness of sunlight at the earth's surface in normal atmosphere at Washington, D.C. (After R. B. Setlow, unpublished.)

autoradiography, "photography") can be accomplished with two-dimensional or three-dimensional solid-state dosimeters based on track etching, radiophotoluminescence, thermoluminescence, and exoelectron emission. Some of these detectors have a higher sensitivity and/or better resolution than conventional photographic films, and some are instantly reusable. Such detectors could permit new approaches to various routine and research problems.

REFERENCES

Allen F. J. and Futtener, A. T., Neutron transmission data, *Nucleonics*, 21(8), 120, 1963.

Ambiger, T. Y., Phantom Measurements to Determine the Most Suitable Site for the Personal Dosimeter, AECL-3379, Chalk River, Ontario, Canada, 1969.

Amelinckx, S., Batz, B., and Strumane, R., *Solid-State Dosimetry*, Gordon and Breach, New York, 1969.

Angino, E. E., Görgler, N., and Call, R., Thermoluminescence Bibliography, TID-3911, 1965.

Attix, F. H. and Roesch, W. C., Eds., *Radiation Dosimetry*, Vol. 1, Fundamentals, Academic Press, New York, 1968.

Attix, F. H., Ed., *Progress in Radiation Dosimetry*, Vol. 1, Academic Press, New York, 1972a.

Attix, F. H., A current look at TLD in personnel monitoring, *Health Phys.*, 22, 287, 1972b.

Barber, D. E., U.S. Department of Health, Education, and Welfare, Public Health Publication No. 999-RH-20, 1966.

Becker, K., Bibliographies on Photographic, Solid-State and Chemical Dosimetry, AEC-C-21-01, 1964, 02, 1964, 03, 1965, 04, 1967, 05, 1967, 06, 1968, 07, 1968, and 08, 1971; Zentralstelle f. Atomkernenergie-Dokumentation, Karlsruhe, Germany.

Becker, K., Photographic, glass, or thermoluminescence dosimetry, *Health Phys.*, 12, 995, 1966.

Becker, K., *Photographic Film Dosimetry*, Focal Press, London, 1966a.

Becker K., Radiophotoluminescence dosimetry — a bibliography, *Health Phys.*, 12, 1367, 1966b.

Becker, K., Radiophotoluminescence dosimetry, *IAEA At. Energ. Rev.*, 5(1), 43, 1967.

Becker, K., Personnel Dosimetry in Large-Scale Nuclear Emergencies, Proc. Conf. on Radiol. Protection of the Public in a Nuclear Mass Disaster, Interlaken, Switzerland, 1968, 175.

Becker, K., Principles and basic dosimetric properties of radiophotoluminescence in silver-activated glasses, in *Solid-State Dosimetry*, Amelinckx, S., Batz, B., and Strumane, E., Eds., Gordon and Breach, New York, 1969a, 153.

Becker, K., Radiophotoluminescence dosimetry bibliography II, *Health Phys.*, 17, 631, 1969b.

Becker, K., Solid-state dosimetry, *CRC Crit. Rev. Radiol. Sci.*, 1(3), 363, 1970b.

Becker, K., Stimulated exoelectron emission from the surface of insulating solids, *CRC Crit. Rev. Solid State Sci.*, 3(1), 39, 1972a.

Becker, K., Dosimetric applications of track etching, in *Topics in Radiation Dosimetry*, Vol. 1, Attix, F. H., Ed., Academic Press, New York, 1972b.

Becker, K., The future of personnel dosimetry, *Health Phys.*, 23, 797, 1972c.

Becker, K., Lu, R. H., and Weng, P. S., Environmental and Personnel Dosimetry in Tropical Countries, *Proc. Third Int. Conf. Luminescence Dosimetry*, Risö, Rep. 249, Vol. 3, Danish AEC, Risö, Roskilde, 1971.

Bennett, B. G., Estimation of gonadal absorbed dose due to environmental gamma radiation, *Health Phys.*, 19, 757, 1970.

Bouwers, A. and van der Tuuk, J. H., X-ray protection, *Br. J. Radiol.*, 3, 503, 1930.

Burckhardt, B. and Piesch, E., Energieabhängigkeit einiger gebräuchlicher Dosisleistungs- und Dosismesser für den Strahlenschutz, KFK 1484, Kernforschungszentrum Karlsruhe, Germany, 1971.

Cameron, J. R., Suntharalingam, N., and Kenney, G. N., *Thermoluminescent Dosimetry*, University of Wisconsin Press, Madison, Wisc., 1968.

Cember, H., *Introduction to Health Physics*, Pergamon Press, New York, 1969.

Chan, F. K. and Burlin, T. E., The energy-size dependence of the response of thermoluminescent dosimeters to photon irradiation, *Health Phys.*, 18, 325, 1970.

Cheka, J. S., Neutron Monitoring by Means of Special Fine Grain Alpha Emulsions Film, U.S. Report M-3685, 1944.

Daniels, F., Thermoluminescence and Related Properties of Crystals, Report 4th Symp. Chem. Phys. Rad. Dosimetry, Part I, Army Chem. Center, Edgewood, Md., 148, 1950.

Delafield, J. H., Gamma Ray Exposure Measurements in a Man Phantom Related to Personnel Film Dosimetry, AERE-R-4430, 1963.

Dessauer, G. and Lennox, E., Photographic Neutron Dosimetry to Date, AECD-2278 (M-1525 A), 944; and AECD-1973 (R-3450), 1944.

Drexler, G. and van Goss, M., Spektren gefilterter Röntgenstrahlung für Kalibrierzwecke, GSF-Report S45, Inst. f. Strahlenschutz, Neuherberg, Germany, 1968.

Dudley, R. A., Dosimetry with photographic emulsions, in *Radiation Dosimetry,* Vol. 2, Attix, F. H. and Roesch, W. C., Eds., Academic Press New York, 1966, 326.

Eggert, J. and Luft, F., Ein neuartiges Röntgen-Filmdosimeter, Parts I and II, *Röntgenpraxis,* 1, 188, 1929.

Evans, H. J., Use of Chromosome Aberration Frequencies for Biological Dosimetry in Man, Proc. Symp. Adv. Phys. Biol. Rad. Detectors, IAEA, Vienna, 1971, 593.

Evans, H. J., Actions of radiations on human chromosomes, *Phys. Med. Biol.,* 17, 1, 1972.

Fowler, J. R. and Attix, F. H., Solid-state integrating dosimeters, in *Radiation Dosimetry,* Attix, F. H. and Roesch, W. C., Eds., Academic Press, New York, 1966, 241.

Frank, M. and Stolz, W., *Festkörperdosimetrie ionisierender Strahlung,* B. G. Teubner Verlagsges., Leipzig, 1969.

Fitzgerald, J. J., Brownell, G. L., and Mahoney, F. J., *Mathematical Theory of Radiation Dosimetry,* Gordon and Breach, New York, 1967,

Fitzgerald, J. J., *Applied Radiation Protection and Control,* Vols. 1 and 2, Gordon and Breach, New York, 1969–1970.

Fricke, H. and Hart, E. J., Chemical dosimetry, in *Radiation Dosimetry,* Vol. 2, Academic Press, New York, 1966, 167.

Frigerio, N. A., Coley, R. F., and Branson, M. H., Depth dose determination. I., *Phys. Med. Biol.,* 17, 792, 1972.

Frigerio, N. A., Coley, R. F., and Branson, M. H., Depth dose determination. II., *Phys. Med. Biol.,* 18, 53, 1973.

Frigerio, N. A. and Coley, R. F., Depth dose determination. III., *Phys. Med. Biol.,* 18, 187, 1973.

Ginther, R. J. and Kirk, R. D., The thermoluminscence of CaF_2:Mn, *J. Electrochem. Soc.,* 104, 365, 1957.

Goldstein, E., *Z. Phys.,* 13, 188, 1912.

Gorbics, S. G., and Attix, F. H., Thermoluminescent dosimeters for high-dose applications, *Health Phys.,* in press.

Gourgé, G. and Hanle, W., Neue Ergebnisse uber Exoelektronen-Emission an Nichtmetallen, *Acta Phys. Austr.,* 10, 427, 1959.

Grant, W. H., Shalek, R. J., Cundiff, J. H., and Golden, R., A Review of the Activities of the AAPM Radiological Physics Center in Interinstitutional Trials Involving Radiation Therapy, Paper 38.1, Third Int. Conf. Medical Physics, Göteborg, 1972.

Hart, J. C., Legal and administrative aspects of personnel dosimetry, *Health Phys.,* in press.

Hendee, W. R., *Medical Radiation Physics,* Year Book Medical Publishers, Chicago, 1970.

Herz, R. H., *The Photographic Action of Ionizing Radiation* Interscience, New York, 1969.

Holm, N. W. and Berry, R. J., Eds., *Manual on Radiation Dosimetry,* Marcel Dekker, New York, 1970.

Hoy, J. E., Personnel Albedo Neutron Dosimeter with Thermoluminescent ^{6}LiF and ^{7}LiF, DP-1277, Savannah River Laboratory, Aiken, S.C., 1972.

Hurst, G. S. and Turner, J. E., *Elementary Radiation Physics,* John Wiley & Sons, New York, 1970.

IAEA, Personnel Dosimetry for Radiation Accidents, IAEA, Vienna, STI/PUB/99, 1965.

IAEA, Neutron Detectors, Bibliograph., Series No. 18, IAEA, Vienna, STI/PUB/21/18, 1966.

IAEA, Solid-State and Chemical Radiation Dosimetry in Medicine and Biology, IAEA, Vienna, STI/PUB/138, 1967.

IAEA, Radiation Protection Monitoring, IAEA, Vienna, STI/PUB/199, 1969.

IAEA, Personnel Dosimetry Systems for External Radiation Exposure, Technical Report Series No. 109, IAEA, Vienna, STI/DOC/10/109, 1970a.

IAEA, Nuclear Accident Dosimetry Systems, Panel Proc. Ser., IAEA, Vienna, STI/PUB/241, 1970b.

IAEA, Advances in Physical and Biological Radiation Detectors, IAEA, Vienna, STI/PUB/269, 1971.

IAEA, Dosimetry Techniques Applied to Agriculture, Industry, Biology and Medicine, IAEA, Vienna, in press.

ICRP, *General Principles of Monitoring for Radiation Protection of Workers,* ICRP Publication No. 12, Pergamon Press, Oxford, 1969.

ICRU, Radiobiological Dosimetry, Recommendations of ICRU Report 10e, National Bureau of Standards Handbook 88, 1962.

ICRU, Measurement of Absorbed Dose in a Phantom Irradiated by a Single Beam of X or Gamma Rays, ICRU Report no. 23, 1973.

Isabelle, D. and Monnin, M., *Proceedings of the International Topical Conference on Nuclear Track Registration in Insulating Solids and Applications,* Clermont-Ferrand, 1969.

Joffre, H., Personnel Dosimetry Techniques for External Radiations, Proc. of Madrid Symposium, ENEA, Paris, 1963.

Johns, H. E. and Cunningham, J. R., *The Physics of Radiology,* 3rd ed., Charles C Thomas, Springfield, Ill., 1969.

Jones, A. R., Proposed calibration factors for various dosimeters at different energies, *Health Phys.,* 12, 663, 1966.

Jones, D. E., Lindeken, C. L., and McMillen, R. E., Natural Radiation Background Dose Measurements with CaF_2:Dy TLD, *Proc. Third Int. Conf. Luminescence Dosimetry Risö-Rep. 249,* Vol. 2, Danish AEC, Risö, Roskilde, 985, 1971.

Jones, T. D., Auxier, J. A., Snyder, W. S., and Warner, G. G., Dose to standard reference man from external sources of monoenergetic photons, *Health Phys.,* 24, 241, 1973.

Kossel, W., Meyer, V., and Wolf, H. C., Simultandosimetrie von Strahlenfeldern am lebenden Objekt, *Naturwissenschaften,* 41, 209, 1954.

Kramer, J., Anwendungen von Exoelektronen, *Acta Phys. Austr.,* 10, 392, 1957.

Langmead, W. A. and Farmer, F. T., Film badge position for x-ray workers, *Br. J. Radiol.,* 44, 480, 1971.

LeVan, J. H., A Survey of Medical Dosimetry Technology in Developing Countries, Paper IAEA-SM-160/90, Proc. Symp. Dosimetry Techniques Applied to Agriculture, Industry, Biology and Medicine, IAEA, Vienna, 1972.

Lin, F. M. and Cameron, J. R., A bibliography of thermoluminescent dosimetry, *Health Phys.,* 14, 495, 1968.

Loevinger, R., Current Trends in Radiation Dosimetry, Paper, Third Int. Conf. Medical Physics, Göteborg, 1972.

Luminescence Dosimetry, Proceedings of International Conference on Luminescence Dosimetry, Stanford 1965, Attix, F. H., Ed., CONF-650637, USAEC Symp. Ser. 8, 1967.

Luminescence Dosimetry, Proc. Second International Conference on Luminescence Dosimetry, Gatlinburg 1968, Auxier, J. A., Becker, K., and Robinson, E. M., Eds., CONF-680920, USAEC Symp. Ser., 1968.

Luminescence Dosimetry, *Proceedings of the Third International Conference on Luminescence Dosimetry, Risö 1971,* Vols. 1–3, Mejdahl, V., Ed., Danish AEC, Risö Rep. 249, 1971.

Markus, B., Die Messung von Tiefendosiskurven im Weichstrahlbereich, *Strahlentherapie,* 135, 25, 1967.

McLaughlin, W. L., Films, dyes and photographic systems, in *Manual on Radiation Dosimetry,* Holm, N. W. and Berry, R. J., Eds., Marcel Dekker, New York, 1970, 129.

McLennan, J. C., On a kind of radioactivity imparted to certain salts by cathode rays, *Phil. Mag.,* 3, 195, 1902.

Meredith, W. J. and Massey, J. B., *Fundamental Physics of Radiology,* Williams & Wilkins, Baltimore, 1972.

Morgan, K. Z. and Turner, J. E., *Principles of Radiation Protection,* John Wiley & Sons, New York, 1967.

Nachtigall, D., *Basic Physics for Dosimetry and Radiation Protection,* Verlag Karl Thiemig, München, 1971.

Nash, A. E. and Attix, F. H., Use of LiF-Teflon disks in laminated identification cards for personnel accident dosimetry, *Health Phys.,* 21, 435, 1971.

NCRP, Report No. 38 on Protection Against Neutron Radiation, National Council on Radiation Protection and Measurements, 1971.

Piesch, E., The indication of adsorbed dose in critical organs by energy independent personnel dosimeters, *Health Phys.,* 15, 145, 1968.

Piesch, E., Developments in RPL dosimetry, in *Topics in Radiation Dosimetry,* Vol. 1, Attix, F. H., Ed., Academic Press, New York, 1972a.

Piesch, E., Zukunftstendenzen der Personendosimetrie, Proc. Ann. Meet. Fachverband für Strahlenschutz, Karlsruhe, 1972b, 207.

Pretre, S., Quantities for the description of neutron irradiations in man. A look at the mechanism of over-conservatism, *Health Phys.,* 23, 169, 1972.

Sagan, L. A., Human costs of nuclear power, *Science,* 177, 487, 1972.

Sasaki, M. S., *Biological Aspects of Radiation Protection,* Sugakara, T. and Hug, O., Eds., 1971, 81.

Schulman, J. H., Ginther, R. J., Klick, C. C., Alger, R. S., and Levy, R. A., Dosimetry of x-rays by radiophotoluminescence, *J. Appl. Phys.,* 22, 1479, 1951.

Schulman, J. H., Shurcliff, W., Ginther, R. J., and Attix, F. H., Radiophotoluminescent dosimetry system of the U.S. Navy, *Nucleonics,* 11(10), 52, 1953.

Schwarz, K. K., Grant, A., Meshs, T. K., and Grube, M. M., *Termolumineszenzaia Dosimetria,* Riga, 1968.

Selman, J., *The Fundamentals of X-Ray and Radium Physics,* 5th ed., Charles C Thomas, Springfield, Ill., 1972.

Shukovsky, L. J., Dose, time volume relationship of squamous cell carcinoma, *Am. J. Roentgenol.,* 108, 27, 1970.

Spurny, Z., Thermoluminescent dosimetry, *IAEA At. Energ. Rev.,* 3, 61, 1965.

Spurny, Z., Solid-State Dosimetry, IAEA Bibliography Ser. No. 23, Vienna, 1967.

Spurny, Z., Additional bibliography of thermoluminescent dosimetry, *Health Phys.,* 17, 349, 1969.

Spurny, Z. and Sulcova, J., Bibliography of thermoluminescence dosimetery (1968–1972), *Health Phys.,* 24, 573, 1973.

Stolz, W., *Strahlensterilisation – Grundlagen and Anwendung in Medizin und Pharmazie,* J. A. Barth, Leipzig, Germany, 1972.

Storm, E. and Shlaer, S., Development of energy-independent film badges with multielement filters, *Health Phys.,* 11, 1127, 1965.

Webb, G. A. M., Dauch, J. E., Slane, D. K., and Phykitt, H. P., Information Content of Personnel Dosimetry Systems, Paper 17th Ann. Meet. Health Phys. Soc., Las Vegas, 1972.

Webster, E. W. and Tsien, K. C., Atlas of Radiation Dose Distributions, IAEA, Vienna, 1965.

Weiss, M. M. and Bernstein, W., Degradation of gamma rays in water, *Phys. Rev.,* 92, 1264, 1953.

Weyl, W. A., Schulman, J. H., Ginther, R. J., and Evans, L. W., On the fluorescence of atomic silver in glasses and crystals, *J. Electrochem. Soc.,* 95, 70, 1949.

United Nations, *Ionizing Radiation: Levels and Effects – A Report of the United Nations Scientific Committee on the Effects of Atomic Radiation,* Vols. 1 and 2, United Nations, New York, 1972.

Wiedemann, E. and Schmidt, G. C., Uber Lumineszenz, *Ann. Phys. Chem.,* 54, 604, 1895.

Willis, C. A. and Handloser, J., Eds., *Health Physics Operational Monitoring,* Vols. 1 to 3, Gordon and Breach, New York, 1973.

Wintle, A. G., Aitken, M. J., and Huxtable, J., Abnormal Thermoluminescence Fading Characteristics, Proc. Third Int. Conf. on Luminescence Dosimetry, Danish AEC, Risö-Rep. 249, 1971.

Young, D. A., Etching of radiation damage in lithium fluoride, *Nature* (London), 182, 375, 1958.

Young, M. E. J., *Radiological Physics,* Charles C Thomas, Springfield, Ill., 1967.

2. THERMOLUMINESCENCE

2.1 Definition and History

When a dielectric solid is exposed to ionizing radiation or ultraviolet light, or when it undergoes certain chemical reactions or mechanical stress, a varying percentage of the liberated electrons (or holes) may become trapped at certain imperfections in the lattice which are called electron (or hole) traps. If the temperature is low and/or the traps are deep enough, they may remain trapped for hundreds or thousands of years before they are released by a sufficient "stimulation," which increases their probability of escaping from the traps. Stimulation usually consists of the transfer of optical (light) or thermal energy to the dielectric.

The return of the electrons (or trapped holes) to the stable state, which is called annealing (or fading if it proceeds slowly), is associated with a release of energy, most of which is thermal and results in a minute temperature increase during annealing. A small fraction of the energy, however, may be released as visible or ultraviolet light, which can be observed as a transient "glow" of the pre-irradiated crystal or glass at a characteristic temperature during heating. This effect is called thermoluminescence (TL).* If several traps of different depths are involved in the process, the recording of the light emission as a function of heating time (or temperature) results in a "glow curve" consisting of several "glow peaks." Figure 2-1 presents the glow curves of some common TL phosphors.

The peak location is fairly constant for a given TL crystal or glass (with or without added "activators") and the light sum under the peak can be made proportional to the radiation dose. The basic readout procedure for a TL phosphor is, therefore, rather simple. The TL material that is in the form of a loose powder, a plastic compound, glass, sintered disk, or single crystal (or encapsulated permanently in a glass capillary, glass bulb, etc.) is heated by electrical resistance heating or exposure to a hot gas. Usually the temperature of the phosphor carrier is monitored during the heating. After passing through infrared and/or additional optical filters, the emitted light is detected by a photomultiplier and the light output is recorded either on a current-vs.-time or (which is preferable for research purposes) a current-vs.-temperature chart, with a peak-height-recording microammeter, or with a charge-integrating circuit.

Varying dose, LET, phosphor pretreatment, heating rate, and reader design (PM tube, filters, etc.) can all modify peak height and location. These parameters are discussed later in this chapter.

Thermoluminescence is a widespread phenomenon. Of 3,000 natural minerals, for example, three quarters exhibit this effect; some minerals, such as fluorite (CaF_2) from certain areas in Iceland, Brazil, and India, are among the most sensitive phosphors known. The effect is not restricted to inorganic crystals and glasses, but TL from organic compounds such as various polymers (Mozisek, 1966 and 1968; Fleming, 1969; Tochin and Nikolskii, 1969; Partridge, 1972; and others) including Teflon (Tomita, 1970), polyethylene, and polypropylene (Mozisek, 1970) and alanine

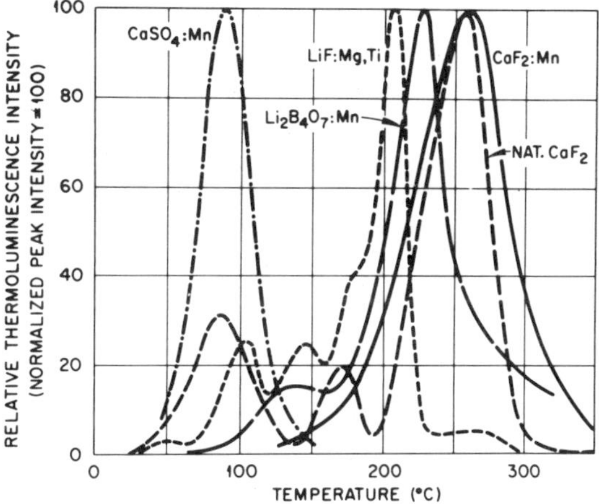

FIGURE 2-1. Thermoluminescence "glow curves" of some common phosphors, normalized for maximum peak height; glow curve and peak temperature vary as a function of the phosphor pretreatment, activator concentration, heating rate, LET of the radiation, etc.; curves as presented here should, therefore, be taken *cum grano salis*. (After Fowler and Attix, 1966.)

---

*For its dosimetric applications, both the terms "thermoluminescen*t* dosimetry" and "thermoluminescen*ce* dosimetry" (TLD) are widely used. The author prefers (and uses in this book) the latter.

(even irradiated tomato seeds are known to exhibit TL at low temperatures, Chadwick and Oosterheert, 1967) usually occurs at temperatures below room temperature and is, therefore, not of interest in practical dosimetry (concerning the search of potential· organic TL materials see, for example, Oster and Gabor, 1967). Only recently, a few complex metallo-organic compounds (cobalt chelates) with glow peaks in the 100 to 245°C range, dependent on LET (Gupta and Sarup, 1971), have been described. Their response is linear in the 10^4 to 10^6 rad dose-range.

Early investigations of artificially activated TL phosphors such as $CaSO_4$:Mn date back to the turn of the century (Wiedemann and Schmidt, 1895). This material was used in 1935, and again since 1949 in rocket experiments for UV dosimetry (Lyman, 1935; Tousey et al., 1951; and Watanabe, 1951), and in 1954 was employed in the first medical dosimetry studies with TLD (Kossel et al., 1954). The effect of X-radiation on synthetic CaF_2 was first studied in 1928 (Wick and Slattery, 1928). Manganese-activated CaF_2 was introduced in the mid-1950's as a sensitive, stable material which has only one major glow peak (Ginther, 1954 and Ginther and Kirk, 1957).

The success story of lithium fluoride dates back to the late 1940's and early 1950's, when Daniels and his students at the University of Wisconsin studied the TL of alkali halides (see Boyd, 1949; Morehead and Daniels, 1952; and Daniels et al., 1953). They used pressed pellets of a lithium fluoride powder manufactured by the Harshaw Chemical Co., Cleveland, Ohio for dosimetry experiments, but abandoned this material in favor of Al_3O_3 because of a disturbing low-temperature peak in their LiF (for an account of the early TL work on LiF, see Daniels, 1965). The TL research in Madison, Wisc. was discontinued from 1956 until 1960, when Cameron et al. revived it with new samples of LiF from Harshaw which happened to contain activating impurities, resulting in more desirable dosimetric properties. A lithium fluoride that is activated with magnesium and titanium, called "TLD-100[®]," dominated the low Z detector market for the following years adn is still the most widely used TL phosphor. Originally available as a crystalline powder, it was later manufactured in the form of extruded polycrystalline ribbons and imbedded in Teflon or silicon rubber.

Early work on optical (infrared) stimulated luminescence of phosphors such as SrS:Eu,Mg (Antonov-Romanovsky et al., 1955) did not lead to practical dosimeters. More successful have been attempts to replace LiF:Mg,Ti, which has a number of undesirable dosimetric properties such as supralinear response at higher dose-levels and complicated annealing characteristics, with other low Z phosphors. Manganese-activated lithium borate (Schulman et al., 1967) and beryllium oxide (Mandeville and Albrecht, 1954; Tochilin et al., 1969; Becker et al., 1970; Scarpa, 1970; Yamashita, 1971; and Spurny, 1972) have been the most successful more recent low Z materials so far.

Considerable efforts have also been devoted during the past 15 years to the improvement of sensitivity and stability of high Z materials. Natural fluorites (Luchner, 1957 and Schayes et al., 1963) turned out to be not too successful, and TL glasses (Botshvar et al., 1963 and Ginther, 1965) could never seriously compete with crystalline phosphors. Some other, highly sensitive phosphors such as $CaSO_4$:Mn and CaF_2:Dy turned out to be not stable enough for long-term experiments, but other, equally stable and more sensitive materials ($CaSO_4$:Dy, $CaSO_4$:Tm — Yamashita et al., 1968 and $MgSiO_4$:Tb — Hashizume et al., 1971) became available. The field is still progressing rapidly with new detector materials, or modified old ones, being suggested at a rate of five to ten each year. Different preparation and read-out techniques improved the annealing characteristics of LiF:Mg,Ti. It is presently difficult to predict which particular phosphors will dominate the scene several years from now.

Rapid developments also took place in the TLD reader field since the first TL personnel dosimetry system based on sealed CaF_2:Mn detectors was developed about ten years ago (Schulman et al., 1963). A first generation of not very versatile and not always reliable laboratory readers such as the one produced by ConRad (now Teledyne/ Isotopes, Westwood, N. J.) was replaced in recent years by a multitude of TL readers manufactured in numerous countries. They can now be classified into three groups: versatile research instruments; inexpensive routine readers for hospital use; and semiautomatic or automatic readers for personnel dosimetry services (these instruments are discussed in Chapter 2.5).

A vast volume of information has been published on thermoluminescence in general and specific applications, including dosimetry. There are bibliographies of early literature on TL in general (Angino and Grögler, 1965); a list with almost 400 titles on geological applications covering the pre-1966 literature (Grögler, 1968); bibliographies on the dosimetric aspects of TLD (Spurny, 1967; Lin and Cameron, 1968; and Becker, 1964, 1965, 1967, 1968, and 1971), as well as review articles and books on various aspects of TL which contain a large number of references (Spurny, 1965; Cameron et al., 1968a; Schwarz et al., 1968; McDougall, 1968; Frank and Stolz, 1969). Good state-of-the-art summaries are provided by the proceedings of three International Conferences on Luminescence Dosimetry in Stanford 1965 (Attix, 1967), Gatlinburg 1968 (Auxier et al., 1968), and Risö 1971 (Mejdahl, 1971; for a summary, see Becker, 1972), as well as some other international meetings on this subject (see, for example, IAEA, 1967 and Amelinckx et al., 1969).

The total number of publications on thermoluminescence dosimetry (TLD) is approaching 2,000 and is increasing at a rate of over 200 per year. Almost half of it is devoted to lithium fluoride (this compares to less than ten publications per year in the whole field of solid-state dosimetry ten years ago — Attix, 1971). Obviously, it is impossible to cover the whole volume of accumulated information in this chapter. Instead, emphasis will be placed on the developments since 1967, when three widely known books on TLD (Cameron et al., 1968; Schwarz et al., 1968; and Frank and Stolz, 1969) were written. The list of references is restricted to publications of particular interest.

2.2 Theory

The theory of thermoluminescence and some closely associated phenomena that are also based on trapping processes, including phosphorescence, thermally stimulated current, and thermally stimulated exoelectron emission, are, despite considerable efforts, not yet completely understood. Usually, discrete energy levels and certain transitions are assumed, and the parameters are adjusted to fit experimental glow curves and fading rates. It is, however, not yet possible to determine trapping parameters of imperfections in crystals on the basis of TL measurements or to calculate the TL characteristics of a phosphor from the knowledge of its crystal lattice and activators.

The following simplified treatment does not nearly reflect the complexity of the processes and the various explanations which have been put forward, but attempts to illustrate a few of the basic principles involved on the basis of two types of traps and a few possible electron transitions only. More complex models are more complicated in the treatment of the kinetic equations and increasingly difficult to verify experimentally. (For a more detailed treatment of the theory of TL, see Randall and Wilkins, 1945; Garlick, 1949; Schön, 1958; Curie, 1963; Bräunlich, 1968; Riehl, and Bräunlich and Kelly, 1971.)

As illustrated in Figure 2-2A, interaction of ionizing radiation with the phosphor results in the transfer of sufficient energy to electrons in the valence (V) band, for transferring them (δ) into the conduction band C. This process, which requires about 10 to 15 eV, is usually due to the effect of secondary electrons which are produced in the environment of primary photoelectron or charged particle tracks. A varying percentage of the thus "liberated" electrons recombines within a very short time with activators A (γ), releasing some of its energy in the form of light

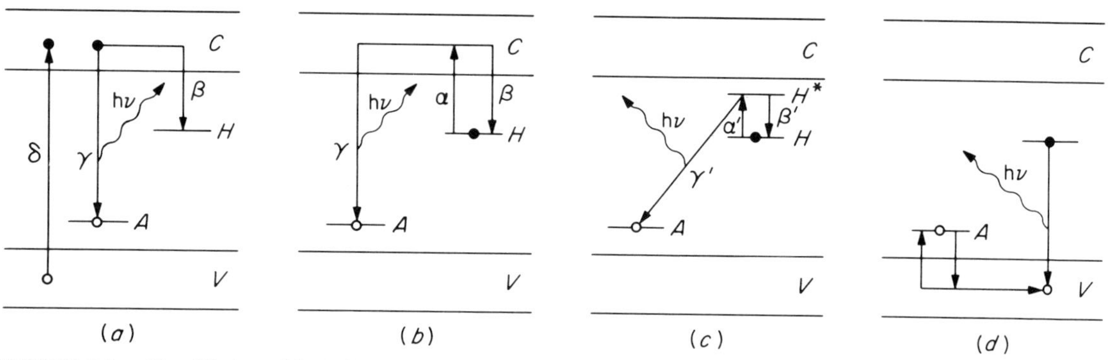

FIGURE 2-2. Simplified model of the processes in a TL phosphor during exposure to ionizing radiation (a), and during heating according to the bimolecular model (b), and the monomolecular model (c); if the hole trap (defect electron trap) is stable, it can be released and recombine with an electron (d).

(fluorescence), while the remaining electrons are captured in electron traps H (β).

If the traps are not very deep, detrapping and recombination may already occur at a substantial rate around room temperature, resulting in a short halflife of the stored energy. This effect is usually called phosphorescence. Only if the traps are deep enough to result in sufficient storage stability at room temperature (halflives of trapped electrons of at least several weeks or months, normally corresponding to glow peaks at temperatures exceeding 100 to 120°C), the effect is called thermoluminescence and becomes of dosimetric interest.

In case of true "thermoluminescence," the electrons are liberated (a) by the transfer of thermal energy (heating), as indicated in Figure 2-2B, representing a "bimolecular" model (Schön, 1958). The electrons either are retrapped (β) or recombine with the activator (γ), emitting part of their energy ($<1\%$) as visible or ultraviolet light. Before retrapping or recombination takes place, the electron population in the conduction band can be observed as a thermally stimulated electrical conductivity.* Such conductivity curves closely parallel the TL glow curves under certain conditions (see, for example, Böhm and Scharmann, 1969 and Bräunlich and Kelly, 1970). According to a "monomolecular" model (Figure 2-2C), electron trap and activator are associated and the radiative transfer γ occurs from the excited state of the electron trap H.

The two models can be described by a set of kinetic equations if a number of simplifying assumptions (no retrapping occurs, the heating rate is constant, and the rate-determining process is the escape rate of the electron from the trap) are made:

Bimolecular model	Monomolecular model
$\dfrac{dn}{dt} = ah - \beta n(H-h) - \gamma nf,$	$\dfrac{dh^*}{dt} = a'h - \beta'h^* - \gamma'h^*.$
$\dfrac{dh}{dt} = -ah + \beta n(H-h),$	$\dfrac{dh}{dt} = -a'h + \beta'h^*,$
$\dfrac{df}{dt} = -\gamma nf,$	$\dfrac{df}{dt} = -\gamma'h^*,$
$f = n + h,$	$f = h + h^*.$ (2.1)

In these equations, n is the concentration of electrons in the conduction band, h,h* is the concentration of electrons in the electron traps H, or excited traps H*, f is the concentration of defect electrons (holes), t is time, and $a,\beta,\gamma,a',\beta',\gamma'$ transfer probabilities of the processes indicated in Figure 2-2.

Equations 2.1 can be simplified because the mean lifetime of trapped electrons (τ) is assumed to be large:

$$h \gg n,\ h \gg h^*,$$
and consequently:

$$f = h, \frac{df}{dt} = \frac{dh}{dt}. \tag{2.2}$$

For a (and a'), the following temperature dependence is assumed:

$$a = a_o e^{-E/kT} \tag{2.3}$$

    E = difference between electron trap and conduction band
    K = Boltzman's constant
    a_o = frequency factor.**

In the following, only one type of trap of a uniform depth E (single glow peak) will be considered. For multipeak phosphors, each trapping level has to be treated separately. The above equations can be used to illustrate the relationship between peak maximum T_m, shape and size of the glow peak, the effect of heating rate q on the peak shape and T_m, and the thermal stability (fading kinetics).

The light emission intensity Φ of the phosphor during heating can be described as

$$\Phi = -c\frac{dh}{dt} \ \ (c = \text{proportionality factor}) \tag{2.4}$$

and the integrated light sum during the heating (or area under glow peak Q) as

$$Q = -\int_{t_1}^{t_2} c\ \frac{dh}{dt'} = ch(t_1), \tag{2.5}$$

---

*Attempts have been made to substantially enhance the sensitivity of TL phosphors by applying a strong electric field of $\sim 10^5$ V/cm during heating. The idea is to accelerate thermally released electrons so that they produce impact ionization (Miyashita and Henisch, 1967). This effect has, however, not found any practical applications yet.

**If the trap is regarded as a potential box, a_o will express the product of the frequency of the electron oscillation and the reflection coefficient. It should be less than the vibrational frequency of the crystal, which is in the order of 10^{-12} sec, but may appear to be much higher.

when t_1 is the starting time and t_2 the end of the heating period, and $h(t_2) = 0$.

If one assumes that no retrapping occurs, the glow curve $\Phi(T)$ can be calculated from 2.4, and $h(t)$ results from the integration of 2.1:

$$\frac{dh}{dt} = -a(T)h \qquad (2.6)$$

and assuming further that the linear heating starts at t_1

$$T = q(t - t_1) + T_A, \quad dt = qdt, \qquad (2.7)$$

integration of 2.6 results in

$$h(T) = h(t_1)e^{-\frac{1}{q}\int_{T_A}^{T} a(T')d\,T'} \qquad (2.8)$$

This equation describes the process of emptying occupied electron traps. The analytical shape of the glow curve can be deduced by

$$\Phi = -c\,\frac{dh}{dt} = -qc\,\frac{dh}{dT} \qquad (2.9)$$

$$\Phi(T) = ch(t_1)a(T)e^{-\frac{1}{q}\int_{T_A}^{T} a(T')dT'} \qquad (2.10)$$

This equation is well known as the Randall-Wilkins formula.* From 2.10, the peak temperature T_m results as $d\,\Phi/dT = 0$, and

$$a'(T_m) = a^2(T_m)\frac{1}{q}. \qquad (2.11)$$

From Equations 2.3 and 2.11 it follows that

$$\frac{E}{kT_m^2} = \frac{a_0}{q}\,e^{-\frac{E}{kT_m}} \qquad (2.12)$$

The implications of Equation 2.10 are presented in Figure 2-3 for various values of E and a_0 at a constant q. Obviously, T_m is shifted toward higher temperatures when E increases or a_0 decreases. (This relationship between E, a_0 and T_m is given in a general form in Figure 2-4A.) The area under each glow peak, however, is proportional to the number of trapped electrons and, therefore, to the dose of a given type of radiation. For constant E, a_0, and q, the peak height (Φ at T_m) can also be used as a measure of dose.

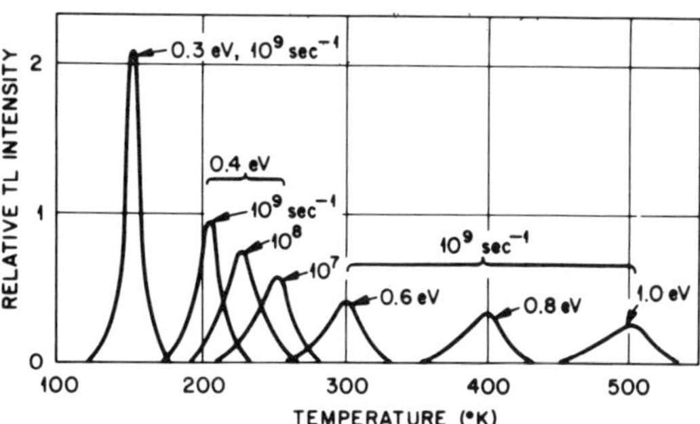

FIGURE 2-3. Theoretical TL glow curves for phosphors with a single trap depth and no retrapping, heated at a constant rate of 2.5°C/sec. (After Garlick, 1949.)

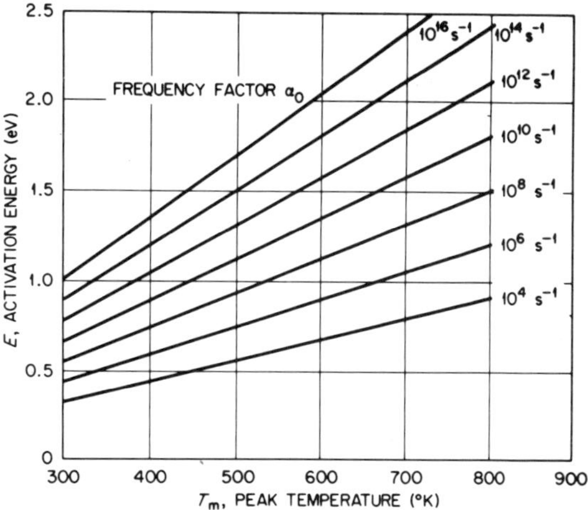

FIGURE 2-4A. Diagram displaying the relationship between frequency factor a_0, activation energy E, and peak temperature T_m, normalized for a heating rate q of 1°K/sec, according to the Randall-Wilkins equation. (After Holzapfel, 1972.)

The peak temperature T_m increases with increasing q. Equation 2.12 forms the basis for experimental methods to determine E and a_0 by measuring, for example, T_m and the half-width of glow peaks for different values of q (Grant et al., 1968 and Pohlit, 1969). For a graphical display of the relationship between peak temperature, peak width, frequency factor, and activation energy, see Figure 2-4B. As a crude approximation, a simple relationship can be used for the estimation of E, assuming an average a_0 of about $10^9/sec^{-1}$ and q $= 1°C/sec$:

$$E = T_m\,(°K)/500 \qquad (2.13)$$

---

*One of its authors, M. H. F. Wilkins, later won a Nobel prize for his work on the structure of DNA.

For the determination of $\Phi(T_m)$ based on 2.10, another approximation can be useful (Witzmann and Buhrow, 1960):

$$\int_{T_A}^{T_E} e^{-\frac{E}{kT'}}\, dT' = \left[T e^{-\frac{E}{kT}} \left\{ \left(\frac{kT}{E}\right) - 2\left(\frac{kT}{E}\right)^2 + 6\left(\frac{kT}{E}\right)^3 + \ldots \right\} \right]_{T_A}^{T_E} \tag{2.14}$$

From Equations 2.14 and 2.12, the glow maximum can be estimated as follows:

$$\Phi(T_m) = c h(t_1) \frac{Eq}{kT_m^2} e^{-\left(1 - 2\frac{kT_m}{E} + \ldots\right)} \tag{2.15}$$

The thermal fading of the stored energy in the phosphor is described as the leakage of electrons from traps h(t) at a temperature T_A. It can be calculated from 2.1. Some solutions, resulting in somewhat different shapes of the theoretical fading curves, are given in Table 2.1. As a rule, the thermal stability increases rapidly with increasing peak temperature, with a peak location of 100°C corresponding to a halflife of the trapped electrons or holes of about 3 hr at 20°C; at 200°C, the halflife at this storage temperature would be about 1 year; at 300°C, 3,000 years; at 400°C, 10 million years; and at 500°C 3 x 10^{10} years. Obviously, the parameters E and a_o can also be determined from isothermal annealing curves. Disagreement between the values as obtained by these different methods usually indicates the presence of complicating secondary processes such as retrapping or trap transformation.

This simple model has been subject to numerous modifications in attempts to explain the complexities of experimental findings. Phosphors rarely show only one peak, but several peaks, which can be described by attributing different values of E and/or a_o to each type of trap. Less than proportional response at high dose-levels, leading to saturation usually in the 10^5 to 10^7 rad gamma dose-range, is easily explained on the basis of the complete filling (saturation) of the relevant traps (it should be noted that nonlinearity and saturation may occur at quite different dose-levels for different traps in the same phosphor). Other models take into account that the traps that give rise to a given glow peak rarely have a uniform, well-defined E, but assume a continuous Gaussian trap depth distribution (Watanabe and Morato, 1971). Other refinements take into consideration the role of holes (defect electrons) in TL processes.

In the probably most profound recent treatment of the phenomenological theory of TL and

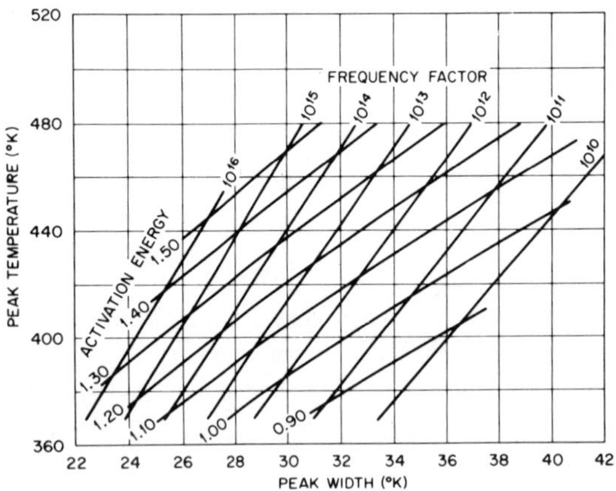

FIGURE 2-4B. Diagram showing the theoretical relationship between peak temperature, peak width, activation energy, and frequency factor for a heating rate of 1°K/sec, according to the Randall-Wilkins equation. (After Grant et al., 1968.)

TABLE 2-1

Solutions for h(t) at Constant T from Equation 2.1

Model	Conditions for parameters	h(t)
Biomolecular	$\dfrac{\beta}{\gamma} = 0$	$h(0)\, e^{-at}$
	$\dfrac{\beta}{\gamma} = 1$	$\dfrac{h(0)}{1 + \dfrac{h(0)}{H} a t}$
	$\dfrac{\beta}{\gamma} = \text{discretionary}$	$\dfrac{h(0)}{(1 + at)^b}$ *
Monomolecular	$\dfrac{\beta}{\gamma} = \text{discretionary}$	$h(0) e^{-\dfrac{a}{1 + \beta/\gamma} t}$

*Approximation, a and b are constants

(After Frank and Stolz, 1969.)

its correlation with thermally stimulated current (TSC) and exoelectron emission, and the "exact" solutions of the kinetic equations which govern these processes (Kelly and Bräunlich, 1970; Bräunlich and Kelly, 1970; Kelly et al., 1971; and Kelly, 1971), it is concluded that even sophisticated experiments will unlikely yield unique values of the trapping parameters. It is, however, possible from correlated TL and TSC experiments to determine whether there is a charge carrier transport associated with the release and recombination of charge carriers and whether there is a large concentration of "thermally disconnected" traps involved. They also show that in a model with a given trap depth in the presence of other deep traps and a single recombination center, conventional approximations become inadequate if the number of active traps is less than 10^{15} cm^{-3}.

Even on the basis of the most sophisticated theoretical considerations it is, however, still impossible to properly describe or predict a number of other, more or less undesirable properties of TL phosphors such as tribo- and chemoluminescence (which is more pronounced in some phosphors than in others and may seriously interfere with low dose measurements); "abnormal" rapid fading in some materials (which is probably due to nonradiative recombination processes); sensitization of phosphors by preheating and/or preirradiation; and a more than proportional ("supralinear") response which has been observed in a number of materials, mostly at higher dose-levels.

Supralinearity, in particular of lithium fluoride whose oversensitivity decreases with increasing LET, has been the subject of intense studies, but no completely satisfactory explanation is available yet. One mathematical model for the supralinearity of LiF:Mg,Ti (Cameron et al., 1967b) proposed the creation of new traps at higher exposure levels. This model was later modified by the assumption that not electron traps, but new recombination centers were formed by the radiation exposure. There was, however, strong experimental evidence against both models. Still another modification of this model (Zimmerman and Cameron, 1968) proposed that the photon emission probability for the same electronic processes increases with increasing radiation exposure; this also is not very likely.

Of the two supralinearity theories that are currently popular, the "overlapping track model" (Claffy et al., 1968 and Dobson and Midkiff, 1970) is based on geometrical and carrier lifetime considerations in the increasingly overlapping "spheres of influence" of the delta rays in the phosphor with increasing dose. Although modified since and supported by some experimental evidence, this theory conflicts with some other results.

Also unable to explain all aspects of supralinearity is the "competing trap model" (Cameron et al., 1968c), which assumes the presence of a trap with a larger cross section for electrons than the ones contributing to the detected TL. Only after filling these competing traps, all (or a constant percentage of) additional liberated electrons can be trapped in the TL traps.

Increasing LET in most phosphors results in a decrease in response per rad of absorbed dose. This effect, which is also observed in radiophotoluminescence and exoelectrons dosimeters, can be explained largely in terms of saturation effects (overexposure) in a narrow channel along heavy charged tracks in which most of the delta electron energy is deposited (Katz et al., 1972). In a recent study (Zimmerman, 1972) of different natural fluorites, CaF$_2$:Mn, CaF$_2$:Tb, CaSO$_4$:Mn, and quartz as TLD phosphors, for example, the TL response per rad of alpha radiation was less than that per rad of beta radiation for all glow peak studies. The 3.7 MeV alpha particle efficiencies ranged from 53% for CaF$_2$:Mn to 2% for the 110°C peak in quartz, and decreased with decreasing alpha energy (= increasing LET): At doses in the saturation region, however, the responses became equal. As a rule, the higher the beta dose at which a peak saturates, the higher is its alpha efficiency at low doses.

Another contributing factor to the LET dependence may be partial thermal annealing along the track (thermal spikes). There is no explanation yet for the fact that in some cases the response actually increases with increasing LET (Figure 2-5). According to one (unconfirmed) report, in some phosphors the sensitivity to alpha particles exceeds the gamma radiation sensitivity by a factor of more than 100 (Kuzmin et al., 1967).

2.3. Phosphors
2.3.1. General Considerations
Many inorganic compounds are known to exhibit TL. However, a TL phosphor for dosimetric use should combine several properties which limit the choice to only a handful of

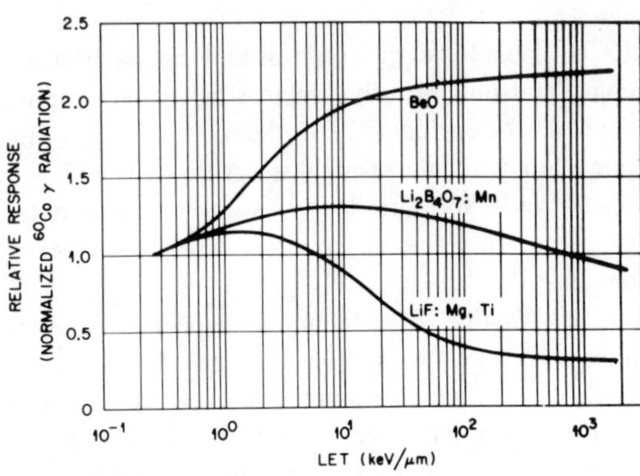

FIGURE 2-5. Relative response (normalized for the response to gamma radiation) of three low Z TLD phosphors as a function of LET. (After Tochilin et al., 1968.)

compounds to be discussed in some detail here. A list of some of the materials that have attracted attention as phosphors for TL dosimetry is given in Table 2-2. Because of the continuing process of introducing more or less promising new materials, replacing some of the older phosphors, this list should by no means be considered comprehensive.

Some of the more desirable properties of a **TLD** phosphor are

a. A high concentration of electron (or hole) traps and a high efficiency in the light emission associated with recombination processes;

b. A sufficient storage stability of the electrons (or holes) to cause no disturbing fading even during extended storage at ambient or slightly increased temperatures such as tropical climates or, in case of biomedical implantation studies, the body temperature. This usually is the case when the T_m of the main peak is between about 180 and 250°C (at higher temperatures, the infrared emission from the hot sample and sample holder increasingly interferes with the measurement of low doses);

c. A spectrum of the emitted TL light to which the detector system (photomultiplier/filter combination) responds well, and which interferes as little as possible with the incandescent (infrared) emission of the heated phosphor and its immediate surroundings (wavelengths between ∿300 and 500 nm are most desirable unless specially UV sensitive systems are used);

TABLE 2-2A

Some Commercially Available Low Z* TLD Phosphors

Host crystal	Activators	Trade name	Manufacturer	Reference
BeO	Na, Li		Matsushita, Osaka, Japan	Yasuno and Yamashita, 1971
BeO	(Si, Fe, Mg, etc.)	Thermalox 995	Brush Beryll. Co., Elmore, Ohio	Scarpa et al., 1971
BeO	?		Consolidated Beryllium Ltd., Milford Haven, Wales, England	Scarpa, 1970
$Li_2B_4O_7$	Mn		Harshaw Chem., Cleveland, Ohio	Schulman et al., 1967
$Li_2B_4O_7$	Mn, Si		Danish AEC, Risö Roskilde, Denmark	Bötter-Jensen and Christensen, 1971
MgB_4O_7	Tb		Dai Nippon Toryo Co., Chigasaki, Japan	
LiF	Mg, Ti	"Throw-away"	Radiat. Detect. Co., Mountainview, Calif.	La Riviere, 1967
LiF	Mg, Ti	200-Ti**	Dohna-lum, VEB Fluorwerke Dohna, East Germany	Frank and Edelmann, 1966
LiF	Mg, Na	PTL 710	CEC, Montrouge, France	Portal et al., 1971
^{7}LiF	Mg, Na	PTL 717	CEC, Montrouge, France	Portal et al., 1971
LiF	Mg, Ti	TLD-100	Harshaw Chemical Co., Cleveland, Ohio	Cameron et al., 1964
^{7}LiF	Mg, Ti	TLD-700	Harshaw Chemical Co., Cleveland, Ohio	Cameron et al., 1964
^{6}LiF	Mg, Ti	TLD-600	Harshaw Chemical Co., Cleveland, Ohio	Cameron et al., 1964

*Maximum inherent photon energy dependence of response +40 and −20% between 10 keV and 2 MeV (for details, see text).
**Other modifications also available.

TABLE 2-2B

Some Commercially Available High Z TLD Phosphors

Host lattice	Activators	Manufacturer	Reference
Mg_2SiO_4	Tb	Dai Nippon Toryo Co. Chigasaki, Japan	Hashizume et al., 1971
Al_2O_3		CEC, Montrouge, France	Francois et al., 1968
CaF_2	Dy	Harshaw Chemical Co., Cleveland, Ohio	Binder et al., 1968
CaF_2	Mn	Harshaw Chemical Co., Cleveland, Ohio	Schulman et al., 1960, 1963
$CaSO_4$	Tm	Matsushita, Osaka, Japan	Yamashita et al., 1968
$CaSO_4$	Dy	Harshaw Chemical Co., Cleveland, Ohio	Yamashita et al., 1968
$CaSO_4$	Mn	Harshaw Chemical Co., Cleveland, Ohio	Bjärngard, 1963 and 1967

TABLE 2-2C

Some Physical Properties of Host Crystals

Crystal	Density (g/cm^3)	Melting point (°C)	Solubility	Energy response*	Effective atomic number
LiF	2.60	870	0.027% cold H_2O	1.35	8.2
$Li_2B_4O_7$	2.3	$\sim$800	Soluble in H_2O	0.8	7.4
BeO	3.01	2530	Conc. H_2SO_4, H_3PO_4	0.86	7.5
CaF_2	3.18	1360	0.00016% cold H_2O	13	16.3
$CaSO_4$	2.96	1450	0.02% cold H_2O, conc. H_2SO_4	12	15.3
Al_2O_3	3.97	2050	—	3.5	

*Sensitivity at $\sim$20 to 30 keV relative to sensitivity at $\sim$1 MeV; measured values usually differ somewhat from calculated.

d. A trap distribution that does not complicate the evaluation procedure by the presence of additional, rapidly fading low temperature peaks, or high-temperature peaks which may be difficult to anneal, give rise to retrapping processes, etc.; the phosphor should preferably be completely annealed by the readout procedure without permanent changes in its sensitivity and/or background reading or linearity;

e. Resistance against potentially disturbing environmental factors including light, humidity, organic solvents, and common fumes and gases unless the detector is permanently sealed into a glass bulb;

f. Low photon energy response (low Z of the host lattice) and a linear response over a wide dose-range are not always, but for most applications are desirable features; depending on the particular use, a high or low thermal neutron sensitivity, or a certain LET response may also be desirable characteristics;

g. The phosphor should, of course, also not be exceedingly expensive (it may be easier to work with a one-way dosimeter), nontoxic, it should not deteriorate during extended storage, should be as little as possible subject to spurious effects such as tribo- or chemoluminescence, oxygen effects, etc.; it may have advantages if the material can easily be prepared with reproducible properties in a normally equipped chemical laboratory.

Obviously, it is difficult to meet all these conditions in a single detector material, and the detectors listed in Table 2-2 represent, at best, a reasonable compromise between the requirements.

2.3.2 Lithium Fluoride

Lithium fluoride with its low Z and simple cubic lattice is not only by far the best-studied and presently most widely used TL phosphor, but also the best-known solid-state detector. For many health and hospital physicists it became almost synonymous with solid-state dosimetry.

Currently more than 100 publications are written annually on the properties and applications of this material, in particular on "Mr. Harshaw's mysterious crystals," LiF:Mg,Ti (trade name TLD-100®, 600® and 700®). Being the first low Z phosphor that became commercially available, its momentum and widespread fame are, unfortunately, not matched by particularly desirable performance characteristics. Its supralinearity at higher dose-levels, the complicated annealing behavior, and relatively poor sensitivity make it, in fact, less suitable for many applications than other, more recently developed and less widely known phosphors. It seems likely that it will soon be replaced in many areas of application.

2.3.2.1. Phosphor Preparation

The original LiF used by Daniels and his students (Morehead and Daniels, 1952) was made by the Harshaw Chemical Co., Cleveland, Ohio, prior to 1954 and exhibited a TL peak around 190°C. LiF made between 1954 and 1960 was probably purer, and consequently much less sensitive. Systematic studies of various activators and activator combinations (U.S. Patent 322248,7,11, 1963) led to a material which has since been widely known in three modifications, namely, TLD-100 (natural isotopic ratio of 7.5% ^{6}Li and 92.5% ^{7}Li), TLD-600 (approximately 95.6% ^{6}LiF, 4.4% ^{7}LiF), and TLD-700 (∿99.99% ^{7}LiF, ∿0.01% ^{6}LiF). The activators have not been officially disclosed, but among the various impurities in this material (Table 2-3), magnesium and titanium are generally believed to be those of dosimetric importance (see, for example, Edelmann, 1968; Rossiter el al., 1971; and Robertson and Gilboy, 1971).

Apparently Mg is the more important coactivator, which can be replaced by other elements only to a limited extent. The optimum sensitivity results from the addition of 0.0013 mol % of Mg

TABLE 2-3

Impurities and TL Sensitivities of Some Commercial Lithium Fluorides

Material	Approximate impurity content (ppm)* Al	Ca	Mg	Si	Ti	Relative response to γ radiation**
Harshaw TLD-100, 600, and 700	20	6	300	40	5	1
Dohna-lum 200-Ti	40	100	400	400	10	0.84
Radiation Detection Co. "throwaway" (E. Leitz)	10	4	40	100	10	0.48
E. Merck "ultrarein" (analytical grade)	3	4	10	40	<3	<0.001
J. T. Baker "analyzed reagent"	7	7	10	100	<4	0.003

*According to semiquantitative spectrochemical analysis, and some mass spectrometric results
**Integral measurement with a Radiation Detection Mark IV reader

(After Becker et al., 1970.)

to pure LiF (Figure 2-6). As the second co-activator, numerous cations have been studied, in particular europium (Jones and Burt, 1967), aluminum, calcium, titanium, and copper (Edelmann, 1967). Several other impurities have either no effect or an adverse (quenching) effect on the TL intensity. Iron and nickel belong to the latter category.

As can be seen in Figure 2-5, a combination of 0.0013 mol % Mg and 0.001 mol % Ti results in a very sensitive material. These numbers should, however, be taken *cum grano salis*. According to one recent study (Rossiter et al., 1971), many supposedly "pure" LiF samples already contain initially about 5 ppm of Ti, and only $\sim$8 ppm (in addition to 80 ppm of Mg) are required for optimum TL sensitivity (at $\sim$3 and $\sim$20 ppm Ti, the main TL peak height is reduced to half of its maximum). Other recommended activator combinations for LiF are 0.03 mol % Mg plus 0.6 mol % Eu (Jones and Burt, 1967), which also results in a main glow peak at $\sim$200°C. Still others have used Cu and Ag, resulting in a rather rapidly fading glow peak at $\sim$120°C (Niewiadomski et al., 1971), or 200 ppm Mg and 1% of Na, resulting in a material for which a simplified annealing procedure is claimed (Portal et al., 1971). In activation studies, it is sometimes difficult to avoid problems due to small impurities of either the host crystal or the activator. An apparent activation of LiF with MnF_2 could, for example, be attributed to Ti traces in the MnF_2 (Rossiter et al., 1972).

Many details in the manufacturing of TL-grade lithium fluorides are surrounded with commercial secrecy, and as a rule they cannot be as easily prepared in reproducible, large quantities as some of the other phosphors. There are, in principle, three methods which are used. The simplest is to select from the wide choice of more or less pure commercial lithium fluorides (which are available for various technical, chemical, and optical purposes), one which happens to contain the proper amounts of activators as impurities. The sensitivity of some materials that can be purchased inexpensively from companies like E. Leitz, Wetzlar, permits their use as one-way ("throwaway") dosimeters (La Riviere, 1968), in particular after sensitization by a pre-irradiation treatment.

Pre-irradiation to a high dose followed by an annealing treatment (for example $>10^6$ rad ^{60}Co irradiation, followed by 4 hr at 300°C and 1 hr at 350°C, or several hours at 290°C) increases the sensitivity by a factor of about five. Pre-irradiation also improves the linearity of response, probably due to the filling of competing high-temperature traps. Recent comparisons of the TL, photo-luminescence, radioluminescence, and thermally stimulated exoelectron emission in LiF:Mg,Ti (Zimmerman, 1971b) confirmed that the radiation-induced sensitization is caused by the removal of "competitors" with the active TL centers.

The second method consists of growing large LiF crystals from a melt to which the dopants are added. A protective atmosphere (N_2 or A) helps to produce optimum sensitivity. The crystals are then powdered to a grain size of about 0.2 mm. TLD-100 is made by this method. Various attempts to cut small crystalline pieces from large crystals for dosimetric purposes have been un-successful, mainly because there are considerable sensitivity fluctuations ("hot spots"). Recently methods for growing large crystals with uniform TL response that can be sliced into useable smaller units have been studied.

Finally, LiF phosphors can be prepared from a lithium chloride solution in water by addition of a mixture of HF and the dopants. The resulting powder is heated for about 1 hr to 630°C. Powders that have been prepared by either method can be used

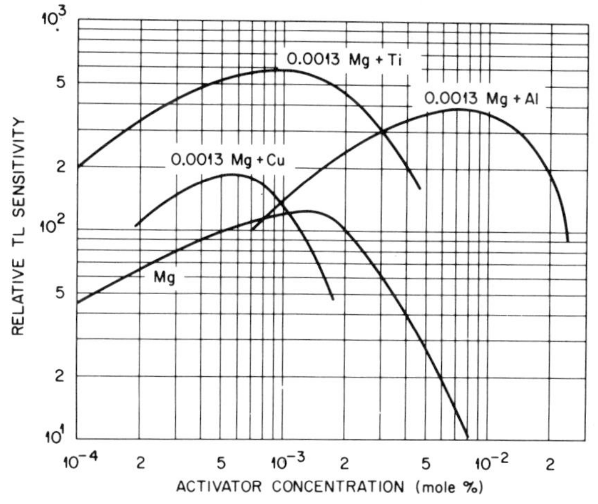

FIGURE 2-6. Effect of varying amounts of activators on the TL response of lithium fluoride. (After Edelmann, 1967.)

a. in amounts of $\sim$8 to 25 mg per reading as a loose powder,

b. permanently encapsulated in potassium-free glass capillaries or attached to a heatable carrier in a light-bulb type of dosimeter,

c. incorporated into high-melting, inert substrates such as Teflon or silicon rubber,

d. hot-pressed into an almost transparent poly-crystalling material ("extruded ribbons") that is commercially available as rods or small squares of varying dimension (Cox, 1968), or

e. for forming pressed (Shambon and Condon, 1968) and/or sintered (Portal et al., 1971 and Niewiadomski et al., 1971) pellets of uniform dimensions, which are less sensitive but easier to handle than powder.

The main disadvantage of most LiF phosphors is that they can only be reused after a relatively complicated and time-consuming annealing consisting of a three-step procedure:

1. 1 hr heating to 400°C (or, in the case of embedded materials which would melt at this temperature, ∿300°C);

2. cooling at a constant rate; and

3. 20 hr storage at 80°C (Zimmerman et al., 1966).

Slight modifications of this "magic treatment" are tolerable. For example, some authors have replaced the 20 hr at 80°C by 2 hr at 100°C. On the other hand, variations in the cooling rate may result in differences in the light output of the irradiated material by a factor of two.

If the detectors are only used a few times for the measurement of comparable and relatively low doses, the reading cycle may provide sufficient annealing, and the remaining few percent of stored signal from the previous exposure(s) as well as the induction of low-temperature peaks which give rise to a short-term fading may not disturb consecutive measurements. It should, however, be noted that a marked growth of the main dosimetry peak with storage time, either before or after exposure to radiation, takes place in samples which have been annealed by the readout only (Booth et al., 1972). This sensitivity increase, which amounts to 40% after one month at 22°C (the exact magnitude depends on the readout heating rate and the population of low-temperature traps), is probably caused by the migration and aggregation of simple, low-temperature trapping centers into more complex, deeper trapping centers. If, therefore, multiple reuse, precise measurements, and/or the

measurement of low doses after high-dose use of the detectors are desirable, complete annealing without changes in the phosphor characteristics according to the above-described procedure is mandatory.

Recently, two approaches to avoid the awkward annealing procedures have apparently been successful. One technique is based on a different preparation of the phosphor (Portal et al., 1971), another on a special postreading annealing step which is built into the reading cycle of the TL reader (Webb and Phykitt, 1971).

Another disadvantage of LiF:Mg,Ti is a permanent loss in sensitivity after exposure to a high radiation dose, which was first reported by Marrone and Attix (1964), and studied in some more detail by Doppke and Cameron (1966). The effect is small for gamma doses of less than a few thousand rads (Figure 2-7), but amounts to a loss of two thirds of the initial sensitivity after pre-exposures exceeding 0.2 Mrad. Both the irradiation dose and annealing temperature have a strong influence on the degree of radiation damage which is produced (for a detailed discussion, see Cameron, 1968).

The desensitizing effect of high gamma radiation doses is superimposed by the previously mentioned, thermally less stable sensitizing effect of pre-irradiation. Optimum sensitization by about a factor of five can be obtained by exposure to 3 x 10^4 to 10^5 rad of gamma radiation, followed by 1 to 4 hr of heating at a temperature of 280 to 300°C (Figure 2-8). The practical value of this procedure is, however, limited because the sensitization also increases the "background" of the phosphor, thus making it more difficult to measure small exposures. A positive effect of radiation sensitization is, however, the reduction in supralinearity which it produces.

2.3.2.2. Glow Curve and Mechanism

The dose response characteristics of LiF:Mg,Ti are rather complex, and affected by numerous parameters such as LET and dose-level of the radiation, thermal treatment (annealing) of the phosphor, pre-irradiation to high doses, pretreatment with certain chemicals, etc. No attempt will be made here to present a comprehensive account of the hundreds of published investigations which deal with the various aspects of these disturbing effects. Only some of the more important factors are discussed in the light of more recent published data.

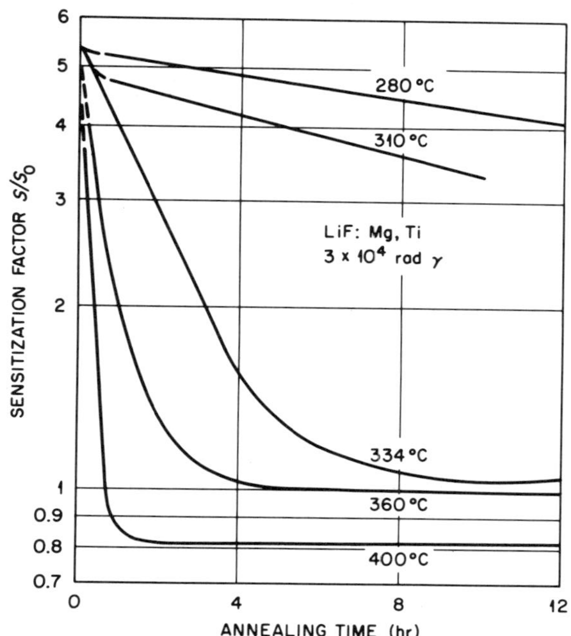

FIGURE 2-7.   TL sensitivity loss of LiF:Mg,Ti as a function of preexposure of the phosphor to gamma radiation. (After Doppke and Cameron, 1966.)

FIGURE 2-8. Sensitization factor of LiF:Mg,Ti after exposure to 3 x 10⁴ rad gamma radiation, as a function of annealing time and temperature. (After Wilson et al., 1966.)

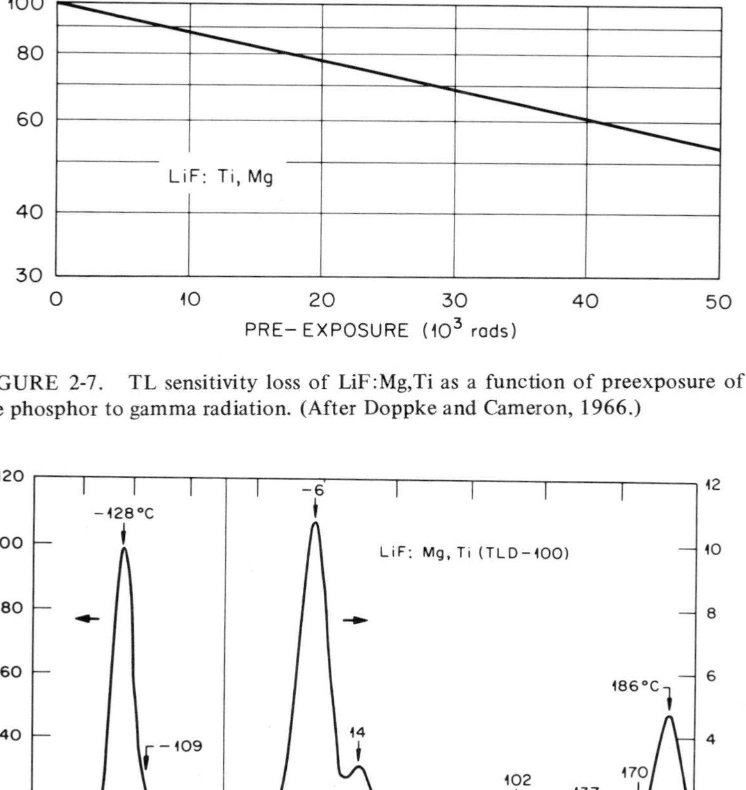

FIGURE 2-9.   Glow curve of a LiF:Mg,Ti crystal during heating from liquid nitrogen temperature to ~200°C (exposure to ~10⁴ rad of X-radiation, heating rate ~35°C/sec). (After Podgorsky et al., 1971.)

A typical "glow curve" of LiF:Mg,Ti, as recorded in a commercial reader (Figure 2-1), will show a main peak at a temperature of about 180 to 220°C (depending mostly on the heating rate and the accuracy of the temperature measurement), as well as one or more smaller peaks between ∿80 and 180°C. A more careful analysis of the glow curve resolves a considerably more complicated glow spectrum (Figure 2-9). Between liquid nitrogen temperature and the main dosimetry peak, at least ten TL peaks occur (Podgorsky et al., 1971 and Nollman and Smolko, 1972). Some of their characteristics are listed in Table 2.4.

It should, however, be noted that different results are obtained with LiF samples from other sources. In "pure" LiF, for example, TL peaks at about −166°C and −135°C have been observed which are very similar in shape and location to peaks in thermally stimulated electrical con-

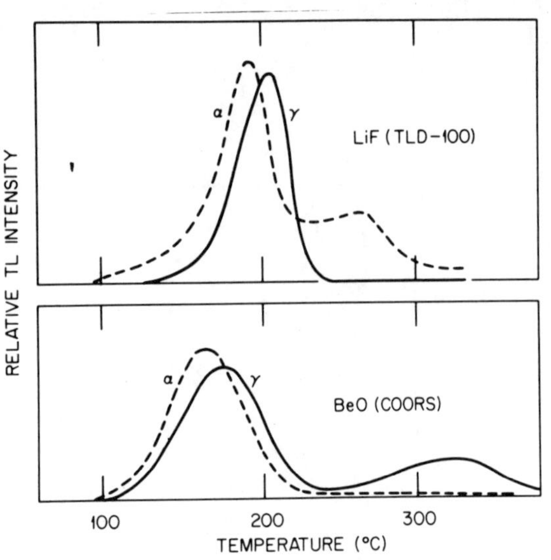

FIGURE 2-10. Glow curves of LiF (TLD-100) and BeO after exposure to alpha and gamma radiation. (After Becker et al., 1970.)

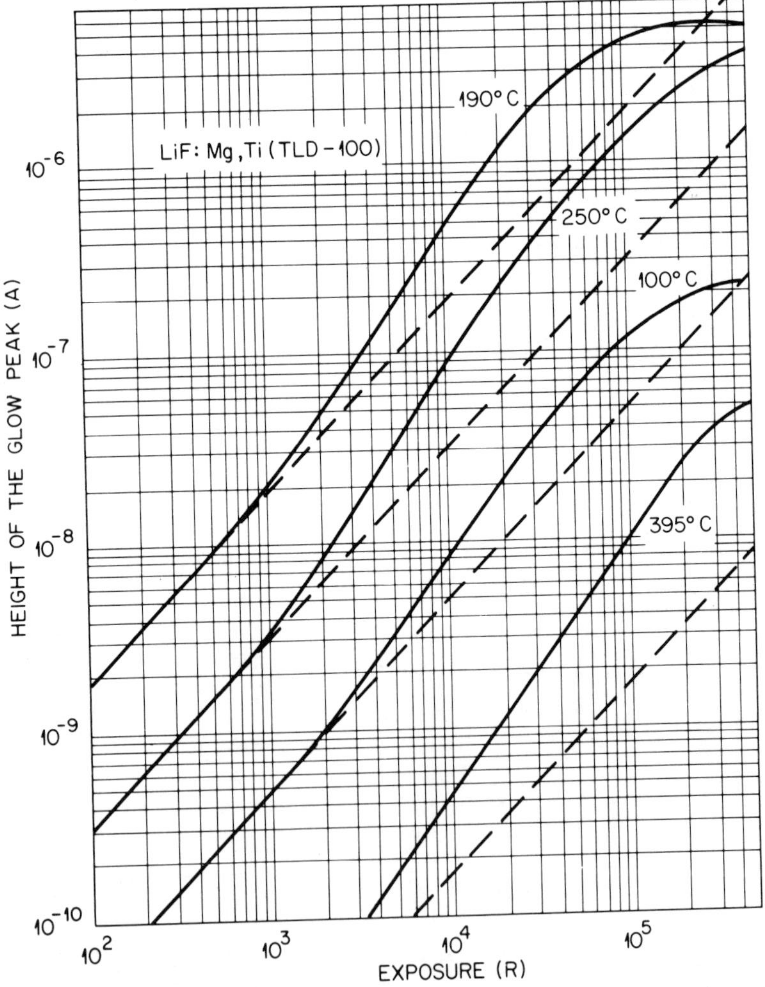

FIGURE 2-11. Height of various TL peaks in LiF:Mg,Ti as a function of exposure. (After Sunta et al., 1971.)

TABLE 2-4

Some Characteristics of Low-temperature TL Peaks in LiF:Mg,Ti

Peak temperature (°C)	Maximum supralinearity factor	Peak of optical emission (nm)	Charge carrier
−128	1	250	h+
−109			e−
− 88			e−
− 6	1.61	400	
14	1.85		
59			e−
102	2.85	400	e−
137		400	e−
164		400	e−
186	4.62	400	e−

(After Podgorsak et al., 1971.)

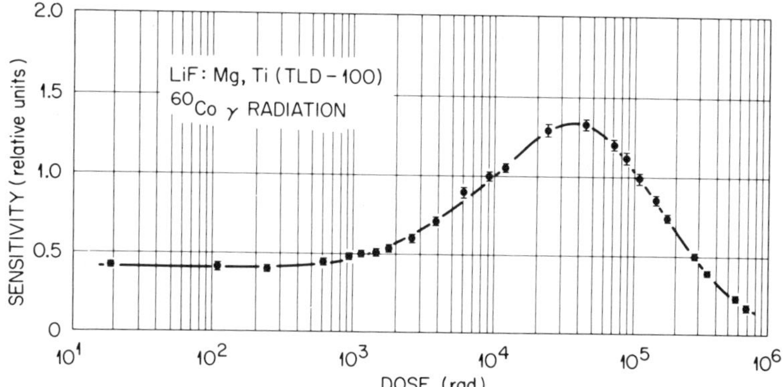

FIGURE 2-12.   Relative response of the main TL peak in unmodified LiF:Mg,Ti (TLD-100) to gamma radiation from ^{60}Co. (After Eggermont et al., 1971.)

ductivity (Böhm and Scharmann, 1969). The decay of the low-temperature glow peaks results in a "phosphorescence" of freshly irradiated LiF: Mg,Ti. In TLD-100, at least two decay times with halflives of 2.7 and 25 min at 21°C have been identified (Figure 2-10 and Burch, 1968).

There are also several peaks at higher temperatures, but increasing infrared emission makes the practical use of these peaks difficult unless special precautions are taken. However, one high-temperature peak which is usually found around 300°C has been studied in some detail and occasionally has been used in dosimetric studies (for some more recent work on deep trapping levels, see Becker et al., 1970 and Sunta et al., 1971); some authors even used a peak between 425 and 475°C for high level dosimetry (Goldstein et al., 1968). The response of one LiF sample made by Isomet Corp., Palisades Park, N. J., was almost linear up to $\sim 10^8$ R of ^{60}Co gamma exposure.

Not only the absolute sensitivity (Figure 2-4) but also the peak ratio in LiF:Mg,Ti depends, as is the case in several other phosphors, on the LET of the radiation (Figure 2-10). If the phosphor is exposed to protons, alpha radiation, or thermal neutrons, the high-temperature peak becomes more pronounced (Wingate et al., 1967; Lucas and Rainbolt, 1968; Ayyangar et al., 1968; and Becker et al., 1970). As can be seen in Table 2-4 and Figure 2-11, supralinearity of response increases with increasing peak temperature. For the common "dosimetry peak" at ~ 200°C and gamma radiation, maximum oversensitivity due to supralinearity by a factor of about 3 to 7 is obtained around 3×10^4 R (Figure 2-12).

As can be seen in Figure 2-12, supralinearity that complicates the dose evaluation somewhat does not occur below gamma or x-ray exposures of about 300 to 1,000 R (the exact value depends somewhat on several experimental factors). Much less or no supralinearity also at high exposure levels occurs under various conditions, in particular if

a. the phosphor has been preexposed to a high radiation dose and the deep traps have not been annealed (La Riviere, 1968);

b. the LET of the radiation is high, for example, for exposure of alpha radiation, thermal neutrons, or low energy x-rays and electrons (Figures 2-29 and 30; see also Naylor, 1965; Wingate et al., 1967; and Suntharalingam and Cameron, 1969); or

c. the LiF:Mg,Ti has been chemically modified, for instance by diffusion of hydroxide ions into the lattice (Figure 2-13). The effect of OH^- is assumed to be due to the formation of complexes with Mg^{2+} cation dipoles (De Werd and Stoebe, 1971).

At least in TLD-100, supralinearity does not depend on the grain size of the phosphor, as it does in a Japanese LiF material (Shiragai, 1970 and Zanelli, 1972).

The fading stability of a TL peak can be described by the equation

$$1 - \frac{h(t)}{h(o)} = 1 - e^{-t} \qquad (2.16)$$

The trap parameters for the main TL peak in LiF:Mg,Ti at $\sim 200°C$ have been determined by

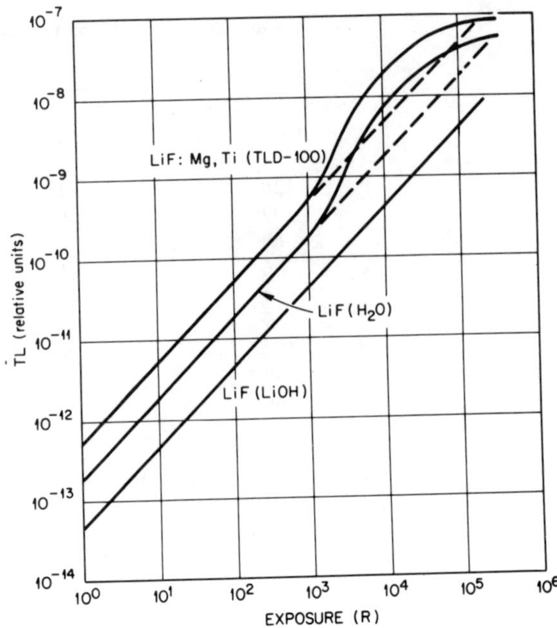

FIGURE 2-13. Relative TL intensity in LiF:Mg,Ti before and after diffusion of water and LiOH into its lattice at 700°C. (After DeWerd and Stoebe, 1971.)

isothermal annealing studies at elevated temperatures to be about E = 1.3 eV, and $a_o = 10^{12}$ sec^{-1} (Zimmerman et al., 1966; Frank and Stolz, 1969; and Johnson, 1972). Extrapolation of the experimental fading data at T=100°C down to ambient or slightly elevated temperatures would result in very low fading rates (calculated value for

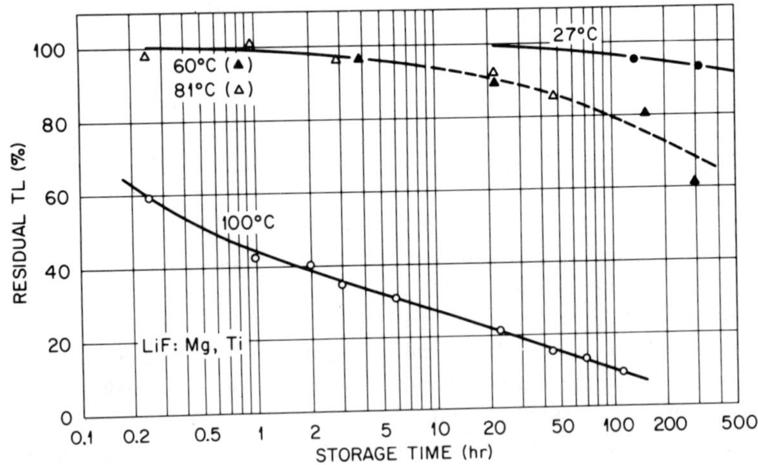

FIGURE 2-14. Fading of TL in LiF:Mg,Ti (TLD-100) powder at various storage temperatures as measured in a TL reader with a preheating cycle (more initial fading would be observed in a reader without such a heating cycle). (After Becker et al., 1971.)

two months at 50°C:3%). Actually, higher rates have been observed, ranging from ∿5% during three months at room temperature (Gorbics et al., 1967 and Suntharalingam et al., 1968) to ∿12% during six weeks at 32°C and ∿10% during one day at 60°C (Becker et al., 1971 and Figure 2-14). The fading of the radiation-induced signal in LiF:Mg,Ti should not be ignored if the phosphor is being used over extended periods of time in warm climates. In one recent study (Becker, 1973), for example, between 10 and 23% fading was observed during three months of storage at tropical or semitropical locations such as Buenos Aires, Santiago de Chile, Bogota, or Guayaquil.

As the low-temperature peaks obviously fade faster than the high-temperature peaks, the measured fading rate will depend on the readout procedure, being less pronounced if a preheating cycle (excluding the low-temperature peaks from integration) is used. It has been suggested to make use of the storage time dependence of the peak ratio for determination of the approximate time elapsed since the exposure occurred (Spurny, 1971). Other factors that complicate the inter-

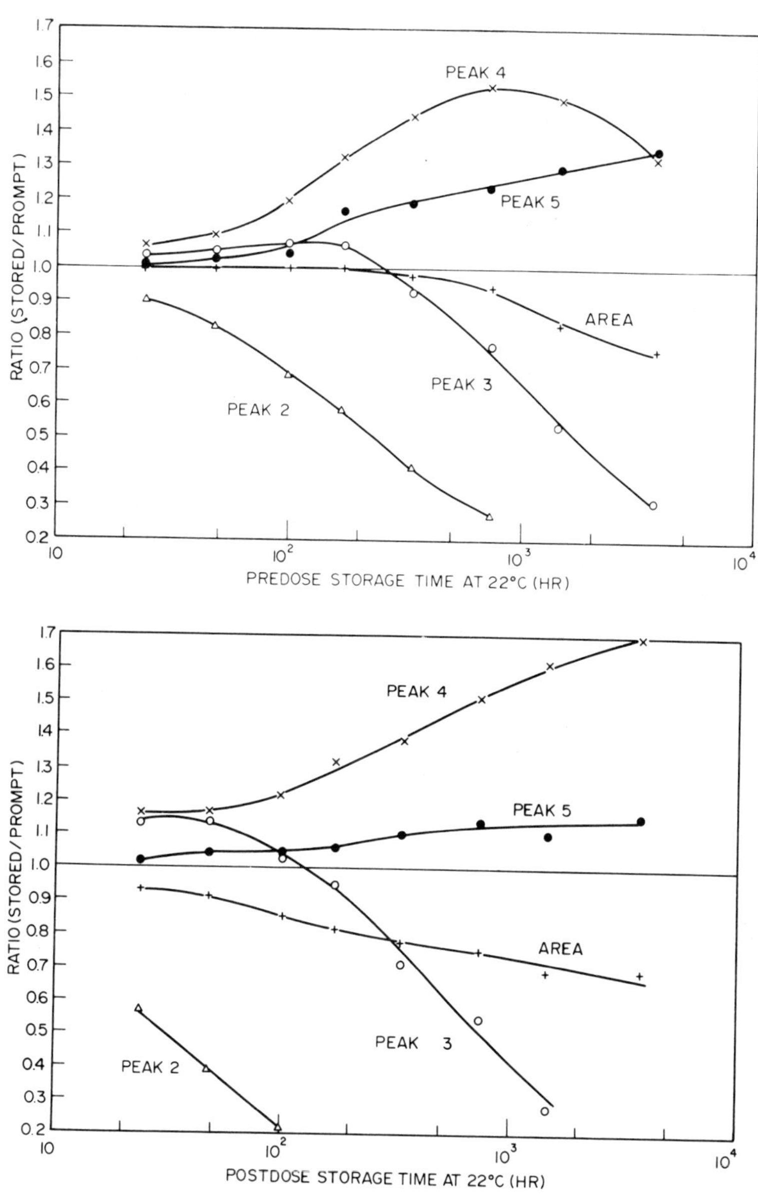

FIGURE 2-15. TL peak height ratios in LiF:Mg,Ti (Harshaw TLD-700 chips) as a function of storage time after annealing (a) and after exposure (b) to 10 rad gamma radiation, with the "prompt" readings immediately following the "stored" readings. (After Johnson, 1972.)

pretation of the fading kinetics are perhaps "abnormal" fading (nonradiative leakage of trapped electrons as observed in a number of other TL phosphors), and especially the complex trap transformations and/or retrapping processes which occur during storage after annealing as well as after radiation exposure (Johnson, 1972). As illustrated in Figure 2-15, both the peak heights of the various peaks as well as the area under the main TL peaks (4 and 5) undergo increases and decreases during storage. The rate with which these changes occur is accelerated by storage at elevated temperatures. Not surprisingly, the values of E and a as obtained by isothermal annealing do not agree with those from the heating-rate dependence and shape of the TL peaks (Johnson, 1972), and apparent frequency factors of up to 10^{42} have been deducted from the half-width of the peaks (Pohlit, 1969).

The relative intensity of the instable and therefore undesirable TL peaks around 100°C can be reduced by a postannealing storage at about 80 to 100°C (Zimmerman et al., 1966). As can be seen in Figure 2-16, approximately 20 hr of storage at 80°C, or somewhat less at 100°C (after an initial annealing for 1 hr at 400°C) leads to a substantial reduction of the traps which gives rise to the undesirable low-temperature peaks. This treatment has to be repeated after every use of the material. It also cannot be accelerated, unlike most other temperature-dependent changes in solid-state detectors, by simply increasing the storage temperature. At temperatures exceeding $\sim$100°C for the second heat treatment, a quenching of the main TL peak(s) and an increase in the low-temperature peaks are observed (Figure 2-16). It is, however, possible to avoid the second

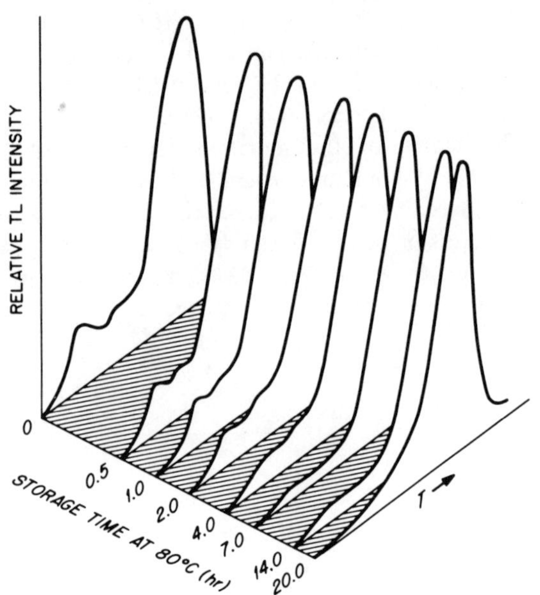

FIGURE 2-16. Influence of postannealing, preexposure storage at 80°C on the low-temperature TL peak height in LiF:Mg,Ti. (After Zimmerman et al., 1966.)

temperature treatment by either a post-irradiation, prereading annealing of the LiF for about 10 min at $\sim$100°C, or by using a TLD reader which does not integrate any light below a preset temperature of $\sim$120°C.

Other authors who also tried to optimize the characteristics of LiF:Mg,Ti by a postannealing heat treatment (see, for example, Shiragai, 1967 and Harris and Jackson, 1968) essentially confirmed the earlier results by Zimmerman et al. (1966). They recommend 10 to 100 min at 500 $\pm$ 25°C and a rapid cooling, followed by 10 to 50 hr at 75 $\pm$ 5°C (16 hr at 80°C). In a high-purity

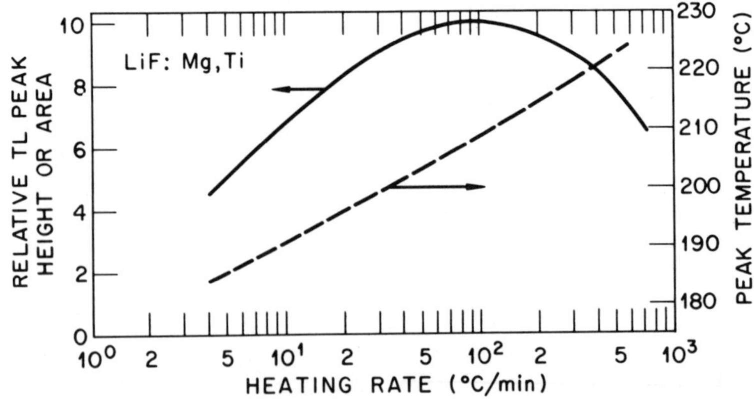

FIGURE 2-17. Relative peak height and area under the main TL peak of LiF:Mg,Ti (TLD-100) as well as main peak temperature (right scale) as a function of heating rate of the irradiated phosphor. (After Gorbics et al., 1968b.)

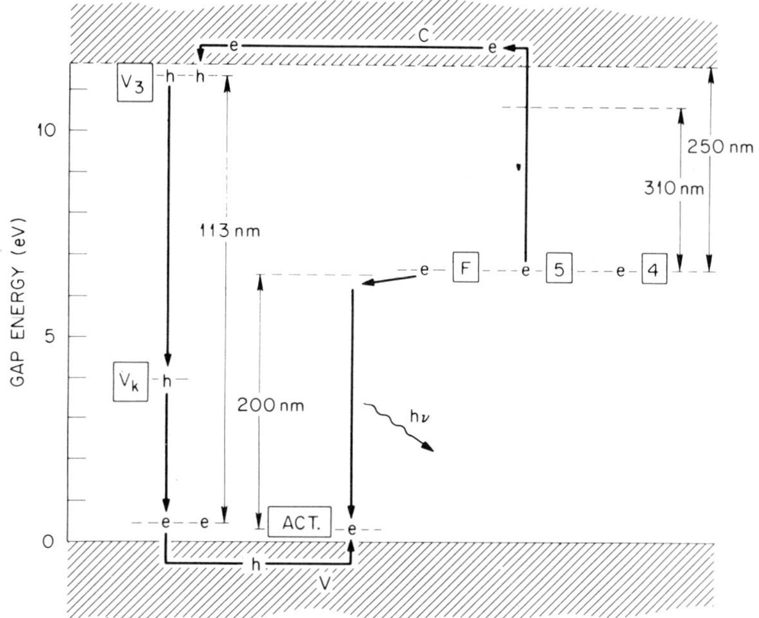

FIGURE 2-18.   Simplified band model of TL in LiF:Mg,Ti. (After Mayhugh, 1970.)

LiF, rapid cooling (1 to 15°C/sec) from 500°C also gave the least pronounced peak at $\sim$100°C, without affecting the main peak at $\sim$200°C (Guilmet et al., 1970).

LiF/Teflon compounds should not be annealed at temperatures above $\sim$300°C because this would result in irreversible changes in their optical properties related to the softening and melting of the Teflon. Studying the annealing of LiF/Teflon discs, some investigators (Carlsson, 1969 and Martensson, 1969) came to the conclusions that "thermal memory" problems are minimized and the reproducibility optimized if the high-temperature annealing is as short as possible, the cooling-down rate constant, and the pre-irradiation storage temperature low. Standard deviations as low as 0.2% have been reported under idealized conditions. Others (Docherty and Marshall, 1970) find less impressive standard deviations of 4 to 12%. A comparison of three annealing procedures for LiF/Teflon discs as recommended by different authors, consisting of several seconds, 20 min, and 16 hr at 300°C, followed by 24 hr at 80°C (Linsley and Mason, 1971) shows no significant difference in the results: In each case, the annealing is incomplete (preexposure to 1,000 rad results in $\sim$30% sensitization) and the standard deviations are comparable. Another comparison (Marshall et al., 1971) indicated that for low dose levels ($\lesssim$1 rad) sufficient annealing can be obtained by the readout cycle itself, but that the additional light sensitivity of the Teflon may disturb low-dose readings.

Another important property of LiF:Mg,Ti is the strong dependence of its TL intensity (peak) height or area) on the heating rate. As can be seen in Figure 2-17, not only the main peak temperature increases, as in all other phosphors, with increasing heating rate and a thermal quenching occurs (also similar to some other phosphors including CaF_2:Mn and $CaSO_4$:Mn) above $\sim$10^2 °C/min, but there is also an unusual decrease at low heating rates resulting in a peak "sensitivity" at the heating rate of $\sim$10^2 °C/min. In most routine applications using a TLD reader with a constant heating rate and reproducible sample geometry, heating-rate effects can, however, be safely ignored unless modified by a flow of cool N_2 gas during readout or other factors.

Despite substantial efforts, no comprehensive explanation of the TL response characteristics of LiF:Mg,Ti has been given yet. One of the more likely models for the description of the TL and optical properties that has been proposed in recent years is presented in Figure 2-18. The primary processes are described as follows (Christy et al., 1967 and Mayhugh, 1968): Ionizing radiation creates highly mobile electrons and less mobile holes in the LiF lattice. The holes are trapped in double-hole centers (V_3) with an optical absorption at 113 nm, the electrons at F centers (250 nm), and thermally instable centers which are associated with Mg (310 or 380 nm). Thermal stimulation releases electrons from the Mg centers, the main dosimetry peaks around 200°C (designated according to a widely used arbitrary nomenclature, peaks 4 and 5) being associated with the electrons from the 310 nm center. When thermally untrapped, these electrons can combine with one of the holes in the V_3 center, thus releasing the remaining hole V_K. V_K may then be captured at the luminescence site and combine with a tunneling F electron or with another free electron (the luminescence site is probably an impurity such as Ti and/or Al).

According to this model, radiation sensitization produces a large density of centers with an absorption at 225 nm, probably resulting in the decrease of radiationless transition sites and/or an increase in the number of luminescence sites. This model is supported by further experimental evidence (see, for example, Booth et al., 1972). Other characteristics of LiF:Mg,Ti remain, however, unexplained by this as well as various other models, including several unusual non-TL effects of potential dosimetric interest such as a radiation-induced, thermally activated depolarization (Fields and Moran, 1973) which occurs at very low temperatures (peaks at ∿160 and 180°K) and is related to the electret polarization which has also been observed in some polymers.

2.3.2.3. General Dosimetric Characteristics

In addition to those already discussed, some other properties of LiF deserve attention relative to their proper use in dosimetry. No dose-rate dependence of the TL response has been detected up to dose-rates of at least 2×10^{11} rad/sec in experiments with pulsed x-rays of 30 μsec duration (Tochilin and Goldstein, 1966). Previous experiments with 15 MeV electron pulses of 1.3 μsec duration up to 2×10^8 rad/sec also did not show any detectable dose-rate effects (Karzmark et al., 1964). At high dose-levels (50,000 R), however, there is a substantial loss of sensitivity (∿30%) in LiF, but not for the TL in $Li_2B_4O_7$:Mn, CaF_2:Mn, and BeO if exposed with pulsed x-rays at more than 15,000 R/pulse (Goldstein, 1972).

The photon energy dependence can easily be calculated from the elementary composition as given in comparison with some other low-Z TL phosphors in Figure 2-19, if some simplifying assumptions (large phosphor particles, narrow-beam geometry) are made. An oversensitivity of 30 to 40% in the 10 to 30 keV range, as compared to the response to gamma radiation, has been experimentally confirmed by many authors (see, for example, Karzmark and Geisselroder, 1968; Almond et al., 1968; and Jayachandran, 1970), but others report slight deviations from the calculated response, in particular a sensitivity drop below 20 keV (Tochilin et al., 1968; Endres et al.,

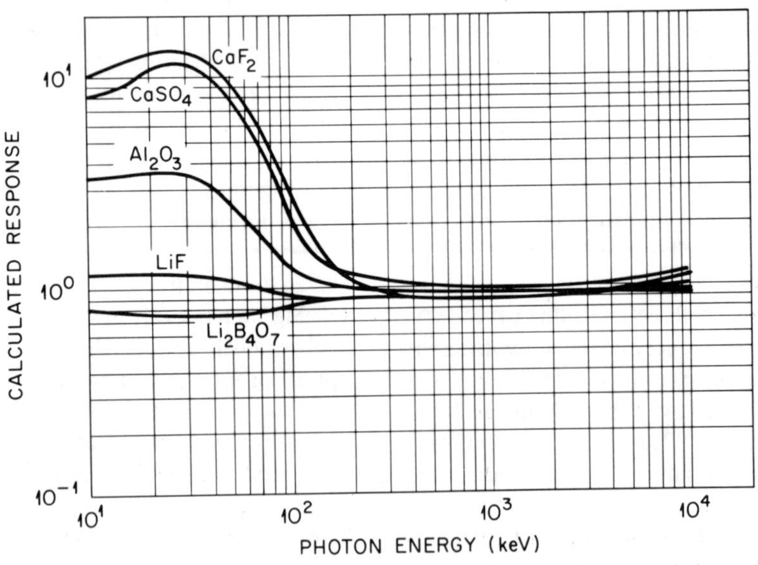

FIGURE 2-19.   Photon energy dependence of the sensitivity of different TL phosphors, calculated from the ratios of the mass energy absorption coefficient in the phosphors and in air. (After Becker, 1970.)

1970; Pendurkar et al., 1971; and Marshall and Docherty, 1971).

This may be due to one or more of several factors including absorption of the very soft x-rays in the sample and/or its encapsulation, or grain-size effects. Indeed, at least one author (Law, 1973) found perfect agreement between calculated and measured sensitivity in the 11 to 26 keV photon energy range within the accuracy limits of the measurement. For the calculated mass attenuation and mass energy absorption coefficients of LiF as well as $Li_2B_4O_7$:Mn in the interesting 1 to 150 keV photon energy range, see Greening et al. (1972); for the calculated effective atomic numbers and kerma values, see Jayachandran (1971). There also have been reports concerning a dose dependence of the energy response in LiF (Hendee et al., 1968), and about a 10% sensitivity decrease for 35 to 65 MeV x-rays relative to the ^{60}Co gamma radiation response (Turner and Anderson, 1973).

Very fine-grained LiF phosphors are markedly less sensitive than coarser samples. It has been speculated that this may be due to an "inactive layer" on the surface of the grains (Zanelli, 1968). Chan and Burlin (1970) calculated the effect of grain size on the photon energy dependence of various TL phosphors which are imbedded in different media including air and Teflon, which is widely employed as an easy-to-handle matrix for the phosphor (Bjärngard and Jones, 1967). Figure 2-20 gives an example of the results.

Obviously, at photon energies around 10 keV the range of the primarily released photoelectric electrons is too short to permit much escape even from very small grains. With increasing photon energy, however, the range of the photoelectrons grows rapidly, and electrons that are released in the surrounding air penetrate the LiF grains to an increasing depth, while a larger and larger fraction of the escaping photoelectrons from the LiF deposit their energy in air. As the effective atomic number of LiF is higher than that of air, the "escaping" energy exceeds the "entering" energy. This results in a minimum between (depending on the grain size) 30 and 60 keV. With further increasing photon energy, the Compton effect begins to dominate. At 200 keV, for example, about 90% of the electrons are Compton electrons with an average energy of 10 keV. The response becomes again fairly independent of grain size.

Naturally, grain size and "environmental"

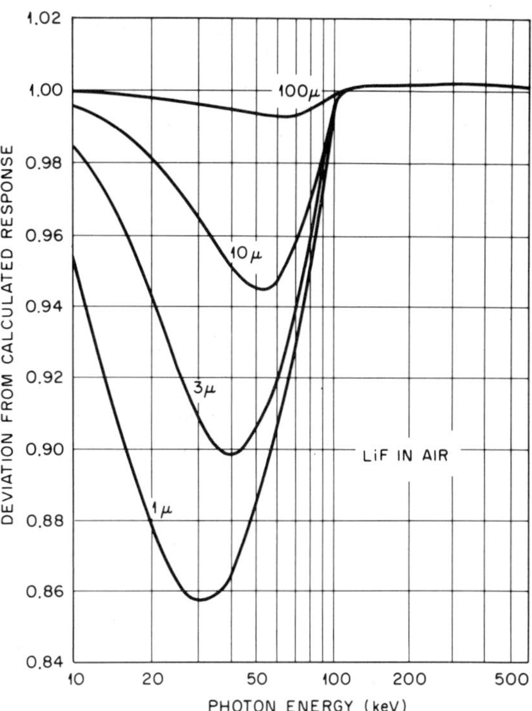

FIGURE 2-20. Calculated deviation of the actual energy response of LiF in air from the calculated values in Figure 2-19, taking into consideration the effect of grain size. (After Chan and Burlin, 1970.)

effects in mixed systems become more pronounced when the Z of either the phosphor or the matrix deviates more from that of air. For LiF powder imbedded in Teflon, the oversensitivity at 60 keV for grains of 10 μm diameter amounts to 10%, and for 1 μm grains at 50 keV to 22% (Chan and Burlin, 1970). This effect is more pronounced if the LiF is imbedded in silicon rubber, which has an inherent oversensitivity for 25 keV photons by a factor of 3.2. The calculated response deviations in this material may amount to a factor of two (Bassi et al., 1971).

Considerable efforts have been devoted to the increase of the precision with which small radiation doses can be measured with LiF:Mg,Ti. In the early years of TLD, various spurious effects and relatively primitive readout equipment made it rather difficult to read doses below 0.1 to 1 rad with any degree of accuracy, mostly because spurious signals (chemo- and triboluminescence) produced a background signal in the unexposed phosphor amounting to as much as several rad. The signal exhibited a large degree of fluctuation and was probably caused by surface trapping states which are known to be affected in a poorly reproducible way by friction, adsorbed gases, electric discharges, and ultraviolet light.

Permanent encapsulation of the phosphor in glass capillaries filled with an inert gas helped to reduce these disturbing effects and made the detection of $\sim$10 mrad possible (Schulman et al., 1963). Also, larger single crystals or extruded ribbons are less subject to this source of errors. Much of the spurious signal can also be suppressed by reading of the samples in an oxygen-free atmosphere (McCall and Fix, 1964). Flowing dry, pure N_2 at a constant rate through the chamber in which the phosphor is heated is now a routinely used technique in most commercial TLD readers. Even small O_2 impurities in the nitrogen are known to affect the reliability of low-dose measurements.

Other limiting factors for low-dose measurements not only with LiF:Mg,Ti are the dark-current of the photomultiplier tube, which can be reduced by cooling, and the black-body radiation of the hot phosphor and the heated parts of the reader (Burch, 1967). The latter can be reduced by the proper choice of filters (Karzmark et al., 1966), the use of low-emissivity heating pans made of (or coated with) silver, or by heating of the phosphor by a hot gas stream.

Another important improvement on the reader side has been the introduction of photon counters (Schlesinger et al., 1971 and Niewiadomski, 1972). Figure 2-21 gives an impression of the current limits in low-dose measurements with LiF, namely, 5 mrad gamma dose readings with a standard deviation of 12% with extruded Harshaw TLD-100 ribbons. Using much larger amounts of the phosphors (500 mg per reading), doses as low as 1 mrad can be read with a similar or better reproducibility (Lippert and Mejdahl, 1967).

The other extreme of the dynamic range has also been the subject of much attention. Saturation and quasipermanent damage of the phosphor takes place around 10^5 to 10^6 rad (most of this radiation damage can, however, be annealed by heating of the phosphor for 2.5 hr at 650°C). For measurements at higher dose-levels, several methods have been suggested. The more important ones are

    a. A TL peak at high temperature (around 450°C) in certain non-TL, optical grade materials

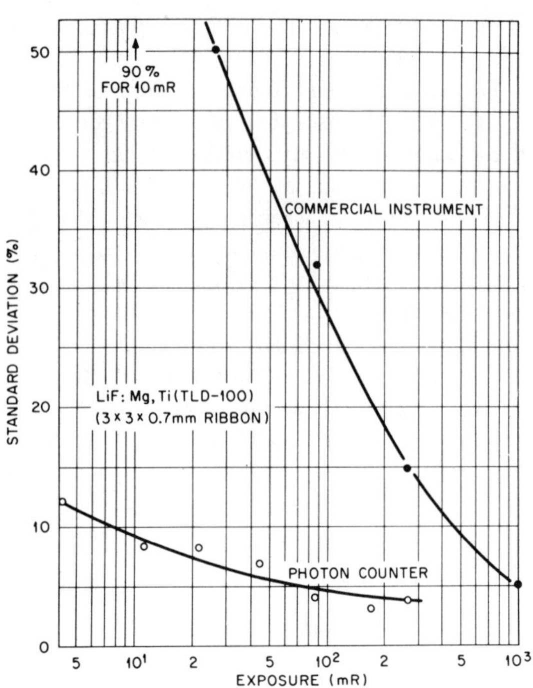

FIGURE 2-21. Standard deviation of low gamma radiation dose measurements, using LiF TLD-100 ribbons as a function of dose, for a typical commercial instrument and for a special reader employing a photon counter. (After Schlesinger et al., 1971.)

can be used for measurements up to $\sim 10^8$ rad (Goldstein et al., 1968).

    b. The ratio in the peak heights at $\sim$200°C and $\sim$280°C decreases with increasing dose and can be used for measurements up to at least 10^7 rad (Figure 2-22).

    c. Optical density measurements of an

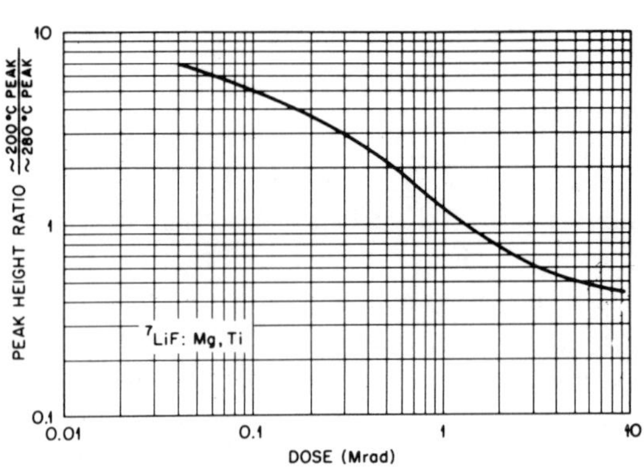

FIGURE 2-22. Ratio of TL main peak height at $\sim$200°C to the peak height at $\sim$280°C in ^{7}LiF:Mg,Ti (TLD-700). (After Jones and Martin, 1968.)

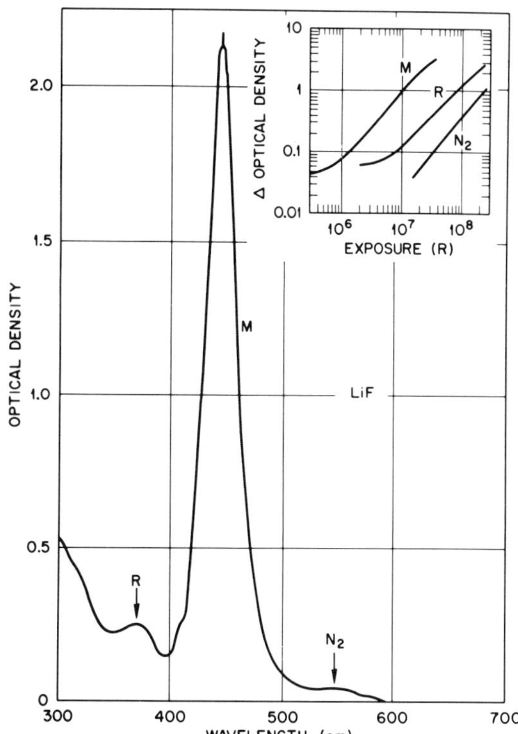

FIGURE 2-23. Optical absorption spectrum of optical-grade (Isomet) LiF after exposure to 2 x 10⁷ rad, and (insert) increase of optical density at the absorption peak as a function of dose. (After Vaughan and Miller, 1970.)

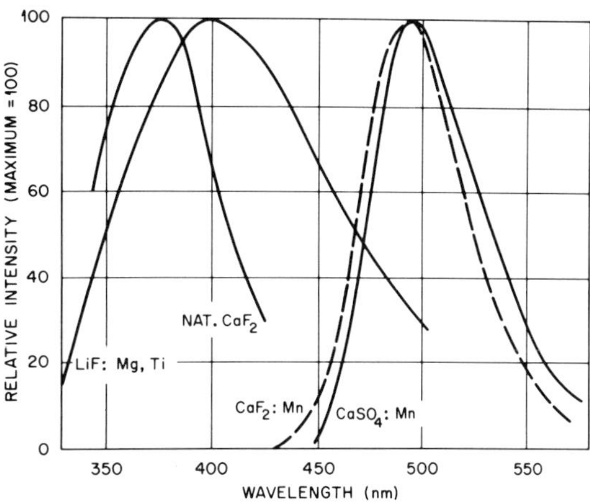

FIGURE 2-24. Thermoluminescence light emission spectrum of the main TL peaks in various phosphors. (Compiled by Fowler and Attix, 1966.)

absorption band that is induced in optical-grade LiF (Isomet Corp., Palisades, N.J. or E. Leitz, Wetzlar, Germany) permit to cover a dose-range of about 10^6 to 2×10^8 rad (Figure 2-23). The peak at 450 nm is thermally more stable than other peaks (Claffy et al., 1971).

d. Radiophotoluminescence (RPL) is observed in TL grade LiF (Regulla, 1972) as well as in optical-grade LiF (Claffy et al., 1971) under excitation with 450 to 455 nm light, with two broad RPL bands at 520 nm, and in the 630 to 670 nm range. The RPL is an almost linear function of dose from ~ 10 to $\sim 10^6$ rad (Figure 4-8).

The spectral energy distribution of the TL light that is emitted by the main peak(s) at $\sim 200°C$ in LiF:Mg,Ti has been compared (Figure 2-24) to that of some other phosphors. Other peaks in this material, in particular some of the spurious signals, are known to have a different spectral distribution. For example, the emission spectrum shifts to shorter wavelengths for the high-temperature glow

peaks (DeWerd and Stoebe, 1972). Variations in the spectral response of the photomultiplier and/or the optical filters in different TL readers may, therefore, explain at least in part the difference in the reported glow curves by different authors. For the same reason, it is impossible to compare absolute sensitivities of phosphors having different emission spectra without normalizing the response characteristics of the readout systems; comparisons of sensitivities as obtained in one type of reader (such as Spurny, 1968) may yield totally different results in another readout system; (see Table 2-5).

A search for a suspected strong TL emission in the ultraviolet did not lead to positive results (Kastner, J., personal communication). Several authors, using different methods, confirmed the ~ 400 nm peak in the emission of the main TL maxima. Unlike some other phosphors such as CaF_2 (Konschak et al., 1971), there appears to be no effect of the dose-level on the TL light spectrum between at least 10^2 and 10^4 rad (Strash, 1969 and DeWerd and Stoebe, 1972), but Oltman et al. (1968) noted an interesting LET dependence (Figure 2-25). The TL light spectrum of LiF:Mg,Ti is, as it is in most other phosphors, very similar to the spectrum of the radio-luminescence which is observed during irradiation (Gorbics, 1966).

The intrinsic efficiency (ratio of light energy emitted during heating to the energy absorbed during gamma or x-ray exposure) was found to be 0.10 to 0.15% by one author (Strash, 1969) and

Relative Sensitivities of Powdered TL Phosphors, Measured with Photomultipliers with Different Spectral Response

Phosphor	Photomultiplier				
	S-20	S-10	S-11	Bi-Alkali	Commercial readers*
$Li_2B_4O_7$:Mn	1	1	1	1	1
LiF:Mg,Ti	2.2	3.3	5.3	8.3	40
CaF_2:Mn	11.8	15.3	26	56	248
CaF_2:Dy	60	73	125	179	590

*With infrared filter

(After Binder et al., 1968.)

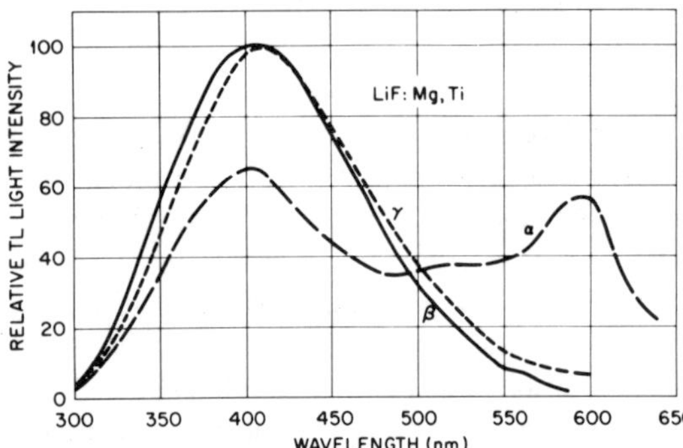

FIGURE 2-25. Optical emission spectrum of the main TL peak of LiF:Mg,Ti after exposure to different types of radiation. (After Oltman et al., 1968.)

0.04% by another (Lucke, 1970). This is substantially below some other phosphors such as CaF_2:Mn (0.44%) and $CaSO_4$:Mn (1.2%).

It is also important to know for all long-term applications of commercial LiF:Mg,Ti that re-ordered materials may not have exactly the same sensitivity or composition. Batch-to-batch variations have been somewhat reduced with improved quality control in recent years, but it is still advisable to calibrate every new batch for photon as well as (in the case of ^{7}LiF) thermal neutron response (in TLD-700 purchased from Harshaw between 1966 and 1968, for example, the ^{6}Li content fluctuated by a factor of 2.7 — Ayyangar et al., 1968).

Neither should it be assumed that the response parameters of ^{6}LiF, ^{7}LiF, and natural LiF are identical for photons. Both the sensitivity and the glow curves of these materials exhibit differences after x- or gamma-radiation exposure. For example, the gamma radiation sensitivity of TLD-600 was found by some investigators to be almost twice as high as TLD-100 (Wingate et al., 1967), while others observed smaller differences (TLD-600 8% less, TLD-700 24% more than TLD-100; Becker et al., 1970). For differences in the glow curves see Sunta et al. (1971). Of course, the differences in the mass energy-absorption coefficients and the mass stopping powers of ^{6}Li and ^{7}Li also result in differences in their photon energy response as well as in the rad to Roentgen ratio, which is 0.835 for TLD-600 and 0.805 for TLD-700 in the case of ^{60}Co gamma radiation (Attix, 1969).

Although one well-known text on TLD (Cameron et al., 1968a) states that there is "essentially no" light sensitivity of LiF, this should not necessarily be taken for granted in case of low-dose measurements. Since the early days of TLD, there have been many reports on a background increase due to extended exposure to sunlight or fluorescent light sources. For example, two days of exposure to a normal fluorescent laboratory light produced a pseudo-dose of 25 mrad gamma equivalent in LiF-Teflon disks (Bjärngard and Jones, 1968 and Freeswick and Shambon, 1970) and 2-hr exposure of powdered LiF:Mg,Ti to daylight ∿300 mR gamma radiation equivalent (Lippert and Mejdahl, 1967). The magnitude of the effect depends strongly on the ultraviolet component of the light source (most TLD materials are sensitive to ultraviolet light). Teflon itself may, incidentally, exhibit some phosphorescence. If virgin ^{7}LiF (TLD-700) is exposed to ultraviolet light from a bactericidal lamp (97% of radiant output at 257 nm), the saturation level of a broad, induced TL peak at about 90°C corresponds to approximately 220 mrad of gamma radiation. As the TL in this case originates mostly from the grain surface, smaller grains will be more sensitive. If the phosphor had been previously exposed to higher gamma radiation levels, peaks at $\sim$120, 170, and 210°C will be observed in addition to the $\sim$90°C peak (Mason, 1971). Others (Gower et al., 1969) report peaks at $\sim$200 and 230°C in gamma-irradiated, UV-exposed LiF:Mg,Ti. It has been suggested to use this increase of the UV response due to previous exposures (Figure 2-26) for possible reassessments of the gamma radiation history, for example, for the remeasurement of accidental exposures exceeding $\sim$10 rad in personnel dosimetry.

Care should also be taken in experiments involving the implantation of unprotected LiF dosimeters into animals. Matrix-imbedded LiF dosimeters exhibit a loss in sensitivity, and extruded LiF reacts with the tissue fluid, resulting initially in a loss of sensitivity and later, after three weeks *in situ,* in a complete dissolution of the detectors (Galkin et al., 1968). In this connection, one should keep in mind the toxicity of LiF. It is quite soluble in water (0.27 g/l at 15°C) and most rats lost weight and died after drinking such a solution for several weeks (Dettmer and Galkin,

1968). Special types of LiF crystals with a protective coating have been developed for in vivo studies (Sachdev et al., 1972).

Despite the common use of a nitrogen flow during readout, there also remains some pressure-sensitivity of the detectors. In LiF powder, chips, or Teflon mixtures, a more or less pronounced signal can be induced by shaking, crushing, or pressing. More intense pressure that leads to a plastic deformation in TLD-100 crystals is also known to change the glow curve (increase of peak 1, decrease of peak 5 – Petralia, 1972).

If loose powder is used, multiple use of the common phosphors (which is mandatory due to its high price) leads to organic contaminations (dust) which cause a yellowish-brownish discoloration of the phosphor during heating and an increase in its TL light self-absorption. Careful washing of the phosphor prior to annealing with pure alcohol and/or acetone helps to avoid this problem. Furthermore, crystals are broken during use and a reduction in average grain size may result in a change in sensitivity. On the other hand, high-temperature annealing tends to fuse smaller crystals. Proper sieving after annealing guarantees the proper size distribution, but leads to material losses of $\sim$10% for each use cycle. (For other common problems in LiF work, see Kartha, 1971.)

Only a few of the numerous potential sources of error in TLD work in general and the use of LiF in particular have been listed here to show the uninitiated reader that TLD is, contrary to some claims of commercial manufacturers, by no means a routinely easy matter to be done by un-experienced technicians, with reliable data to be expected within hours of the delivery of phosphors and readout instrument. Much is still left to the trouble-shooting capabilities, intelligence, and relevant knowledge of the experimenter, particularly if the measurement problems are somewhat unusual.

2.3.2.4. Neutron, Charged Particle, and Electron Response

Compared to the effect that ^{6}Li has on the thermal neutron response of LiF, the contribution of its other constituents can be normally ignored. The reaction ^{6}Li $(n,\alpha)^3$H has a cross section of 945 barn and produces short-range alpha particles (2.07 MeV) and tritons (2.74 MeV). The corresponding reaction in the ^{7}Li$(n,\gamma)^8$Li has a cross section of only 0.033 barn and produces gamma

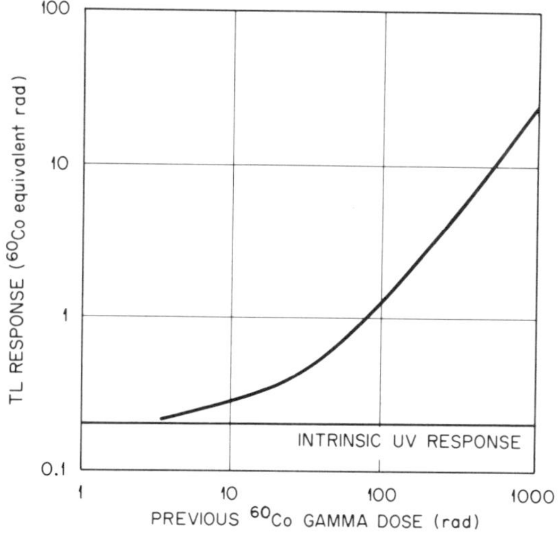

FIGURE 2-26. Increase of the intrinsic response of ^{7}LiF:Mg,Ti powder (TLD-700) to 257 nm ultraviolet light as a function of previous exposure of the phosphor to ^{60}Co gamma radiation. (After Mason, 1971.)

TABLE 2-6

Thermal Neutron Response of Some TL Phosphors

	Sensitivity[1]		
Phosphor	Tochilin et al., 1969[2]	Scarpa, 1970[3]	Ayyangar et al., 1968[4]
LiF, TLD-100	610	260	440
LiF, TLD-700	4.0	4.0	3.8
CaF_2, natural		0.64	
CaF_2:Mn	2.2		0.8
CaF_2:Dy			2.8
BeO	0.76	0.68−2.8	
$Li_2B_4O_7$:Mn			220

[1] Gamma radiation tissue dose from ^{60}Co in rad required to give same TL response as 1 rad (4×10^{10} n_{th}/cm^2) of thermal neutrons in a small tissue volume.

[2] TL peak height measurement

[3] TL integrated light up to 400°C

[4] TL integrated light up to 300°C

radiation of which only a small fraction is absorbed in the phosphor. Trace impurities of thermal neutron absorbers in TL-grade LiF (about 1.5×10^{-3} ppm Dy, 7×10^{-5} ppm Eu, and 1.5×10^{-2} Mn — Ayyangar et al., 1968) also usually have a negligible effect.

In LiF with the normal isotopic composition (7.4% ^{6}Li, 92.6% Li), 2.7×10^7 thermal neutrons/cm^2 produce an effect corresponding to that of 1 R gamma radiation (Cameron et al., 1964). Reported differences between this and other measurements (according to Wingate et al., 1967, for example, 10^{10} n/cm^2 produce the equivalent of 210 R gamma radiation) are probably explainable on the basis of different measuring conditions (peak height vs. area measurement, different maximum heating temperatures), and by neutron self-absorption in samples of different size and geometry.

In the commercially available TLD-600 (Harshaw Chemical Co.) with 95.6% ^{6}Li and 4.4% ^{7}Li, 10^{10} n/cm^2 produce the equivalent of about 600 R gamma radiation. This is the highest thermal neutron response of any known TL phosphor (for a comparison, see Table 2-6). Even more than in TLD-100, self-absorption is a source of potential error in this material. In a layer of 0.1 mm thickness of TLD-100, for example, 5 to 10% of all incident thermal neutrons are absorbed. In TLD-600, 50% are absorbed.

Among the other factors that one should keep in mind when using LiF for thermal neutron measurements are the LET-dependence of the response characteristics (thermal neutrons interact by creating high LET radiation, resulting in differences in the glow curve, the spectrum of the emitted light, reduced supralinearity, etc. — see Figures 2-10, 2-25, and 2-29); the fact that ^{6}Li is not a 1/E neutron absorber, but has a resonance peak at 0.25 MeV (Kastner et al., 1966a); and that TLD-700 is by no means free from ^{6}Li (its ^{6}Li content has been found to vary between 0.034 and 0.087%, resulting in a response to 10^{10} n/cm^2 varying between the equivalent of 0.87 and 2.3 rad of ^{60}Co gamma radiation — Ayyangar et al., 1968). In Figure 2-27, some typical response data are given. It has been suggested to make use of the differences in the TL glow curve of LiF (TLD-100) after exposure to low-LET radiation and thermal neutrons for distinguishing between those types of radiation in mixed fields (Mason, 1970; Busuoli et al., 1970; and Furuta and Tanaka, 1972), but it seems unlikely that this method has much practical value. Naturally, exposure of LiF to high thermal neutron fluxes also creates some permanent structural changes which are reflected in differences in its response characteristics (Mason, 1970).

The inherent fast-neutron sensitivity is small for TLD-100, TLD-600, and TLD-700. As all fast-neutron sources are contaminated with some photon radiation, the precise neutron response is

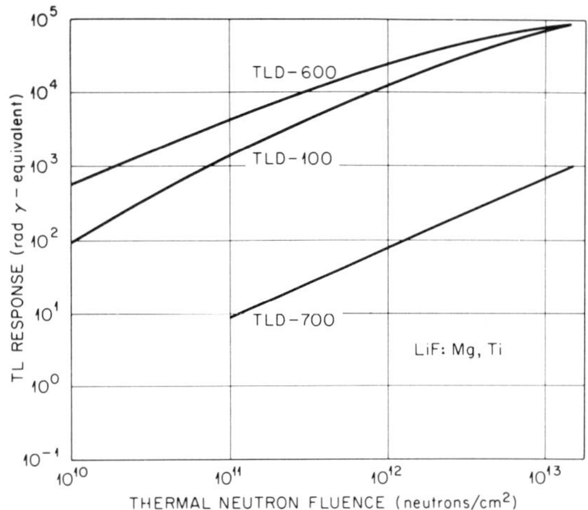

FIGURE 2-27. Response of different types of LiF:Mg,Ti to thermal neutrons. (After Ayyangar et al., 1968.)

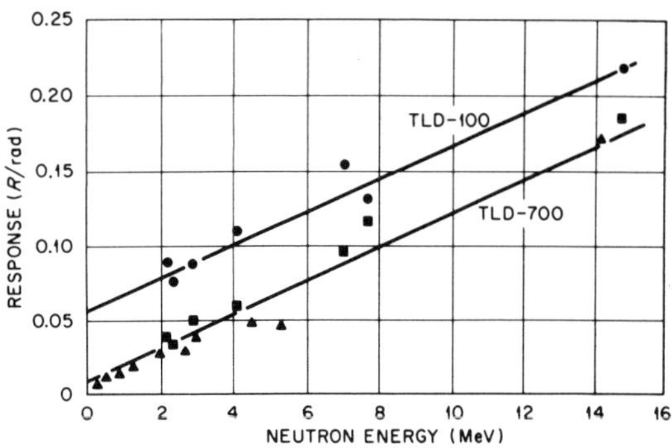

FIGURE 2-28. Relative fast neutron response of LiF:Mg,Ti (TLD-100 and 700), expressed in the TL response produced by a neutron field which would deposit 1 rad in tissue, divided by the TL response produced by exposure to 1 R of ^{60}Co gamma radiation. (After Wingate et al., 1967 ▲, and McGinley, 1972 ●, ■.)

not easy to measure, and scattered thermal neutrons may further falsify the results unless the detectors are well shielded in lithium during irradiation. Some early results on the fast-neutron response of ^{7}LiF by Wingate et al. (1967) have recently been confirmed by Goldstein et al. (1970), Furuta and Tanaka (1972), and McGinley (1972).

A good agreement between experimental and calculated values in the 17 keV to 14 MeV neutron energy range was found if the fluorine recoils are not considered in the calculation (Furuta and Tanaka, 1972). As can be seen in Figure 2-28, the response obtained per 100 erg/g neutron kerma in tissue, expressed as equivalent ^{60}Co gamma radiation exposure, increases from about 0.005 (0.5% of the gamma sensitivity) as 0.25 MeV to 0.16 at 14 MeV. LiF is, therefore, a good gamma radiation detector in mixed radiation fields if the neutron energy does not exceed approximately 1 MeV (Attix et al., 1972).

The fast-neutron response of all inorganic TLD phosphors can be increased by intimately mixing them with hydrogeneous materials. Karzmark et al. (1964) first used this principle by mixing the phosphor during irradiation with alcohol, which can easily be evaporated prior to the reading. Further studies of the energy response of the system (Wingate et al., 1967) were only partially encouraging. First, the system required continuous stirring or shaking during exposure. Second, its response decreased rather rapidly with neutron energy (0.17 per 100 erg/g neutron kerma in tissue

at 2.9 MeV, 0.66 at 14.9 MeV) because of the relationship between recoil proton range and grain diameter.

A better response at low-neutron energies could be expected with exceedingly fine-grained LiF, but this introduces other problems such as a drop in inherent radiation sensitivity. Naturally, mixing LiF with alcohol or other liquids also affects its photon energy dependence (Endres et al., 1970). Mixing fine LiF powder with water is not recommended because it induces a rapid fading (Spurny et al., 1971).

The response of LiF and some other TLD phosphors in a 2 mm polyethylene radiator to 14 MeV neutrons has only recently been studied by Sunta et al. (1972), and the neutron to gamma radiation response ratio (at a 40 rad dose-level) was reported 1.24 for TLD-100. The response of the other phosphors varied between 0.88 for natural CaF_2 and 2.84 for $Li_2B_4O_7$:Mn. It has also been attempted to measure the range of the recoil protons in polyethylene sandwiches as an indicator of the energy of the incident neutrons (Facey, 1968).

Another method for increasing the fast-neutron response of LiF is to surround it with sufficient amounts of moderating or backscattering media. A fairly energy-independent response can be obtained with polyethylene spheres of 15 to 30 cm diameter (Distenfeld et al., 1967 and Lazanoff and McLaughlin, 1969). These and the related "albedo" methods are discussed in more detail in Chapter 2.4.1. A third technique for measuring fast neutrons with LiF, based on its activation by the reaction $^{19}F(n,2n)^{18}F$, has been proposed (Pearson et al., 1972), but the high threshold of this reaction (11 MeV) restricts its usefulness for most practical situations.

It had been reported (Oltman et al., 1967) that the gamma radiation response of ^{6}LiF is reduced by 5 to 12% if the sample was previously, or simultaneously, exposed to fast neutrons. Goldstein et al. (1970) studied the additivity of gamma and fast-neutron radiation not only in LiF (TLD-100), but also in CaF_2:Mn, BeO and $Li_2B_4O_7$:Mn, and did not find such a reduction within the experimental errors of their measurements (±3%). This difference in results was resolved recently (Wallace et al., 1971) in a somewhat surprising way: As it turned out, a pseudo-"fading" had been induced by heat-sealing the samples in polyethylene, which apparently released vapors with a TL-quenching effect. Some medical-type polyethylene tubing seems not to exhibit this effect, but there are other reports on a spurious TL build-up of samples in polyethylene holders.

If LiF is to be used for the dosimetry of alpha radiation or other low energy charged particles, the short range of these particles, being frequently shorter than the average grain diameter, has to be considered. Thin layers of finely powdered phosphor, with or without additional absorber foils to permit extrapolation to zero thickness, have been employed in such studies (Harvey and Townsend, 1971).

Another important factor is the previously mentioned (Figure 2-5) LET dependence of response. According to recent experiments with ^{7}LiF (Jähnert, 1971), the response (normalized for ^{60}Co gamma radiation) decreases from 87% to 58% with decreasing proton energy from 13.3 to 3.7 MeV, and amounts to 16% for alpha particles in the 1.7 to 3.7 MeV range. As the LET dependence is small below $\sim$10 keV/μm, high energy protons and other charged particles are registered with approximately the same sensitivity as gamma, beta, or x-radiation. For the response of LiF:Mg,Ti to protons in the 5 to 137 MeV range, see Parker (1968) and Momeni et al. (1973).

The shape of the LET curve is usually explained on the basis of saturation effects along the track of the densely ionizing particle. Several observations are in agreement with this theory, in particular the decrease in supralinearity with increasing LET

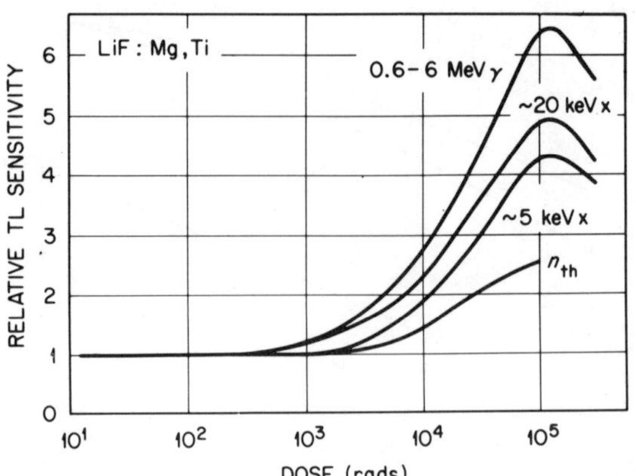

FIGURE 2-29. Relative sensitivity increase of radiation-sensitized LiF:Mg,Ti (TLD-100) powder for radiations of different LET; the sensitivity increase would be somewhat less pronounced if the integrated light output would be used instead of the main TL peak height. (After Suntharalingam and Cameron, 1969.)

(Figure 2-29). The fact that there is a small remaining supralinearity even for alpha particle exposures amounting to an oversensitivity of about 25% at 10^5 rad may, however, indicate the presence of other processes such as a possible interference between the delta rays from different high LET particle tracks. On the basis of an activator saturation model, Davy and O'Brien (1969) postulate that at least one of the trapping sites in LiF must have a LET dependent hole capture cross section.

The response of various TL phosphors including LiF:Mg,Ti to high energy electrons and x-rays has been subject to some controversy. With 6 to 18 MeV electrons and x-radiation of 18.5 and 11 MV, Almond and McCray (1970) reported a 6 to 10% drop in the sensitivity with increasing electron energy in LiF, compared to 3% in CaF_2 and $Li_2B_4O_7$:Mn. They explain this on the basis of Burlin's cavity theory, but Burlin (1970) doubts that his theory can be applied in this case. The results of other authors seemed to confirm that LiF is 10 to 12% less sensitive on a rad/rad basis for high energy electrons than for ^{60}Co gamma radiation, but found no decrease between 10 and 13 MeV (Pinkerton and Laughlin, 1966; Binks, 1969; and others).

Still other authors (Nakajima et al., 1968; Suntharalingam and Cameron, 1969; Rassow and Strüter, 1969; Ehrlich, 1970; Degner et al., 1972; and others) find, however (as one would expect for radiation with a LET well below $\sim$10 keV/

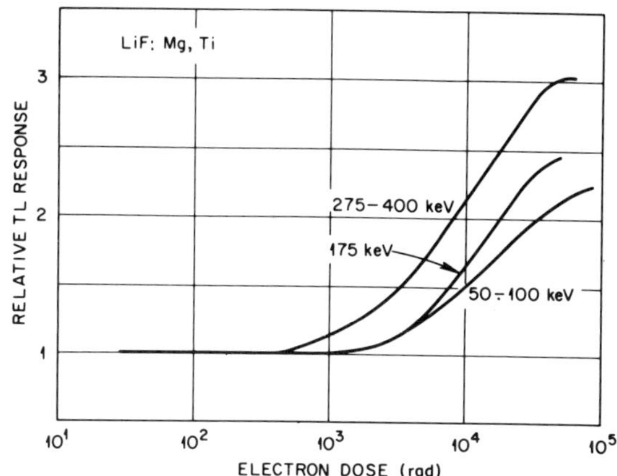

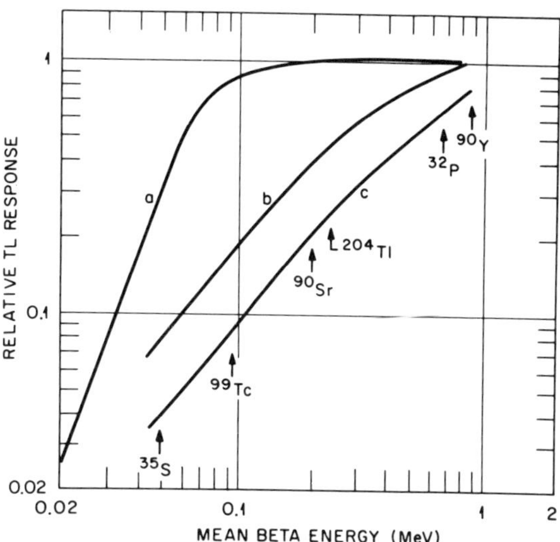

FIGURE 2-30. The relative TL response (in TL units per rad) of an extruded TLD-100 ribbon as a function of dose for various low energy electrons. (After Suntharalingam and Cameron, 1969.)

FIGURE 2-31. Relative response of TL in LiF:Mg,Ti as a function of the mean beta energy of various emitters, (a) with powder (0.15 mm average grain diameter) exposed at the surface of a gelatine mold (after Kastner et al., 1967); (b) with 1 mm diameter LiF/Teflon rods, at the surface of a container with the emitting solution; and (c) submersed in the solution of the nuclide. (After Greitz and Rudén, 1972.)

μm), no difference in the response to high energy electrons and ^{60}Co gamma radiation on a rad/rad basis within the precision of the measurement ($\sim$5%). Assuming a constant TL response over a wide electron energy range, it has been calculated that the response to 15 MeV electrons should be about 2% less than to ^{60}Co in the case of small LiF volumina, and 4% less for large LiF volumina (in CaF_2:Dy, the corresponding values are 4 and 6% — Ehrlich, 1970).

Depth-dose measurements in lucite that was externally exposed to ^{90}Sr/^{90}Y beta radiation resulted in an excellent agreement between LiF and extrapolation chamber measurements (Kastner et al., 1967). The supralinearity for higher energy electrons corresponds to that of high energy x-rays and gamma radiation, but it begins to approach that of high LET types of radiation with decreasing electron energy (Figure 2-30).

Kastner et al. (1967) exposed thin layers of commercial TLD-100 powder (average grain diameter 0.15 mm) to gelatin molds containing various nuclides. The response initially increased with increasing average beta energy, and leveled off when the electron range exceeded the grain diameter (Figure 2-31). A thin water layer on the grain surface did not affect the response characteristics, but there was a less-than-calculated sensitivity to very low beta energies. In the tritium exposed sample, only 5.5% of the calculated response was observed. Perhaps this is due to the hypothetical "insensitive layer" surrounding each grain which has been postulated on the basis of

results with extremely fine ($<$10 μm) LiF powder in trabecular bone (Spiers and Zanelli, 1966). It has also been speculated (Kastner, personal communication) that an electric dipole layer is built up in the highly insulating LiF crystals, which "repels" further incoming low energy electrons.

Obviously, the energy response changes when the geometry is changed. Also given in Figure 2-31 are data for the relative response of LiF-impregnated Teflon rods 1 mm in diameter, either submersed in the solution of a beta emitter (this case would approximately correspond to implanted detectors in in vivo studies), or pressed against the wall of a container with the beta emitting solution. On the other hand, a flat response down to somewhat lower beta energies would be obtainable with finer powders. For the practical beta dosimetry of the human skin, for example, in fingertip dosimetry, the depth of the basal layer has to be considered. The depth of this layer, reported to be between 2.5 and 50 mg/cm^2 (an average value of 7 mg/cm^2 is usually used), can reasonably well be approximated with 8.9 mg/cm^2 thick LiF/Teflon disks in mixed beta and gamma radiation fields (Gibson et al., 1971a and b; Marshall and Docherty and others, 1971).

2.3.3. Lithium Borate

Among the various other low Z TLD phosphors that have been studied over the years, manganese-activated lithium tetraborate, which was introduced around 1965, is the oldest (Schulman et al., 1967 and Kirk et al., 1967). One of its most obvious advantages is the fact that it can be prepared fairly easily in the laboratory: 32.5 g reagent-grade Li_2CO_3 and 108.96 g H_3BO_3 are mechanically mixed, the mixture is moistened with 5 ml of a solution containing 75 mg of $MnCl_2 \cdot 4 H_2O$, and dried for 12 hr at $100^\circ C$. The product is fired in a 200 ml platinum dish at $950^\circ C$ until molten, stirred, fired another 20 min, stirred again, and poured into a second platinum crucible. The cooled crystalline mass is ground and sieved to yield a grain size of 75 to 175 μm.

The optimum conditions for preparing lithium borate phosphors have been studied by several investigators. Among the parameters that affect its properties are

1. The Li_2O/B_2O_3 ratio (Kirk et al., 1967 and Christensen, 1968) which can be varied within rather wide limits. As lithium borates are quite soluble in water and tend to be hygroscopic at high relative humidities, attempts have also been made to "stabilize" it by additions of BeO (Becker, 1969) or, more successfully, 0.25% by weight of silicon as SiO_2 (Bötter-Jensen and Christensen, 1972). Pressed and sintered (at $900^\circ C$) tablets of this composition, which are now used in routine personnel dosimetry at the Danish AEC Establishment in Risö, are apparently not hygroscopic.

2. Type and concentration of the activator. Besides Mn^{++}, only silver is a good activator for $Li_2B_4O_7$ (Moreno et al., 1971; Archundia et al., 1972; and Thompson and Ziemer, 1972a). Its maximum of TL light emission is in the near ultraviolet and high sensitivities can be obtained with concentrations of about 0.1 to 0.5% and an UV sensitive photomultiplier. The standard Mn concentration is 0.1%, which leads to a slight "undersensitivity" for photons in the 10 to 100 keV range (Christensen, 1968; Jayachandran et al., 1968, 1971, and 1972; Thompson and Ziemer, 1972b; and others). The energy response of the system (Li_2O, B_2O_3, BeO, $MnCl_2$) can be adjusted very closely to that of air (Figure 2-32) and also to muscle, water, and soft tissue (Becker, 1969). Without the addition of BeO, a similar effect can

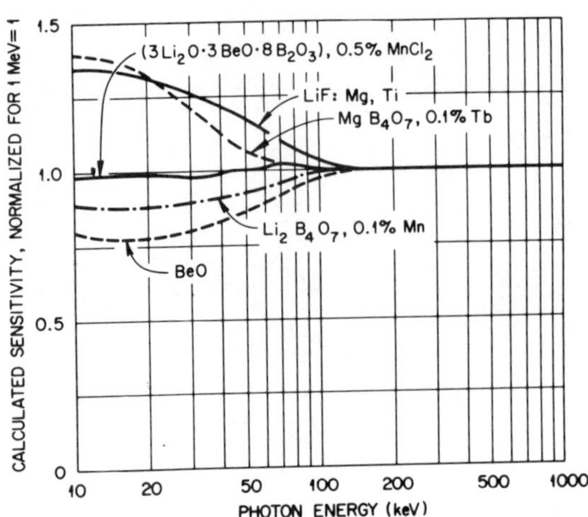

FIGURE 2-32. Calculated photon energy response, normalized for the response at 1 MeV, for different low Z TLD phosphors. (After Becker and Cheka, unpublished.)

be obtained by varying the Mn concentration: 0.34% by weight Mn brings $Li_2B_4O_7$ very close to the energy response of air (Jayachandran et al., 1968 and Jayachandran, 1971). With increasing Mn content, the temperature of the main glow peak shifts to higher temperatures.

3. Melting and cooling-down procedures. Best results were obtained with 15 min of melting at $950^\circ C$, and a slow cooling to room temperature (Jayachandran et al., 1968).

The dosimetric properties of lithium borate are much simpler than those of LiF:Mg,Ti, and many of them have been studied during the past few years (see the papers quoted above, Wilson and Cameron, 1968; and Wallace and Ziemer, 1968). Only a few will be discussed here briefly. As in LiF, there is a supralinearity on the response to low LET radiation at dose-levels exceeding about 100 rad, the degree depending somewhat on the procedure of measurement (Figure 2-33).

The "main peak" at $\sim$200 to $250^\circ C$ actually consists of at least two peaks, whose ratio depends on such parameters as phosphor preparation, photon energy, and dose (Figure 2-34). $Li_2B_4O_7$ can be glassy as well as crystalline. Both forms may, in principle, be used for dosimetry. The crystalline powder has, however, been found to have more desirable properties. As the TL light emission spectrum of Mn- or Ag-activated lithium borate is quite different from that of lithium fluoride, their relative sensitivities will depend strongly on the spectral response of the reader

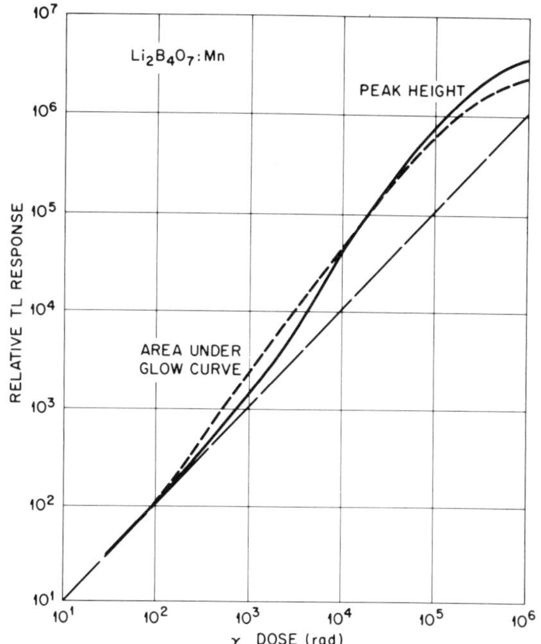

FIGURE 2-33. Relative TL response, expressed as main TL peak height and area under the glow curve, of commercial $Li_2B_4O_7$ activated with 0.1% Mn (Harshaw) as a function of gamma radiation dose. (After Wilson and Cameron, 1968.)

(Table 2-5). Most modern commercial readers are adjusted to LiF:Mg,Ti, with $Li_2B_4O_7$ consequently appearing to be less sensitive. For example, the borate exhibited one eighth of the fluoride sensitivity if measured with a S11 photocathode, but one half with a S20 photocathode.

The precise glow peak locations, peak ratios, and consequently also the fading rates depend strongly on the details of the preparation of the phosphor and the readout procedure. About 10% fading has been found between 4 and 50 days (Wilson and Cameron, 1968) and 37% during one year (Christensen, 1968) at room temperature. Evidently the peak at $\sim 100°C$, which is quite pronounced in the unactivated material and more or less pronounced in the TLD phosphor, causes a short-term fading after which the phosphor is quite stable. In one recent study (Christensen et al., 1973) no fading was observed in the lithium borate powder during three months of storage in dry air at 25°C. Due to its hygroscopic nature, however, 30% fading occurred during two months at the same temperature in moist air. The addition of 0.5% SiO_2 stabilized this material considerably.

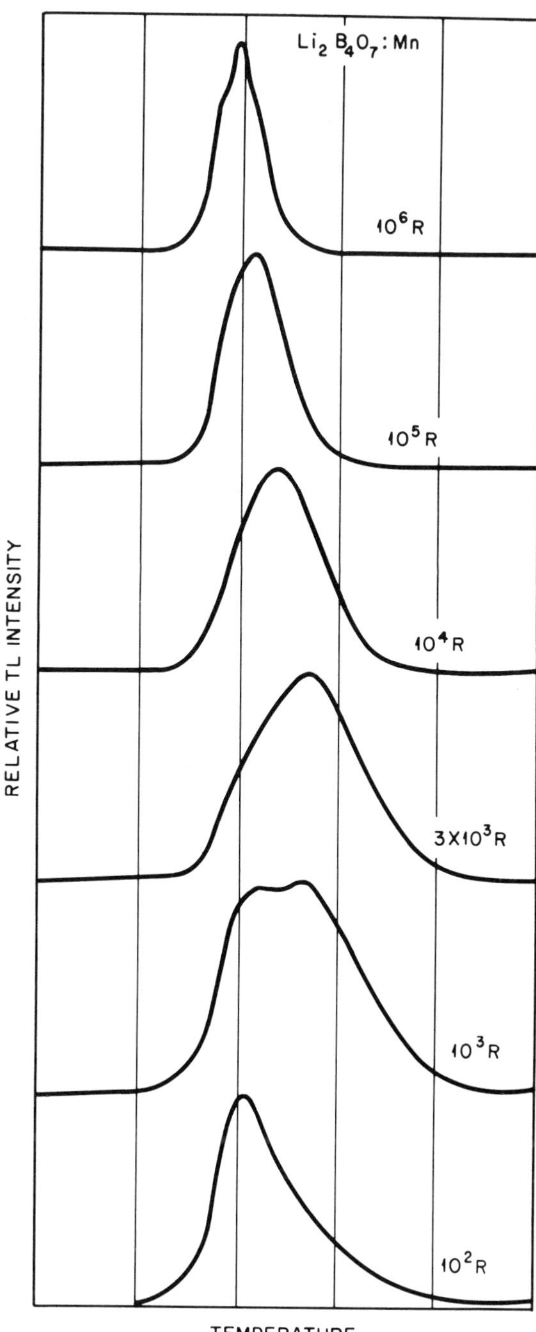

FIGURE 2-34. Glow curve of a batch of $Li_2B_4O_7$:Mn for different gamma radiation levels. (After Wilson and Cameron, 1968.)

In a sintered mixture of $Li_2B_4O_7$:Mn and $CaSO_4$:Dy, there was even an increase of the TL by about 17% during storage at 25°C in moist air. As in LiF:Mg,Ti, spurious signals are to a large degree suppressed by flowing dry N_2 through the reader during readout, but unlike LiF, lithium borate requires only a simple annealing procedure ($\sim$15 min at $\sim$300°C). For the approximate glow curve of $Li_2B_4O_7$:Mn, see Figure 2-1 for its LET dependence of response which is much less pronounced than in LiF (Figure 2-5).

In a recent comparison of the response of various TLD phosphors to recoil protons induced by 14 MeV neutrons (Sunta et al., 1972), it was found that the neutron to gamma response ratio was higher in $Li_2B_4O_7$:Mn (2.84) than in TLD-100 (1.24) and $CaSO_4$:Dy or Tm (1.6). Furthermore, lithium borate exhibited a pronounced high-temperature peak at about 350°C after exposure to the high LET radiation.

2.3.4. Beryllium Oxide

Another promising low Z TLD phosphor attracting much attention in recent years is beryllium oxide. It is particularly promising because of its chemical, mechanical, and thermal inertness, and its extremely low thermal neutron response of $\sim$0.20 R gamma equivalent per 10^{10} n/cm^2 (Tochilin et al., 1969 and Spurny, 1972). With commercial TL-grade materials now becoming available, it can be expected to play a major role in TLD soon.

Early investigations of the luminescent properties of BeO (Mandeville and Albrecht, 1954; Albrecht and Mandeville, 1956; and Moore, 1957) had shown glow peaks at 200°C and 270°C, the possibility of optical stimulation, and the fact that the light emission occurs mostly in the ultraviolet region, but its dosimetric properties have not been studied until quite recently (Tochilin et al., 1969; Rhyner and Miller, 1970; Becker et al., 1970; Scarpa, 1970; Scarpa et al., 1971; Yasuno and Yamashita, 1971; Spurny, 1972; and Scarpa and Benincase, 1972).

A deterring factor was perhaps the well-known high toxicity of finely disperse ("unfired") BeO powders which, if inhaled, induce a potential lethal reaction in allergic persons. However, no such reactions occur after BeO has been compacted (sintered, "fired") at high temperatures; the extremely hard ceramics can be considered safe unless they are machined, crushed, or otherwise transformed into the reactive, disperse form. Another somewhat complicating factor is the need

for a UV-sensitive TL reader if a good sensitivity is to be obtained.

Of the powdered and ceramic samples from various manufacturers that have been studied so far, the high-fired, large-grained powders and sintered ceramics not only happen to be least toxic, but also most sensitive (Tochilin et al., 1969 and Becker et al., 1970). Besides a more or less pronounced glow peak at $\sim$180 to 220°C, a high-temperature peak at $\sim$350°C was observed in most samples. In at least some materials, the main peak is a composite peak: At low dose-levels, T_m is at $\sim$180°C, but at higher exposure levels a second peak at $\sim$220°C overtakes this peak. There is also an LET dependence of the glow curve (Figure 2-10).

The calculated photon energy response of BeO is shown in Figure 2-32. For unknown reasons (perhaps high Z impurities of the commercial BeO's), experimental determinations of the energy response yielded a slight oversensitivity instead of the expected undersensitivity in the <100 keV energy range (Tochilin, 1969 and Scarpa, 1970). The photon response of a sintered mixture of 94% BeO and 6% SiO_2 has been calculated to be within less than ±1% of energy independence between <10 keV and several MeV (Becker, unpublished).

The somewhat unusual LET response of BeO is given in Figure 2-5. Otherwise, the response characteristics are similar to those of other low Z phosphors. There is a supralinearity of the response to low LET radiation above $\sim$10^2 R, which is much less pronounced for high LET radiation (Figure 2-35). The high melting point ($\sim$2,500°C) of BeO permits rather high annealing temperatures (5 min at 600°C has been recommended).

The TL characteristics of small ceramic BeO discs, which are commercially manufactured as heat sinks for electronic instruments by various manufacturers,* have been studied rather carefully, mostly because such small discs are very easy to handle, and because they also attracted much attention in connection with stimulated exoelectron emission dosimetry (see Chapter 3.2.2.). The dosimetric characteristics (glow curve, sensitivities) of such sintered units are distinctly different, suggesting a strong influence of pretreatment and/or impurity content (Scarpa, 1970; Scarpa et al., 1971; Nash et al., 1971; and Crase et al.,

---

*Consolidated Beryllium Ltd., Milford Haven, Wales, England (Slip Cast[®], Hot pressed[®], Nuclear Quality[®]) and Brush Beryllium Comp., Elmore, Ohio (Thermalox 995[®], Thermalox 998[®]).

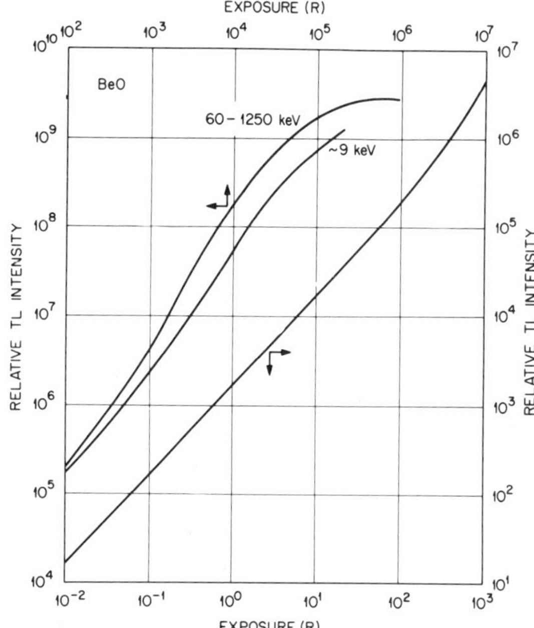

FIGURE 2-35.   Relative intensity of the BeO glow peak height as a function of exposure to ^{60}Co gamma radiation, 60 keV X-radiation, and $\sim$9 keV X-radiation. (After Tochilin et al., 1969.)

unpublished). Another complicating factor in such comparisons is the thermal gradient in the samples which makes it sometimes difficult to determine the actual temperature of the TL-emitting layer.

Scarpa et al. report a main peak in Thermalox 995® whose location shifts between 220 and 245°C depending on the dose-level, with a second peak at higher temperatures. Both peak area and height of the first peak become supralinear at $\sim$100 R, while the high-temperature peak is slightly sublinear and saturates already at 10^4 rad. Consequently, the peak ratio depends strongly on dose (4 at 1 R, 3,500 at 10^5 R) as well as photon energy. Crase et al. (ORNL, unpublished) found lower peak locations in the same material ($\sim$160 and 250°C for a heating rate of 16°C sec); both report standard deviations for multiple measurements around 1 to 3% for higher dose-levels.

During storage of exposed samples at room temperature in the dark, a fading of $\sim$7% was observed after two months by some authors and 5% during one month (3% during the first day) by others. The intrinsic response to UV light levels off at a value corresponding to <100 mR of gamma radiation, but it is higher in samples which had previously undergone gamma radiation exposure

and readout. The response of such UV-irradiated samples during a second readout is actually a more linear function of dose than the conventional first readout. During multiple UV exposures followed by readout of the same sample, there is naturally a decrease in response (Scarpa et al., 1971). In BeO:Na, UV stimulation of pre-irradiated and annealed samples limited the second readout to initial gamma dose-levels of >10 R. As the effect is due to retrapping from deeper traps, it disappears with increasing annealing temperature (Yasuno and Yamashita, 1971).

Optical stimulation of the luminescence in the visible region during reading of the luminescence in the UV also has the advantage of permitting multiple readout of the same samples. With another modification of optical stimulation, namely, light exposure followed by measurement of the phosphorescence which decays with a halflife of $\sim$10 sec, it was possible to measure doses of >1 R in the BeO powder and >10 R in the ceramic disks. After 40 readouts, the signal was reduced to one half (Rhyner and Miller, 1970).

Little is known about the mechanism of TL in BeO. In the pure material it has been associated with oxygen vacancies (Albrecht and Mandeville, 1956). The samples discussed so far contained arbitrary amounts of various impurities, mostly Al, Mg, and Si, with the "dirtier" materials apparently being more sensitive than the very pure ones (Becker et al., 1970). The effect of high-temperature doping with various monovalent (Li, Na, K), trivalent (B, Al) and tetravalent ions (Si, Ge) has recently been studied systematically by Yasuno and Yamashita (1971). With all activators, a main peak at $\sim$180°C was found, in some cases also additional peaks at 190 to 230°C which become more pronounced in the supralinear region and at 320°C. Sintering of purified BeO powder with Na at 1800°C resulted in a material with the highest sensitivity. A shift in the peak location during isothermal annealing from 180 to 215°C indicated a trap depth distribution.

Exposure of ceramic BeO:Na to a fluorescent lamp greatly accelerates fading due to optical stimulation, and 90% of the radiation-induced effect can be bleached out during one day exposure to a 330 lux fluorescent lamp. If kept in the dark, however, BeO:Na can be used for the reliable measurement of gamma exposures as small as 1 mrad.

2.3.5. Calcium Fluoride

Of the various high Z thermoluminescent phosphors with their less desirable photon energy response, calcium fluoride has been most carefully investigated. Either natural fluorite which contains, depending on its origin, a wide variety of different impurities acting as activators, or synthetic CaF_2 as a host lattice of "artificial" dopants such as manganese or dysprosium has been used.

The TL of natural fluorite, a rather widespread mineral, has been the subject of investigations as far back as 1903 (for bibliographies of the early work, see Angino and Grögler, 1956 and Ginther and Kirk, 1957). During the early years of modern solid-state dosimetry, Wölsendorf fluorite was studied under the dosimetric point of view in Germany and Switzerland (Luchner, 1957 and Grögler et al., 1958). A complex glow curve with rapidly fading peaks around 80 to $100°C$, reasonably stable peaks at 150 to $200°C$, and very stable peaks at $\sim$250 to $300°C$ was reported.

In the early 1960's a major effort was made in Belgium by M.B.L.E., a Brussels affiliate of Philips, Eindhoven, to introduce a commercial system based on a fluorite of undisclosed origin. This material has been the subject of intense research efforts not only in the manufacturer's laboratories (see, for example, Schayes et al., 1963 and 1967; Brooke and Schayes, 1967; van Espen, 1968; and Brooke, 1969), but also by various independent researchers (Harvey, 1967; Aitken, 1968; Fleming, 1968; Adam and Katriel, 1971; and Burgkhardt and Piesch, 1972). Since the commercial production of this system was discontinued in 1970, however, these results are of limited interest and will be discussed only briefly. More recently, the declining interest in natural fluorite was somewhat revived in developing countries, where a local supply of large quantities of an inexpensive TL phosphor has obvious advantages even if its dosimetric properties are not ideal (Nambi et al., 1969; Sunta, 1971; and Okuno and Watanabe, 1971).

As can be seen in Figure 2-36, fluorites from different sources may exhibit rather different glow curves (for similar data on Indian fluorites, see Sunta, 1971). The response of all peaks is fairly linear in all fluorites, with saturation of the peaks occurring at slightly different dose levels in the 10^4 to 10^5 rad gamma radiation range.

Attempts to correlate the various peaks with certain impurities and to understand the mechanism of TL have been partially successful. It seems that the release of holes from lattice centers and their recombination at the sites of rare earth impurities (Gd, Dy, Er, Tb) play a major role (Sunta, 1971). For CaF_2:Mn, it has been suggested that yttrium and other trivalent rare earth impurities act as electron traps and Mn^{++} acts as a hole trap (Puite, 1971 and Puite and Arends, 1971).

ESR studies of gamma irradiated natural CaF_2 (Nambi and Higashimura, 1971) also strongly indicate that trivalent rare earth impurities, possibly through a charge conversion process (RE^{3+} RE^{2+}), are involved in the TL of this material. For a "continuous" instead of a single trap depth model in Brazilian fluorite, describing the isothermal decay characteristics of the various peaks better than the old model, see Watanabe and Marato, 1971. An analysis of the glow curve of the M.B.L.E. fluorite (Adam and Katriel, 1971) resulted in activation energies E of 1.20 eV for the peak at $\sim$110°C, E = 1.65 for the peak at $\sim$175°C, and E = 1.71 for the main dosimetry peak at $\sim$263°C (the apparent frequency factors were found to be 4 x 10^{15}, 6 x 10^{18}, and 4 x 10^{18}, respectively), but other authors report different values such as E = 1.34 eV and a_o = $10^{11.4}$ sec^{-1} for the main peak (Frank and Stolz, 1969).

The inherent sensitivity of some natural CaF_2 samples for gamma radiation is, depending on the spectral response of the reader, up to 50 times higher than that of LiF:Mg,Ti (Schayes et al., 1967). Gamma radiation doses of less than 0.1 mrad have been measured with optimized sealed dosimeters and sophisticated readers (Brooke and Schayes, 1967 and Burgkhardt and Piesch, 1967). Simple open detectors have a lower limit of $\sim$10 mrad (Nambi et al., 1969).

The TL light spectrum from the main TL peak has a maximum at $\sim$370 nm (Figure 2-24) and shifts to $\sim$330 nm for very high doses. A similar, less pronounced shift of the $\sim$515 nm emission maximum takes place in CaF_2:Mn with increasing dose (Konschak et al., 1971). Heating above 500 to $600°C$ permanently damages the phosphor. A pressure annealing for 5 to 15 min at 400 to $450°C$ is suggested, which leaves some of the initially filled very deep traps unannealed.

Natural CaF_2 is quite sensitive to visible ultraviolet light. As in most other TL phosphors,

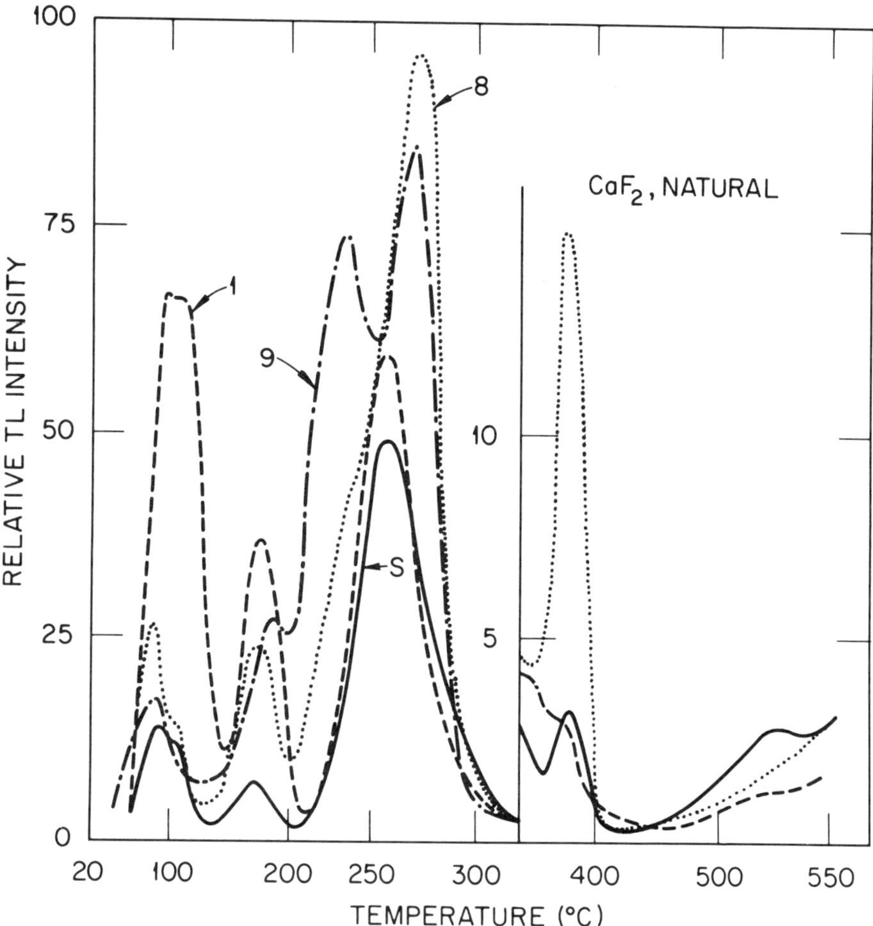

FIGURE 2-36.   TL glow curves of different natural fluorite samples. (After Schayes et al., 1967.) (For the simplified glow curve of a commercial TLD fluorite, see Figure 2-1.)

basically three effects occur, which are super-imposed to a varying degree:

1. In pre-irradiated samples, light-induced bleaching (optical stimulation) occurs which is particularly pronounced in the low-temperature peaks; during 1 hr of exposure to 500 lux daylight there is ∿10% fading, increasing to ∿25% after 4 hr (Brooke and Schayes, 1967).

2. In unirradiated samples, ultraviolet light induces a TL signal; as the deep traps are usually not completely annealed at a high temperature prior to the first use, there is also some UV-induced trapped electron transfer from deep to shallow traps, as discussed in point 3.

3. In pre-irradiated samples that have been conventionally evaluated by heating briefly to a temperature somewhat above the main TL peak, ultraviolet light exposure causes the acceleration

of the normally very slow transfer of electrons from deep, unannealed traps into the shallower traps which are annealed during standard readout. Thus, a repeated rereading of a previously "read" detector at a reduced sensitivity level becomes possible by the proper use of ultraviolet exposure (Brooke and Schayes, 1967 and van Espen, 1968); previous integrated gamma exposures above ∿5 rad can be repeatedly reread.

In Figure 2-37, the competing effects of UV-induced TL build-up, fading, and transfer are illustrated; Figure 2-38 shows the loss of stored TL information during such repeated transfer and readout cycles. The strong effect of ultraviolet light on natural CaF_2 has also been studied as a possible method for the integrating measurement of ultraviolet light (Konschak, 1972). Saturation, followed by a decrease of the TL signal, occurs in a Brazilian fluorite during exposure to 365 nm

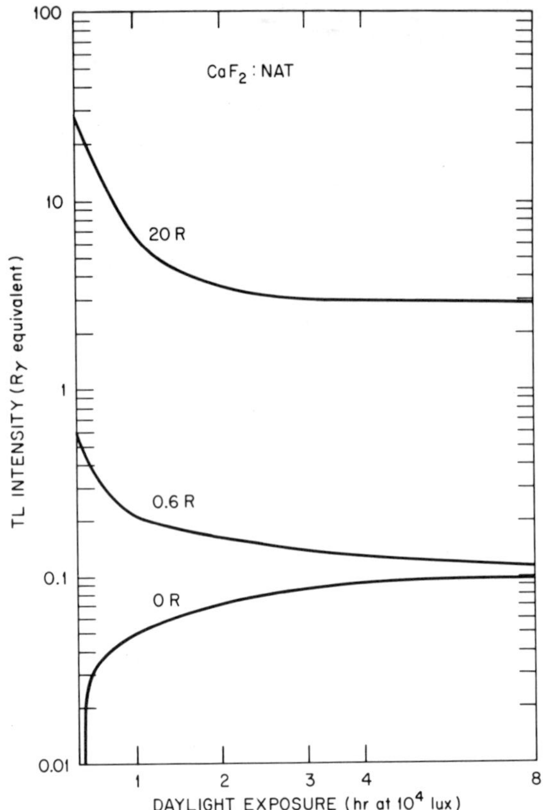

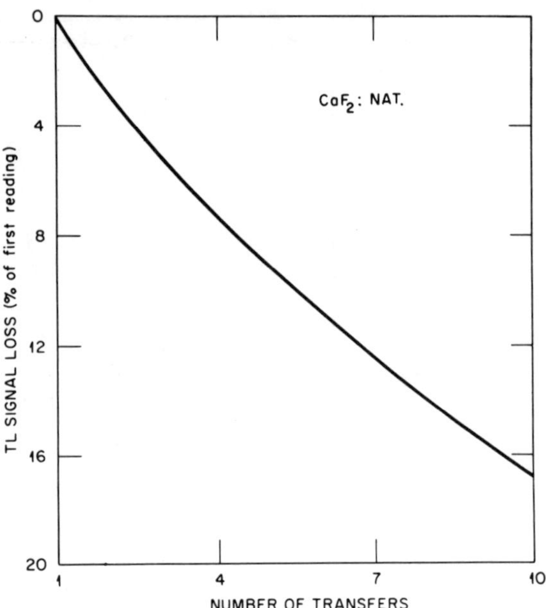

FIGURE 2-38. Loss of stored information in M.B.L.E. natural fluorite after exposure to 20 R gamma radiation and conventional readout during repetitive UV-induced electron transfers from deeper traps and multiple readout. (After Brooke, 1969).

FIGURE 2-37. Effect of diffuse daylight on commercial (M.B.L.E. PNE 007) fluorite dosimeter: (a) build-up of UV-induced TL in an unirradiated sample (lower curve); (b) light-induced bleaching and superimposed UV-induced signal build-up in sample preexposed to a low gamma radiation dose (middle); and (c) bleaching and UV-induced electron transfer during UV irradiation of sample preexposed to a high gamma radiation dose (upper curve). (After Brooke, 1969.)

light (Okuno and Watanabe, 1971). The M.B.L.E. fluorite was found to be most sensitive at 300 nm (McCullough et al., 1971), but it exhibits a similar nonlinearity due to the competing reactions.

Although the response of fluorite is usually quite linear, a slight supralinearity has been observed in some samples at high dose-levels (in an Indian fluorite, for example, between 1 and 4 x 10^4 rad). Pre-irradiation to ~ 6 x 10^4 rad, followed by annealing for 1 hr at 450°C, increased the sensitivity by a factor of ~ 4 (for a detailed study of residual TL, see Sunta and Kathuria, 1971). If individually calibrated, reproducibilities of better than ±1% have been reported for higher dose-levels and commercial, sealed dosimeters.

The photon energy dependence of CaF_2, which can be calculated to amount to an oversensitivity by about a factor of 12 at ~ 25 keV (Figure 2-19), is found in experiments to be slightly less (~ 9) and can be compensated with a good metal filter down to ~ 35 keV within approximately ±15% (Burgkhardt and Piesch, 1972). The effect of grain size on the calculated energy response becomes rather pronounced for very small grains and may amount to a deviation by a factor of six from the value which has been calculated on the basis of the mass energy absorption coefficients only (Figure 2-39). At higher photon energies (^{60}Co), the thickness of the CaF_2 layer may also affect the response due to build-up and interface effects (Ehrlich, 1971).

The thermal neutron sensitivity of natural CaF_2 depends on the impurities and type of encapsulation, but is usually small. The fast-neutron sensitivity is certainly less than 1 x 10^{-10} R (gamma equivalent) cm^2/neutron for 2.5 MeV neutrons, and <4 x 10^{-10} R-cm^2/neutron at 14.5 MeV, corresponding to less than 3 to 6.5% of the gamma radiation response (Harvey, 1967). In CaF_2:Mn, the sensitivity in R gamma equivalent (see Table 2-5) has been calculated to be 0.77 per 10^{10} n/cm^2 (measured value 1.05) for thermal neutrons

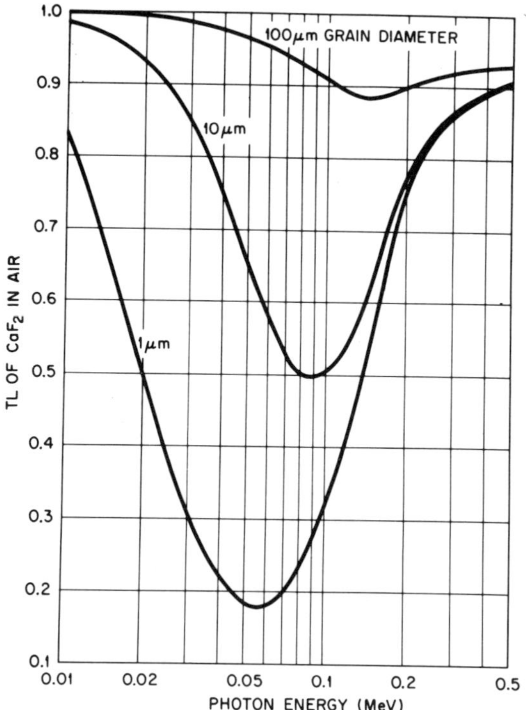

FIGURE 2-39. Deviation of the actual TL energy response of fine-grained CaF_2 in air from the photon energy response as calculated on the basis of the mass energy absorption coefficients, for different grain diameters. (After Chan and Burlin, 1970.)

and 3.54 (measured 2.30) for 14 MeV neutrons (Prokic, 1971).

According to the most recent published data (Attix et al., 1972), the fast-neutron sensitivity of CaF_2:Mn amounts to 0.6 of that of ^{7}LiF:Mg,Ti. However, Attix (personal communication) now believes that this figure probably applies best in the vicinity of 15 MeV and rises with decreasing neutron energy. Blum et al. (1973) have used CaF_2:Mn imbedded in Teflon for fast-neutron measurements in phantoms, with one of two detector discs surrounded by polyethylene and the other by lead. Obviously, this method can only be used for rather high neutron energies. A lower energy threshold of $\sim$0.7 MeV, which is the minimum energy required for the displacement of F^- ions, has been postulated by Puite (1971a). No permanent change in the gamma radiation response characteristics due to thermal neutron fluences up to 4×10^{12} n/cm^2 and fast-neutron fluences up to 8×10^{13} n/cm^2 has been found (Puite, 1971b).

The thermal neutron response can, of course,

be increased by mixing (or covering) CaF_2 with materials which undergo (n, α) reactions; the fast-neutron response is increased if the material is in intimate contact with organic substances during irradiation (Puite, 1969; Mishev et al., 1972; and others). An interesting method for selective thermal neutron measurements without the interference of gamma, beta, or X-radiation in a mixed radiation field is based on the neutron-induced activation of the phosphor (Mayhugh et al., 1971). In natural fluorite, mostly ^{45}Ca (165 d halflife) is formed; in CaF_2:Dy activation results in the formation of ^{165}Dy (2.3h) which "internally" irradiate the phosphor after exposure.

If the exposed sample is annealed, rereading after a sufficient delay time gives a measure of this self-dose. In CaF_2:Dy, a thermal neutron fluence as low as 5×10^7 n/cm^2 is thus detectable (in natural fluorite the lower limit is about 5×10^{10} n/cm^2). CaF_2 responds, as do all other TLD phosphors, to neutron-induced fission-fragments from fissionable materials. The high radioactivity of such materials prevents, however, the use of this principle in practical dosimetry (Ettinger et al., 1970).

For the response of CaF_2:Mn to protons in the 5 to 137 MeV range, see Parker, 1968. The relative sensitivity of this material to alpha particles in the 2 to 5 MeV range is about twice as high as that of LiF:Mg,Ti and amounts to 20 to 32% of the gamma radiation response (Lucas and Rainbolt, 1968). According to a more recent study (Zimmerman, 1972), the efficiency for 3.7 MeV alpha radiation in CaF_2:Mn is 53% of that for beta radiation on a rad per rad basis in the <100 rad dose-range, but becomes 100% at high doses (10^6 to 10^7 rad). Natural CaF_2 has been used for measurements of the alpha and beta radioactivity in soil (Fleming, 1968).

A strong interest of the U.S. Naval Research Laboratory (Ginther and Kirk, 1957) in artificially prepared, manganese-activated CaF_2 with only one major TL peak at $\sim$260°C (Figure 2-1) and an emission at $\sim$500 nm (Figure 2-24) led to the development of one of the first reliable TLD systems in the late 1950's (Schulman et al., 1960 and 1963). An early version of the dosimeter consisted of $\sim$90 mg of the phosphor cemented to a carbon heater plate with silicone cement and sealed into a glass bulb which was evacuated or filled with an inert gas. In later versions, the phosphor was cemented on a large-surface metallic heater spiral.

Miniature (1 x 14 mm) dosimeters were also made by sealing $\sim$8 mg of CaF_2:Mn into a glass capillary, with or without an additional sealed-in heating wire. A recent version (Figure 2-52) consists of two extruded ribbons of CaF_2:Mn, sandwiching the metal heater strip in a sealed glass tube. The gamma radiation response of the larger version was linear between 10^{-3} and 2×10^5 R (the smaller version was by a factor of 10 less sensitive). Recently, it has been reported (Gorbics et al., 1973) that thin layers of CaF_2:Mn exhibit up to about 40% supralinearity in terms of the light-sum response, but only half that much in terms of peak height. This effect is "swamped out" by optical absorption (discoloration of the phosphor) when thicker samples, e.g., 0.9-mm thick extruded chips, are employed.

Somewhat surprisingly, 6 to 10% of the signal faded during the first 16 hr of storage in the dark at room temperature. The amount of fading depends on the heating rate: No fading was observed at very low rates around 1°C/sec. There is also a strong effect of the heating rate on the sensitivity (no matter whether it is expressed as peak height or area under the peak) in CaF_2:Mn, but not in fluorite (Figure 2-40): At heating rates exceeding 0.5°C/sec, the response begins to drop and is reduced by a factor of two at 10°C/sec (Gorbics et al., 1968b). This is explained by preferential thermal quenching of the more stable components of the TL glow curve (Schulman et al., 1969).

Synthetic CaF_2:Mn (Mn content usually 3 mol %) is generally prepared by coprecipitation of CaF_2 and MnF_2 from a chloride solution with ammonium fluoride, followed by drying and firing at temperatures up to 1,200°C in an inert atmosphere, grinding of the sintered mass, and sieving to a convenient particle size of $\sim$100 μm (Ginther and Kirk, 1957; Frank and Herforth, 1962; Palmer et al., 1965; and Niewiadomski, 1967). It can then be formed into "extruded ribbons" (polycrystalline, almost transparent squares or rods) or sintered pellets, with or without the addition of binders (Uran et al., 1971). Enamel-covered CaF_2 has also been considered as a dosimetric material (Mihailovic et al., 1967).

What frequently has been claimed to be the single glow peak of CaF_2:Mn represents in reality a superposition of trapping levels of different depth. There are also small high-temperature peaks in this phosphor which limit its reusability after exposure to high radiation doses (Figure 2-41). By optimizing readout time and temperature, however, the residual TL can be reduced to 10^{-5} of the original signal (Lucas and Kapsar, 1972). There is no effect of temperature during irradiation between at least -50°C and the temperature when fading becomes significant in CaF_2:Mn as well as LiF:Mg,Ti (Gorbics et al., 1967).

A key problem in the use of CaF_2 dosimeters for the measurement of low doses has been the

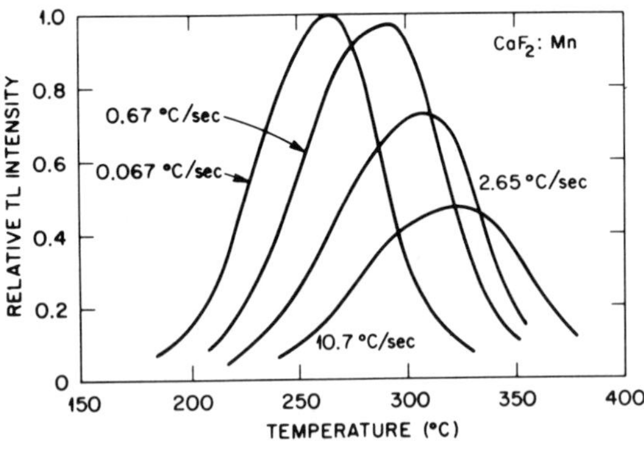

FIGURE 2-40. Glow curves of CaF_2:Mn for different heating rates, with the exposure adjusted to the reciprocal heating rate. (After Gorbics et al., 1968b.)

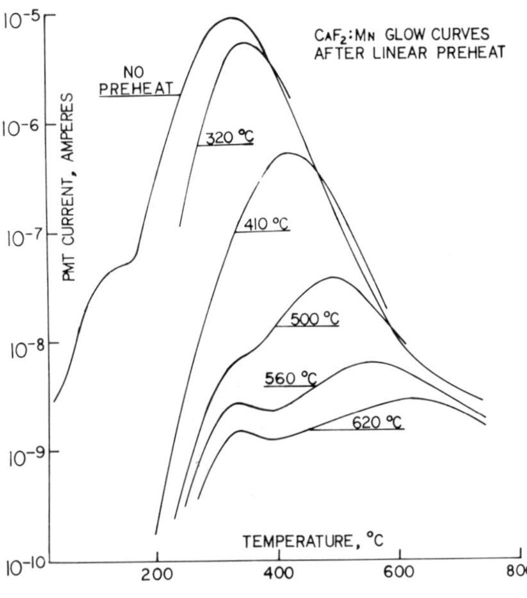

FIGURE 2-41. Glow curve of irradiated CaF_2:Mn without preheating, and after preheating to various indicated temperatures. (After Lucas and Kapsar, 1972.)

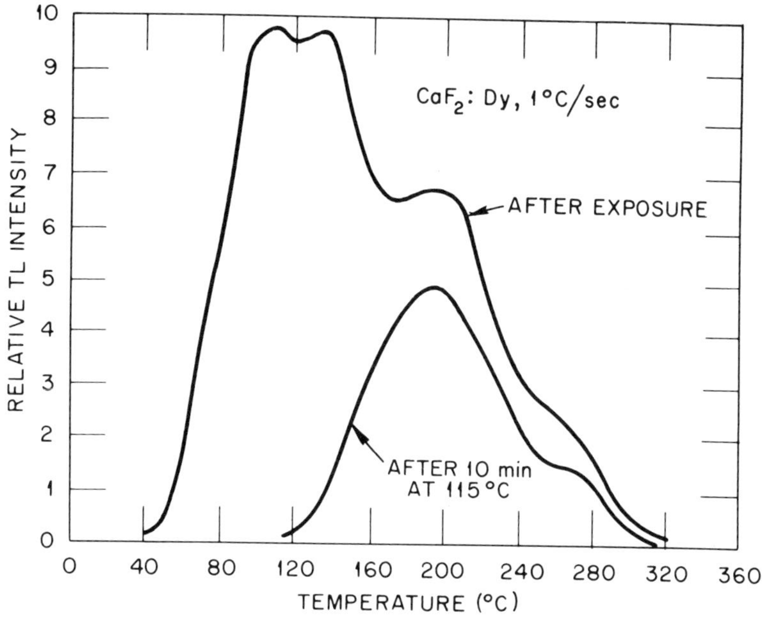

FIGURE 2-42. Glow curve of CaF_2:Dy immediately after exposure to gamma radiation, and after a "stabilizing" prereading heat treatment for 10 min at 115°C. (After McCurdy et al., 1969.)

"self-dosing," namely, a build-up of a TL signal without external radiation exposure either in the case of natural fluorites because of the inherent radioactivity of the phosphor (8 to 12 mrad/year self-dosing occur in the M.B.L.E. phosphor — Aitken, 1968) or in the case of glass-encapsulated CaF_2:Mn due to the potassium content of the glass envelope and/or cement. In some commercial dosimeters such as the early types made by EG & G of Santa Barbara, Calif., the background build-up due to potassium in the dosimeter by far exceeded the external background radiation dose (even in the improved new EG & G Tl-035 dosimeter, widely scattering self-irradiation rates between 7 and 21 μR/hr have been reported by Burke and Shambon, 1972).

Dysprosium-activated CaF_2, commercially available as TLD-200® from the Harshaw Chemical Co. and manufactured similar to CaF_2:Mn, is substantially more sensitive to gamma radiation than CaF_2:Mn in standard TLD readers (Table 2-5). It is, however, not more sensitive than LiF:Mg,Ti to high energy protons or alpha particles. A response of 3×10^{-7} rad gamma equivalent per 735 MeV proton, and of 2.1×10^{-6} rad per 930 MeV alpha particle has been reported (Jones et al., 1971).

The phosphor has a rather complicated glow curve with at least four peaks between 110 and ∿270°C (Figure 2-42), as well as two high-temperature peaks at ∿340 and ∿400°C. The TL emission spectrum exhibits main peaks at 483.5 and 576.5 nm. The gamma radiation response is rather complex: Supralinearity begins at ∿800 R, peaks at $\sim 10^4$ R, and begins to show saturation between 10^5 and 10^6 R. The response can, however, be made linear by pre-irradiation and annealing for 2 hr at 600°C. The glow curve undergoes complicated changes as the dose increases and/or during fading (Binder et al., 1968).

The photon energy dependence of CaF_2:Dy was measured to amount to a factor of about 12 to 17 between 30 keV and 1 MeV. The peak ratio in the glow curve changes, as in most other phosphors, as a function of LET. Due to the presence of low-temperature peaks, there is a 12% fading during one day at room temperature in the dark; if the sample is kept in diffuse room light, it increases to 30% during one day and 40% in two days. For example, a post-irradiation annealing for 10 min at 80°C reduces the sensitivity by 20%, but improves the long-term storage stability to 12% fading during one month (Binder et al., 1968). Other authors (McCurdy et al., 1969) report 30% fading during ten days and suggest a post-irradiation annealing treatment for 10 min at 115°C (Figure 2-42). Obviously, such a "stabilizing" prereadout annealing substantially reduces the sensitivity.

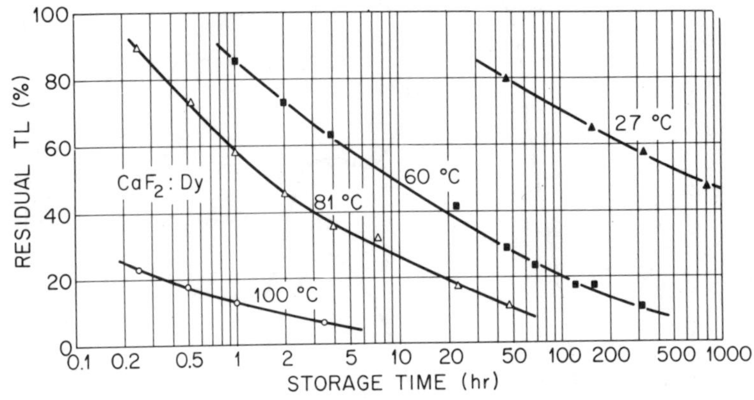

FIGURE 2-43. Fading of gamma radiation induced TL in CaF_2:Dy at various storage temperatures. (After Becker et al., 1971.)

Due to its high sensitivity, CaF_2:Dy has attracted some attention as a possible long-term integrating detector for measurements of the environmental (background) radiation. Its value for this purpose has, however, been subject to some controversy because of the necessity of large fading corrections. Some investigators simply ignore the fading (Hoy, 1971), or wait for at least one day prior to readout to avoid the $\sim$25% fading they find during the first day after exposure (Jones et al., 1971 and Lindeken et al., 1973) or they apply, in addition to the one-day delayed reading, a prereadout annealing procedure, as discussed above (Denham et al., 1972).

Others (Becker et al., 1971) conclude that the fading rate increases too rapidly with increasing temperature (Figure 2-43), thus making it advisable not to use it at least in a hot climate, especially considering that other equally or more sensitive phosphors are now available which are thermally much more stable, including $CaSO_4$:Dy and Mg_2SiO_4:Tb (Figure 2-49). Furthermore, the properties of CaF_2:Dy are apparently subject to rather large batch-to-batch fluctuations (Sukis, 1971).

Some other activators for CaF_2 have also been tested, among them erbium (Batygov and Voronko, 1969) and terbium (Batsanov et al., 1971). Apparently they offer no substantial advantages. A supposedly "pure" synthetic CaF_2 sample exhibited peaks at 60 and 360°C, the ratio between both changing during fading, for example, by a factor of two during five days at 10°C. It is, in principle, possible to use this ratio as an indicator of the time of exposure.

2.3.6. Calcium Sulfate

Manganese-activated calcium sulfate, with its long history of use mostly for ultraviolet measurements (see Chapter 2.1), attracted some attention in the early 1960's because of its outstanding sensitivity (Krasnaja et al., 1961 and Bjärngard, 1963a and b). Employing a large amount of the phosphor, a cooled photomultiplier, and special amplification of the modulated TL signal, gamma doses as low as 5 μrad have been measured with a 10% standard deviation (Lippert and Mejdahl, 1967).

With a reemission of $\sim$1.5% of the absorbed dose as TL light, it is one of the most sensitive TL phosphors in absolute terms (Lucke, 1970). Its capability to measure very low doses is, however, not only due to the high inherent sensitivity, but also to the facts that the incandescent background signal which is emitted by heater and phosphor is very small at the low temperature of its glow peak (Figure 2-1), and its convenient TL spectrum peaking is near 500 nm (Figure 2-14).

The low temperature of the main TL peak implies a rapid fading at room temperature, which various authors have found to amount to between 14 and 40% (most likely $\sim$30%) after 10 hr, $\sim$40% after one day, and 43 to 85% after 3 days, or after an initial period of $\sim$10 hr of more rapid fading, a fading rate of 6% per hour (Fowler and Attix, 1966; Bjärngard, 1967; and Lippert and Mejdahl 1967; see also Figure 2-44).

The practical use of $CaSO_4$:Mn is, therefore, limited to laboratory measurements of a short duration and/or at a low temperature (see, for example, Adam, 1972). If mixed with a suitable

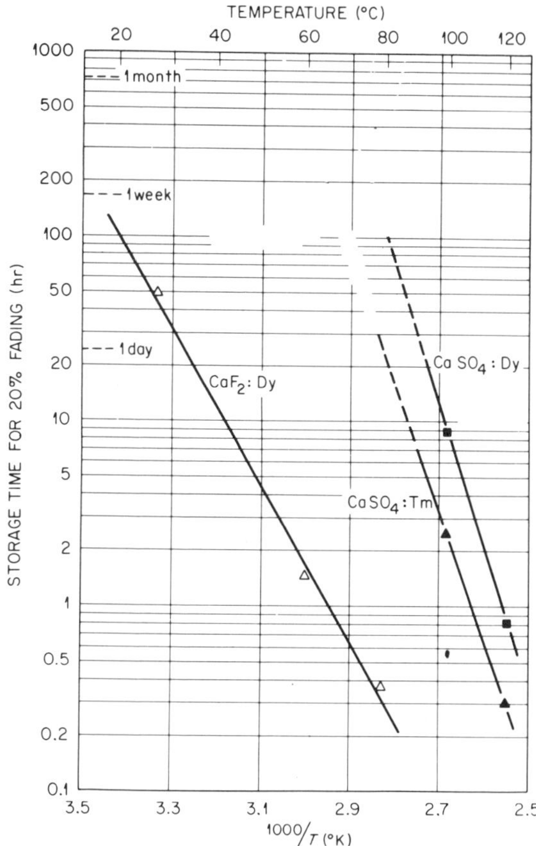

FIGURE 2-44. Simplified schematic diagram of effects of the storage temperatures on the fading rates in three sensitive modern phosphors. (After Becker et al., 1971.)

radioactive material and warmed to >50°C, its "permanently excited thermoluminescence" can actually be used as a weak light source (Spurny, 1969). It has been used for some dental x-ray experiments and for checking the validity of the Bragg-Gray cavity chamber principle for solid-state detectors (Nakajima, 1968).

The phosphor can easily be prepared, usually by dissolving $CaSO_4 \cdot 2 H_2O$ and the appropriate amount ($\sim$1 mol %) of $MnSO_4 \cdot 2 H_2O$ in sulfuric acid, slowly evaporating the acid, followed by heating and powdering of the phosphor (Watanabe, 1951). One version of this method (Bjärngard, 1963b) starts with a diluted (25%) H_2SO_4, with which the powders are mixed and left standing for one day at room temperature. The liquid is driven off by heating for 3 to 4 hr at 90°C and then 3 to 4 hr at $\sim$300°C. The residue is heated for about 2 hr at 900°C and the resulting white powder is ground and sieved.

$CaSO_4$:Mn can, however, also be prepared by sintering the finely powdered constituents at 900 to 1,000°C (Ivanii, 1967). It was determined in EPR studies that the Mn^{2+} centers have to be highly disperse for optimum sensitivity (Ikeya and Itoh, 1969). $CaSO_4$:Mn is quite sensitive to visible and UV light (daylight causes $\sim$15 μR/min gamma equivalent). Its response becomes nonlinear in the 10^3 to 10^4 R dose range. Its fast-neutron sensitivity is, as in all other inorganic TL materials, rather low. The thermal neutron response can be increased, for example, by mixing it with Li_2SO_4 (Ikeya et al., 1971).

Because of the rapid fading of TL in $CaSO_4$:Mn, there has been a search of other activators producing a more stable TL peak. The first alternate activator that was investigated in some detail was samarium (Krasnaja et al., 1962 and Bjärngard, 1964 and 1967). It was reported to have a small glow peak at $\sim$95°C and a larger one at $\sim$400°C. The response of the high-temperature peak becomes supralinear at $\sim$100 R, that of the low-temperature peak (less) supralinear at about 1,000 R. Although other authors report a more desirable glow curve for $CaSO_4$:Sm (Yamashita et al., 1968, found a main peak at $\sim$240°C, see Figure 2-45), this activator offers no great advantages over other rare earth activators.

A combination of high sensitivity (TL yield 4%, minimum detectable dose 0.05 mR ± 25%) and good stability (glow peaks at 160 and 190°C, with a fading rate of 5%/week at room temperature) has been reported for a $CaSO_4$ activated with 0.2 mol % lead and 3 mol % Mn (Yamashita et al., 1968). The lead content has, however, the effect of further increasing the photon energy response of $CaSO_4$.

By far, the most interesting and useful phosphors among the calcium sulfates are those activated with dysprosium or thulium (Yamashita et al., 1968). They can easily be prepared by a method similar to $CaSO_4$:Mn (Becker, 1972): 75 mg of powdered Dy_2O_3 (or 80 mg of Tm_2O_3) are dissolved in a few milliliters of diluted H_2SO_4 and the solution is added to 250 cm^3 of concentrated H_2SO_4, which is heated to about 250°C in an open dish; 34.4 g of powdered $CaSO_4 \cdot 2 H_2O$ are then dissolved in the hot acid, which is slowly (within several hours) evaporated in a fumehood at $\sim$300°C.

The resulting $CaSO_4$ crystals are ground and sieved (the size fraction of 100 to 300 μm flows

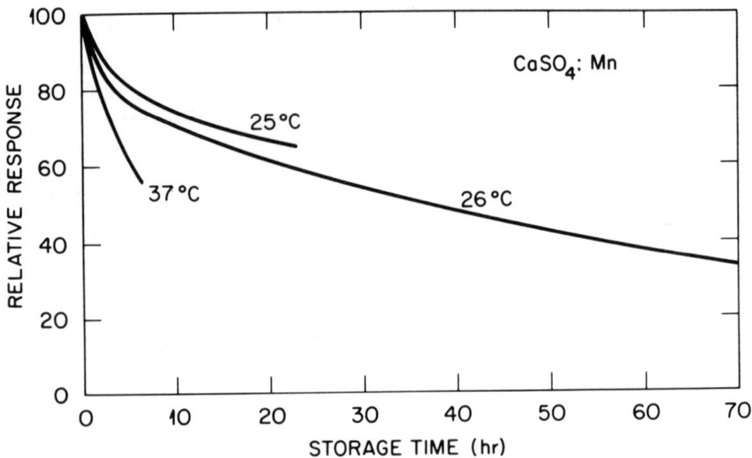

FIGURE 2-45.   Fading of TL in CaSO$_4$:Mn at 25°C and 37°C (after Bjarngard, 1967) and at 26°C. (After Lippert and Mejdahl, 1967.)

easily from automatic dispensers). The crystal powder is heated for 1 to 2 hr to 700 to 750°C in an open crucible to remove residual H_2SO_4, burn traces of organic residues, and reduce a low-temperature glow peak (some investigators report a pronounced peak at $\sim$100°C in their self-made material — Shambon, 1972 and Niewiadomski, personal communication).

Both dysprosium- and thulium-activated $CaSO_4$ have essentially the same single glow peak at $\sim$205°C, similar to the one given as Yamashita's glow curve for $CaSO_4$:Sm in Figure 2-45. The Dy-activated phosphor is slightly more sensitive to gamma radiation, but $CaSO_4$:Tm is less subject to thermal neutron response in mixed radiation fields. The TL emission spectrum consists of main peaks at 450 nm and some smaller peaks for $CaSO_4$:Tm and two main peaks at 480 and 570 nm for $CaSO_4$:Dy. The response is linear up to 3 x 10^3 R, supralinear between 3 x 10^3 and 10^4 R, and reaches saturation around 10^5 R. Only a small light sensitivity of $CaSO_4$:Tm has been observed (Nakajima, 1972).

The lower limit for gamma measurements has been of particular interest in connection with background monitoring studies. It also depends on the readout instrument. Using powdered samples of $\sim$15 mg each in commercial readers such as the Teledyne/Isotopes 7710[®] or the Harshaw 2000[®], $CaSO_4$:Dy is found to be 20 to 30 times as sensitive as LiF:Mg,Ti. Gamma radiation doses as low as 0.3 mrad can be detected. The precision is $\sim$4% for one standard deviation at 10 mrad and

$\sim$2% at 20 mrad and above (Becker, 1972). Others report σ = 20% at 0.5 mrad (Yamashita et al., 1968) and σ = 1.5 – 3% for measurements in the 5 to 10 mrad range (Mejdahl, 1970). With a special low-noise hot-air-jet reader, the reproducibility of readings with $CaSO_4$:Tm sealed in glass capillaries is better than ±0.1 mrad (Oonishi et al., 1971). $CaSO_4$:Dy has been used successfully in environmental dose measurements in tropical (Becker et al., 1971) as well as arctic countries (Nielsen and Mejdahl, 1970).

Only little fading (about 1 to 2% during one month and 5 to 8% during six months) is detected at room temperature due to the disappearance of minor low-temperature peaks. The fading rate increases slightly to around 3%/month at 30°C. For the stability at higher temperatures, see Figure 2-46. In recent laboratory tests with three months of storage at 25°C and 95% relative humidity, no fading has been observed in $CaSO_4$:Dy powder (Christensen et al., 1973). Even under harsh field conditions in various tropical countries where LiF:Mg,Ti faded up to 23%, the fading during three months amounted to not more than 5% (Becker, 1973). There have, however, also been some reports about higher fading rates in $CaSO_4$:Dy. Mejdahl (1972), for example, found almost 30% fading during the first month at 25°C and another $\sim$8% during the next 6 months of storage at this temperature. Apparently, a low-temperature peak has been more pronounced in the material used in this study.

Annealing is simple (several minutes at 400°C)

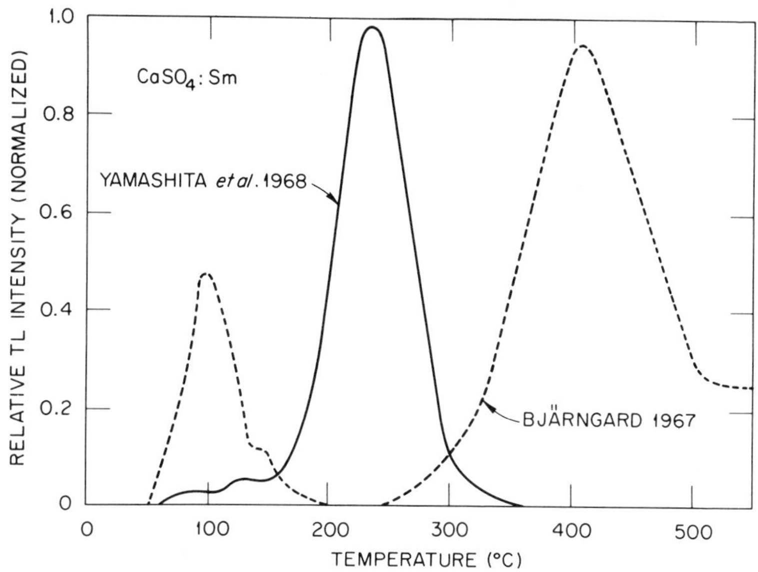

FIGURE 2-46. TL glow curve of CaSO$_4$:Sm. (After Bjarngard, 1967 and Yamashita et al., 1968.)

without changes of the response during multiple reuse and there is only minor interference from tribo effects and light. CaSO$_4$:Dy also became available as a dispersion in polytrifluorethylene (Teflon®) in the shape of small discs (Webb et al., 1972b). Their standard deviation is ±15% at 1 mR and 3.5% at 100 mR, but there is 10% fading during the first day and 20% during the first week after exposure. Various environmental effects such as UV light, mechanical vibration, and ultrasound can induce spurious signals in the order of a few mR in these detectors.

The inherent fast-neutron sensitivity of CaSO$_4$ is, independently of the activator, small. An interesting method for increasing the fast-neutron response has recently been suggested (Blum et al., 1972). Finely powdered CaSO$_4$:Tm is intimately mixed with glucose, enclosed in a Teflon tube, and exposed together with another Teflon tube filled with CaSO$_4$:Tm powder. For evaluation, the glucose is dissolved in water, the phosphor dried, and the response difference of the two phosphors related to the fast-neutron dose. At ORNL, a similar system is being studied, but the TLD phosphors are pressed into permanent, reheatable pellets with high-melting hydrogenous organics such as p-hexaphenyl. Pairs of such and Teflon imbedded detectors are used for fast-neutron dosimetry in a way similar to ^{6}LiF/^{7}LiF pairs in thermal neutron measurements (Becker et al., 1973). Such systems are also being studied in the Soviet Union (Kazanskaya et al., 1970).

2.3.7. Other TL Phosphors

Of the wide variety of potential TL phosphors that have been considered during the past ten years, some, such as thorium oxide (Angino, 1967) or Kapis shells that grow in muddy waters around the Philippine Islands (Bustamente et al., 1971) are obviously too exotic to be discussed here. Only a few that attracted more than passing attention for one reason or another will be mentioned.

Aluminum oxide — A well-known TL emitter (for some early work see Rieke and Daniels, 1957), aluminum oxide is of some interest because it is cheap and less energy-dependent (Figure 2-19) than CaF$_2$ or CaSO$_4$. It was found that ruby (Al$_2$O$_3$ with 0.05% Cr$_2$O$_3$) gave a prominent TL peak at 350 to 400°C, with some shift occurring as a function of dose. Its light emission peaks around 700 nm and a fairly linear response is observed between 30 and 10^5 rad (Philbrick et al., 1967).

Other investigators claim that doses as low as 100 mrad can also be measured with unactivated Al$_2$O$_3$ (McDougall and Rudin, 1970), but Buckman (1971) points out that even "pure" Al$_2$O$_3$ normally contains activating traces of Cr^{3+}. Other activators such as Ti or Na may, however, be involved in some of the samples which have been studied. There is a strong UV-induced fading (Buckman et al., 1968) and a pronounced dependence of the thermal fading on the Cr^{3+} concentration in ruby. In France, bricks made from Al$_2$O$_3$ have been used as "accident" dosi-

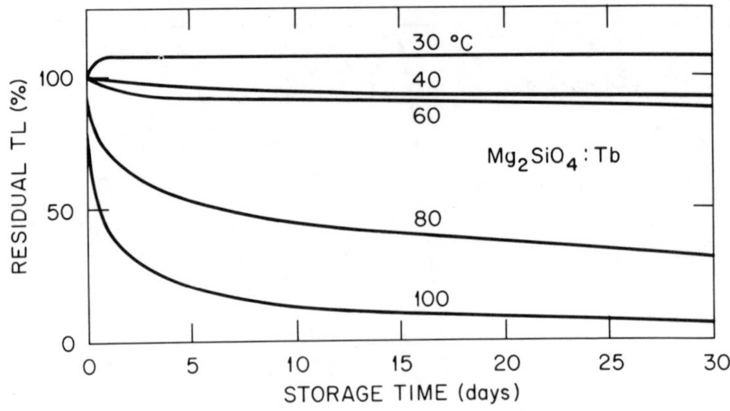

FIGURE 2-47.   Fading characteristics of Mg_2SiO_4:Tb at various storage temperatures. (After Hashizume et al., 1971.)

meters which are built into the walls of shelters (Portal et al., 1968).

Boron nitride — As a low Z and highly heat-resistant material, boron nitride (BN) was investigated as a potential TLD phosphor only quite recently. According to a study by Lubyanskii et al. (1971) of pure BN as well as samples activated with Mn, Eu, and Sm, there are two main peaks between 110 and 150°C, between 425 and 490°C, and an additional peak at 250 to 270°C for the Mn-activated material. The high-temperature peak response was linear up to 10^5 rad, the Mn peak up to 10^6 rad. No tribo-luminescence was observed, but there is a strong light sensitivity (20 hr of daylight resulted in a TL signal corresponding to 12 rad). The glow curve changed as a function of dose level and LET.

Quartz — It is the second most abundant mineral at the solid surface of the earth. With a material prepared by heating natural alpha quartz for 18 hr at 700°C, doses as low as 0.3 rad can be detected. There is a slight disagreement with regard to the type of temperature and/or irradia-tion pretreatment which produces optimum dosimetric characteristics (Fleming and Thompson, 1970 and McDougall, 1971). It is interesting to note that the TL in quartz is rather strongly affected by subjecting it to shocks, for example, in explosions.

In a rather detailed study on the $\sim$100°C TL peak of natural alpha quartz preheated at 700°C (Zimmerman, 1971a), a pronounced sensitivity increase due to pre-irradiation (e.g., by a factor of 10 after exposure to 1,000 rad of beta radiation and heating to 500°C) has been observed. From comparisons between the TL, radioluminescence, and the thermally stimulated exoelectron emis-

sion, it is concluded that the TL sensitivity increase results primarily from an increase in the number of activated luminescence centers. The TL sensitivity increase is reduced by UV exposure, but can be restored by heating to 500°C. Other properties of the three TL peaks at about 111, 325, and 375° that can be found in natural quartz grains as contained in ancient pottery have been the subject of intense studies in connection with TL dating studies (Aitken and Fleming, 1972; Mejdahl, 1972; and others).

Magnesium silicate — Terbium-activated mag-nesium orthosilicate (2 $MgO \cdot SiO_2$ with $\sim$0.001 g-at/mol Tb) is a very promising phosphor because of its high sensitivity and stability. It has a glow curve peaking at $\sim$195°C and a complex optical emission spectrum with main peaks around 380 and 552 nm. The response is linear between 0.3 mrad and several hundred rad. It becomes supra-linear at higher dose-levels. As it is about 50 times as sensitive as LiF:Mg,Ti in standard commercial readers, gamma doses as low as $\sim$0.2 mrad can be detected. Its fading stability is quite good up to $\sim$60°C. At 30°C, only 3% fading occurs during two months (Figure 2-47). At ambient temper-ature, there is a slight increase in TL after exposure. A maximum TL that exceeds the reading immediately after exposure by $\sim$14% is reached after a storage time of 10 min at 40°C and 10 days at 18°C (Nakajima, 1972 and Becker, unpublished data).

The phosphor is annealed for reuse for 20 min at 450°C. Its energy dependence (oversensitivity by a factor of 4.5 around 30 keV) can be compensated with a tin ribbon of semicircular or trapezoidal cross section wound into a helical coil to obtain an almost flat response down to $\sim$40 keV (Hashizume et al., 1971 and Nakajima et al., 1971). The phosphor can be further sensitized by pre-irradiation (Figure 2-48), for example, by a factor of 4 with 10^5 R gamma radiation followed by 1 hr at 300°C (Nakajima, 1971).

Mg_2SiO_4:Tb is rather sensitive to light, both with regard to light-induced fading and to the build-up of a TL signal in the pre-annealed phosphor (Nakajima, 1972). It should, therefore, be used only in light-tight encapsulation. The activation of magnesium ^{24}Mg (n,p) ^{24}Na in magnesium silicate has been used for fast-neutron measurements, with the gamma radiation effect being annealed after exposure and the neutron-induced effect measured after $\sim$3 days delay (Pearson et al., 1972).

A close "relative" of Mg_2SiO_4:Tb is MgB_4O_7:Tb. Its energy response for photons is comparable to that of LiF (Figure 2-32), but the sensitivity is smaller.

Glasses — The first TLD glass was developed in the Soviet Union in the early 1960's and named "system IKS" (Botshvar et al., 1963). It consisted of a magnesium phosphate/aluminum phosphate

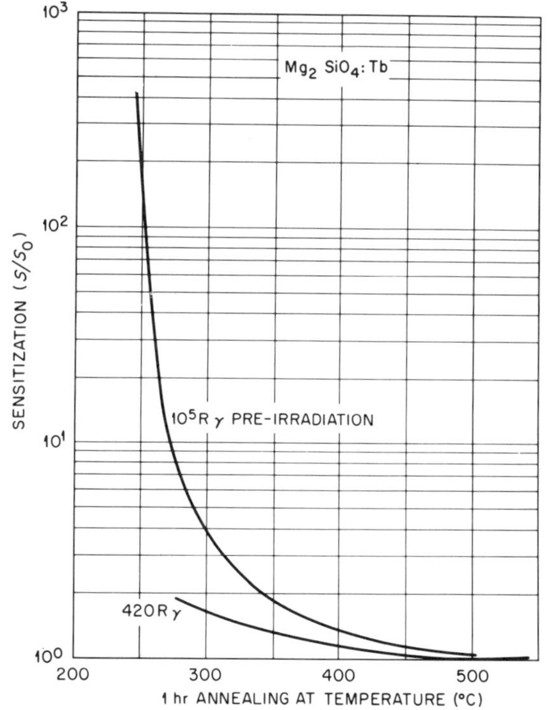

FIGURE 2-48. Residual sensitivity increase in Mg_2SiO_4:Tb after exposure to different gamma radiation doses, followed by 1 hr of annealing at various temperatures. (After Nakajima, 1971.)

glass base activated with manganese, had a range from 20 mrad to 10^6 rad, and no fading and light sensitivity under "normal" conditions. Ginther (1967) studied another glass base and arrived at an optimum composition of 29% Li_2O, 12% Al_2O_3, 59% SiO_2, and 0.002% Tb. This glass had a stable peak at $\sim$260°C, an oversensitivity to 25 keV photons by a factor of 2.7, and a complicated, partly sublinear and partly supralinear dose response between $\sim$30 mrad and 10^5 rad. Exposure to UV light produced a different glow curve in this material.

For a manganese-activated aluminophosphate glass with added MgO and SrO similar to that of Botshvar et al. (1963), Spurny (1968) reported a main TL peak at $\sim$275°C, an emission maximum of 640 nm, an oversensitivity by a factor of 7 at $\sim$30 keV, and a better sensitivity than LiF:Mg,Ti. A similar Mn-activated phosphate glass was studied by Regulla (1972). It has several wide, merging glow peaks between about 60 and 210°C, which makes peak height measurements obligatory for evaluation. The TL emission spectrum peaks at 600 nm and there is an oversensitivity by a factor of 4 around 40 keV. Other authors focused on various other systems (Bettinali et al., 1967 and Sidorov et al., 1968) including boralumino silicate "enamels" with or without added lithium silicate (Mihailovic and Kosi, 1971). Despite some obvious advantages such as transparency, TLD glasses never became very popular.

Others — An interesting system with a sharp TL peak at $\sim$160°C, a main TL emission at $\sim$550 nm, and a linear response between a few mrad and several times 10^4 rad of 40 keV x-rays with the composition $(Al_2O_3 \cdot 2 Y_2O_3, 2 \times 10^{-4}$ g-at Tb) was studied by Sakamoto et al. (1968). As a high Z material with an oversensitivity by a factor of 100 at $\sim$65 keV, pure BaF_2 has been suggested for the determination of the effective energy of x-rays (Dixon and Watts, 1972). There is, however, a very pronounced fading in the undoped material. (For the TL in $MgCO_3$, $SrCO_3$, $CaCO_3$, $BaCO_3$, and $BaTiO_3$, see Angino, 1967.)

In MgF_2 with 0.1% MnF_2, small peaks at 68°C and 218°C, and a main peak at 131°C have been reported (Bräunlich et al., 1961). Y_2O_3:Eu, with and without additions of LiF, CaF_2, ZnF_2, and NaF, has been studied by Huzimura and Ato (1968). In Ag-activated NaCl, a peak at $\sim$110°C was found in addition to various peaks at lower temperatures (Rothermel and Scharmann, 1966).

In KCl:Ca, use can be made of a peak at $\sim$85°C in the 5 mR to 50,000 R range (Mehendru et al., 1970 and Kumar et al., 1971). In the complex system ZnS/CdS:Ag,Ni,Co, a linear response in the 1 to 500 rad range and 4% fading during 48 hr have been observed (Elmanharawy and Mostafa, 1972). The TL in the concrete of walls has been suggested as a method for evaluating dose distributions after radiation accidents (de Franceschi et al., 1973). Numerous other compounds have also been considered.

2.4. Dosimeter Design

2.4.1. Open Phosphors, Chips, and Plastic Compounds

Several aspects of dosimeter design have already been touched upon while discussing various phosphors, but there are some more general aspects that have to be mentioned. As it is difficult to grow single large crystals with exactly reproducible properties, their use of TLD phosphors has only rarely been considered. Loose crystalline phosphor powders, however, are still finding widespread application.

The grain size distribution should be fairly uniform and fit the particular requirements of the dosimetry system. If the powder is too fine, it tends to stick in a humid atmosphere and not to flow easily and uniformly from the standard, commercially available vibrating powder dispensers. Depending on the type of dispenser and the density of the powder, uniform quantities of the phosphor of about 8 to 25 mg each are normally used for each readout. Increasing the amount of phosphor per reading does not necessarily increase the total TL light output proportionally because absorption of the TL light which is emitted by the lower layers occurs. The use of the excessive amounts of phosphor during readout also leads to a nonuniform heating of the sample and a distortion of the glow curve.

Very fine powders of some phosphors such as LiF:Mg,Ti are also known to be less sensitive than coarser samples. Too coarsely grained phosphors, on the other hand, introduce larger statistical fluctuations into the dispensing procedure. An average grain size of 75 to 200 μm usually represents a good compromise for most phosphors and dispensers. The fluctuations due to the dis-

penser should, if instructions are followed, not exceed 0.5 to 1% for one standard deviation and, therefore, not be the limiting factor in overall reproducibility of TL readings. If large numbers of powdered samples are to be used, the individual dispensing of the powder can be avoided by using a simple device. It permits preparation of larger quantities in each batch with about 1% standard deviation (Gorbics and Attix, 1973).

Working with loose powders has some obvious disadvantages: It may be quite cumbersome if large amounts of detectors have to be read; fractions of the phosphor are easily lost; the phosphor is exposed for relatively long periods of time, and with a relatively large surface area, to atmospheric humidity and pollution; it is in many cases mandatory to perform low dose readings with phosphor powders in an inert gas (N_2) flow; and, if different types of TL phosphors are used in the same laboratory, there is a chance of contamination due to accidental mixing or insufficient cleaning of the dispenser and reader. It is particularly unpleasant to "poison" unknowingly ^{7}LiF with natural LiF or ^{6}LiF in neutron dosimetry experiments or contaminate an insensitive phosphor with a highly sensitive one.

There are, however, also several advantages. It is, for example, easy to homogenize the properties of a larger amount of a phosphor by simply shaking the bottle, without having to worry about the individual thermal and/or irradiation histories of single detector units. Powders can be filled into containers of any desired shape* including polyethylene tubes or bags to be heat-sealed with a soldering iron, if one takes care that gases from the heated plastic do not negatively affect the TL properties (Wallace et al., 1971). Some polyethylene tubes for medical purposes such as one sold by Clay Adams, Parsippany, N.J., apparently permit heat-sealing without affecting the TL. It is also more economical to use enough powder for permitting multiple readings in personnel or environmental dosimetry than the corresponding number of impregnated or extruded detectors costing about $1/unit (sealed detectors are priced even higher).

In a good TL reader, an experienced technician can easily read medium and high gamma radiation doses with a standard deviation of 2 to 3%.

*A treasured curiosity in the author's collection of personnel dosimeters is a religious amulet filled with LiF powder, which has been used by Dr. Cullen et al., Catholic University of Rio de Janeiro, for measuring the long-term dose to the natives in "black sand" coastal areas in Brazil which have an increased natural background.

Normally, the powder is annealed (and, if necessary, washed and sieved) for reuse, resulting in a $\sim 10\%$ material loss for each use cycle. To reduce these losses and simplify handling, various investigators produced sintered, polycrystalline pellets of "rods" from several phosphors including LiF:Mg,Ti (Shambon and Condon, 1968; Portal et al., 1971; Niewiadomski et al., 1971; Koczynski et al., 1972; and others), BeO:Na (Yasuno and Yamashita, 1971), $Li_2B_4O_7$:Mn, Si (Christensen, 1971), and Al_2O_3 (Portal et al., 1968). There is, however, some loss of sensitivity due to light absorption and scattering in such pellets; and some of them break or rub off easily.

The powdered TLD phosphor can also be permanently attached to an open heatable carrier. In India, for example, strips of Kanthal metal (an alloy consisting of 72% Fe, 23% Cr, and 2% Co) with the dimensions 40 x 12.5 x 0.025 mm have been coated in their central portion with a few drops of a Dow Corning Silicone Resin (805)® prior to sieving ~ 100 mg of natural CaF_2 over it and burning it in for 6 hr at $300°C$ (Nambi et al., 1969). The dosimeter is then used in a sealed black PVC bag.

As has been mentioned before, LiF is quite soluble in water as well as toxic and should, therefore, not be used without protection for in vivo implantation studies (Galkin et al., 1968 and Dettmer and Galkin, 1968). Furthermore, etching of the surface changes the light output from the samples, making it necessary to recalibrate the dosimeters after retrieval. Some of those difficulties can be overcome by a protective coating of the LiF crystals which is sufficiently stable in body fluids. Hukkoo et al. (1972) described a technique for producing a thin SnO_2 film which meets the requirements quite well.

Small "extruded ribbons" of polycrystalline and transparent TL phosphors became commercially available in recent years (Cox, 1968). The following materials are currently available either as rods (1 x 6 mm) or as square chips (3.175 x 3.175 x 0.89 mm) from the Harshaw Chemical Co.: natural LiF:Mg,Ti; [7]LiF:Mg,Ti; [6]LiF:Mg,Ti; and CaF_2:Dy. The latter is also available in a larger size (6.3 x 6.3 x 0.89 mm). Some typical samples are pictured in Figure 2-49. They exhibit a good sensitivity per unit weight and are much easier to handle than loose powder in most situations. If individually calibrated and read with great care with an optimized counter, doses as low as 5 mrad

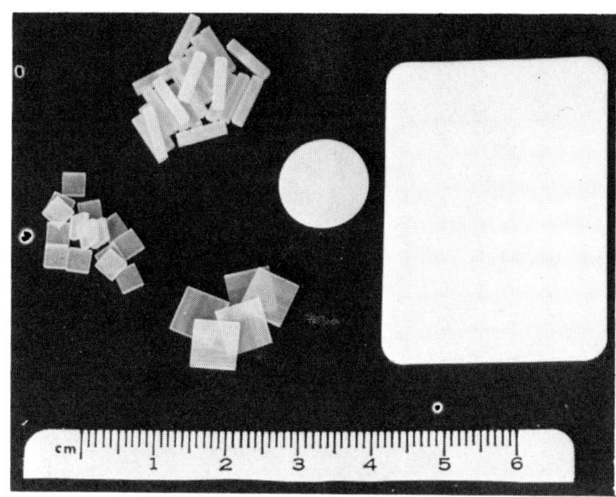

FIGURE 2-49. Some common types of TL detectors: "extruded ribbon" type polycrystalline chips of different sizes made by Harshaw (left), phosphor-impregnated Teflon disk (center), and film-sized phosphor-Teflon foil (right) made by Teledyne/Isotopes.

have been read with a 12% standard deviation in LiF:Mg,Ti chips (Schlesinger et al., 1971). For higher dose-levels, standard deviations of ~ 1 to 5% are typical (see, for example, Rassow and Strüter, 1969).

The extruded chips and rods also have some disadvantages: They are difficult to mark and to distinguish. Various users have drilled holes into them or rounded one or more edges to simplify identification, for example, of different phosphor types. They also are rather expensive; depending on the size of the order and type of chips, they cost currently between $0.90 and 4.00 per unit. Their individual sensitivity depends, of course, on each chip's irradiation and/or thermal history. The most serious drawback in the use of extruded LiF:Mg,Ti is, however, the large fluctuations in sensitivity which have been observed both within one batch (typical standard deviation 15% with variations of single detectors up to a factor of 3), and even more so from batch to batch. (Regulla (1972) reports a drop in average sensitivity of TLD 100 needles delivered between 1967 and 1970 by a factor of 5.) Therefore, if large numbers of such detectors are ordered for a personnel dosimetry program, it is very advisable to check the sensitivities of at least a representative sample (such as 5 to 10% of each package) and reject all those not meeting the usual $\pm 10\%$ specifications.

Another approach to simplify handling and reduce disturbing environmental effects consists of the intimate mixing of phosphors with heat-

TABLE 2-7

Some Commercial TLD Phosphor Teflon Dosimeters

Phosphor	Shape	Dimensions (mm)	Phosphor % by weight	Phosphor content (mg)
LiF:Mg,Ti nat.[1]	disc	12 x 0.4	30	30
[7]LiF:Mg,Ti[1]	disc	12.7 x 0.4	30	30
	disc	8 x 0.5	5	3
	disc	12.7 x 0.13	30	10
	disc	6 x 0.02	30	0.2
	rod	1 x 6	4	0.4
	long rod	1 x 150	4	
	tape	25 wide x 0.4	30	
[6]LiF:Mg,Ti[1]	disc	11.3 x 0.4	30	30
	rod	1 x 6	4	0.4
	long rod	1 x 150	4	
$Li_2B_4O_7$:Mn[1]	disc	9.5 x 0.13	5	1
	disc	9.5 x 0.4	5	3
	rod	1 x 6	4	0.4
	long rod	1 x 150	4	
$CaSO_4$:Dy[1]	disc	6 x 0.02	30	0.2
	rod	1 x 6	4	0.4
	long rod	1 x 150	4	
CaF_2:Mn[1]	disc	6 x 0.13	5	0.3
	disc	6 x 0.4	5	1
	disc	6 x 0.02	30	0.2
Mg_2SiO_4:Tb[2]	rod	6 x 1		
	disc	5 x 0.3		

[1] Teledyne/Isotopes, Westwood, N.J.
[2] Dai Nippon Toryo, Chiyoda-ku, Tokyo, Japan

resistant polymers such as Teflon (Bjärngard and Jones, 1967) or silicone rubber (Nakajima, 1968). The Teflon-incorporated materials are available in various sizes, shapes, and phosphor contents from Teledyne/Isotopes, Westwood, N.J. As can be seen in Table 2-7, the different phosphors can easily be distinguished by the different dimensions of the detector. Prices are slightly lower than those of the chips (depending on the type of detector and size of the order between $0.10 and 2.50 per unit, with the ones suitable for personnel dosimetry in the $0.85 to 2.50 range); the sensitivity obviously depends mostly on the type of phosphor and its total amount in the detector.

As the chips, the Teflon dosimeters can be cleaned when dirty.* The key problem in their use is the fact that the Teflon softens, changes its optical properties, and finally melts and disintegrates at temperatures around 320°C, making it difficult to anneal the phosphors, as desirable, at temperatures around 400°C (Preston et al., 1970). Furthermore, the Teflon foil bends and changes shape if not squeezed between metal plates during annealing (for this reason, a special holder has to be used during readout which keeps the foil flat in a constant position). Multiple use is, however, possible if the foils are not exposed to very high doses and always annealed with the same, well-reproducible annealing cycle (see, for example, Carlsson, 1969; Martensson, 1969; Docherty and Marshall, 1970; Linsley and Mason, 1971; and Marshal et al., 1971).

The recommended annealing cycle for LiF-Teflon consists of several hours at 300°C, followed by cooling at a constant rate to 80°C and 24 hr of storage at 80°C. There are several commercial readers such as the Teledyne/Isotopes TLD-7300 and the Victoreen 2800 which provide a simplified

---

*There are, however, exceptions to that rule. In one instance (Robertson and Gilboy, 1972), a high and erratic background of Teflon discs, which could not be removed by ultrasonic cleaning with alcohol, was attributed to fumes from a plastic (Araldite®). The annealing oven should not be used for other purposes which could result in such a contamination.

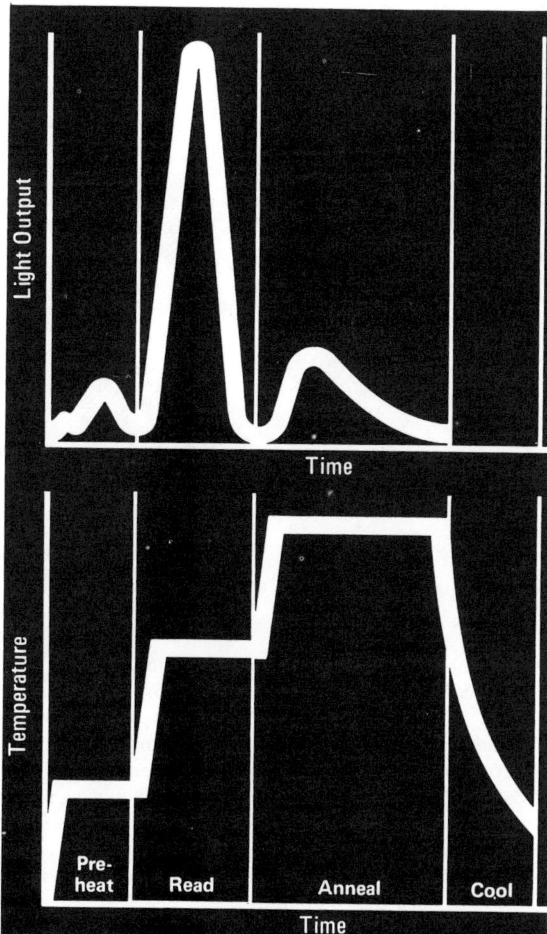

FIGURE 2-50A. Typical programmed readout cycle in modern TLD reader for evaluation of LiF:Mg,Ti, consisting of preheating without light integration to discriminate against low-temperature peaks, actual readout of the "dosimetry peak(s)" around 200°C; annealing at elevated temperature, and rapid cooling of the sample. (Courtesy of Victoreen, Cleveland, Ohio.)

annealing procedure as part of the readout cycle (Figure 2-50A). Such an annealing cycle has been found to anneal all but $\sim$0.03% of the signal (Webb and Phykitt, 1971). It may, for example, consist of $\sim$12 sec of preheating at $\sim$130°C during which no TL light is recorded and a readout for 12 sec up to $\sim$270°C, followed by annealing for 18 sec at 300°C.

The sensitivities of individual dosimeters, which may vary substantially from batch to batch, may be brought to within ±5% by cycling once through such a readout and annealing procedure. One should, however, keep in mind that Teflon is a poor heat conductor; for example, in a disc 0.4 mm thick at a heating rate of 20°C/sec a temperature difference of more than 50°C can be observed between the two faces of the disc (Perry and Preston, 1970).

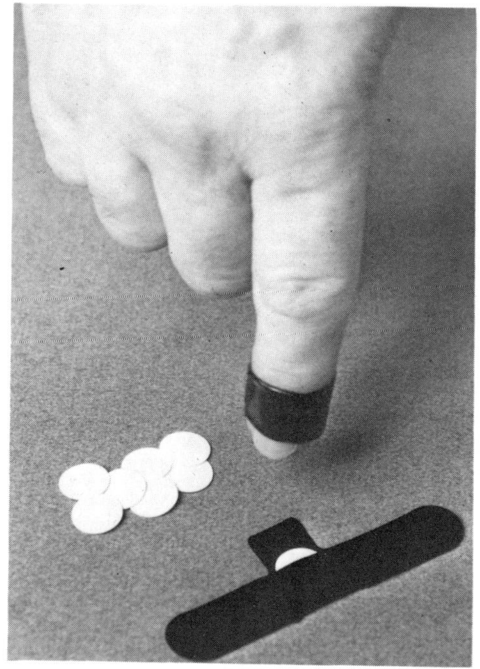

FIGURE 2-50. Teflon-LiF-discs as fingertip dosimeter. (Courtesy D. A. Pitman, Weybridge, Surrey, U.K.)

As can be seen in Table 2-7, foils with a thickness of only 20 μm are available as Bragg-Gray type solid-state cavity chambers for the dosimetry at interfaces, beta dosimetry, etc. By using a liquid molding technique and a silicone binder, ultrathin foils with 30% phosphor at 15 μm thickness have been prepared (Webb and Bodin, 1971) with a standard deviation of 3 to 4%. In foils containing 0.27 mg $CaSO_4$:Dy with an effective thickness of 2.1 mg/cm^2, doses as low as 20 mR can be detected. The sensitivity is a linear function of the phosphor content.

The Teflon discs can be used without much additional protection for many purposes such as measurements at interfaces (Schulz, 1967) or phantom depth-dose studies. For the dosimetry of very short-range radiations, stacks of the ultrathin foils have been employed (see, for example, Gibson et al., 1971). If placed in a pocket of a thin opaque protective plastic, they can be used as fingertip dosimeters (Figure 2-50B) or they may be incorporated in simple holders or identification cards for personnel dosimetry (an example is given in Figure 2-51).

Usually more than one disc is used in the latter application. Either ^{6}LiF and ^{7}LiF pairs are employed to gain some information on neutron

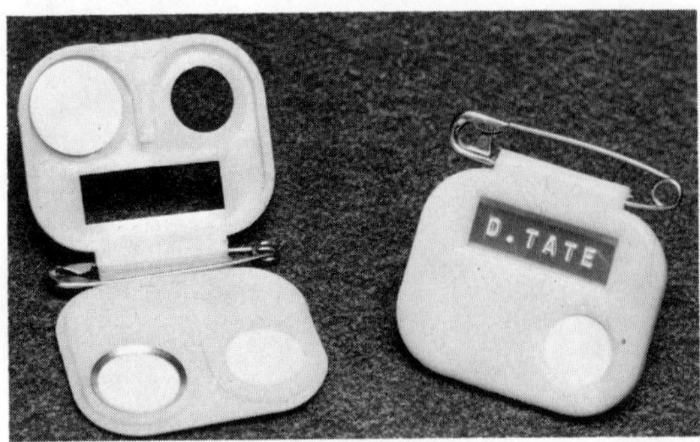

FIGURE 2-51. Example of a very simple personnel dosimeter with Teflon-LiF-discs. (Courtesy D. A. Pitman, Weybridge, Surrey, U.K.)

exposures (Kocher et al., 1970; Nash and Attix, 1971; Johns et al., 1971; and others) and/or the discs are located behind different filters to permit distinguishing between photon and beta radiation or estimating the effective photon energy. The filter-analytical principle can also be applied to a single, larger sheet of the phosphor-Teflon foil, if the areas behind different filters are read out separately by localized heating. Such a system (to be discussed in more detail later) has been developed by Teledyne/Isotopes.

There are, of course, an almost unlimited number of possibilities for badge design. The details depend on the choice of detectors; their combination with other dosimeters, such as photographic film or RPL glasses; the type of radiation(s) to be measured; the number of persons to be monitored; the evaluation system; and many other parameters. (For a variety of recent designs, see Gangadharan et al., 1969; Kocher et al., 1970; Nash and Attix, 1971; Johns et al., 1971; Christensen, 1971; Oonishi et al., 1971; Jones, 1971; Rich and Samardzich, 1971; Webb et al., 1972b; and many others.)

Extruded ribbon chips usually have to be removed from a badge-type holder for evaluation, and various types of such badges are in use. In very simple badges, there is only one chip (see, for example, Gollnick, 1972). Normally the badge provides room for two or three units, for example, behind 10 mg/cm^2 plastic and 285 mg/cm^2 in a commercial badge (Eberline, Santa Fe, N.M.). In some cases such as the badge of the Lawrence Livermore Laboratory (Figure 2-52, left), three chips are located behind differently colored plastic films, from where they can be removed, read, and replaced automatically.

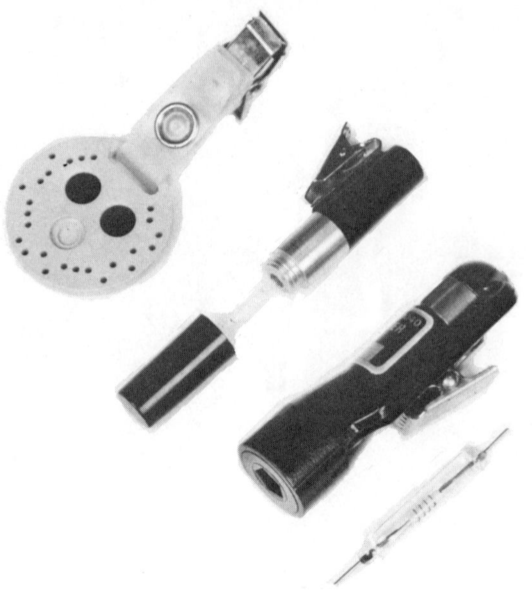

FIGURE 2-52. Three typical examples of more advanced TLD dosimeters: a circular disc with room for three TLD chips to be evaluated in a fully automatic reader as used in the Lawrence Livermore Laboratory (left); a dosimeter with two capillaries containing $CaSO_4$:Tm and/or BeO:Na, to be read by blowing hot air against the needles as manufactured by Matsushita, Osaka (middle); and the new U.S. Navy dosimeter developed by Harshaw, with two extruded CaF_2:Mn pieces sandwiching the heater in a glass bulb (left).

The evaluation of dosimeters containing TLD chips, and the problem of their fast and reliable identification can both be simplified by attaching them permanently to the carrier. One example of such a device, where two LiF chips are enclosed between transparent thin Teflon foils into holes of a credit card-type identification card, is given in Figure 2-53. A small area constant temperature heater permits one to read and essentially anneal one or both of the detectors without removing them from the coded card. A similar system was developed at the Chalk River Nuclear Laboratory (Jones, 1971). However, adhesive-coated Teflon tapes that have also been studied for this purpose exhibited some "thermoluminescence" of their own (Shipley et al., 1972).

Both Teflon foil and extruded ribbon pairs, usually with ^{6}LiF and ^{7}LiF, have been used widely in thermal neutron dosimetry. Numerous attempts have been made to modify the response characteristics of such LiF pairs with different thermal neutron response, but similar gamma response in such a way that they can be used in

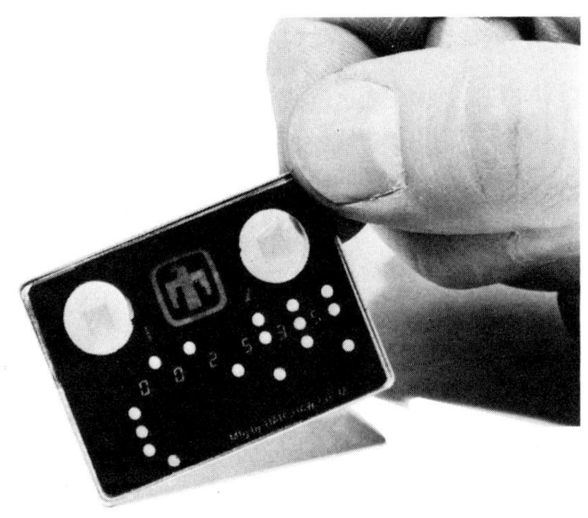

FIGURE 2-53. Credit card-type simple personnel dosimeter with two LiF:Mg,Ti extruded ribbons between Teflon foils, where they can be evaluated automatically without being removed. (Courtesy Harshaw Chemical Co., Solon, Ohio.)

neutron personnel dosimetry. As significant personnel exposures to thermal neutrons are extremely rare, their measurement is of little interest per se.

The problem is to make use of the moderating and backscattering effect of the human body which produces a thermal neutron field near the surface of a person being exposed to intermediate or fast neutrons. The detector's oversensitivity to thermal neutrons originating outside the body has to be reduced; only those should be measured that originate in the body. To some extent this can be done by asymmetrical neutron absorbers, covering only the detector side opposite the body surface.

As this technique is essentially the only one available for measuring personnel exposures to intermediate energy neutrons, substantial efforts have been made in recent years in several countries, such as the U.S. (Hoy, 1972; Hankins, 1973; and others), Germany (Piesch and Burghardt, 1972 and 1973), England (Brunskill and Sellars, 1973; Harvey et al., 1973; and Knight et al., 1973), and Denmark (Bötter-Jensen et al., 1973) to optimize its response characteristics. Unfortunately, such "albedo" types of intermediate and fast-neutron personnel dosimeters are rather sensitive to changes in the neutron spectrum, with rapidly decreasing sensitivity to neutron energies above $\sim$0.1 to 1 MeV due to the energy dependence of the neutron albedo (Figure

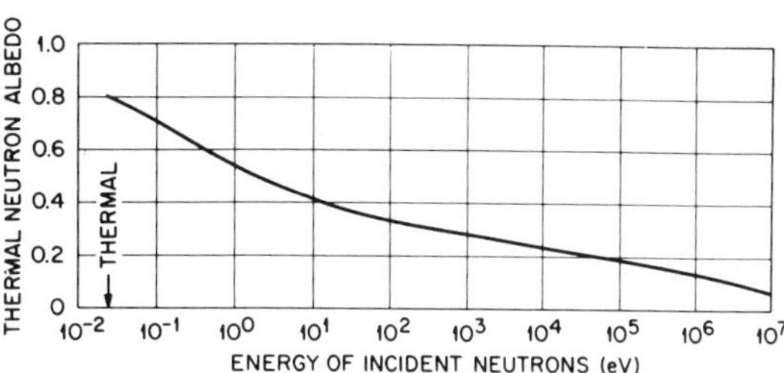

FIGURE 2-54. Thermal neutron albedo from the human trunk as a function of incident neutron energy. (After Harvey, 1967.)

2-54), as well as variations in the detector-body distance (a few additional cm distance may reduce the sensitivity by a large factor).

The results of numerous studies on this subject (Brown et al., 1967; Mejdahl, 1967; Endres and Kocher, 1968; Preston, 1968; Hall and Hudd, 1969; Korba and Hoy, 1970; Kalk, 1971; Attix, 1971; Majborn et al., 1972; Bötter-Jensen and Christensen, 1972; Tymons et al., 1973; and others) can be summarized as follows: It is unlikely that more than semiquantitative dose information will ever be obtainable from albedo neutron personnel dosimeters. Even the best-adjusted system will remain oversensitive to thermal neutrons and undersensitive to fast and relativistic neutrons. There is, however, at least one sophisticated device (Hoy, 1972) which, if worn at a small and constant distance from the body, can serve as an adequate dosimeter in the intermediate neutron range as encountered around swimming pool reactors, plutonium glove boxes, etc. and which is not accessible by other personnel dosimeters.

This system (Figure 2-55) consists of two pairs of TLD-600 and 700 chips in a polyethylene-filled

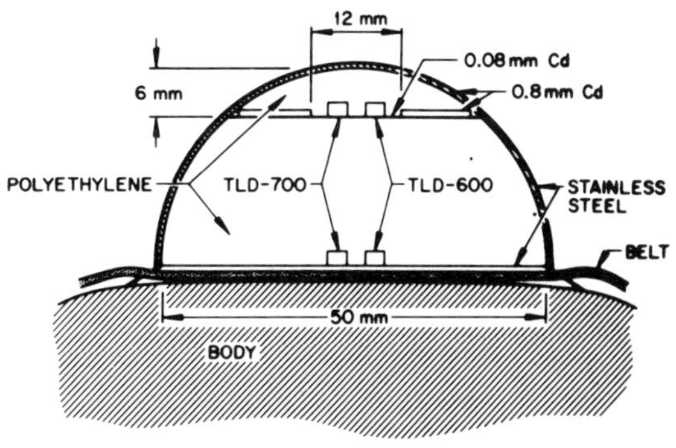

FIGURE 2-55. Schematic diagram of albedo neutron personnel dosimeter. (After Hoy, 1972.)

hemispherical unit of 5 cm diameter worn tightly on the body surface by a belt. The hemisphere surface and bottom are covered with stainless steel, below which an additional 0.8 mm Cd shield is located, on the hemispherical side only. One pair of dosimeters (B_{600} and B_{700}) is located at the bottom of the device close to the body, another pair (S_{600} and S_{700}) is behind a 6 mm thick polyethylene segment which is separated by an additional Cd layer in the upper part of the device. From the TL readings of the four chips, the dose equivalent DE is calculated by the formula

$$DE = (B_{600} - B_{700}) - (S_{600}R - S_{700}) \times CF \times \frac{\ln T}{-1.52} 2.16$$

where

R = gamma response ratio of TLD-700/TLD-600,

CF = calibration factor, which depends somewhat on the exposure conditions, and

$$T = \frac{(S_{600}R - S_{700})}{(B_{600}R - B_{700})} \qquad 2.17$$

Using this somewhat complicated evaluation procedure, an energy response to unscattered monoenergetic neutrons, as shown in Figure 2-56, is obtained. These data are in good agreement with the results of response calculations (Alsmiller and Barish, 1972). As in most monitoring situations, a large fraction of the neutrons is scattered and degraded in energy; however, the actual deviations between neutron dose and meter reading have frequently been less pronounced.

A simpler device also based on the albedo

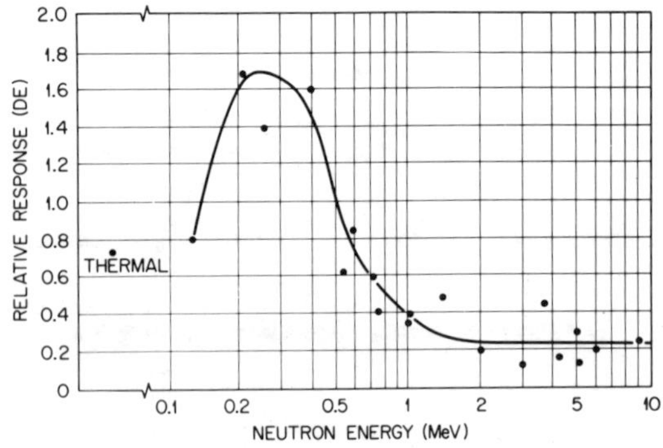

FIGURE 2-56. Relative response of the Hoy-type albedo dosimeter as a function of incident fast neutron energy. (After Hoy, 1972.)

principle, with a ^{6}LiF/^{7}LiF pair shielded only on the side opposite the body surface by a phenolic plastic loaded with boron carbide, was developed and tested at the Berkeley Nuclear Laboratories in England (Harvey et al., 1973 and Knight et al., 1973). It shows a satisfactorily neutron energy independent response up to 10 keV. At the Lawrence Livermore Laboratory, a ^{10}B-loaded plastic is being studied for the same purpose (Griffith, 1973). Similar devices are also under investigation in Winfrith, England (Preston and Peabody, 1973) and in Karlsruhe, Germany (Piesch and Burgkhardt, 1973), where two albedo dosimeters with ^{7}LiF/^{6}LiF pairs in a boron shield are to be worn at the front and the back of the thorax to reduce the directional response (Figure 1-9B).

When stationary measurements over a wide neutron energy range are required without limitations to the size and weight requirements of the detector, a simpler approach can be utilized (Distenfeld et al., 1967 and Engelke, 1969). A pair of TLD-600 and TLD-700 detectors is placed in the center of a large moderator sphere about 15 to 34 cm in diameter, for a varying energy response between thermal and approximately 10 MeV. There is usually some oversensitivity in the intermediate energy region. They are made of polyethylene, pressed wood, or paraffine.

The difference in the reading of the detectors is related to the neutron dose on the basis of calibration data. With spheres of different diameters, even some "poor man's neutron spectroscopy" is possible, for example, around reactors or accelerators. One can use essentially several spheres in one by placing several ^{6}LiF/^{7}LiF pairs at different locations throughout a spherical moderator. Machining of such devices may be simplified by using other shapes which approximate a sphere sufficiently (Lazanoff and McLaughlin, 1969 and Piltingsrud and Engelke, 1973).

2.4.2. Sealed Detectors

The simplest permanently encapsulated TLD dosimeters are those with the phosphor sealed in a glass capillary without built-in heater, preferably tightly packed, with a protective gas in the capillary, and a potassium-free glass to avoid spurious effects and self-dosing. Some commercially available dosimeters of this type are listed in

TABLE 2-8

Some Commercial TLD's with the Phosphor Sealed in Glass Capillaries

Manufacturer	Phosphor	Dimensions (mm)
EG & G, Salem, Mass.	CaF_2:Mn	12 x 1.4 (Mini)
		6 x 1.4 (½ Mini)
		6 x 0.9 (Micro)
	LiF (nat.)	12 x 1.4
		6 x 1.4
		6 x 0.9
	[6]LiF	12 x 1.4
		6 x 1.4
		6 x 0.9
	[7]LiF	12 x 1.4
		6 x 1.4
		6 x 0.9
Dai Nippon Toryo, Chiyoda-ku, Tokyo, Japan	Mg_2SiO_4:Tb	10 x 2
Matsushita, Osaka, Japan	BeO:Na	14 x 2[1]
	$CaSO_4$:Tm	14 x 2

[1] Permanently attached to holder as indicated in Figure 2-51.

Table 2-8. Uniformity of response within a batch is ±3 to 6% for higher dose-levels, but reproducibilities of multiple readings of the same dosimeter may be as low as $\sigma = 1\%$. The highest sensitivity (0.5 mR detectable) is available with the Mg_2SiO_4:Tb dosimeter.

Obviously, dosimeters of this type to be heated externally are limited to a fairly small size. Larger sealed dosimeters all have internal resistance heating which may, in the simplest case, consist of a metal wire through a powder-filled capillary (Schulman et al., 1963). The design of most dosimeters is more sophisticated: Early bulb-type dosimeters based on CaF_2:Mn had the powdered phosphor cemented to a square, 16 x 16 mm carbon heater plate with Dow Corning 805 Silicone Cement®.

Later, under a contract with the U.S. Navy Bureau of Ships, EG & G developed a cylindrical bulb ∿10 mm in diameter containing a spiral metal filament coated with the phosphor. Attached to the base was a 6 cm long handle carrying eight cams whose setting established the serial number of the dosimeter. In the reader, this number was sensed by an appropriate mechanism and pointed out, together with dose indication between 2 mR and 10^4 R, on a digital recorder (Palmer et al., 1966). The energy dependence was ±20% between 80 keV and 1.3 MeV, the directional dependence for gamma radiation ±15%. This system is not in production anymore.

In a similar system that was developed during the early 1960's by M.B.L.E. in Belgium, the phosphor (normally natural CaF_2) is firmly bonded onto a metal cylinder containing a filament as heating element (Figure 2-57). This bulb was used in pen-shaped aluminum or plastic holders into which the energy dependence compensation filters were incorporated. An optimized filter consisting of 0.5 mm lead with 10% of its area perforated flattened the energy response within about ±10% above ∿40 keV, but there remained a strong directional response (Brooke and Schayes, 1967).

The reproducibility of these dosimeters was about ±2%. A modification of this dosimeter provided a possibility to read the same detector twice. It consisted of two layers of the same phosphor, one being heated and read directly, the second after a delayed time due to the effect of a metal sleeve acting as a heat-sink. The production of M.B.L.E. dosimeters has also been discontinued.

The double dosimeter idea has also been used for a different purpose, namely, the estimation of the effective photon energy (Gorbics and Attix, 1968 and Spurny et al., 1973). A small extruded CaF_2:Mn chip was bonded to the top of an extruded LiF:Mg,Ti dosimeter, resulting in a good

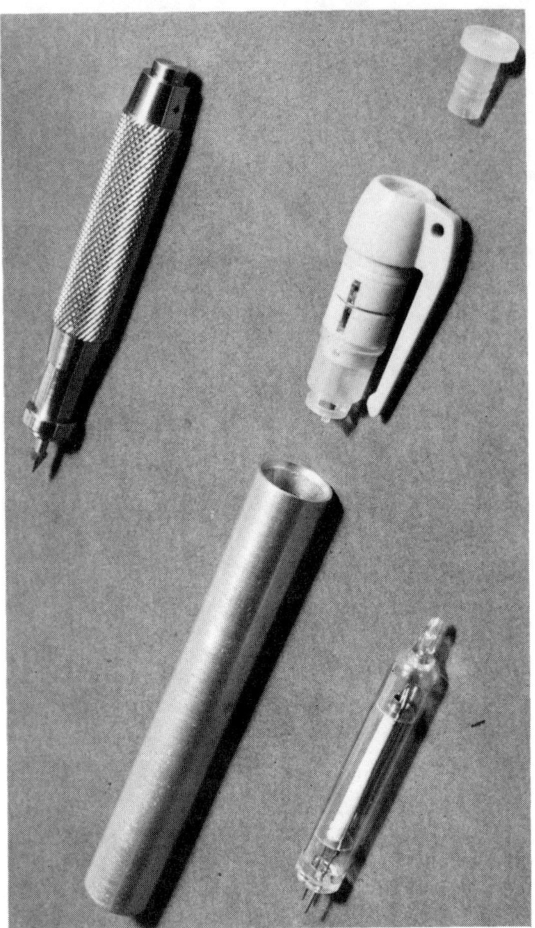

FIGURE 2-57. Sealed dosimeter containing natural calcium fluoride on a heatable metal cylinder, opened holder with built-in energy compensation filter, and opener for the tamperproof unit. (Courtesy M.B.L.E., Brussels, Belgium.)

separation between the LiF peaks and the peak of the CaF_2:Mn, which appears later because of delayed heating (Figure 2-58). The peak ratio obviously depends on the effective photon energy. This system was, however, never commercially manufactured.

Two currently available commercial bulb systems (EG & G and Harshaw) are very similar in design. In this dosimeter, extruded CaF_2:Mn chips sandwich a heater ribbon. This system (see Figure 2-52) is to be introduced in U.S. naval shipyards. Some proposals for less conventional systems can also be found in the literature. Brunskill and Langmead (1971), for example, describe a two-element dosimeter incorporating graphite trays which are contained in a Teflon body. The TLD phosphor is attached to the graphite in a solid

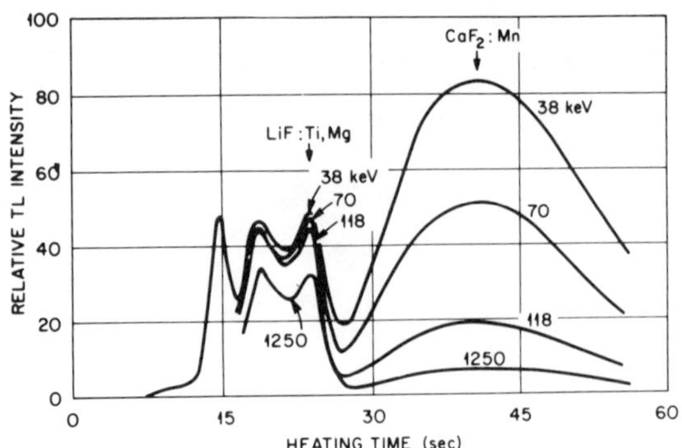

FIGURE 2-58. Glow curves of a LiF:Mg,Ti/CaF_2:Mn "tandem" dosimeter with delayed heating of the CaF_2:Mn for different photon energies. (After Gorbics and Attix, 1968.)

matrix of nonluminescent adhesive and covered with a transparent Teflon film. The dosimeter is read out by R.F. eddy current heating.

2.5. Readout Instruments

Thermoluminescence can easily be observed visually, for example, by pouring the irradiated phosphor on a hot electric plate in the dark. Instrumentation is, consequently, basically very simple: It consists of a heat source and a quantitative light detector. During the early years of TLD, most researchers in the field built their own readout equipment, usually based on electrical resistance heaters, photomultipliers, and more or less sophisticated filter systems for discrimination against spurious signals and the infrared emission from heated components.

This was partly due to the nonavailability of the type of instrument that may have been required for a certain purpose and partly to the high cost and poor performance of some of the earlier commercial readers. Even today, when a wide variety of readers are offered in many countries (Table 2-9), there are clearly situations when a self-designed, home-built reader may serve a given purpose better and/or will be much less expensive than a commercial instrument. Examples of this type of instruments are those that permit a spectral analysis of the TL light, a simultaneous recording of TL and thermally stimulated exoelectron emission (see Chapter 3.3), or automatic systems for the reading of large quantities of detectors.

A simple, easy-to-build TLD reader has, for example, been described by Cameron et al. (1967).

TABLE 2-9

List of Some Commercial TLD Readers[1]

Country	Manufacturer	Model	Price (U.S. $)[2]	Type of dosimeter[5]	Remarks
U.S.A.	Eberline Instr. Corp., Santa Fe, N.M.	TLR-5	~2000	Best for extruded ribbons, powder reading possible	Low-priced routine reader
	EG & G, Salem, Mass.	TL-3 A, B, or C	4325 to 4675	For extruded ribbons, glass capillaries, tube	Range 5 mR to 500 R (A), 50 mR to 5000 R (B), or 0.5 to 50,000 R (C)
		2005 A	4945	Powder, chips, ribbon, tube	Wide variability of heating cycle
		8506 B	2950	Glass capillaries	0.1 to 18,000 R, for medical physics
	Harshaw Chemical Co., Solon, Ohio	2000 A and B	~6000	Powder, ribbons	Versatile research instrument
		3000	~3000	Powder, ribbons	
		2271		Ribbons in special cards	Automated system for personnel dosimetry
	Victoreen, Cleveland, Ohio	2800	2800	Powder, chip, rod, or bulb	Simple routine reader
	Teledyne/Isotopes, Westwood, N.J.	TLD-7300	4000	Powder, discs, ribbons, rods	New instrument with pre- and post-reading annealing
		7100 TS	~10.000 (including accessories)	Powder, discs, ribbons, rods	Research type instrument, production discontinued
		9100	~25.000	Large Teflon badge dosimeter	Automatic reader for personnel dosimetry, under development
		8300	~7000	Large Teflon badge	Simple, semiautomatic version of 9100

[1] Not to be considered complete; reflects information available to the author by October 1972.

[2] Approximate prices F.O.B. factory without accessories such as printing-out equipment, glow curve recorders, powder dispensers, annealing ovens, etc., which many suppliers also offer.

[3] Price in U.S. (represented by Teledyne/Isotopes).

[4] U.S. price (represented in the U.S. by Amperex, Hicksville, L.I.).

[5] Some readers can also be adopted for other dosimeter types.

TABLE 2-9 (Continued)

List of Some Commercial TLD Readers[1]

Country	Manufacturer	Model	Price (U.S. $)[2]	Type of dosimeter[5]	Remarks
U.S.A.	Madison Research, Middleton, Wisc.	E-IV		Powder	Production discontinued
	Radiation Detect. Co., Mountainview, Calif.	Mark IV, Model 100	1610	Powder, ribbons	Production to be discontinued
		Model 1100	2240	Powder, ribbons	Production discontinued
Poland	Institute Nucl. Physics, Cracow			Sintered pellets	
Germany (West)	PTW Dr. Pychlau, Freiburg i.Br.			TLD powder in glass or capillaries	Production to be discontinued
Germany (East)	VEB Vacutronik, Dresden			LiF powder	
England	D. A. Pitman Ltd., Weybridge, Surrey	205 C		LiF-Teflon discs	Mainly for personnel dosimetry
		205 B		LiF-Teflon discs	Portable rugged version of 205 B for nuclear submarines, etc.
		605			Automated system for personnel dosimetry
		654			High-precision instrument for clinical dosimetry, under development
	Dynatron Electron., Maidenhead, Berks.	Harwell 2000	∿3600	Powder	Production apparently discontinued
France	Saphymo-Srat, 51 rue Mouchez, Paris 13	LDT-20	∿6000.-	Powders or pellets	Includes preheating cycle
Japan	Dai Nippon Toryo Co., Chiyoda-ku	Kyokko TLD-1200	$Mg_2 SiO_4$:Tb in glass	0.5 mR to 2000 R capillary, as rod, or pellet	

TABLE 2-9 (Continued)

List of Some Commercial TLD Readers[1]

Country	Manufacturer	Model	Price (U.S. \$)[2]	Type of dosimeter[5]	Remarks
Japan	Matsushita, Osaka	502	9500[3]	$CaSO_4$:Dy or BeO in glass capillaries	Research model with glow-curve recorder, etc.
		505 A	4800[3]	$CaSO_4$:Dy or BeO in glass capillaries	Simplified versions, hot air heater
		503	$\sim$3000[4]	$CaSO_4$:Dy or BeO in glass capillaries	Simplified versions, hot air heater
Belgium	M.B.L.E., Brussels	PNH 001	3600[4]	Nat. CaF_2 bulbs	Production discontinued
		PNH 651	3600[4]	Nat. CaF_2 bulbs	Production discontinued
		PNH 151 and others	5900[4]	Nat. CaF_2 bulbs	Production discontinued

[1] Not to be considered complete; reflects information available to the author by October 1972.

[2] Approximate prices F.O.B. factory without accessories such as printing-out equipment, glow curve recorders, powder dispensers, annealing ovens, etc., which many suppliers also offer.

[3] Price in U.S. (represented by Teledyne/Isotopes).

[4] U.S. price (represented in the U.S. by Amperex, Hicksville, L.I.).

[5] Some readers can also be adopted for other dosimeter types.

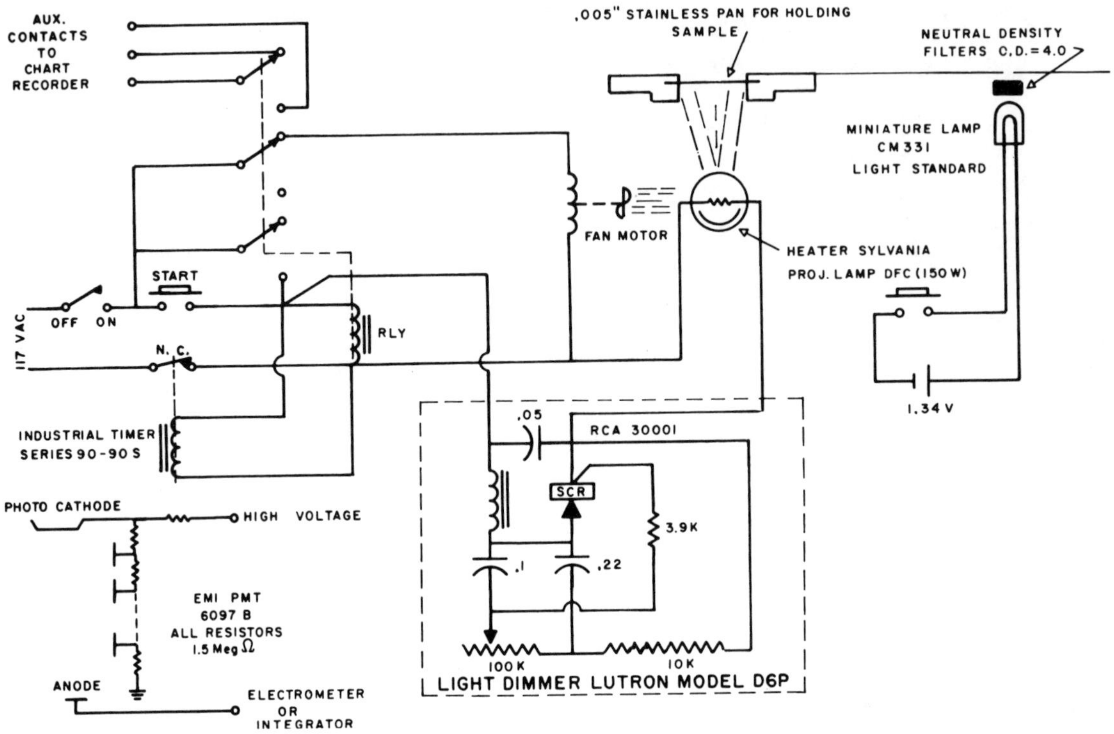

The circuit diagram is Figure 2-59.

FIGURE 2-59. Operating circuit of an easy-to-build TLD reader. (After Cameron et al., 1967a.)

It costs less than \$200 in parts, requires about 20 hr of construction time, has variable heating rate and time, and accepts a large variety of sample sizes. Shutter action is accomplished by moving the PM tube mounted on a metal sheet, and the planchette is heated by a standard projection light bulb. The circuit for this reader is given in Figure 2-59.

Only a few simple, general rules for the design of a TLD reader can be given here. For the heating of the phosphor one could, in principle, use various methods including infrared irradiation and ultrasound. In practice, however, only three methods have been studied, of which only two found widespread application. These are electric resistance heating, either by placing the sample in a pan or planchette, or by bonding the phosphor to the heating element, and hot-gas heating. As most of the heat transfer to a loose powder occurs through the heated gas and not through conductive or radiative heat transfer, both methods are, in principle, identical, but differ in some important practical details: Hot air (Onishi et al., 1971 and Hiraki et al., 1972) or nitrogen (Petrock and Jones, 1968 and Botter-Jensen, 1971) heating substantially reduces the infrared-induced back-ground signal from the heated planchette, but it does not permit the use of loose powder.

Only one commercial instrument made by Matsushita uses this heating principle (Figure 2-60), but it appears to have several advantages for automatic reader systems. Heating by eddy currents of a conductive graphite carrier sealed in

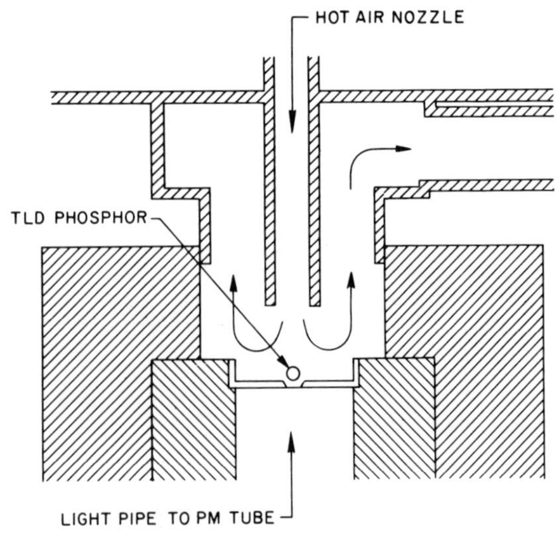

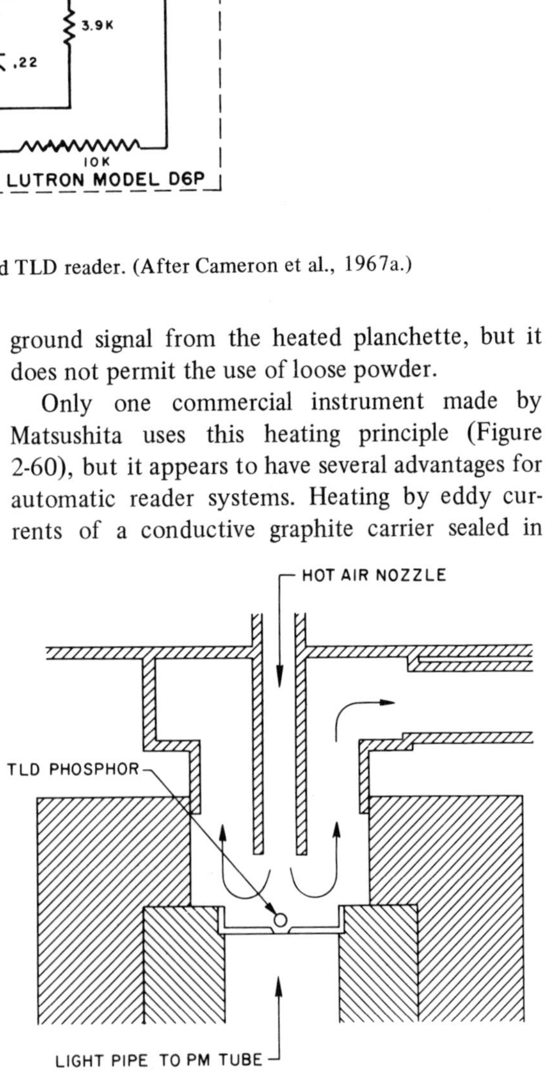

FIGURE 2-60. Principle of TLD sealed capillary heating with hot air stream. (After Oonishi et al., 1971.)

Teflon, with the R.F. energy being fed to a small water-cooled coil below the dosimeter, has also been studied (Brunskill and Langmead, 1971). Heating by magnetic induction of the sample on a copper ring in the gap of a special transformer was also considered (Frank and Stolz, 1969).

In research instruments, the heating rate should be well reproducible. For a good resolution and analysis of glow curves, linear heating rates as low as $\sim1°C/sec$ may be desirable and the peak location may have to be studied as a function of the heating rate. Various types of controllers have been described for this purpose (Gall and Mason, 1969). The fast evaluation of large quantities of TLD's in routine studies, on the other hand, may call for heating rates as high as $30°C/sec$ to minimize the time required for each reading, and nonlinear heating (fast heat-up, holding a constant temperature for readout) has certain advantages in routine readers.

As a rule, such a reader should be able to anneal at least 95% of the stored TL signal in ~10 to 20 sec. Many modern readers provide for an adjustable heating cycle which may, in the case of LiF:Mg,Ti, consist of a rapid preheating for several seconds to ~100 to $130°C$ without light recording to destroy low-temperature peaks which are subject to rapid fading, a linear heating at 10 to $25°C/sec$ to ~250 to $270°C$, during which the TL output is measured and an "annealing" for ~10 to 20 sec at 300 to $400°C$ without light measurement (Webb and Phykitt, 1971).

The heated metal strip serving as a carrier for "open" TLD phosphors (powder, chip, Teflon compound) or small capillaries may itself affect the results, for instance, by changes in its reflectivity or emissivity which occur during oxidation or discoloration. Changing or cleaning planchettes may result in changes of the apparent TL by up to 10%. Metals do not exhibit any TL or phosphorescence, but some metal oxides such as Al_2O_3 do. The Al_2O_3 layer that covers all aluminum surfaces can give a sizable TL signal after exposure to room light. To minimize the energy requirements and have a rapid cooling between reading cycles, the mass of the heated parts should be small. Unless the heater is cooled down completely after each reading, the final temperature will depend on the "starting" temperature.

Minimizing the infrared emission of the heated parts is another important factor. Pans made of (or coated with) silver offer very satisfactory properties and can easily be cleaned with diluted HNO_3. Furthermore, the reader should be designed in such a way that the PM tube "sees" as much of the phosphor, and as little of the other heated areas as possible. There are many other possible sources of error. According to a recent report (Saunders, 1971), for instance, variations up to 9% were observed in a commercial reader which had been caused by the effect of residual magnetic fields from the AC heater current on the sensitivity of the photomultiplier. A ~10 cm light pipe between sample and PM tube was suggested in this case as a possible remedy.

It is not easy to establish the precise temperature of the phosphor during readout. Some readers have a thermocouple touching the bottom of the planchette. The temperature indication consequently refers to this location and not to the center of the emitting sample where it may be substantially lower. There are several methods for temperature calibration including infrared measurement, observing the boiling of liquids (alcohol, water, glycerine) on the planchette, and test runs with TLD phosphors which have well-established TL peaks.

Because of the difficulties in establishing the true temperature of the phosphor, many authors have published glow curves with time instead of temperature as the measured parameter. Although no exact knowledge of the glow curve is required for most routine applications, this confusing habit should be discouraged in research papers.

As the portion of a sample that is not in direct contact with the heater is heated with some delay, its size, shape, and thermal conductivity as well as the heating rate will affect the shape of the resulting glow curve. Very thinly and uniformly spread powders heated at a low constant rate ($\sim1°C/sec$) result in the most reliable and reproducible data. The type of the glow curve recorder may also have an influence. A potentiometric recorder with a full-scale pen speed of 0.25 sec is usually sufficient.

Another factor of major importance is the choice of the optimum light detector, usually a PM tube. The use of photodiodes is not recommended, at least not for low dose-levels; for the exclusive measurement of high doses, they may be superior to the PM tube (Au and Moran, 1972). The spectral response of the photomultiplier should match that of the TL light to be measured. This requires a quartz window* if the emission occurs, as in BeO, in the ultraviolet. It should be highly sensitive (good quantum efficiencies are obtainable with Sb-K-Na-Cs, and oxygen-activated Sb-Cs photocathodes). Its dark current should be low and constant. Typical dark currents at ambient temperature are 10^{-8} to 10^{-10} A, which can be reduced by a factor of 10 to 100 by cooling with dry ice to $-78°C$; some readers provide a built-in cooling device for the tube (Carlson and Richey, 1968). It should also be stable over extended periods of time. Most PM tubes exhibit fatigue effects. If one tries to measure a low intensity after exposing the tube to intense light, it may take several hours before it has completely recovered its "normal" response. The PM tube is also subject to aging.

To check the long-term stability of the PM tube and the electronic system, pre-irradiated "standard" TLD phosphors have been used in the early days of TLD. Due to the fading of these, nowadays a standard light source is inserted into a built-in component of a TL reader. It usually consists of a mixture of a beta or alpha emitter with a long halflife (^{14}C, ^{85}Kr, ^{226}Ra) and an inorganic scintillator such as NaJ:Tl or ZnS. If measurements are done in a wide dose-range, several such sources with different light intensities have advantages. It should be noted that the high energy beta radiation from some light sources containing ^{90}Sr/^{90}Y may induce luminescence and TL effects in the glass of the optical system (Spanne, 1973). Typical fluctuations of light source readings are slightly better than 2% (Harvey et al., 1973b). An external light source has the disadvantage that it takes up to 30 min before a light-induced phosphorescence has completely decayed away. With sources that are kept continuously in the dark, this problem is avoided, but low-intensity sources may still exhibit inherent fluctuations of ±0.4 to 0.6% (McCall, 1969). Furthermore, the light output of radioactive sources is not completely independent of ambient temperature and age; it is unlikely that the long-term performance of detectors or readers can be assessed with an accuracy better than ±1% due to such changes. To reduce errors that are related to the "tarnishing" of the planchets, it has been suggested to place the light source exactly on the location of the TLD powder or chip (Robertson, 1970). Less conventional approaches to the light source problem include the use of a light-emitting diode (Lindskoug and Bengtsson, 1972).

Infrared absorbing filters between the sample and PM tube are a disadvantage since they also absorb some of the TL light. The Corning filters 1-69 or 4-97 and $CuSO_4$ solutions have been recommended. Obviously, discrimination of the infrared becomes more difficult at higher phosphor (peak) temperatures and long wavelengths of the TL light.

The complex electronic circuitry of modern TLD readers cannot be discussed here in any detail. Most readers nowadays employ an integrator because for various reasons the integrated TL light is more closely related to the dose and easier to measure reproducibly than the TL peak height. The instruments usually provide a digital data display with automatic switch of range. Most of the more expensive instruments also have additional provisions for an X-Y glow curve recorder and/or a data printout system. The existing reader systems can be roughly classified as follows:

1. Moderately priced (about $2,000 to 4,000), rugged, simple-to-operate readers for routine use — frequently for one type of detector only — by little-trained technicians, mostly in medical physics departments, but also for training or routine personnel dosimetry in small installations. Some typical examples of this category are the Eberline Model TLR-5® reader (Figure 2-61), the Victoreen 2800® instrument (Figure 2-62), whose block diagram is given in Figure 2-63, the Harshaw Model 3000® reader (Figure 2-64), the Pitman 205® reader (Figure 2-65), the Kyokko 1200® instrument, the Teledyne TLD-7300®, and the Matsushita 503® instrument. Some even more ruggedized versions of such readers have been designed for military purposes. One such portable TLD reader developed for the British Royal Navy is pictured in Figure 2-66, another (Belgian) model in Figure 2-67. None of these portable readers are, however, battery-operated because of the high energy requirements of the heater.

2. In the second category are the more sophisticated, versatile, and expensive ($\sim$\$6,000 to

---

*PM tubes with a normal glass window can be made sensitive to UV light by painting the window with a sodium salicylate solution; this substance converts UV into visible light.

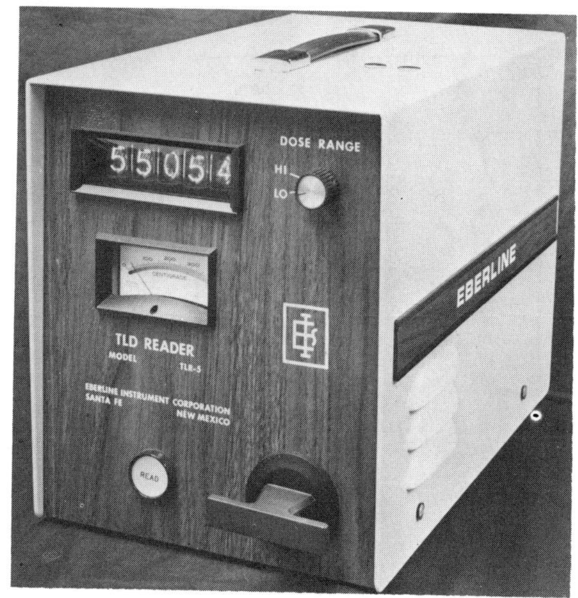

FIGURE 2-61. Small routine readout instrument, mostly for extruded ribbons. (Courtesy Eberline Instrument Corp., Santa Fe, N.M.)

FIGURE 2-62. Small TLD reader for routine use, with open and closed door of the control panel. (Courtesy Victoreen Instrument Co., Cleveland, Ohio.)

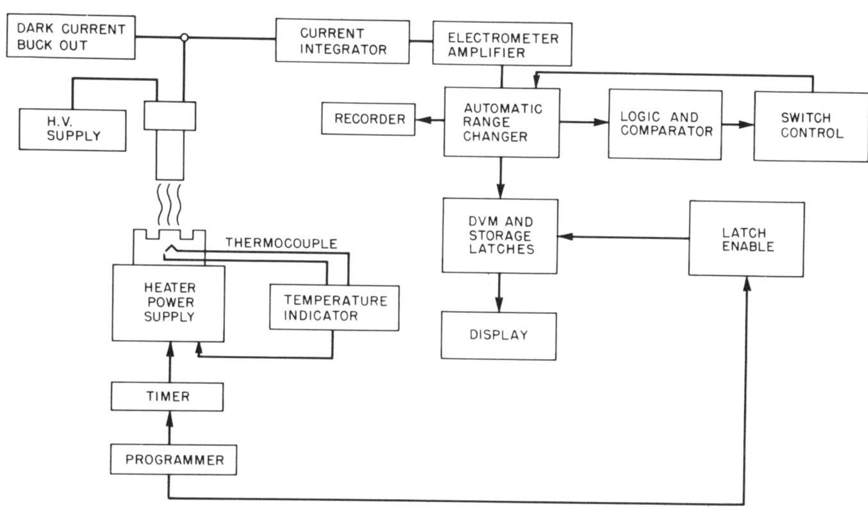

FIGURE 2-63. Block diagram of the reader pictured in Figure 2-62.

10,000) laboratory-type instruments which can be used for a wide variety of dosimeter types, and which normally offer adjustable heating programs, cooled PM tubes, precise temperature control, and higher stability. Their main use is in research, and for special problems such as the accurate measurement of low doses. Some typical examples are the Harshaw 2000 reader (Figure 2-68) and an EG & G system consisting of three units (Figure 2-69) with a sophisticated programmer for the heating cycle. With the generally high level of development in commercial readers available, it is normally not recommended to attempt the development of a new reader unless this is absolutely unavoidable for technical or economic reasons. Nevertheless, some descriptions of homemade readers in this category can be found in the literature (see, for example, Gangadharan et al., 1969). Among various devices for special studies is one for studying TL at very low temperatures (Land and Wysong, 1968) and various "photon counter" readers (Schlesinger et al., 1971; Niewiadomski, 1972; and others) for very sensitive measurements.

3. Finally, there are more or less automatic

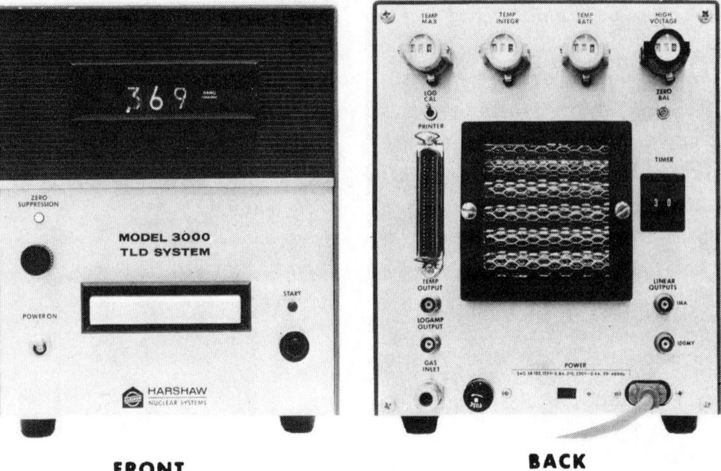

FIGURE 2-64.   Front and back view (with controls) of Harshaw Model 3000 system. (Courtesy Harshaw Chemical Co., Solon, Ohio.)

FIGURE 2-65.   Portable readout instrument for LiF/Teflon discs with built-in nitrogen gas-flow meter. (Courtesy D. A. Pitman, Weybridge, Surrey, U.K.)

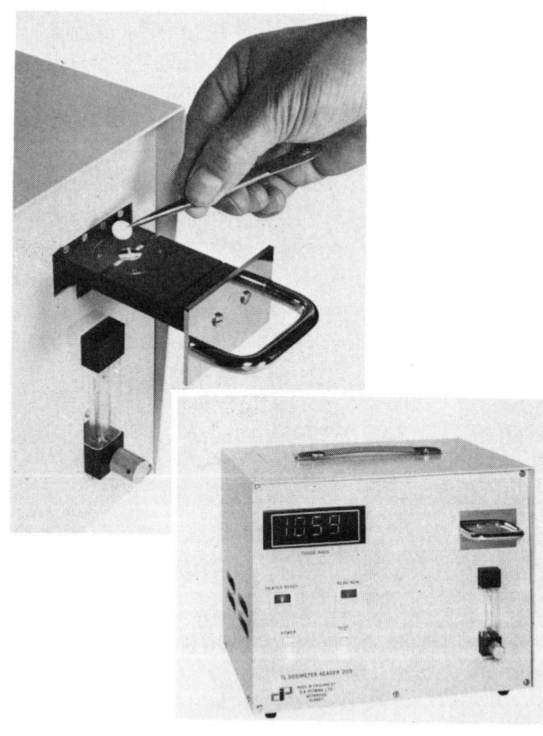

FIGURE 2-66.   Ruggesized version of a portable reader for LiF/Teflon discs (Model 205 B) for the British Navy. (Courtesy D. A. Pitman, Weybridge, Surrey, U.K.)

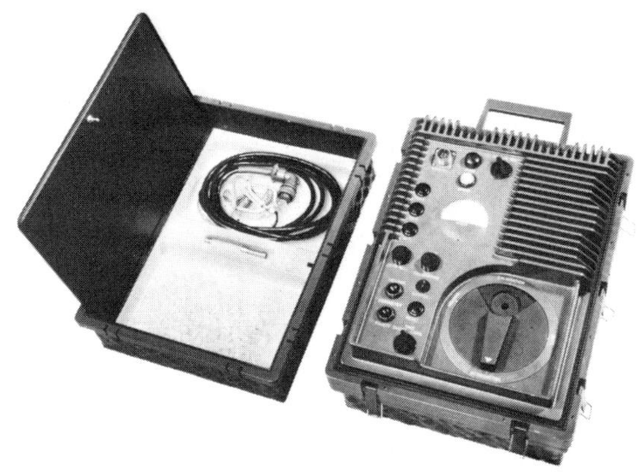 

FIGURE 2-67. Opened case of portable military-type TLD reader for the dosimeter type pictured in Figure 2-57. (Courtesy of M.B.L.E., Brussels, Belgium.)

FIGURE 2-68. The Harshaw 2000 System, consisting of the thermoluminescence detector (left), and the integrating picoammeter (right). (Courtesy Harshaw Chemical Co., Solon, Ohio.)

FIGURE 2-69. TLD heating head (upper right), temperature cycle programmer (upper left), and readout unit of a sophisticated research instrument. (Courtesy EG & G, Goleta, Calif.)

systems for large-scale TLD personnel dosimetry, to be used only for evaluation of a special type of badges. A U.S. system of this type with computer interface is pictured in Figure 2-70 and a simpler British instrument with dose printout in Figure 2-71. Another automatic reader is under development at Teledyne/Isotopes (Model 9100). It is based on a film-sized LiF/Teflon foil with four differently filtered areas (Figure 2-72), each of which can be read twice (once, for example, during routine readouts, and the second time as a back-up in accidents, after instrument mal-

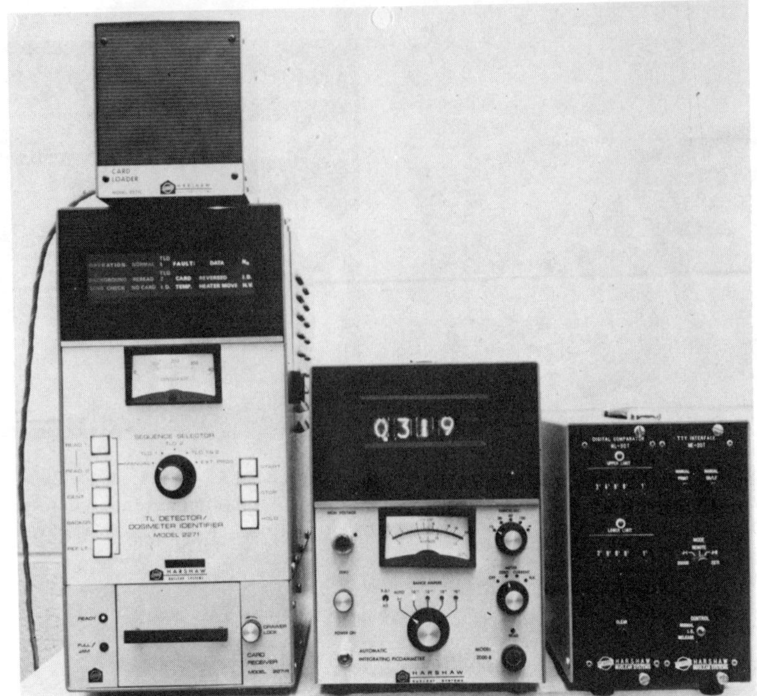

FIGURE 2-70. Automatic reader for sealed, credit card-type TLD chip personnel dosimeters (Figure 2-53) with digital comparator. (Courtesy Harshaw Chemical Co., Solon, Ohio.)

FIGURE 2-71. Simple reader with printout for personnel dosimeters. (Courtesy D. A. Pitman, Weybridge, Surrey, U.K.)

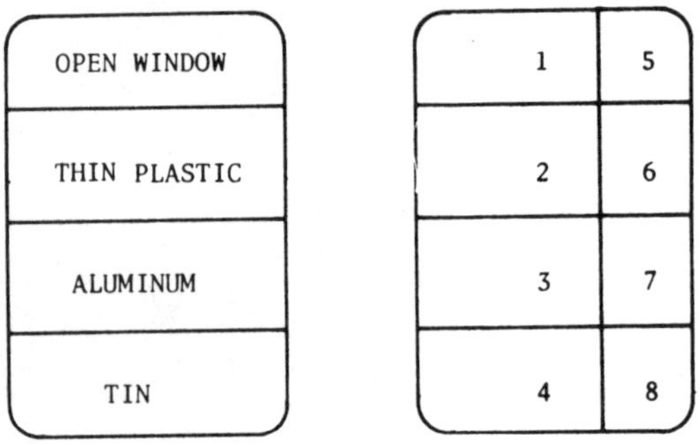

OPEN WINDOW		1	5
THIN PLASTIC		2	6
ALUMINUM		3	7
TIN		4	8

(a) Shielding on irradiation

(b) Areas read out separately

FIGURE 2-72. Film-sized TLD dosimeter (see Figure 2-49) with four differently filtered areas and eight readout possibilities in fully automatic reader. (After Webb et al., 1972a.)

functions, or for long-term integration). The heater consists of an anvil, heating only part of the indicated area.

Such readers, whose use may become more widespread in the future in countries with high labor costs, usually also provide for automatic reading and print-out of the number of the individual dosimeter. Of course, they are rather expensive ($\sim$\$25,000) and may require a back-up system in case of malfunction.

Several larger installations built their own automatic readers partly because they are less expensive and partly because they can be made to meet better the specific requirements (for a compilation, see Table 2-10). The number of such

TABLE 2-10

Some Personnel Dosimetry Services Employing TLD[1]

Country	Institution	Detector material	Automatic readout	Approx. number of monitored persons
Canada	Chalk River Nuclear Laboratories	$LiF:Mg,Ti$ extruded ribbons	Yes	4,000
Denmark	Danish AEC Research Establishment Risö, Roskilde	$Li_2B_4O_7:Mn,Si$ and $^7LiF:Mg,Ti$	Yes	600
France	CEA	Al_2O_3 powder	No	30,000[2]
Great Britain	UKAEA Windscale and Harwell Atomic Energy Establishment Winfrith	LiF-Teflon $LiF:Mg,Ti$ powder	No	
Netherlands	Radiological Service Unit TNO, Arnhem	$LiF:Mg,Ti$ extruded ribbons	Yes	7,000
U.S.	Battelle Northwest, Richland, Wash.	$LiF:Mg,Ti$ extruded ribbons	Yes	1,500
	Bureau of Radiolog. Health, Rockville, Md.	$LiF:Mg,Ti$	No	2,700
	Dow Chemical Co., Rock Flats, Calif.	$LiF:Mg,Ti$ extruded ribbons	No	2,500
	G.E. Knolls Atomic Power Lab., Schenectady, N.Y.	$LiF:Mg,Ti$ extruded ribbons	Yes	400
	Lawrence Livermore Lab., Livermore, Calif.	$CaF_2:Dy$ and $LiF:Mg,Ti$ extruded ribbons	Yes	7,100
	Nat. Reactor Testing Station, Idaho Falls, Idaho	LiF-Teflon	Semi	2,000
	Naval Medical Center, Washington, D.C.	$LiF:Mg,Ti$ extruded ribbons	Yes	2,000
	Naval Research Lab., Washington, D.C.	$LiF:Mg,Ti$ extruded ribbons	Yes	3,600
	Naval Shipyards	$CaF_2:Mn$	Yes	10,000
	Oak Ridge Nat. Lab., Oak Ridge, Tenn.	$LiF:Mg,Ti$ extruded ribbons	No	4,000
	Savannah River Lab., Aitken, S.C.	$LiF:Mg,Ti$ extruded ribbons	Yes	5,500
	Stanford Linear Accelerator, Stanford, Calif.	$LiF:Mg,Ti$ extruded ribbons	No	1,200
	Univ. of North Carolina, Chapel Hill, N.C.	$LiF:Mg,Ti$ extruded ribbons	No	400

[1] This list, which is largely based on a compilation prepared in 1971 (Attix, 1972), is subject to rapid change and not to be considered complete. For descriptions of larger TLD services, see Cusimano and Cipperly, 1968; Jones, 1968 and 1971; Brunskill and Langmead, 1971; Johns et al., 1971; Bötter-Jensen and Christensen, 1971; and others.
[2] Back-up for accident dosimetry only.

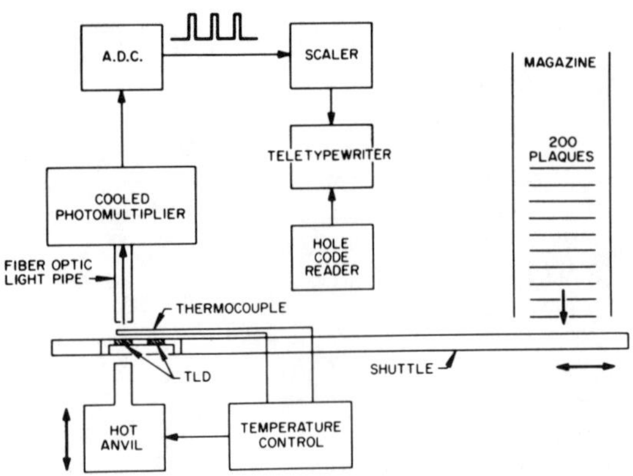

FIGURE 2-73. Schematic diagram of an automatic TLD personnel dosimeter reader used at the Chalk River Nuclear Laboratories, Canada. (After Jones, 1971.)

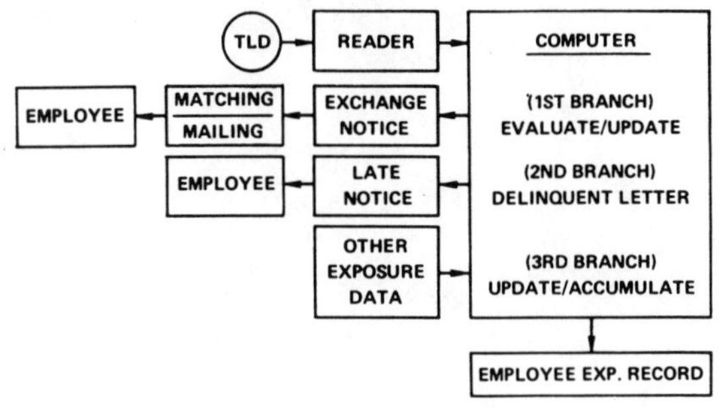

FIGURE 2-74. Block diagram of computer system for employee exposure record keeping. (After Rich and Samardzich, 1971.)

systems described in the literature is rapidly increasing. One at the Chalk River Nuclear Laboratories in Canada, for example (Jones, 1971), employs a semiautomatic reader for finger-tip dosimeters and a fully automatic reader for the normal personnel dosimeters with two "kapton" sealed LiF chips. A diagram of this reader, in whose tray 200 badge inserts can be loaded for readout, is given in Figure 2-73.

Another system for the automatic reading of personnel dosimeters (Christensen, 1971) with two detectors, one $Li_2B_4O_7$:Mn,Si pellet and one ^7LiF:Mg,Ti unit, is based on hot nitrogen heating. It is being developed for the Danish AEC Research Establishment at Risö (Bötter-Jensen and Bechmann, 1968 and Bötter-Jensen and Christensen, 1971). A somewhat similar hot nitrogen system that is used at the Lawrence Livermore Laboratory (Rich and Samardzich, 1971) accepts 60 dosimeters, as pictured in Figure 2-52, removes the plastic cap and then puts the LiF chip with a vacuum needle into the hot nitrogen stream for readout, decodes the identification number, and finally transmits the decoded data to a summary punch. For a description of a British automatic system, see Perry and Preston (1970). A semiautomatic instrument with a circular loading device for 23 dosimeters has been developed for a smaller facility (Kartha and MacDonald, 1971).

The organizational aspects of TLD personnel monitoring services are mentioned in several of the reports describing automatic systems discussed above. In particular, a detailed analysis of the experience in the Livermore Radiation Laboratory (Rich and Samardzich, 1971) points out some economic advantages of the TLD system for a larger installation (5,000 to 6,000 employees and 19,000 visitors per year). The annual costs for the film badge service were $75,600, but for the TLD service $59,000 (for example, the exchange costs have been reduced by a factor of 8 by using the in-plant mail). The block diagram for the computer system for employee records is given in Figure 2-74.

Such organizational schemes can, however, hardly be transferred without modifications to other installations. Instead, a system should be tailored by an expert to fit exactly the particular requirements of a given situation, with enough flexibility to adjust to changing technical or administrative conditions. In particular, the question of automation should be carefully considered. Although automatic dosimeter handling, readout, and data processing greatly reduce labor and drudgery, and significant savings can be expected in large services, there is an equilibrium point where further automation increases costs and technical effort instead of decreasing them.

One has to keep in mind that increased complexity frequently implies increased breakdown probability and maintenance costs and that not even a "fully automatic" system should be left without some supervision by a qualified technician over extended periods of time. Besides convenience and costs, increased accuracy and reliability can be important factors favoring automation. No practical experience is available yet with new commercial automatic systems.

Both the basic aspects and the technical performance of TLD vs. film in personnel dosimetry have been the subject of numerous discussions (see, for example, Becker, 1966, 1967b, and 1972c and Webb et al., 1972a) and the trend is clearly in favor of TLD (Attix, 1972). The film still has a higher information content regarding direction of radiation incidence, detection of contaminations, detection of single exposures, etc., but the gap is closing due to possible increases in the TLD's information content (Webb et al., 1972a).

As far as the accuracy of dose readings is concerned, no single case has yet been reported in which TLD turned out to be inferior. The exact degree by which TLD was found to be more accurate in studies (Eastes, 1968; Cameron and Suntharalingam, 1968; Johnson and Attix, 1968; Busuoli and Cavallini, 1969; Scott, 1970; Preston et al., 1970; Geiger, 1971; Fitzsimmons et al., 1972; Tatsuta et al., 1972; and Crosby, 1972, to name only a few more recent ones) depends, of course, on various factors, including the quality of both the TLD and the film badge service, the dose-level, and other conditions of the experiment.

TLD has also found many applications in studies of the more basic aspects of personnel dosimetry. Typical examples are the classical experiments by Jones (1964) on the relationship between surface and organ dose for different photon energies between $\sim$27 and 1,250 keV, Ambiger's (1969) phantom measurements to determine the most suitable site for the personal dosimeter, and studies on the depth-dose distribution in a phantom exposed in a ^{60}Co field (Beck et al., 1968).

2.6. Applications other than Personnel Dosimetry

There are such a large number of applications and so many publications describing experiments in which TLD has been used as an experimental tool that no attempt will be made here to cover this field comprehensively. The most important applications of TLD are in personnel dosimetry, hospital physics, and biomedical research. Many aspects of these applications, particularly in personnel dosimetry, have already been mentioned. As it would take too much space to review all the procedural type of papers which have been published over the years, only a few of the more important and/or most recent contributions to the TLD applications field that have not been treated in previous sections of this chapter will be discussed briefly.

There is an extensive literature on phantom and in vivo TLD measurements during diagnostic or therapeutic procedures. In fact, monitoring of radiation doses in body cavities such as the mouth, the esophagus, the stomach, the rectum, the bladder, the uterus, and the cervix with external as well as internal radiation sources was among the earliest applications of TLD. For example, LiF powder was put in a rectal tube with steel balls as separator and radiographic markers. Later, TLD was used for measuring the entrance and exit dose of patients undergoing radiation therapy, particularly for beams traversing the lung (for a review of the early work, see Cameron et al., 1968a).

To give a few more recent examples: The dose distribution in the diagnostic beam (Vacirca et al., 1972) as well as the scattered radiation outside the beam (Vacirca and Thompson, 1972) during thorax radiography has been measured; the radiation dose to critical organs during petrous tomography was determined (Chan et al., 1970); there have been several studies concerning the patient exposure during dental radiography (Regulla et al., 1972; Forsythe et al., 1973; Weissman, 1973; Alcox, 1973) and of the radiation exposure of infants (Muffelman, 1973); three-dimensional dose distribution measurements in the head and neck have been carried out (Mansfield et al., 1968); dose gradient studies on bone-tissue interfaces (Schulz, 1968 and Carlsson and Carlsson, 1971) and air-tissue interfaces (Hvolby, 1972) have been made; the transit dose in extracorporeal blood irradiation devices has been measured (Bötter-Jensen and Christensen, 1972); measurements in trabecular bone have been reported (Zenelli and Spiers, 1968 and Spiers et al., 1972); and the equipping of intrauterine contraception devices with TLD's for the determination of the dose to female gonads was considered (Parker et al., 1972). Nowadays, the use of TLD in clinical dosimetry is a routine procedure; TLD readers can be found even in small hospitals in remote areas of the world. (For reports on experiences and progress in this field, see Rassow, 1970; Rudén, 1971; Gooden and Brickner, 1971; Suntharalingam and Mansfield, 1971; Castro, 1972; Lindskoug and Bengtsson, 1972; and others.) There have also been some rather unusual approaches, such as attempts (so far unsuccessful – J. Kastner, personal communication) to use the dental enamel as an integrating TL detector. The field is still rapidly expanding.

Another area of widespread application of TLD's (mostly LiF:Mg,Ti) has been the intercomparison of radiation sources, particularly radiation therapy equipment, on a national or an international scale (see, for example, Pohlit, 1972). The best-known of these studies has been a joint worldwide project of the International Atomic Energy Agency (IAEA) and the World Health Organization (WHO) carried out by mailing LiF:Mg,Ti powder in small capsules to the participating institutions with detailed instructions on how to expose them in a special holder (Figure 2-75), which supplied together with the dosimeters, at a location is exactly 5 cm below the "phantom" water level. The results (Classen et al., 1971) clearly show that the standard deviations in the reported dose values (average of all participat-

ing countries 7.5%) are less in more advanced countries ($\sim$3.5% in two European countries).

The present situation in clinical dosimetry is, however, less than ideal even in highly developed countries: According to one recent study of 71 institutions in the U.S., only 80% of all therapy units delivered doses within ±5% of the prescribed dose, and the extremes were as much as −29% and ±22% off (Grant et al., 1972). In a similar study involving radiobiological research laboratories in 8 European countries using 80 to 130 keV X-radiation (Puite et al., 1972), only 6 out of 14 institutions agreed within ±5% of the "standard" value.

There are a wide variety of other, more or less specialized applications of TLD. For example, such detectors have been used as personnel dosimeters in all space flights and several biosatellite experiments. Use has been made of TLD for dose measurement in reactors at temperatures up to 100°C (see, for example, Tobias, 1972). They also have been used extensively in reactor shielding studies (see, for example, Woodsum, 1972). In the NASA Skylab Program, LiF and CaF_2 will be used as passive detectors (Schneider et al., 1972). And TLD materials have been employed as alpha particle detectors in radon progeny personnel dosimeters for uranium miners, but this system is, despite considerable efforts, still plagued by mechanical and other problems (see Schiager and Savignac, 1972 and Shambon and White, 1972).

The TL in roof tiles in Hiroshima and Nagasaki helped to establish the precise dose distribution around point zero of those infamous nuclear explosions. In lunar materials, the accumulated TL signal due to their inherent radioactivity and solar and cosmic radiation helped to establish age, radiation and/or thermal history, and turn-over rate of the lunar suface. Thermoluminescence techniques have also been used for dating ancient pottery and other mineralic, once heated artifacts (there are, for example, fascinating stories on the identification of art forgeries by TL measurements — for excellent reviews, see Aitken and Fleming, 1972 and Mejdahl, 1972).

At this time, only five laboratories in Great Britain, Denmark, and the U.S. are active in this field but worldwide interest in the method is rapidly increasing. For example, the age of ancient potsherds and terracottas is determined by dividing the measured accumulated TL by the

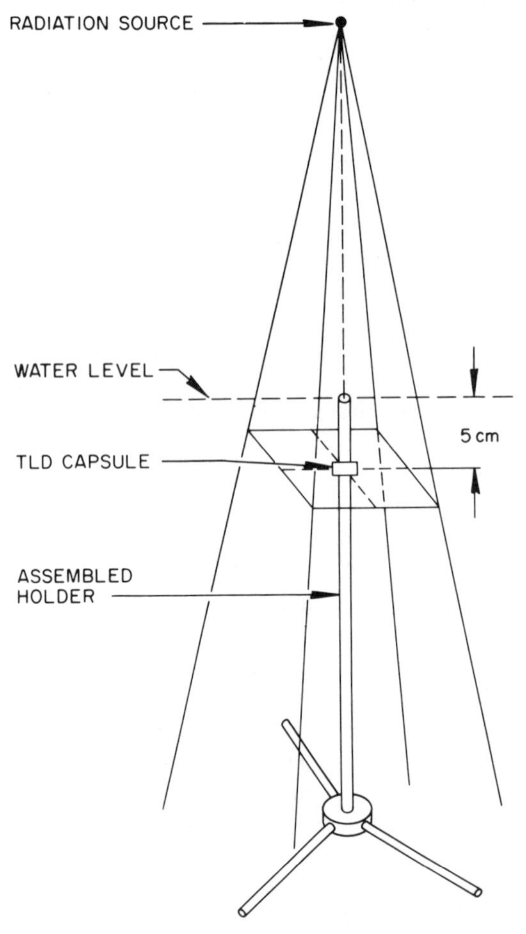

FIGURE 2-75. TLD exposure set-up for worldwide IAEA/WHO intercomparison study of radiotherapy units.

sensitivity of the clay constituents (as determined by a calibration exposure to a known radiation dose) multiplied with the dose-rate to which it was exposed (as determined by local dose-rate measurements with sensitive TL phosphors and/or determination of the natural U, Th, and K content of the ceramic). There are, in principle, three fairly independent methods that can be used for such measurements:

1. Measurement of the TL signal in the coarse ($\sim$0.1 to 0.2 mm) quartz grains that respond mostly to long-range gamma and beta radiation at a dose-rate of $\sim$0.1 rad/year (quartz has two stable peaks at 325°C with a halflife at 20°C of $\sim$3,000 year) and at 375°C, halflife 4 x 10^7 year to be used for this purpose; the alpha-exposed surface of the grains can be removed by etching with HF;

2. fine-grain ($<$10 μm particles) measurements, emphasizing the larger contribution of "internal" alpha radiation ($\sim$1 rad/year); and

3. measurements of radiation-induced changes in a $\sim$110°C peak in quartz ("predose dating").

Unfortunately, several complicating factors, such as abnormal fading, spurious signals, and difficulties in accurately measuring the very small signals involved against a strong infrared emission, limit the total accuracy of the method to ±5 to 10%. If the age of the samples is known, the same techniques can be used for determination of the average dose-rate at a given location over very long periods of time to be compared with today's dose-rates (Becker, unpublished).

Environmental dose measurements with TL have been mentioned before. With the increasing concern about population exposures, they are rapidly gaining in importance. Using LiF:Mg,Ti and CaF_2:Mn, pertubations in the natural radiation background have been related to snow cover and rain which reduce the radon exhalation (Burke, 1972). Today most experts agree that other phosphors such as $CaSO_4$:Dy (Mejdahl, 1970 and Becker, 1972b) or Mg_2SiO_4:Tb are superior detector materials for this purpose. One month of exposure of these powdered phosphors encapsulated in plastic and protected from direct sunlight yields measurements in commercial readers with an accuracy of ±3% without serious fading problems and with no need for energy dependence corrections. According to Mejdahl (1970), photon energies of $<$150 keV contribute only $\sim$4% to the environmental dose; shielding of $CaSO_4$:Dy with 1 mm Cu did not noticeably affect the results as compared with the unshielded, plastic-encapsulated phosphor (Becker, unpublished).

Screens containing TL phosphors have been used for the image storage of x-ray or thermal neutron radiographs (see, for example, Kastner et al., 1966b). There have even been attempts to identify ball lightning as the annihilation of small particles of antimatter from outer space by the increased TL from bricks in hallways where it occurred. Although interesting or amusing, these and many other applications are, however, only remotely related to the everyday problems in radiation dosimetry and cannot be discussed in any detail here.

REFERENCES

Adam, G. and Katriel, J., Analysis of Thermoluminescence Kinetics of CaF_2:Mn Dosimeters, *Proc. Third Int. Conf. Luminescence Dosimetry, Risö-Rep. 249,* Vol. 1, Danish AEC, Risö, Roskilde, 1971, 9.

Adam, G., Thermoluminescence dating and the sensitization method, *Health Phys.,* 23, 579, 1972.

Aitken, M. J., Low-level Environmental Radiation Measurements Using Natural Calcium Fluoride, Proc. Sec. Int. Conf. Luminescence Dosimetry, CONF-680920, 1968, 281.

Aitken, M. J. and Fleming, S. J., Thermoluminescence dosimetry in archaeological dating, in *Topics in Radiation Dosimetry,* Attix, F. H., Ed., Academic Press, London, 1972, Chap. 1.

Albrecht, H. O. and Mandeville, C. E., Storage of energy in beryllium oxide, *Phys. Rev.,* 101, 1250, 1956.

Alcox, R. W., A Dosimetry Study of Dental Exposures from Intraoral Radiography, *Proc. HPS 7th Midyear Symp. on Health Physics in the Healing Arts,* U.S. Public Health Service, Washington, D.C., 1973.

Almond, P. R., McCray, K., Espejo, D., and Watanabe, S., The Energy Response of LiF, CaF_2, and $Li_2B_4O_7$:Mn from 26 keV to 22 MeV, Proc. Sec. Int. Conf. Luminescence Dosimetry, CONF-680920, 1968, 411.

Almond, P. R. and McCray, K., The energy response of LiF, CaF_2, and $Li_2B_4O_7$ Mn to high energy radiations, *Phys. Med. Biol.,* 15, 335, 1970.

Alsmiller, R. G. and Barish, J., The Calculated Response of the TLD Personnel Dosimeter to Neutrons with Energies below 400 MeV, *Trans. Am. Nucl. Soc.,* 15, 984, 1972. (Detailed version ORNL-TM-3984, Oak Ridge Nat. Lab., Oak Ridge, Tenn.)

Ambiger, T. Y., Phantom Measurements to Determine the Most Suitable Site for Personal Dosimeter, AECL-Report 3379, Chalk River, Ontario, Canada, 1969.

Amelinckx, S., Batz, B., and Strumane, R., *Solid State Dosimetry,* Gordon and Breach, New York, 1969.

Angino, E. E. and Grögler, N., Thermoluminescence Bibliography, Report TID 3911 (Rev. 2), USAEC Tech. Inf. Series, 1965.

Angino, E. E., Thermoluminescence of some Carbonates, Thorium Dioxide, and Barium Titanate, Luminescence Dosimetry, CONF-650637, 1967, 158.

Antonov-Romanovsky, V. V., Keirim-Marcus, I. B., Poroshina, M. S., and Trapeznikova, Z. A., Dosimetry of Ionizing Radiation with the Aid of Infrared Sensitive Phosphors, Conf. Acad. Sci. USSR on Peaceful Uses Atomic Energy, Moscow, Sess. Div. Phys. Math. Sci. (Engl. transl.), U.S. Atomic Energy Comm., 1956, 239.

Archundia, C., Moreno, A., and Navarrete, M., Optimization of the Method of Obtaining Manganese-Activated Lithium Borate for Use in Thermoluminescence Dosimetry, Paper IAEA/SM-160/36, Proc. Symp. Dosimetry Techniques Applied to Agriculture, Industry, Biology, and Medicine, IAEA, Vienna, 1972.

Attix, F. H., Ed., Luminescence Dosimetry, CONF-650637, USAEC Symp. Ser. No. 8, 1967.

Attix, F. H., Isotopic effect in lithium fluoride thermoluminescent dosimeters, *Phys. Med. Biol.,* 14, 147, 1969.

Attix, F. H., Luminescence and Exoelectron Dosimetry in Personnel Monitoring, Proc. Symp. Advances in Radiation Detectors, IAEA, Vienna, 1971, 3.

Attix, F. H., A current look at TLD in personnel monitoring, *Health Phys.,* 22, 287, 1972.

Attix, F. H., Theus, R. B., Shapiro, P., Surratt, R. E., and Nash, A. E., Neutron Beam Dosimetry at the NRL Cyclotron, Paper 6.3, Third Int. Conf. Medical Physics, Gothenburg, Sweden, 1972.

Au, Y. F. and Moran, P. R., Evaluation of the UDT-500 Photodiode-Operational Amplifier Pair as Detector for Thermoluminescence, COO-1105-177, University of Wisconsin, Madison, Wisc., 1972.

Auxier, J. A., Becker, K., and Robinson, E. M., Eds., Proc. Sec. Int. Conf. Luminescence Dosimetry, CONF-680920, 1968.

Ayyangar, K., Reddy, A. R., and Brownell, G. L., Some Studies on Thermoluminescence from Lithium Fluoride and Other Materials Exposed to Neutrons and other Radiations, Proc. Sec. Int. Symp. Luminescence Dosimetry, CONF-680920, 1968, 525.

Bassi, P., Busuoli, G., Cavallini, A., Lembo, L., and Rimondi, O., Dependence of the Response of LiF TLD-100 Powders, Incorporated in Silicone Rubber, on Grain Size, *Proc. Third Int. Conf. Luminescence Dosimetry, Risö-Rep. 249,* Vol. 1, Danish AEC, Risö, Roskilde, 1971, 504.

Batsanov, S. S., Korobeinikova, V. N., Kazakov, V. P., and Kobets, L. I., Thermoluminescence of CaF_2:Tb^{3+} crystals, *Opt. Spektrosk.* (Engl. transl.), 30, 265, 1971.

Batygov, S. K. and Voronko, H. K., Thermoluminescence of CaF_2:Er crystals, *Izv. Akad. Nauk SSSR, Neorg. Mater.,* 5, 1048, 1969.

Beck, W. L., Callis, E. L., and Cloutier, R. J., Phantom Depth-Dose Measurements with Extruded LiF in a Low-Exposure-Rate Total-Body Irradiator, Proc. Sec. Int. Conf. Luminescence Dosimetry, CONF-680920, 1968, 976.

Becker, K., *Chemical and Solid-State Dosimetry Bibliography,* AED-C-21-02, ZAED, Karlsruhe, Germany, 1964.

Becker, K., *Photographic, Chemical and Solid-State Dosimetry Bibliography,* AED-C-21-03, ZAED, Karlsruhe, Germany, 1965.

Becker, K., Photographic, glass or thermoluminescence dosimetry, *Health Phys.,* 12, 995, 1966.

Becker, K., *Photographic, Chemical and Solid-State Dosimetry Bibliography,* AED-C-21-04, ZAED, Karlsruhe, Germany, 1967.

Becker, K., *Photographic, Chemical and Solid-State Dosimetry Bibliography,* AED-C-21-05, ZAED, Karlsruhe, Germany, 1967a.

Becker, K., Film dosimetry, a thing of the past? *Euratom-Bull.,* 6, 78, 1967.

Becker, K., *Photographic, Chemical and Solid-State Dosimetry Bibliography,* AED-C-21-06, ZAED, Karlsruhe, Germany, 1968.

Becker, K., *Photographic, Chemical and Solid-State Dosimetry Bibliography,* AED-C-21-07, ZAED, Karlsruhe, Germany, 1968.

Becker, K., Radiophotoluminescence and thermoluminescence, ORNL—4446, Oak Ridge Nat. Lab., Oak Ridge, Tenn., 1969, 258.

Becker, K. and Auxier, J. A., Luminescence dosimetry, *Science,* 164, 974, 1969.

Becker, K., Solid-state dosimetry, *CRC Crit. Rev. Radiol. Sci.,* 1(3), 363, 1970.

Becker, K., Cheka, J. S., and Oberhofer, M., Thermally stimulated exoelectron emission, thermoluminescence and impurities in LiF and BeO, *Health Phys.,* 19, 391, 1970.

Becker, K., *Photographic, Chemical and Solid-State Dosimetry Bibliography,* AED-C-21-08, ZAED, Karlsruhe, Germany, 1971.

Becker, K., Lu, R. H., and Weng, P. S., Environmental and Personal Dosimetry in Tropical Countries, *Proc. Third Int. Conf. Luminescence Dosimetry, Risö-Rep. 249,* Vol. 2, Danish AEC, Risö, Roskilde, 1971, 960.

Becker, K., Progress in luminescence dosimetry, *Science,* 177, 539, 1972a.

Becker, K., Environmental dosimetry with TLD, *Nucl. Instr. Meth.,* 104, 405, 19, 2b.

Becker, K., The future of personnel dosimetry, *Health Phys.,* 23, 729, 1972c.

Becker, K., Long-term stability of film, TLD, and other integrating dosimeters in warm and humid climates, ORNL–TM–4297, Oak Ridge Nat. Lab., Oak Ridge, Tenn., 1973.

Becker, K., Tham, T. D., and Haywood, F. F., Compounds of TLDs and High-Melting Organics for Fast Neutron Personal Dosimetry, 3rd Congr. IRPA, Washington, D.C., 1973.

Bekes, M., Feher, I., Deme, S., and Suha, Z., Thermoluminescent Glass Dosimeter, *KFKI Kozlem,* 17, 179, 1969.

Bettinali, C., Gottardi, V., and Ferraresso, G., Thermoluminescence in vitreous sodium alumino silicates and germanates, in *Proceedings Solid-State Conference,* Bishay, A., Ed., Plenum Press, New York, 1967, 281.

Binder, W., Disterhoft, S., and Cameron, J. R., Dosimetric Properties of CaF_2:Dy, Proc. Sec. Int. Conf. Luminescence Dosimetry, CONF-680920, 1968, 43.

Binks, C., Energy dependence of lithium fluoride dosemeters at electron energies from 10 to 35 MeV, *Phys. Med. Biol.,* 14, 327, 1969.

Bjärngard, B., The Properties of $CaSO_4$:Mn Thermoluminescent Dosimeters, Report AE-109, AB Atomenergi, Stockholm, Sweden, 1963a.

Bjärngard, B., Measurement of Small Exposures of Gamma Radiation with $CaSO_4$:Mn Radiothermoluminescence, Aktiebolaget Atomenergi, Stockholm AE-118, 1963b.

Bjärngard, B., The Radiothermoluminescence of $CaSO_4$:Sm and its Use in Dosimetry, Report AE-167, AB Atomenergi, Stockholm, Sweden, 1964.

Bjärngard, B., Use of Manganese and Samarium-activated Calcium Sulfate in Thermoluminescence Dosimetry, Luminescence Dosimetry CONF-650637, 1967, 195.

Bjärngard, G. and Jones, D., Thermoluminescent Dosimeters of LiF and CaF_2:Mn Incorporated in Teflon, Proc. Int. Symp. Solid-State and Chem. Rad. Dosimetry, IAEA, Vienna, SM-78/24, 1967, 99.

Bjärngard, B. E. and Jones, D., Experience with a new Thermoluminescence Method for Finger and Hand Dosimetry Employing Lithium Fluoride-Teflon Dosimeters, *Proceedings First International Congress of Radiation Protection,* Vol. 1, Pergamon Press, Oxford, 1968, 473.

Blum, E., Bewley, D. K., and Heather, J. D., The use of $CaSO_4$:Tm powder for fast neutron dosimetry, *Phys. Med. Biol.,* 17, 661, 1972.

Blum, E., Bewley, D. K., and Heather, J. D., Thermoluminescent dosimetry for fast neutron beams using CaF_2:Mn, *Phys. Med. Biol.,* 18, 226, 1973.

Böhm, M. and Scharmann, A., Zur Thermolumineszenz and thermisch stimulierten Leitfähigkeit von LiF, *Z. Phys.,* 225, 44, 1969.

Booth, L. F., Johnson, T. L., and Attix, F. H., Lithium fluoride glow-peak growth due to annealing, *Health Phys.,* 23, 137, 1972.

Bötter-Jensen, L. and Bechmann, P., A Versatile Automatic Sample Charger for Reading of Thermoluminescence Dosimeters and Phosphors, Proc. Sec. Int. Conf. Luminescence Dosimetry, CONF-680920, 1968, 640.

Bötter-Jensen, J. and Christensen, P., Progress Towards Automatic TLD Processing for Large-Scale Routine Monitoring at Risö, *Proc. Third. Int. Conf. Luminescence Dosimetry, Risö-Rep. 249,* Vol. 2, Danish AEC, Risö, Roskilde, 1971, 851.

Bötter-Jensen, L. and Christensen, P., Transit dose in extracorporeal blood irradiation devices measured by means of thermoluminescence, *Acta Radiol. Suppl.,* 313, 247, 1972.

Bötter-Jensen, L., Christensen, P., and Majborn, P., Interpretation of Monitoring Data from a Personnel TLD Badge Exposed to Mixed Neutron and Gamma Radiations, Paper IAEA/SM-167/46, Proc. Symp. Neutr. Monitor. Radiat. Protect., IAEA, Vienna, 1973.

Botshvar, I.A., Keirim-Markus, I. B., Vasileve, A. A., Proskina, T. I., Syritskaya, Z. M., and Yakubik, V. V., Dosimetry of Ionizing Radiation by the Aid of Thermoluminescent Alumino-Phosphate Glasses, *At. Energ.,* 15, 48, 1963.

Boyd, C. A., A kinetic study of thermoluminescence of lithium fluoride, *J. Chem. Phys.,* 17, 1221, 1949.

Bräunlich, P., Hanle, W., and Scharmann, A., Zur Thermolumineszenz von MgF_2/Mn, *Z. Naturforsch.,* 16a, 869, 1961.

Bräunlich, P., Thermoluminescence and Thermally Stimulated Current Tools for the Determination of Trapping Parameters, in *Thermoluminescence of Geological Materials,* McDougall, D. J., Ed., Academic Press, New York, 1968, 61.

Bräunlich, P. and Kelley, P., Correlations between thermoluminescence and thermally stimulated conductivity, *Phys. Rev. B,* 1, 1596, 1970.

Brooke, C. and Schayes, R., Recent Development in Thermoluminescent Dosimetry; Extensions in the Range of Applications, Proc. Int. Symp. Solid-State and Chem. Radiat. Dosimetry, IAEA, Vienna, 1967, 31.

Brooke, C., Properties of Different Phosphors as used in Packaged TLD System; Multiple Readings of Dose by U.V. Light Transfer, in *Solid State Dosimetry,* Amelinckx, S., Batz, B., and Strumane, R., Eds., Gordon and Breach, New York, 1969, 645.

Brown, J. B. C., Recommendations Concerning the Use of the Personnel Neutron Albedo Dosimeter, Report RD/B/R-828, Berkeley Nucl. Lab., Berkeley, England, 1967.

Brunskill, R. T. and Langmead, W. A., A Sealed Thermoluminescent Dosimeter Employing R.F. Heating for Routine Individual Monitoring, Adv. Radiat. Detect., IAEA, Vienna, 1971, 67.

Brunskill, R. T. and Sellars, J., Personal Neutron Dosimetry in a Plutonium Plant, *Proc. Symp. Neutr. Monitor. Radiat. Protect.*, IAEA, Vienna, 1973.

Buckman, W. G., Philbrick, C. R., and Underwood, N., The Characteristics of Ruby as a Thermoluminescent Radiation Dosimeter, Proc. Sec. Int. Conf. Luminescence Dosimetry, CONF-680920, 1968, 82.

Buckman, W. G., Aluminum oxide thermoluminescence properties for detecting radiation, *Health Phys.*, 22, 402, 1972.

Burch, W. M., Thermoluminescence, low radiation dosage and blackbody radiation, *Phys. Med. Biol.*, 12, 523, 1967; see also *Phys. Biol. Med.*, 13, 461, 1968.

Burch, W. M., Two-level isothermal read-out for high-precision thermoluminescence dosimetry with LiF, *Phys. Med. Biol.*, 13, 627, 1968.

Burgkhardt, B. and Piesch, E., Neutronendosimetrie mit LiF Proc. Conf. Strahlenschutz am Arbeitsplatz, 6, Jahrestagung Fachverband für Strahlenschutz. Karlsruhe, 1972a, 195.

Burgkhardt, B. and Piesch, E., Use of CaF_2 thermoluminescent dosimeters for measuring the natural background radiation, *Kerntechnik*, 14, 128, 1972b.

de Burke, G. P., Thermoluminescent Dosimeter Measurements of Pertubations of the Natural Radiation Environment, Proc. Sec. Int. Symp. on the Nat. Radiat. Environ., Houston, 1972.

de Burke, G. P. and Shambon, A., Investigation of Thermoluminescence Dosimeters for Environmental Monitoring, HASL-265, USAEC Health and Safety Lab., New York, 1972.

Burlin, T. E., The energy response of LiF, CaF_2, and $Li_2B_4O_7$ to high energy radiations, *Phys. Med. Biol.*, 15, 558, 1970.

Bustamente, N. T., Petel, R., and Bartolome, Z. M., Kapis as a Thermoluminescent Dosimeter, *Proc. Third Int. Conf. Luminescence Dosimetry, Risö-Rep. 249*, Vol. 3, Danish AEC, Risö, Roskilde, 1971, 1177.

Busuoli, G. and Cavallini, A., Comparison of lithium fluoride and film for personal dosimetry of X- and gamma-radiations, *Minerva Fisiconucl.*, 13, 265, 1969.

Busuoli, G., Cavallini, A., Fasso, A., and Rimondi, O., Mixed radiation dosimetry with LiF (TLD-100), *Phys. Med. Biol.*, 15, 673, 1970.

Cameron, J. R., Vought, C., and Liss, M., A Simple Design for a TLD Reader, Report COO-1105-133, U.S. At. Energy Comm., Washington, D.C., 1967.

Cameron, J. R., Zimmerman, D. W., and Bland, R. W., Thermoluminescence vs. Roentgens in Lithium Fluoride: A Proposed Mathematical Model, in Luminescence Dosimetry, CONF-650637, USAEC Symp. Ser. No. 8, 1967b.

Cameron, J. R., Suntharalingam, N., and Kenney, G. N., *Thermoluminescent Dosimetry,* University of Wisconsin Press, Madison, Wisc., 1968a.

Cameron, J. R. and Suntharalingam, N., A Comparison of TLD and Film for Personnel Dosimetry, *Proceedings First International Congress on Radiation Protectors, Rome 1966*, Vol. 1, Pergamon Press, Oxford, 1968b, 451.

Cameron, J. R., Suntharalingam, N., Wilson, C. R., and Watanabe, S., Supralinearity of Thermoluminescent Phosphors, in Proc. Sec. Int. Symp. Luminescence Dosimetry, CONF-680920, 1968c, 332.

Carlson, R. W. and Richey, J. B., The Harshaw Model 2000 Thermoluminescence Analyzer, Proc. Sec. Int. Conf. Luminescence Dosimetry, CONF-680920, 1968, 706.

Carlsson, C. A., Thermoluminescence of LiF: dependence on thermal history, *Phys. Med. Biol.*, 14, 107, 1969.

Carlsson, G. A. and Carlsson, C. A., Methodological Aspects on Measurements of Steep Dose Gradients at Interfaces between Two Different Media by Means of Thermoluminescent LiF, *Proc. Third Int. Conf. Luminescence Dosimetry, Risö-Rep. 249,* Vol. 3, Danish AEC, Risö, Roskilde, 1971, 1193.

Castro, V. G., *In vivo* Thermoluminescent Dosimetry in Clinical Radiation Therapy, Paper 38.6, Third Int. Conf. Medical Physics, Gothenburg, Sweden, 1972.

Chadwick, K. H. and Oosterheert, W. R., Radiation-Induced Thermoluminescence of Tomato Seeds at Low Temperature, Proc. Symp. on Microdosimetry, Euratom EUR 3747, 1967, 633.

Chan, F. K. and Burlin, T. E., The energy-size dependence of the response of thermoluminescent dosimeters to photon irradiation, *Health Phys.,* 13, 325, 1970.

Chan, F. K., Anderson, W. B., and Gilbertson, J. D., Radiation dose to critical organs during petrous tomography, *Radiology*, 94, 623, 1970.

Christensen, P., Manganese-Activated Lithium Borate Crystals, Glasses, and Sintered Pellets as Thermoluminescence Dosimeters, Proc. Sec. Int. Conf. Luminescence Dosimetry, CONF-680920, 1968, 90.

Christensen, P., A Combined Lithium Borate and Lithium Fluoride Thermoluminescence Dosimeter for Routine Personal Monitoring, Proc. Symp. Adv. Phys. Biol. Radiat. Detectors, IAEA, Vienna, 1971, 101.

Christensen, P., Bötter-Jensen, L., and Majborn, B., Influence of ambient humidity on TL-dosimeters for personnel monitoring, Paper Regional Conf. Radiat. Protect., Jerusalem, 1973.

Christy, R. W., Johnson, N. M., and Wilborg, R. R., Thermoluminescence and color centers in LiF, *J. Appl. Phys.,* 38, 2099, 1967.

Claffy, E. W., Klick, C. C., and Attix, F. H., Thermoluminescence Processes and Color Centers in LiF:Mg, in Proc. Sec. Int. Symp. Luminescence Dosimetry, CONF-680920, 1968, 302.

Claffy, E. W., Gorbics, S. G., and Attix, F. H., Radiation Induced Optical Absorption and Photoluminescence of LiF Powder for High-Level Dosimetry, *Proc. Third Int. Conf. Luminescence Dosimetry, Risö-Rep. 249*, Danish AEC, Risö, Roskilde, 1971, 756.

Classen, I., Seelentag, W., and Waldeskog, B., Joint IAEA/WHO Intercomparison Programme — Evaluation of Present Situation, Paper, IAEA Panel on National and International Dose Intercomparison, Vienna, Austria, 1971.

Cox, F. M., New Solid LiF Thermoluminescent Dosimeter, Proc. Sec. Int. Symp. Luminescence Dosimetry, CONF-680920, 1968, 60.

Crosby, E. H., Comparison of film badges and thermoluminescent dosimeters, *Health Phys.*, 23, 371, 1972.

Curie, D., *Luminescence in Crystals*, John Wiley & Sons, New York, 1963.

Cusimano, J. P. and Cipperly, F. V., Personnel dosimetry using thermoluminescent dosimeters, *Health Phys.*, 14, 339, 1968.

Daniels, F., Boyd, C. A., and Saunders, D. F., Thermoluminescence as a research tool, *Science*, 117, 343, 1953.

Daniels, F., Early Studies of Thermoluminescence Radiation Dosimetry, in Luminescence Dosimetry, CONF-650637, 1965, 34.

Davy, D. R. and O'Brien, B. G., An adapted model for the LET dependence of LiF thermoluminescence, *Health Phys.*, 17, 471, 1969.

De Franceschi, L., Gentili, A., and Sabbatini, V., Thermoluminescence induced in concrete for gamma dosimetry in case of nuclear accident, *Health Phys.*, 24, 287, 1973.

Degner, W., Hegewald, H., and Windelband, R., Dose-dependent sensitivity of LiF-detectors considering the polarization effect with high-energy electrons, *Radiobiol. Radiother.*, 13, 355, 1972.

Denham, D. H., Kathren, R. L., and Corley, J. P., A CaF_2:Dy Thermoluminescent Dosimeter for Environmental Monitoring, Paper 143, 17th Ann. Meet. Health Phys. Soc., Las Vegas, 1972.

Dettmer, C. M. and Galkin, B. M., The Toxicity of Thermoluminescent Phosphors, Proc. Sec. Int. Conf. Luminescence Dosimetry, CONF-680920, 1968, 944.

De Werd, L. A. and Stoebe, T. G., The Influence of Hydroxide Impurities on Thermoluminescence in Lithium Fluoride, *Proc. Third Int. Conf. Luminescence Dosimetry, Risö-Rep. 249*, Vol. 1, Danish AEC, Risö, Roskilde, 1971, 78.

De Werd, L. A. and Stoebe, T. G., The emission spectrum of LiF (TLD-100) at low and high exposures, *Phys. Med. Biol.*, 17, 187, 1972.

Distenfeld, C., Bishop, W., and Colvett, D., Thermoluminescent Neutron-Dosimetry System, in Luminescence Dosimetry, CONF-650637, 1967, 457.

Dixon, R. L. and Watts, F. C., The use of BaF_2 thermoluminescence in determining radiation quality, *Phys. Med. Biol.*, 17, 81, 1972.

Dobson, P. N. and Midkiff, A. A., Explanation of supralinearity in thermoluminescence of LiF in terms of the interacting track model, *Health Phys.*, 18, 571, 1970.

Docherty, J. and Marshall, M., A method for annealing lithium fluoride Teflon disks, *Phys. Med. Biol.*, 15, 739, 1970.

Doppke, K. P. and Cameron, J. R., Desensitization of LiF (TLD-100) Due to Radiation Damage, Report C00-1105-119, USAEC, *Phys. Med. Biol.*, 11, 624, 1966.

Eastes, J. D., Comparison of Thermoluminescent and Film Dosimetry in Routine Radiation Dosimetry, Proc. Sec. Int. Conf. Luminescence Dosimetry, CONF-680920, 1968, 184.

Edelmann, B. U., Ph.D. dissertation, Technische Hochschule Dresden, East Germany, 1967.

Eggermont, G., Jacobs, R., Janssens, A., Segaert, O., and Thielens, G., Dose Relationship, Energy Response and Rate Dependence of LiF-100, LiF-7 and $CaSO_4$:Mn from 8 keV to 30 MeV, *Proc. Third Int. Conf. Luminescence Dosimetry, Risö-Rep. 249*, Danish AEC, Risö, Roskilde, 1971, 444.

Ehrlich, M., Response of thermoluminescent dosimeters to 15 MeV electrons and ^{60}Co gamma rays, *Health Phys.*, 18, 287, 1970.

Ehrlich, M., Influence of Size of CaF_2:Mn Thermoluminescence Dosimeters on [60]Co Gamma-Ray Dosimetry in Extended Media, *Proc. Third Int. Conf. Luminescence Dosimetry, Risö-Rep. 249*, Vol. 2, Danish AEC, Risö, Roskilde, 1971, 550.

Elmanharawy, M. S. and Mostafa, Z. E. A., Radiothermoluminescence of (ZnS/CdS:Ag,Ni, Co) phosphors irradiated with beta rays, *Czech. J. Phys.*, 22, 318, 1972.

Endres, G. W. R. and Kocher, L. F., The Response of Selected Thermoluminescent Materials to Fast Neutron Exposures, Proc. Sec. Int. Conf. Luminescence Dosimetry, CONF-680920, 1968, 486.

Endres, G. W. R., Kathren, R. L., and Kocher, L. F., Thermoluminescence personnel dosimetry at Hanford, *Health Phys.*, 18, 665, 1970.

Engelke, M. J., Neutron Measurement Using Thermoluminescent Dosimeters, LA-4335, Los Alamos, N.M., 1969.

Ettinger, K. V., Durrani, S. A., and Christodoulides, C., Observation of Thermoluminescence Induced by Fission Fragments, *Radiat. Eff.*, 5, 99, 1970.

Facey, R. A., Paper, Annu. Meet. Am. Nucl. Soc., 1968.

Falk, R. B., A Personnel Neutron Dosimeter Using Lithium Fluoride Thermoluminescent Dosimeters, USAEC-Report REP-1581, Rocky Flats, Golden, Col., 1971.

Fields, D. E. and Moran, P. R., Observation of a Radiation Induced Thermally Activated Depolarization in Lithium Fluoride, to be published.

Fitzsimmons, C. K., Horn, W., and Klein, W. L., Comparison of Film Badges and Thermoluminescent Dosimeters Used for Environmental Monitoring, SWRHL-93-R, West. Environment. Res. Lab., Las Vegas, Nev., 1972.

Fleming, S. J., Development and Application of Calcium Fluoride for Evaluation of Dosage within a Radioactive Powder, Proc. Sec. Int. Conf. Luminescence Dosimetry, CONF-680920, 1968, 464.

Fleming, S. J., Thermoluminescence from Irradiated Polymers, *Proc. R. Aust. Chem. Inst.*, 36, 199, 1969.

Fleming, S. J. and Thompson, J., Quartz as a heat-resistant dosimeter, *Health Phys.*, 18, 567, 1970.

Forsythe, R. J., Roessler, C. E., Collett, W. K., Kavanaugh, H. V., and Bolch, W. E., Development of a Dental X-ray Protection Survey Technique Employing Thermoluminescent Dosimeters, *Proc. 7th HPS Midyear Symp. on Health Physics and the Healing Arts,* USPHS, Washington, D.C., 1973.

Fowler, J. R. and Attix, F. H., Solid-state integrating dosimeters, in *Radiation Dosimetry,* Vol. 2, Academic Press, New York, 1966.

Franceschi, L. de, Gentili, A., and Sabbatini, U., Thermoluminescence induced in concrete for gamma dosimetry in case of nuclear accident, *Health Phys.*, 24, 287, 1973.

Francois, H., Pradel, J., Carpentier, S., and Delarue, R., Materiel Dosimetre Thermoluminescent en Ceramique Systeme, C.E.A. Proc. Symp. Protect. Population Nucl. Mass Disasters, Interlaken, 1968.

Frank, M. and Herforth, L., Zur Thermolumineszenzdosimetrie mit CaF_2:Mn, *Kernenergie*, 5, 173, 1962.

Frank, M. and Edelmann, B. U., *Isotopenpraxis*, 2, 371, 1966.

Frank, M. and Stolz, W., *Festkörperdosimetrie ionisierender Strahlung*, G. G. Teubner, Leipzig, East Germany, 1969.

Freeswick, D. C. and Shambon, A., Light sensitivity of LiF thermoluminescent dosimeters, *Health Phys.*, 19, 65, 1970.

Furuta, Y. and Tanaka, S., Response of ^{6}LiF and ^{7}LiF thermoluminescence dosimeters to fast neutrons, *Nucl. Instr. Meth.*, 104, 365, 1972.

Galkin, B. M., Dettmer, C. M., and Suntharalingam, N., Sensitivity Changes in Solid Thermoluminescent Dosimeters after Subcutaneous Implantation, Proc. Sec. Int. Conf. Luminescence Dosimetry, CONF-680920, 1968, 963.

Gall, A. and Mason, E. W., A versatile controller for thermoluminescence read-out, *Phys. Med. Biol.*, 14, 661, 1969.

Gangadharan, P., Balasubramanian, V., Kale, L. R., Sane, S. G., and Pendurkar, H. K., TLD System for Measurement of a Wide Range of X- and Gamma Exposures, Proc. Symp. Radiat. Protect. Monitoring, IAEA, Vienna, 1969, 307.

Garlick, F. F. J., *Luminescent Materials,* Clarendon Press, Oxford, 1949.

Geiger, E. L., *TLD vs. Film, Paper Symp. Recent Dev. Pract. Dosimetry and Standards,* National Bureau of Standards, Washington, D. C., 1971.

Gibson, J. A. B., Marshall, M., and Docherty, J., Comparison of calculated and measured surface dose using LiF discs, *Phys. Med. Biol.*, 16, 283, 1971a.

Gibson, J. A. B., Marshall, M., and Docherty, J., Surface Dosimetry Using Thermoluminescent Discs, *Proc. Symp. Adv. Phys. Biol. Radiat. Detect.*, IAEA, Vienna, 1971b, 45.

Ginther, R. J., Sensitized luminescence of CaF_2:(Ce + Mn), *J. Electrochem. Soc.*, 101, 248, 1954.

Ginther, R. J. and Kirk, R. D., The thermoluminescence of CaF_2:Mn, *J. Electrochem. Soc.*, 104, 365, 1957.

Ginther, R. J., Terbium-Activated Thermoluminescent Glass for Dosimetry, in Luminescence Dosimetry, CONF-650637, 1965, 118.

Goldstein, N., Tochilin, E., and Miller, W. G., Millirad and megarad dosimetry with LiF, *Health Phys.*, 14, 159, 1968.

Goldstein, N., Miller, W. G., and Rago, P. F., Additivity of neutron and gamma ray exposures for TLD dosimeters, *Health Phys.*, 18, 157, 1970.

Goldstein, N., Dose-rate dependence of lithium fluoride for exposures above 15,000 R per pulse, *Health Phys.*, 22, 90, 1972.

Gollnick, D. A., An inexpensive TLD personnel badge, *Health Phys.*, 23, 255, 1972.

Gooden, D. S. and Brickner, T. J., Thermoluminescence Dosimetry for Clinical Use in Radiation Therapy, *Proc. Third Int. Conf. Luminescence Dosimetry, Risö-Rep. 249,* Vol. 2, Danish AEC, Risö, Roskilde, 1971, 793.

Gorbics, S. G., The Emission Spectra of Various Thermoluminescent Materials, NRL Report 6408, Naval Research Lab., Washington, D.C., 1966.

Gorbics, S. G., Attix, F. H., and Pfaff, J. A., Temperature stability of CaF_2:Mn and LiF (TLD-100) thermoluminescence dosimeters, *Int. J. Appl. Radiat. Isotopes,* 18, 625, 1967.

Gorbics, S. G. and Attix, F. H., LiF and CaF_2:Mn thermoluminescence dosimeters in tandem, *Int. J. Appl. Radiat. Isotopes,* 19, 81, 1968a.

Gorbics, S. G., Nash, A. E., and Attix, F. H., Thermal Quenching of Luminescence in Six Thermoluminescent Dosimetry Phosphors, Parts I and II, Proc. Sec. Int. Symp. Luminescence Dosimetry, CONF-680920, 1968a, 568.

Gorbics, S. G., Attix, F. H., and Kerris, K., Thermoluminescent dosimeters for high-dose applications, *Health Phys.*, in press.

Gorbics, S. G. and Attix, F. H., A device for loading thermoluminescent dosimetry powder into capsules, *Rev. Sci. Instrum.*, in press 1973.

Gower, R. G., Hendee, W. R., and Ibbott, G. S., Ultraviolet-induced changes in residual thermoluminescence from gamma-irradiated lithium fluoride, *Health Phys.*, 17, 607, 1969.

Grant, R. M., Stowe, W. S., and Correll, J., Computer-Assisted Theoretical and Experimental Analysis of the Thermoluminescence of LiF:Mg, Proc. Sec. Int. Conf. Luminescence Dosimetry, CONF-680920, 1968, 613.

Grant, W. H., Shalek, R. J., Cundiff, J. H., and Golden, R., A Review of the Activities of the AAPM Radiological Physics Center in Interinstitutional Trials Involving Radiation Therapy, Paper 38.1, Third Int. Conf. Medical Physics, Gothenburg, Sweden, 1972.

Greening, J. R., Law, J., and Redpath, A. T., Mass attenuation and mass energy absorption coefficients for LiF and $Li_2B_4O_7$ for photons from 1 to 150 keV, *Phys. Med. Biol.*, 17, 585, 1972.

Greitz, U. and Rudén, B. I., Calibration of LiF Teflon rods for internal beta-ray dosimetry, *Phys. Med. Biol.*, 17, 193, 1972.

Griffith, R. V., The Use of ^{10}B-loaded Plastic in Personnel Neutron Dosimetry, Paper IAEA/SM-167/64, Proc. Symp. Neutr. Monitor. Radiat. Protect., IAEA, Vienna, 1973.

Grögler, N., Houtermans, F. G., and Stouffer, H., The use of thermoluminescence for dosimetry and in research on the radiation and thermal history of solids, *Proc. 2nd Int. Conf. Peaceful Uses Atomic Energy, Geneva*, Vol. 21, United Nations, New York, 1958, 226.

Grögler, N., Classified bibliographies, in *Thermoluminescence of Geological Materials*, McDougall, D. J., Ed., Academic Press, New York, 1968, 645.

Guilmet, G. M., Stoebe, T. D., and Dawson, H. I., Effect of quenching temperature and rate on thermoluminescence in high purity lithium fluoride, *Health Phys.*, 19, 582, 1970.

Gupta, S. S. and Sarup, S., Thermoluminescence in Cobalt Chelates, Proc. Natl. Symp. Radiat. Phys., Bombay, 1971, 289.

Hall, J. A. and Hudd, W. H. R., The Calibration of the Mark I Personnel Neutron Albedo Dosimeter at Trawsfynydd Nuclear Power Station, Report RD/B/N-1332, Berkeley Nucl. Lab., Berkeley, England, 1969.

Hankins, D. E., Design of Albedo-Neutron Dosimeters, Paper IAEA/SM-167/62, Proc. Symp. Neutr. Monitor. Radiat. Protect., IAEA, Vienna, 1973.

Harris, A. M. and Jackson, J. H., Pre-irradiation annealing of TLD lithium fluoride, *Health Phys.*, 15, 457, 1968.

Harris, A. M. and Jackson, J. H., In the low temperature annealing of TLD-LiF, *Health Phys.*, 18, 162, 1970.

Harvey, J. R., The Energy Dependence of a Personal Neutron Dosimeter Which Utilizes a Thermal Neutron Detector at the Body Surface, Report RD/B/N 827, Berkeley Nucl. Lab., Berkeley, England, 1967.

Harvey, J. R. and Townsend, S., The Measurement of Dose from a Plane Alpha Source, Proc. Third. Int. Conf. Luminescence Dosimetry, Risö, Roskilde, 1971.

Harvey, J. R., Some Applications of Low-dose Thermoluminescence Measurements with Calcium Fluoride, in Luminescence Dosimetry, AEC Symp. Ser. No. 8, CONF-650637, 1972a, 331.

Harvey, J. R., Hudd, W. H. R., and Townsend, S., A Personal Dosimeter Which Measures Dose from Thermal and Intermediate Energy Neutrons and From Gamma and Beta Radiation, Paper IAEA/SM 167/43, Proc. Symp. Neutr. Monitor. Radiat. Protect., Vienna, 1973a.

Harvey, J. R., Tobias, A., and Townsend, S., The effect of tray current on the sensitivity of ConRad 5100 thermoluminescence dosimetry readers, *Health Phys.*, 24, 437, 1973b.

Hashizume, T., Kato, Y., Nakajima, T., Toryu, T., Sakamoto, H., Kotera, N., and Eguchi, S., A New Thermoluminescence Dosimeter of High Sensitivity Using a Magnesium Silicate Phosphor, Proc. Symp. Adv. Radiat. Detectors, IAEA-SM-143/11, Vienna, 1971.

Heinzelmann, M., Erfahrungen im praktischen Strahlenschutz mit Thermolumineszenzdosimetern zur Bestimmung der Dosis bei Teilbestrahlungen, JÜL-640-St, KFA Jülich, Germany, 1970.

Hendee, W. R., Ibbott, G. S., and Gilbert, D. B., Effects of total dose on energy dependence of TLD-100 LiF dosimeters, *Int. J. Appl. Radiat. Isotopes*, 19, 431, 1968.

Hiraki, H., Hasegawa, S., Yamamoto, O., Shirakusa, H., and Kunishige, H., Thermoluminescence Dosimeter. Development of TLD Reader with Digital Display, Natl. Tech. Report, Matsushita, Japan, 18, 181, 1972.

Holzapfel, G., Thermionic emission from electron traps, *Vacuum*, in press.

Hoy, J. E., Environmental radiation monitoring with thermoluminescent dosimeters, *Health Phys.*, 21, 860, 1971.

Hoy, J. E., Personnel Albedo Neutron Dosimeter with Thermoluminescent ^{6}LiF and ^{7}LiF, USAEC-Rep. DP-1277, Savannah River, S. C., 1972.

Hoy, J. E., An albedo-type personnel neutron dosimeter, *Health Phys.*, 23, 385, 1972.

Hukkoo, R. K., Sachdev, M. R., Wadhwani, C. N., and Somasundaram, S., Feasibility Studies of Using Bare LiF Single Crystals for *in-vivo* Radiation Dosimetry, BARC-594, Bhabha Atomic Research Center, Bombay, India, 1972.

Huzimura, R. and Ato, Y., Application of Radiothermoluminescence of Some Phosphors to Accident Dosimetry, Proc. Sec. Int. Symp. Luminescence Dosimetry, CONF-680920, 1968, 54.

Hvolby, J., Depth dose near the surface by fast electron irradiation. Measurements with thermoluminescence dosimetry, *Acta Radiol. Suppl.*, 313, 60, 1972.

Ikeya, M. and Itoh, N., Properties of $CaSO_4$:Mn powder for thermoluminescence dosimeter, *J. Nucl. Sci. Technol.* (Tokyo), 6, 132, 1969.

Ikeya, M., Ishibashi, M., and Itoh, N., Thermoluminescence response of the mixture of $CaSO_4$:Mn and Li_2SO_4 to thermal neutron and gamma-ray fields, *Health Phys.*, 21, 429, 1971.

Ivanii, G. M., Roentgenoluminescence of the $CaSO_4$:Mn phosphor and its stored energy, *Ukr. Fiz. Zh.*, 12 (Trans. AEC-tr-6896/7, 1181).

Jähnert, B., Thermoluminescent Research of Protons and Alpha Particles with LiF (TLD-700), *Proc. Third Int. Conf. Luminescence Dosimetry, Risö-Rep. 249*, Vol. 3, Danish AEC, Risö, Roskilde, 1971, 1031; *Health Phys.*, 23, 112, 1972.

Jayachandran, C. A., West, M., and Shuttleworth, E., The Properties of Lithium Borate as a Solid-State Dosimeter, Proc. Sec. Int. Conf. Luminescence Dosimetry, CONF-680920, 1968, 118.

Jayachandran, C. A., The response of thermoluminescent dosimetric lithium borates equivalent to air, water, and soft tissue and of LiF TLD-100 to low energies, *Phys. Biol. Med.*, 15, 325, 1970.

Jayachandran, C. A., Calculated effective atomic number and kerma values for tissue-equivalent and dosimetry materials, *Phys. Med. Biol.*, 16, 617, 1971.

Johns, T. F., Peabody, C. O., Preston, H. E., and Perry, K. E. G., Development of Thermoluminescence Dosimetry Systems at AEE Winfrith, Advances Radiation Detectors, IAEA, Vienna, 1971, 77.

Johnson, T. L. and Attix, F. H., Pilot Comparison of Two Thermoluminescent Dosimetry Systems with Film Badges in Routine Personnel Dosimetry, *Proceedings First International Congress on Radiation Protection, Rome 1966*, Vol. 1, Pergamon Press, Oxford, 1968, 457.

Johnson, T. L., Trap Parameters of LiF (TLD-700), Paper No. 139, 17th Ann. Meet. Health Phys. Soc., Las Vegas, to be published.

Jones, A. R., Measurement of the Dose Absorbed in Various Organs as a Function of the External Gamma Ray Exposure, AECL Report 2240, Chalk River Nucl. Lab., Chalk River, Canada, 1964.

Jones, A. R., A thermoluminescent Dosimetry System Based on Extruded Lithium Fluoride Dosimeters, Proc. Sec. Int. Symp. Luminescence Dosimetry, CONF-680920, 1968, 757.

Jones, A. R., A Personnel Dosimeter System Based on LiF Thermoluminescent Dosimeters (TLD). *Proc. Third Int. Conf. Luminescence Dosimetry, Risö-Rep. 249,* Vol. 2, Danish AEC, Risö, Roskilde, 1971, 831.

Jones, D. E. and Burt, A. K., Reproducibility Studies of the Preparation of a Lithium Fluoride Dosimeter Material, in Luminescence Dosimetry, CONF-650637, USAEC Symp. Ser. 8, 1967, 103.

Jones, D. E., Lindeken, C. L., and McMillen, R. E., Natural Radiation Background Dose Measurements with CaF_2 :Dy TLD, *Proc. Third Int. Conf. Luminescence Dosimetry, Risö-Rep. 249,* Vol. 2, Danish AEC, Risö, Roskilde, 1971, 985.

Jones, J. L. and Martin, J. A., Use of LiF (TLD)-100 for doses greater than 0.1 Mrad, *Health Phys.*, 14, 521, 1968.

Kartha, M., Some intrinsic inaccuracies of thermoluminescence dosimetry, *Health Phys.*, 20, 431, 1971.

Kartha, M. and MacDonald, J. C. F., Semi-automatic TLD reader, *Phys. Med. Biol.*, 16, 141, 1971a.

Kartha, M. and MacDonald, J. C. F., LiF surface and depth dose measurements of megavoltage photon and electron beams, *Acta Radiol. Ther. Phys. Biol.*, 10, 279, 1971b.

Karzmark, C. J., White, J., and Fowler, J. F., Lithium fluoride thermoluminescence dosimetry, *Phys. Med. Biol.*, 9, 273, 1964.

Karzmark, C. J., Fowler, J. F., and White, J., Problems of reader design and measurement in lithium fluoride thermoluminescent dosimetry, *Int. J. Appl. Radiat. Isotopes*, 17, 161, 1966.

Karzmark, C. J. and Geisselroder, J., Intercomparison of the Fricke Chemical, the LiF Thermoluminescent and Air Ionization Dosimeters for 6 MV, 250 KVCP and 50 KVP X-Rays, Proc. Sec. Int. Symp. Luminescence Dosimetry, CONF-680920, 1968, 400.

Kastner, J., Oltman, B. G., and Tedeschi, P., LiF thermoluminescent response to fast neutrons, *Health Phys.*, 12, 1125, 1966a.

Kastner, J., Berger, H., and Kraska, I. R., LiF thermoluminescence for neutron image storage, *Nucl. Appl.*, 2, 252, 1966b.

Kastner, J., Hukkoo, R., and Oltman, B. G., Lithium Fluoride Thermoluminescent Dosimetry for Beta Rays, in Luminescence Dosimetry, CONF-650637, 1967, 482.

Katz, R., Sharma, S. C., and Homayooufar, M., The structure of particle tracks, in *Progress in Radiation Dosimetry*, Attix, F. H., Ed., Academic Press, New York, 1973, Chap. 6.

Kazanskaya, V. A., Kuzmin, V. V., Makarov, Y. A., et al., Dispersed Detector for the Individual Thermoluminescent Dosimetry of Neutrons, *Yadern. Proborostr.*, 13, 118, 1970.

Kelly, P. and Bräunlich, P., Phenomenological theory of thermoluminescence, *Phys. Rev. B*, 1, 1587, 1970.

Kelly, P., Thermally stimulated exoelectron emission, *Phys. Rev. B*, 5, 749, 1971.

Kelly, P., Laubitz, M. J., and Bräunlich, P., Exact solutions of the kinetic equations governing thermally stimulated luminescence and conductivity, *Phys. Rev. B*, 4, 1960, 1971.

Kirk, R. D., Schulman, J. H., West, E. J., and Nash, A. E., Studies on Thermoluminescent Lithium Borate for Dosimetry, Proc. Symp. Solid-State Chem. Rad. Dosimetry, IAEA, Vienna, 1967, 91.

Knight, A., Marshall, T. O., Harvey, C. L., and Harvey, J. R., The Performance of Nuclear Track Emulsion and Albedo Dosimeters for Monitoring the Neutron Radiation from [252]Cf sources, Proc. Symp. Neutr. Monitor. Radiat. Protect., Paper IAEA/SM-167/40, IAEA, Vienna, 1973.

Kocher, L. F. and Endres, G. W. R., Neutron Dosimetry with Activation Filters and TLD, Proc. Sec. Int. Conf. Luminescence Dosimetry, CONF-680920, 1968, 552.

Kocher, L. F., Kathren, R. L., and Endres, G. W. R., Thermoluminescence personnel dosimetry at Hanford, *Health Phys.*, 18, 311, 1970.

Koczynski, A., Wolska-Witer, M., and Musialowicz, T., Thermoluminescent Finger Gamma Radiation Dosimeter, CLOR-89/D, Central Lab. for Radiat. Protect., Warsaw, Poland, 1972.

Konschak, K., Pulzer, R., and Hübner, K., The Emission Spectra of Various Thermoluminescence Phosphors, *Proc. Third Int. Conf. Luminescence Dosimetry, Risö-Rep. 249,* Vol. 1, Danish AEC, Risö, Roskilde, 1971, 249.

Konschak, K., UV dosimeter based on the thermoluminescence of natural CaF_2, *Radiochem. Radioanal. Lett.,* 10, 177, 1972.

Korba, A. and Hoy, J. E., A thermoluminescent personnel neutron dosimeter, *Health Phys.,* 18, 581, 1970.

Kossel, W., Mayer, V., and Wolf, H. C., Simultandosimetrie von Strahlenfeldern am lebenden Objekt, *Naturwissenschaften,* 41, 209, 1954.

Krasnaya, A. R., Nosenko, B., Revzin, L. S., and Yaskolko, V. Ya., Use of $CaSO_4$:Mn for dosimetry, *At. Energ.,* 10, 630, 1961; Engl. trans. *Sov. J. At. Energ.,* 10, 625, 1962.

Kumar, N., Sootha, G. D., and Mahendru, P. C., Thermoluminescence Dosimeters for Small Dosages, Proc. Natl. Symp. Radiat. Physics, Bombay (BARC), 1971, 353.

Land, P. L. and Wysong, E. D., Apparatus for Thermoluminescence Measurements, Proc. Sec. Int. Conf. Luminescence Dosimetry, CONF-680920, 1968, 689.

La Riviere, P. D., A Unique Throwaway LiF Dosimeter, Proc. Sec. Int. Symp. Luminescence Dosimetry, CONF-680920, 1968, 78.

Law, J., Dosimetry of low energy x-rays using LiF, *Phys. Med. Biol.,* 18, 38, 1973.

Lazanoff, A. S. and McLaughlin, J. E., Feasibility Study of Using LiF Detectors in a Neutron Flux Integrator, HASL-206, USAEC Health and Safety Lab., New York, 1969.

Lin, F. M. and Cameron, J. R., A bibliography of thermoluminescent dosimetry, *Health Phys.,* 14, 495, 1968.

Lindeken, C. L., Jones, D. E., and McMillen, R. E., Environmental background variations between residences, *Health Phys.,* 24, 81, 1973.

Lindskoug, B. and Bengtsson, B. E., A Dosimetry System Using Thermoluminescence Dosimetry for Clinical Measurements during Radiotherapy, Paper 22.6, Third Int. Conf. Medical Physics, Gothenburg, Sweden, 1972.

Linsley, G. S. and Mason, E. W., Sensitization of LiF:Teflon dosimeters, *Phys. Med. Biol.,* 16, 695, 1971.

Lippert, J. and Mejdahl, V., Thermoluminescence Readout Instrument for Measurement of Small Doses, Luminescence Dosimetry, CONF-650637, 1967, 204.

Lubyanskii, G. A., Styrov, V. V., and Sokolov, V. A., Dosimetric properties of boron nitride, *At. Energ.,* 31, 119, 1971.

Lucas, A. C. and Rainbolt, C., Response of Calcium Fluoride and Lithium Fluoride to Alpha Particles, Proc. Sec. Int. Symp. Luminescence Dosimetry, CONF-680920, 1968, 456.

Lucas, A. C. and Kapsar, B. M., Operational Importance of Retrapping in CaF_2:Mn Dosimeters, Paper 142, 17th Ann. Meet. Health Physics Society, Las Vegas, 1972.

Luchner, K., Über die Thermolumineszenz von natürlichem Flußspat, *Ann. Phys.,* 149, 435, 1957.

Lucke, W., Intrinsic Efficiency of Thermoluminescent Dosimetry Phosphors, NRL-Rep. 7104, U.S. Naval Research Lab., Washington D.C., 1970.

Lyman, T., The transparency of air between 1100 and $1300°A$ (using a $CaSO_4$:Mn thermoluminescent detector), *Phys. Rev.,* 48, 149, 1935.

Majborn, B., Bötter-Jensen, L., and Christensen, P., Thermoluminescence Dosimetry Applied to Areas with Mixed Neutron and Gamma Radiation Fields, Paper IAEA/SM-160/25, Proc. Symp. Dosimetry Techniques Applied to Agriculture, Industry, Biology, and Medicine, IAEA, Vienna, 1972.

Mandeville, C. E. and Albrecht, H. O., Luminescence of beryllium oxide, *Phys. Rev.,* 94, 44, 1954.

Mansfield, C. M., Galkin, B. M., Chow, M. C., and Suntharalingam, N., Three-Dimensional Dose Distribution Measurements in the Head and Neck using LiF, Proc. Sec. Int. Conf. Luminescence Dosimetry, CONF-680920, 1968, 990.

Marrone, M. J. and Attix, F. H., Damage effects in CaF_2:Mn and thermoluminescent dosimeters, *Health Phys.,* 10, 431, 1964.

Marshall, M. and Docherty, J., Measurement of skin dose from low energy beta and gamma radiation using thermoluminescent discs, *Phys. Med. Biol.,* 16, 503, 1971.

Marshall, T. O., Shaw, K. B., and Mason, E. W., The Consistency of the Dosimetric Properties of [7]LiF in Teflon Discs over Repeated Cycles of Use, *Proc. Third Int. Conf. Luminescence Dosimetry, Risö-Rep. 249,* Vol. 2, Danish AEC, Risö, Roskilde, 1971, 530.

Martensson, B. K. A., A statistical analysis of the influence of pre-annealing on the precision of measurement, *Phys. Med. Biol.,* 14, 119, 1969.

Mason, E. W., The effect of thermal neutron irradiation on the thermoluminescent response of CON-RAD Type-7 lithium fluoride, *Phys. Med. Biol.,* 15, 79, 1970.

Mason, E. W., Thermoluminescence response of [7]LiF to ultra-violet light, *Phys. Med. Biol.,* 16, 303, 1971.

Mayhugh, M. R., Christy, R. W., and Johnson, N. M., Color Centers and the Thermoluminescence Process in LiF:Mg, Proc. Sec. Int. Conf. Luminescence Dosimetry, CONF-680920, U.S. Atomic Energy Comm., 1968, 294.

Mayhugh, M. R., Watanabe, S., and Mucillo, R., Thermal Neutron Dosimetry by Phosphor Activation, *Proc. Third Int. Conf. Luminescence Dosimetry, Risö-Rep. 249,* Vol. 3, Danish AEC, Risö, Roskilde, 1971, 1040.

McCall, R. C. and Fix, R. C., A sensitive LiF dosimeter for routine β and γ personnel monitoring, *Health Phys.,* 10, 602, 1964.

McCall, R. C., Use of light source for testing thermoluminescent dosimeter readers, *Health Phys.,* 16, 811, 1969.

McCullough, E. C., Fullerton, G. D., and Cameron, J. R., Transferred Thermoluminescence in CaF_2:nat as a Dosimeter of Biomedically Interesting Ultraviolet Radiation, *Proc. Third Int. Conf. Luminescence Dosimetry, Risö-Rep. 249,* Vol. 3, Danish AEC, Risö, Roskilde, 1971, 1118.

McCurdy, D. E., Schiager, K. J., and Flack, E. D., Thermoluminescent dosimetry for personal monitoring of uranium miners, *Health Phys.*, 17, 415, 1969.

McDougall, D. J., Ed., *Thermoluminescence of Geological Materials*, Academic Press, London, 1968.

McDougall, D. J., Some Thermoluminescent Properties of Quartz and its Potential as an "Accident" Radiation Dosimeter, *Proc. Third Int. Conf. Luminescence Dosimetry, Risö-Rep. 249*, Vol. 2, Danish AEC, Risö, Roskilde, 1971, 256; *Health Phys.*, 20, 452, 1971.

McDougall, R. S. and Rudin, S., Thermoluminescent dosimetry of aluminum oxide, *Health Phys.*, 19, 281, 1970.

McGinley, P. H., Response of LiF to fast neutrons, *Health Phys.*, 23, 105, 1972.

Mehendru, P. C., Sootha, G. D., and Humar, N., Alkali halide thermoluminescent dosimeters, *Radiat. Eff.*, 4, 185, 1970.

Mejdahl, V., Thermoluminescence Dosimetry in Fast-Neutron Personnel Monitoring using Body-Moderation, Proc. Symp. Neutron Monitoring, IAEA, Vienna, 1967.

Mejdahl, V., Measurement of environmental radiation intensity with thermoluminescent $CaSO_4$:Dy, *Health Phys.*, 18, 164, 1970.

Mejdahl, V., Ed., *Proc. Third Int. Conf. Luminescence Dosimetry, Risö-Rep. 249*, Vols. 1-3, Danish AEC, Risö, Roskilde, 1971.

Mejdahl, V., *Dosimetry Techniques in Thermoluminescence Dating, Risö-Rep. 261*, Danish AEC, Risö, Roskilde, 1972.

Mihailovic, M., Kosi, V., Mihailovic, M. V., and Milavc, Z., An enamel coated CaF_2:Mn thermoluminescent dosemeter and the reading instrument, *Phys. Med. Biol.*, 12, 395, 1967.

Mihailovic, M. and Kosi, V., Thermoluminescent Enamels, *Proc. Third Int. Conf. Luminescence Dosimetry, Risö-Rep. 249*, Vol. 1, Danish AEC, Risö, Roskilde, 1971, 277.

Mishev, I. T., Radicheva, A. A., and Levi, S. M., The natural fluorite in the dosimetry of the mixed field of gamma rays and thermal neutrons, *Health Phys.*, 23, 367, 1972.

Miyashita, K. and Henisch, H. K., Field-Effect Enhancement of Thermoluminescence Dosimetry (CONF-650637), 1967, 259.

Momeni, M. H., Cahill, T. A., and Horn, P. L., Proton irradiation of beagle eyes. Si(Li) and thermoluminescent dosimetry of 20, 35, and 45 MeV protons, *Radiat. Res.*, 53, 15, 1973.

Moore, L. E., Thermoluminescence of $NaSO_4$, lead sulfate and miscellaneous sulfates, carbonates, and oxides, *J. Phys. Chem.*, 61, 636, 1957.

Morehead, F. F. and Daniels, F., Storage of radiation energy in crystalline lithium fluoride and matamict minerals, *J. Phys. Chem.*, 56, 546, 1952.

Moreno, A., Archundia, C., and Salsbery, L., A Study of Silver, Iron, Cobalt, and Molybdenum as Lithium Borate Activators for its Use in Thermoluminescent Dosimetry, *Proc. Third Int. Conf. Luminescence Dosimetry, Risö-Rep. 249*, Vol. 1, Danish AEC, Risö, Roskilde, 1971, 305.

Mozisek, M., Thermoluminescence of Some Irradiated Polymers, Proc. 2nd Tihany Symp. Radiat. Chem., Dobo, J. and Hedvig, P., Eds., Budapest, 1966.

Mozisek, M., Thermoluminescence of some irradiated polymers, *Jaderna Energ.*, 14, 198, 1968.

Mozisek, M., Radiothermoluminescence of Polyethylene and Polypropylene, Int., *J. Appl. Radiat. Isotopes*, 21, 11, 1970.

Muffelman, D., Scheele, R. V., Wakley, J., and Agarwal, S. K., A Survey of Radiation Exposure of Infants in a New-Born Special Care Unit, *Proc. 7th HPS Midyear Symp. on Health Physics and the Healing Arts*, U.S. Public Health Service, Washington, D.C., 1973.

Nakajima, T., On the applicability of a solid-state Bragg-Gray cavity chamber to thermoluminescent dosimetry, *Int. J. Appl. Radiat. Isotopes*, 19, 789, 1968.

Nakajima, T., Hiaoka, T., and Habu, T., Energy dependence of lithium fluoride dosemeters and high electron energies, *Health Phys.*, 14, 266, 1968.

Nakajima, T., On the Sensitivity Factor Mechanism of Some Thermoluminescence Phosphors, *Proc. Third Int. Conf. Luminescence Dosimetry, Risö-Rep. 249*, Vol. 2, Danish AEC, Risö, Roskilde, 1971, 466.

Nakajima, T., Kato, Y., Kotera, N., and Eguchi, S., Improvements on both energy and direction dependence of the Mg_2SiO_4 thermoluminescence dosimeter, *Health Phys.*, 21, 118, 1971.

Nakajima, T., Optical and thermal effects on thermoluminescence response of Mg_2SiO_4 (Tb) and $CaSO_4$ (Tm) phosphors, *Health Phys.*, 23, 133, 1972.

Nambi, K. S. V., Kathuria, S. P., and Sunta, C. M., Radiation Monitoring with a Natural Calcium Fluoride Thermoluminescent Detector, Radiat. Protect. Monitoring, IAEA, Vienna, 1969, 321.

Nambi, K. S. V. and Higashimura, T., Optical Absorption and ESR Properties of Thermoluminescent Natural CaF_2 after Heavy Gamma Irradiation, *Proc. Third Int. Conference on Luminescence Dosimetry, Risö-Rep. 249*, Vol. 3, Danish AEC, Risö, Roskilde, 1971, 1155.

Nash, A. E. and Attix, F. H., Use of LiF-Teflon discs in laminated identification cards for personnel accident dosimetry, *Health Phys.*, 21, 439, 1971.

Nash, A. E., Ritz, V. H., and Attix, F. H., Storage Stability of TL and TSEE from Six Dosimetry Phosphors, *Proc. Third Int. Conf. Luminescence Dosimetry, Risö-Rep. 249*, Vol. 3, Danish AEC, Risö, Roskilde, 1971, 1123.

Naylor, G. P., Thermoluminescence phosphor: variation of quality response with dose, *Phys. Med. Biol.*, 10, 564, 1965.

Nielsen, B. L. and Mejdahl, V., Gamma-Ray Logging by Means of Thermoluminescence Dosimeters, Risö-Rep. 219, Danish AEC, Risö, Roskilde, 1970.

Niewiadomski, T., Doping method and some thermoluminescent properties of LiF:Ag and CaF$_2$:Mn, *Nukleonika,* 12, 281, 1967.

Niewiadomski, T., Jasinska, M., and Ryba, E., Sintered TL Dosimeters, *Proc. Third Int. Conf. Luminescence Dosimetry, Risö-Rep. 249,* Vol. 1, Danish AEC, Risö, Roskilde, 1971, 332.

Niewiadomski, T., Determination of Optimal Conditions of Photon Counting in TL Measurements, Paper IAEA/SM-160/72, Proc. Symp. Dosimetry Techniques Applied to Agriculture, Industry, Biology, and Medicine, IAEA, Vienna, 1972.

Nollman, C. E. and Smolko, E. E., Thermoluminescence of LiF TLD-100 at Temperatures Above that of Liquid Nitrogen and Extension of its Range into the R region, Paper IAEA/SM-160/1, Proc. Symp. Dosimetry Techniques Applied to Agriculture, Industry, Biology, and Medicine, IAEA, Vienna, 1972.

Okuno, E. and Watanabe, S., UV-Induced Thermoluminescence in Natural Calcium Fluoride, *Proc. Third Int. Conf. Luminescence Dosimetry, Risö-Rep. 249,* Vol. 2, Danish AEC, Risö, Roskilde, 1971, 864.

Oltman, B. G., Kastner, J., Tedeschi, P., and Beggs, J. N., The effects of fast neutron exposure on the [7]LiF thermoluminescence response to gamma rays, *Health Phys.,* 13, 918, 1967.

Oltman, B. G., Kastner, J., and Paden, C., Spectral Analysis of Thermoluminescence Curves, Proc. Sec. Int. Symp. Luminescence Dosimetry, CONF-680920, 1968, 623.

Oonishi, H., Yamamoto, O., Yamashita, T., and Hasegawa, S., Dosimeter and Reader by Hot Air Jet, *Proc. Third Int. Conf. Luminescence Dosimetry, Risö-Rep. 249,* Vol. 1, Danish AEC, Risö, Roskilde, 1971, 237.

Oster, G. and Gabor, G., Organic Phosphorescent Materials, in Luminescence Dosimetry, AEC Symp. Ser. 8, CONF-650637, 1967, 132.

Palmer, R. C., Blase, E. F., and Poirier, V., A coprecipitation technique for the preparation of thermoluminescent manganese-activated calcium fluoride for use in radiation dosimetry, *Int. J. Appl. Radiat. Isotopes,* 16, 737, 1965.

Palmer, R. C., Rutland, D., Lagerquist, R., and Blase, E. F., A Prototype Thermoluminescent Dosimetry System for Personnel Monitoring, *Int. J. Appl. Radiat. Isotopes,* 17, 399, 1966.

Parker, C. V., The Response of Thermoluminescent Dosimeters to Protons with Energies to 137 MeV, Proc. Sec. Int. Conf. Luminescence Dosimetry, CONF-680920, 1968, 438.

Parker, H. M., Endres, G. W. R., and Wheeler, R. G., An intrauterine dosimeter, *Health Phys.,* 23, 389, 1972.

Partridge, R. H., Thermoluminescence in polymers, in *Radiation Chemistry of Polymers,* Vol. 1, Dole, M., Ed., Academic Press, New York, 1972, 194.

Pearson, D., Wagner, J., Moran, P. R., and Cameron, J. R., Fast Neutron Dosimetry Using Activated TL Phosphors, USAEC Report COO-1105-175, 1972.

Pendurkar, H. K., Boulenger, R., Choos, L., Nicasi, W., and Mertens, E., Energy Response of Certain Thermoluminescent Dosimeters and their Application to Dose Measurements, *Proc. Third Int. Conf. Luminescence Dosimetry, Risö-Rep. 249,* Vol. 3, Danish AEC, Risö, Roskilde, 1971, 1089.

Perry, K. E. G. and Preston, H. E., Progress Towards a Thermoluminescent Dosimetry System for Large-Scale Routine Personnel Monitoring, Paper No. 58, Sec. Int. Congr. of the Int. Rad. Protect. Assoc., Brighton, England, 1970.

Petralia, S. and Gnani, G., Thermoluminescence of Plastically Deformed LiF (TLD-100), *Lett. Nuovo Cim.,* 4, 483, 1972.

Petrock, K. F. and Jones, D. E., Hot Nitrogen Gas for Heating Thermoluminescent Dosimeter, Proc. Sec. Int. Conf. Luminescence Dosimetry, CONF-680920, 1968, 652.

Philbrick, C. R., Buckman, W. G., and Underwood, N., Ruby as a thermoluminescent radiation dosimeter, *Health Phys.,* 13, 789, 1967.

Piesch, E. and Burgkhardt, B., Use of LiF Albedo Dosimeter for Personnel Monitoring in the Radiation Field of Fast Neutrons, Paper IAEA/SM-167/8, Proc. Symp. Neutr. Monitor. Radiat. Protect., IAEA, Vienna, 1973.

Piltingsrud, H. V. and Engelke, M. J., A Passive, Broad-Energy Response Neutron Spectrometer-Dosimeter, Paper IAEA/SM-167/51, Proc. Symp. Neutr. Monitor. Radiat. Protect., IAEA, Vienna, 1973.

Pinkerton, A. P. and Laughlin, J. S., Energy dependence of lithium fluoride dosimeters for high electron energies, *Phys. Med. Biol.,* 11, 129, 1966.

Podgorsky, E. B., Moran, P. R., and Cameron, J. R., Thermoluminescent behavior of LiF (TLD-100) from 77° to 500°K, *J. Appl. Phys.,* 42, 2761, 1971.

Pohlit, W., Zur ThermoluminSzenz in Lithiumfluorid, *Biophysik,* 5, 341, 1969.

Pohlit, W., TLD-Probes for Intercomparison of Dosimetric Data, Paper IAEA/SM-160/7, Proc. Symp. Dosimetry Techniques Applied to Agriculture, Industry, Biology, and Medicine, IAEA, Vienna, 1972.

Portal, G., Francois, H., Carpentier, S., and Dajlevic, R., Radiothermoluminescent Aluminum Oxide used as Building Material for Accidental Gamma Radiation Dosimetry, Proc. Sec. Int. Conf. Luminescence Dosimetry, CONF-680920, 1968, 894.

Portal, G., Berman, F., Blanchard, P., and Prigent, R., Improvement of Sensitivity and Linearity of Radiothermoluminescent Lithium Fluoride, *Proc. Third Int. Conf. Luminescence Dosimetry, Risö-Rep. 429,* Vol. 1, Danish AEC, Risö, Roskilde, Denmark, 1971, 410.

Preston, H. E., The Measurement of Personnel Dose in the Energy Region 0.5 eV to 10 MeV with Thermoluminescent Lithium Fluoride Report AEEW-M-801, Atomic Energy Establishment, Winfrith, England, 1968.

Preston, H. E., Clifton, J. J., and Hallett, C. D., Experience in the Use of Thermoluminescent Dosimeters for Radiation Dose Control Purposes, UKAEA Report AEEW-M-962, 1970.

Preston, H. E. and Peabody, C. O., The Measurement of Personnel Neutron Dose in Reactor and Associated Areas, Paper IAEA/SM-167/37, Symp. Neutr. Monitor. Radiat. Protect., IAEA, Vienna, 1973.

Prokic, M., Determination of the Sensitivity of the CaF_2:Mn Thermoluminescent Dosimeter to Neutrons, *Proc. Third Int. Conf. Luminescence Dosimetry, Risö-Rep. 249*, Vol. 3, Danish AEC, Risö, Roskilde, 1971, 1051.

Puite, K. J., Thermoluminescent Response of CaF_2:Mn Mixed with Organic Liquids in Thermal and Fast Neutron Fields, *Health Phys.*, 17, 661, 1969.

Puite, K. J., Thermoluminescent Properties of Manganese Doped Calcium Fluoride, Thesis, University of Groningen, Netherlands, 1971a.

Puite, K. J., Thermoluminescent sensitivity of CaF_2:Mn in a mixed neutron-gamma field, *Health Phys.*, 20, 437, 1971b.

Puite, K. J. and Arends, J., Trapping Centers in CaF_2:Mn from Thermoluminescence and Thermally Stimulated Exoelectron Emission Measurements in Undoped and Mn-doped CaF_2 Samples, *Proc. Third Int. Conf. Luminescence Dosimetry, Risö-Rep. 249*, Vol. 2, Danish AEC, Risö, Roskilde, 1971, 680.

Puite, K. J., Crebolder, D. L. J. M., Hogeweg, B., and Broerse, J. J., Intercomparisons of absorbed dose and dose distribution for X-irradiations using mailed LiF thermoluminescence dosimeters, *Phys. Med. Biol.*, 17, 390, 1972.

Randall, J. T. and Wilkins, M. H. F., Phosphorescence and electron traps, *Proc. R. Soc. Ser. A*, 184, 366, 1945.

Rassow, J. and Strüter, H. D., Messbedingungen und Reproduzierbarkeit von TLD-100-Dosimetern bei energiereichen Photonen-und Elektronenstrahlen, Proc. 4th Ann. Meet., Fachverband für Strahlenschutz, Berlin, 1969, 471.

Rassow, J., Grundlagen und Planung der Elektronentiefentherapie mittels Pendelbestrahlung, Habilitationsschrift, Klinikum Essen d. Ruhr-Universität, Essen, Germany, 1970.

Regulla, D., Differences of Gamma and Neutron Induced Radiophotoluminescence Spectra of Phosphate Glass Dosimeters, Proc. Sec. Int. Symp. Luminescence Dosimetry, CONF-680920, 1968, 518.

Regulla, D. F., Ricci, A., and Sonnabend, E., Dosismessungen bei Panorama-Röntgenaufnahmen des Gebisses mit Thermolumineszenz-Dosimetern, *Dtsch. Zahnärztl. Z.*, 27, 14, 1972.

Regulla, D., Lithium fluoride dosimetry based on radiophotoluminescence, *Health Phys.*, 22, 491, 1972a.

Regulla, D., Some Studies on the Radiothermoluminescence of Manganese-Activated Metaphosphate Glass and Its Application to Low- and High-Level Photon Dosimetry, Paper Sm/160/8, Proc. Agriculture, Industry, Biology, and Medicine, IAEA, Vienna, 1972b.

Rhyner, C. R. and Miller, W. C., Radiation dosimetry by optically-stimulated luminescence of BeO, *Health Phys.*, 18, 681, 1970.

Rich, B. L. and Samardzich, B. G., Organizational Aspects of a Personnel Dosimetry Service, UCRL-73254, Lawrence Livermore Laboratory, 1971; *Health Phys.*, 24, 95, 1973.

Riehl, N., Einführung in die Lumineszenz, No. 35, Thiemig-Taschenbücher, Thiemig Verlag, München.

Rieke, J. K. and Daniels, F., Thermoluminescence studies of aluminum oxide, *J. Phys. Chem.*, 61, 633, 1957.

Robertson, M. K., Use of light source for testing thermoluminescent dosimeter readers, *Health Phys.*, 18, 440, 1970.

Robertson, M. E. A. and Gilboy, W. B., Studies of the Thermoluminescence of Lithium Fluoride Doped with Various Activators, *Proc. Third Int. Conf. Luminescence Dosimetry, Risö-Rep. 249*, Vol. 1, Danish AEC, Risö, Roskilde, Denmark, 1971, 350.

Robertson, M. E. A. and Gilboy, W. B., Reduction of abnormal background levels in LiF/Teflon thermoluminescent dosimeters, *Int. J. Appl. Radiat. Isotopes*, 23, 439, 1972.

Rossiter, M. J., Rees-Evans, D. B., and Ellis, S. C., Impurities and Thermoluminescence in Lithium Fluoride, *Proc. Third Int. Conf. Luminescence Dosimetry, Risö-Rep. 249*, Vol. 3, Danish AEC, Risö, Roskilde, Denmark, 1971, 1002.

Rossiter, M. J., Rees-Evans, D. B., and Ellis, S. C., Thermoluminescence in lithium fluoride: apparent activation by manganese, *J. Phys. D* (London), 5, 1164, 1972.

Rothermel, H. M. and Scharmann, A., Zur Thermolumineszenz von NaCl/Ag, *Z. Phys.*, 192, 1, 1966.

Rudén, B. I., Two Years Experience of Clinical Thermoluminescence Dosimetry at the Radiumhemmet, Stockholm, *Proc. Third Int. Conf. Luminescence Dosimetry, Risö-Rep. 249*, Vol. 2, Danish AEC, Risö, Roskilde, 1971, 781.

Sachdev, M. R., Hukkoo, R. K., Somasundaram, S., and Wadhwani, S. C. N., Development of LiF Single Crystals for *in-vivo* Radiation Dosimetry, Paper IAEA/SM-160/40, Proc. Symp. Dosimetry Techniques Applied to Agriculture, Industry, Biology, and Medicine, IAEA, Vienna, 1972.

Sakamoto, H., Hitomi, T., and Kotera, N., Thermoluminescence of Y_2O_3-Al_2O_3:Tb Phosphors, Proc. Sec. Int. Conf. Luminescence Dosimetry, CONF-680920, 1968, 27.

Saunders, J. E., Significant Changes in TLD Readings Produced by AC Heater Currents, *Proc. Third Int. Conf. Luminescence Dosimetry, Risö-Rep. No. 249*, Vol. 1, Danish AEC, Risö, Roskilde, 1971, 209.

Scarpa, G., The dosimetric use of beryllium oxide as a thermoluminescent material: a preliminary study, *Phys. Med. Biol.*, 15, 667, 1970.

Scarpa, G., Benincasa, G., and Ceravolo, L., Further Studies on the Dosimetric Use of BeO as a Thermoluminescent Material, *Proc. Third Int. Conf. Luminescence Dosimetry, Risö-Rep. 249*, Vol. 1, Danish AEC, Risö, Roskilde, Denmark, 1971, 427.

Schayes, R., Lorthoir, M., and Lheureux, M., La Dosimétrie par Thermoluminescence, Rev. MBLE 6, 33 MBLE, Brussels, Belgium, 1963.

Schayes, R., Brooke, C., Kozlowitz, I., and Lheureux, M., Thermoluminescent Properties of Natural Calcium Fluoride. Luminescence Dosimetry, CONF-650637, 1967, 138.

Schiager, K. J. and Savignac, N. F., Radiation Monitoring of Uranium Miners: A Comparison of Bioassay, TLD, and the Kusnetz Determinations of Current Exposures, Rep. COO-1500-21, Colorado State University, Fort Collins, Col., 1972.

Schlesinger, T., Avni, A., Feige, Y., and Friedland, S. S., Photon counting as applied to thermoluminescence dosimetry, *Proc. Third Int. Conf. Luminescence Dosimetry, Risö-Rep. 249*, Vol. 1, Danish AEC, Risö, Roskilde, 1971, 226.

Schneider, M. F., Jamm, J. F., and Ainsworth, G. C., Experimental Active and Passive Dosimetry Systems for the NASA Skylab Program, NASA-TM-X 2440, Kirtland AFB, N. Mex., 1972, 958.

Schön, M., *Halbleiterprobleme*, Vol. 4, Schottky, W., Ed., Braunschweig, 1958, 282.

Schulman, J. H., Ginther, R. J., Kirk, R. D., and Goulard, H. S., Thermoluminescent dosimeter has storage stability, linearity, *Nucleonics*, 18(3), 92, 1960.

Schulman, J. H., Attix, F. H., West, E. J., and Ginter, R. J., Thermoluminescence Methods in Personnel Dosimetry, Proc. Symp. Personnel Dosimetry Techn., Madrid, OECD/ENEA, Paris, 1963, 319.

Schulman, J. H., Kirk, R. D., and West, E. J., Use of lithium borate for thermoluminescence dosimetry, in *Luminescence Dosimetry*, Attix, F. H., Ed., AEC Symp. Ser. No. 8 (CONF-650637), 1967, 113.

Schulman, J. H., Ginther, R. J., Gorbics, S. G., Nash, A. E., West, E. J., and Attix, F. H., Anomalous fading of CaF_2:Mn thermoluminescent dosimeter, *Int. J. Appl. Radiat. Isotopes*, 20, 523, 1969.

Schulz, R. J., Interface Dosimetry with LiF-Teflon Microdosimeters, Proc. Symp. on Microdosimetry, Euratom EUR 3747, 1967, 643.

Schwarz, K. K., Grant, A., Meshs, T. K., and Grube, M. M., Thermoluminescence Dosimetry, Riga, 1968, in Russian.

Scott, A. G., The Effects of Energy and Angular Dependence on the Response of Film, Quartz-Fibre Electroscope and Thermoluminescent Personnel Dosimeters, AECL-2714, Atomic Energy of Canada Ltd., Pinawa, Manitoba, 1970.

Shambon, A. and Condon, W., LiF pellet dosimeter, *Phys. Med. Biol.*, 13, 654, 1968.

Shambon, A., Evaluation of $CaSO_4$:Dy and $CaSO_4$:Tm TLD for Measurement of Small Gamma-Ray Exposures, Report HASL-270, USAEC Health and Safety Lab., New York, 1972.

Shambon, A. and White, O., Evaluation of a Thin LiF Thermoluminescent Dosimeter for Radon Daughter Personnel Monitoring, HASL-264, USAEC Health and Safety Lab., New York, 1972.

Shearer, D. R., Investigation of Thermoluminescent Lithium Borate Glasses using Electron Spin Resonance, *Proc. Third Int. Conf. Luminescence Dosimetry, Risö-Rep. 249*, Vol. 1, Danish AEC, Risö, Roskilde, 1971, 16.

Shipley, D. B., Nichols, L. L., and Kocher, L. F., Thermoluminescence from adhesive coated Teflon tape, *Health Phys.*, 22, 195, 1972.

Shiragai, A., Annealing of a LiF thermoluminescence dosimeter, *Health Phys.*, 13, 1040, 1967.

Shiragai, A., Effect of grain size and initial trap density on superlinearity of LiF-TLD, *Health Phys.*, 18, 728, 1970.

Sidorov, T. A., Tyulkin, V. A., and Aksenov, V. S., The thermoluminescence, EPR, and electron absorption spectra of irradiated silicate glass, *Dokl. Akad. Nauk (SSSR)*, 175, 1094, 1968.

Spanne, P., Warning against high energy β-emitters in light sources for TLD readers, *Health Phys.*, 24, 568, 1973.

Spiers, F. W. and Zanelli, G., The application of thermoluminescence to problems of radiation dosimetry in bone, *Phys. Med. Biol.*, 11, 148, 1966.

Spiers, F. W., Zanelli, G. D., Poli, V., and Beddoe, A., Application of Thermoluminescence Methods to Bone Dosimetry, Paper IAEA/SM-160/53, Proc. Symp. Dosimetry Techniques Applied to Agriculture, Industry, Biology, and Medicine, IAEA, Vienna, 1972.

Spurny, Z., Thermoluminescence dosimetry, *At. Energ. Rev.*, 3(2), 61, 1965.

Spurny, Z., Solid-State Dosimetry, IAEA Bibliographical Ser. No. 23, STI/PUB/21/23, 1967.

Spurny, Z., Relative sensitivities of thermoluminescent phosphors, *Phys. Med. Biol.*, 13, 465, 1968.

Spurny, Z., A Glass Thermoluminescent Dosimeter, Proc. Sec. Int. Conf. Luminescence Dosimetry, CONF-680920, 1968, 18.

Spurny, Z., Permanently Excited Thermoluminescence Substance, *Int. J. Appl. Radiat. Isotopes*, 20, 395, 1969.

Spurny, Z., Simultaneous estimation of exposure and time elapsed since exposure using multipeaked thermoluminescent phosphors, *Health Phys.*, 21, 755, 1971.

Spurny, Z., Novotny, J., and Hedvicakova, L., Thermoluminescent dosimetry using lithium fluoride in aqueous suspension, *Phys. Med. Biol.*, 16, 295, 1971.

Spurny, Z., Thermalox 995 (BeO) as Thermoluminescent Dosimeter of the Gamma Component in the Field of Neutrons, Paper IAEA/SM-160/52, Proc. Symp. Dosimetry Techniques Applied to Agriculture, Industry, Biology, and Medicine, IAEA, Vienna, 1972.

Spurny, Z., Milu, C., and Racoveanu, N., Comparison of x-ray beams using thermoluminescent dosimeters, *Phys. Biol. Med.*, 18, 276, 1973.

Spurny, Z. and Sulcova, J., Bibliography of thermoluminescent dosimetry 1968—1972, *Health Phys.*, 24, 573, 1973.

Strash, A. M., The Thermoluminescence Spectrum and the Energy Conversion Efficiency of Lithium Fluoride, HASL-219, USAEC Health and Safety Lab., New York, 1969.

Sukis, D. R., Thermoluminescent Properties of CaF_2:Dy TLD's, Trans. Nucl. Sci., NS-18, 185, 1971.

Sunta, C. M., Thermoluminescence of Natural CaF_2 and its Applications, *Proc. Third Int. Conf. Luminescence Dosimetry, Risö-Rep. 249*, Vol. 2, Danish AEC, Risö, Roskilde, 1971, 392.

Sunta, C. M. and Kathuria, S. P., Effects of Residual Thermoluminescence in Fluorite and Lithium Fluoride Thermoluminescent Dosimeters, Proc. IAEA Symp. Adv. Phys. Biol. Radiat. Detect., 1971, 125.

Sunta, C. M., Bapat, V. N., and Kathuria, S. P., Effect of Deep Traps on Supralinearity, Sensitisation and Optical Thermoluminescence in LiF TLD, *Proc. Third Int. Conf. Luminescence Dosimetry, Risö-Rep. 249*, Vol. 1, Danish AEC, Risö, Roskilde, 1971, 146.

Sunta, C. M., Nambi, K. S. V., and Bapat, V. N., Fast Neutron Response of Thermoluminescent Detectors using Proton Radiator Technique, Proc. Symp. Neutr. Monitor. Radiat. Protect., IAEA, Vienna, 1972.

Suntharalingam, N., Cameron, J. R., Shuttleworth, E., West, M., and Fowler, J. F., Fading characteristics of thermoluminescent lithium fluoride, *Phys. Med. Biol.*, 13, 97, 1968.

Suntharalingam, N. and Cameron, J. R., Thermoluminescent response of lithium fluoride to radiations with different LET, *Phys. Med. Biol.*, 14, 397, 1969.

Suntharalingam, N. and Mansfield, C. M., Lithium Fluoride Dosimeters in Clinical Radiation Dose Measurements, *Proc. Third Int. Conf. Luminescence Dosimetry, Risö-Rep. 249*, Vol. 2, Danish AEC, Risö, Roskilde, 1971, 816.

Tatsuta, H., Moriuchi, Y., Katoh, A., Yamachi, I., and Matsumoto, K., Intercomparison of the accuracy of dosimetry systems used in the different facilities, *Hoen Butsuri*, 7, 37, 1972.

Thompson, J. J. and Ziemer, P. L., The Effect of Various Activators on the Thermoluminescence of Lithium Borate, Paper P/141, 17th Ann. Meet. Health Phys. Soc., Las Vegas, 1972a.

Thompson, J. J. and Ziemer, P. L., Energy response of lithium borate thermoluminescent dosimeters, *Health Phys.*, 22, 399, 1972b.

Tobias, A., Irradiation of TLD LiF at Elevated Temperatures, Rep. RD/B/N-2368, Berkeley Nuclear Lab., Berkeley, England, 1972.

Tochilin, E. and Goldstein, N., Dose-rate and spectral measurements from pulsed x-ray generators, *Health Phys.*, 12, 1705, 1966.

Tochilin, E., Goldstein, N., and Lyman, J. T., The Quality and LET Dependence of Three Thermoluminescent Dosimeters and Their Potential Use as Secondary Standards, in Proc. Sec. Int. Symp. Luminescence Dosimetry, CONF-680920, 1968, 424.

Tochilin, E., Goldstein, N., and Miller, W. G., Beryllium oxide as a thermoluminescent dosimeter, *Health Phys.*, 16, 1, 1969.

Tochin, V. A. and Nikolskii, V. G., Concerning the radiothermoluminescence of organic compounds, *Khim. Vys. Energ.*, 3, 281, 1969.

Tomita, A., Thermoluminescence of gamma-irradiated polytetrafluoroethylene, *J. Phys. Soc. Jap.*, 28, 731, 1970.

Tousey, R., Watanabe, K., and Purcell, D., Measurement of solar extreme uv and x rays from rockets by means of a $CaSO_4$:Mn phosphor, *Phys. Rev.*, 83, 792, 1951.

Turner, A. P. and Anderson, D. W., Thermoluminescent response of lithium fluoride to high-energy photons, *Phys. Med. Biol.*, 18, 46, 1973.

Tymons, B. J., Tuyn, J. W. N., and Baarli, J., A System for Personnel Dosimetry in Mixed Radiation Fields, Proc. Symp. Neutr. Monitor. Radiat. Protect., IAEA, Vienna, 1973.

Uran, D., Knezevic, M., Susnik, D., and Kolar, D., Some Dosimetric Properties of Sintered Activated CaF_2 Dosimeters, *Proc. Third Int. Conf. Luminescence Dosimetry, Risö-Rep. 249*, Vol. 3, Danish AEC, Risö, Roskilde, 1971, 1196.

Vacirca, S. J. and Thompson, D. L., Dose outside useful beam in water and realistic phantoms exposed to chest radiographic procedures, *Health Phys.*, 23, 533, 1972.

Vacirca, S. J., Thompson, D. L., Paternack, B. S., and Blatz, H., A film-thermoluminescent dosimetry method for predicting body doses due to diagnostic treatment, *Phys. Med. Biol.*, 17, 71, 1972.

Van Espen, E., Memory Effect with M.B.L.E. CaF_2 Thermoluminescent Dosimeters, *Proceedings First International Congress Radiation Protection, Rome 1966*, Vol. 1, Pergamon Press, Oxford, 1968.

Vaughan, W. J. and Miller, L. O., Dosimetry using optical density changes in LiF, *Health Phys.*, 18, 578, 1970.

Wallace, R. H. and Ziemer, P. L., Studies on the Thermoluminescence of Manganese-Activated Lithium Borate, Proc. Sec. Int. Conf. Luminescence Dosimetry, CONF-680920, 1968, 140.

Wallace, R. H., Ziemer, P. L., Kastner, J., and Oltman, B. G., The relationship between encapsulation and apparent fast neutron induced fading in TLD, *Health Phys.*, 20, 221, 1971.

Watanabe, K., Properties of $CaSO_4$:Mn phosphor under vacuum ultraviolet excitation, *Phys. Rev.*, 83, 785, 1951.

Watanabe, S. and Morato, S. P., Continuous Model for TL Traps, *Proc. Third Int. Conf. Luminescence Dosimetry, Risö-Rep. 249*, Danish AEC, Risö, Roskilde, 1971, 58.

Watanabe, S. and Okuno, E., Thermoluminescent Response of Natural Brazilian Fluorite to [137]Cs Gamma Rays, *Proc. Third Int. Conf. Luminescence Dosimetry, Risö-Rep. 249*, Vol. 1, Danish AEC, Risö, Roskilde, 1971, 380.

Webb, G. A. M. and Bodin, G., Manufacture of Uniform, Extremely Thin Thermoluminescence Dosimeters by a Liquid Moulding Technique, *Proc. Third Int. Conf. Luminescence Dosimetry, Risö-Rep. 249*, Vol. 2, Danish AEC, Risö, Roskilde, 1971, 518.

Webb, G. A. M. and Phykitt, H. P., Possible Elimination of the Annealing Cycle For Thermoluminescent LiF, *Proc. Third Int. Conf. Luminescence Dosimetry, Risö-Rep. 249*, Vol. 1, Danish AEC, Risö, Roskilde, Denmark, 1971, 185.

Webb, G. A. M., Dauch, J. E., Slane, J. K., and Phykitt, H. P., Information Content of Personnel Dosimetry Systems, Paper 17th Ann. Meet. Health Physics Society, Las Vegas (pre-print from Teledyne/Isotopes, Westwood, N. J.), 1972.

Webb, G. A. M., Dauch, D. E., and Bodin, G., Operational evaluation of a new high sensitivity thermoluminescent dosimeter, *Health Phys.*, 23, 89, 1972b.

Weissman, D. D., Comparative Absorbed Doses in Dental Radiography. IV. Pedodontic Radiography, *Proc. 7th HPS Midyear Symp. on Health Physics and the Healing Arts,* U.S. Public Health Service, Washington, D.C., 1973.

Wick, F. G. and Slattery, M. K., Thermoluminescence excited by x-rays; further experiments upon synthetically prepared materials, *J. Opt. Soc. Am.,* 16, 398, 1928.

Wiedemann, E. and Schmidt, G. C., Uber Liminsecenz, *Ann. Phys. Chem.,* 54, 604, 1895.

Wilson, C. R., De Werd, L. A., and Cameron, J. R., Stability of the Increased Sensitivity of LiF (TLD-100) as a Function of Temperature, Report COO-1105-116, USAEC, 1966.

Wilson, C. R. and Cameron, J. R., Dosimetric Properties of $Li_2B_4O_7$:Mn (Harshaw), Proc. Sec. Int. Conf. Luminescence Dosimetry, CONF-680920, 1968, 161.

Wingate, C. L., Tochilin, E., and Goldstein, N., Response of Lithium Fluoride to Neutrons and Charged Particles, Luminescence Dosimetry, AEC Symp. Ser. 8, CONF-650637, 1967, 421.

Witzmann, H. and Buhrow, J., in Zur Physik und Chemie der Kristallphosphore, Berlin, 1960.

Woodsum, H. C., Selection and Use of TLDs for High-Precision NERVA Shielding Measurements, NASA-TM-X-2440, Westinghouse, Large, Pa., 1972, 466.

Yamashita, T., Nada, N., Onishi, H., and Kitamura, S., Calcium Sulfate Phosphor Activated By Rare Earth, Proc. Sec. Int. Symp. Luminescence Dosimetry, CONF-680920, 1968, 4.

Yasuno, Y. and Yamashita, T., Thermoluminescent Phosphor Based on Beryllium Oxide, in *Proc. Third Int. Conf. Luminescence Dosimetry, Risö-Rep. 249,* Danish AEC, Risö, Roskilde, 1971, 290.

Zanelli, G. D., The effect of particle size on the thermoluminescence of lithium fluoride, *Phys. Med. Biol.,* 13, 393, 1968.

Zanelli, G. D. and Spiers, F. W., LiF Thermoluminescence Dosimetry in Trabecular Bone, Proc. Sec. Int. Conf. Luminescence Dosimetry, CONF-680920, 1968, 920.

Zanelli, G. D., Particle size and supralinearity in LiF, *Phys. Med. Biol.,* 17, 99, 1972.

Zimmerman, D. W., Rhyner, C. R., and Cameron, J. R., Thermal annealing effects on the thermoluminescence of lithium fluoride, *Health Phys.,* 12, 525, 1966.

Zimmerman, D. W. and Cameron, J. R., Supralinearity of Thermoluminescence of LiF versus Dose, in *Thermoluminescence of Geological Materials,* McDougall, D., Ed., Academic Press, London, 1968.

Zimmerman, D. W., Relative thermoluminescence effects of alpha- and beta-radiation, *Radiat. Eff.,* 14, 81, 1972.

Zimmerman, J., The radiation-induced increase of the 100°C thermoluminescence sensitivity of fired quartz, *J. Phys. C: Solid-State Phys.,* 4, 3265, 1971a.

Zimmerman, J., The radiation-induced increase of thermoluminescence sensitivity of the dosimetry phosphor LiF (TLD-100), *J. Phys. C: Solid-State Phys.,* 4, 3277, 1971b.

3. STIMULATED EXOELECTRON
EMISSION

3-1. Principles
3-1-1. History

Exoelectron emission is the effect of structure-dependent emission of low energy electrons which occurs from surfaces of many insulating solids at temperatures well below those at which thermionic emission takes place. It has not yet found any widespread practical applications, but is presently the subject of intense and promising investigations in an increasing number of laboratories. It has already been the subject of more than 300 publications and several international meetings, most recently in Braunschweig (Holzapfel and Kaul, 1970); Sverdlovsk (Minz et al., 1970); Braunschweig, 1972; Detroit, 1973; and Liblice near Prague, 1973. It is the purpose of this chapter to provide an introduction to the field in general and to briefly discuss its potentials in radiation dosimetry, summarizing and updating previous reviews on this subject (Becker, 1970 and 1972a) and the results presented at a recent seminar in Braunschweig (1972).

The basic effect of exoelectron emission, following mechanical deformation, physical phase changes, chemical reactions, or exposure of the surface to ionizing radiation, has also been called electron after-emission or "cold emission." The term "Kramer effect" has become most widely accepted because it was Kramer who first systematically investigated it around 1950 and stimulated further activities (Kramer, 1950). He introduced the term "exoelectrons" because it was originally assumed that exothermic processes were involved in the emission.

Several other names and a number of different abbreviations such as TE (for thermo-emission), TSEE (for thermo-stimulated electron emission), EE (for electron exoemission), and EEE (for exoelectron emission) can also be found in the early literature. It is now generally accepted that the term "exoelectrons" should be used for the electrons emitted, and the name "Kramer effect" for the physical process of "cold" electron emission. Spontaneous exoelectron (after)-emission from a material or exoelectron emission

in general is abbreviated EE, optically stimulated exoelectron emission OSEE, and thermally stimulated exoelectron emission TSEE.

The observation of effects which can now be explained with exoelectron emission is by no means new (for a recent review of early observations of EE, see Scharmann, 1970). The first detailed paper on exoelectrons was published by McLennan in 1902, who excited various inorganic salts with electrons and recorded a delayed electron emission. The negative charge and low energy of the emitted particles were established and a relationship with thermoluminescence was suspected.

In the 1930's, after the development of the Geiger-Mueller counter, an abnormally high background count in freshly manufactured G-M counters was observed frequently which decreased on aging of the counters, as well as an increase of the background in G-M counters after gas discharges or after mechanical treatment of the electrodes. However, these early observations did not lead to systematic investigations until Kramer's work.

He studied exoelectron emission from a number of ionic crystals and metal surfaces after mechanical treatment or exposure to heat or light, and found emission maxima at different characteristic temperatures during heating (Kramer, 1952). Kramer's results soon were confirmed by other investigators. Gobrecht and Barthow (1956) demonstrated that the electron emission from metal surfaces depends on the presence of oxygen and that the electrons originate not in the metal but in the metal oxide layer on its surface.* Bohun (1954) and others pointed out similarities between thermoluminescence and TSEE; a kinetic model of TL was used to explain TSEE (Nassenstein, 1955).

An international symposium on exoelectrons in 1956 (*Acta Phys. Austr.*, 10, 1957) first summarized the state of the art; new contributions such as the first measurements in vacuum were reported. A period of steady progress (Bohun, 1963) culminated in a Second Exoelectron Conference (1966). Much of the information in

---

*The term "surface" is used here not only for the two-dimensional boundary between a solid and its environment, but also for its thin uppermost layer of $\sim$10 to 100 Å thickness.

the early literature is, however, only of historical interest.

Recent years have seen a revival of the interest in exoelectrons. Although several aspects are still unclear, much progress has been made in understanding the physical nature of exoelectron emission and its relation to other trapping and emission phenomena in ionic crystals. Many practical applications were discovered, such as the use of EE as an imaging device (exoelectron "photography"), as an ultrasensitive analytical method, in metallographic or catalytic studies, for investigations on solid-state chemical reactions, in biophysics, in geology, and in lunar research. Of these applications only those integrating radiation dosimetry are discussed briefly in this chapter.

3-1-2. Emission Mechanism

Exoelectron emission has been observed in connection with numerous physical and chemical processes which have in common that they result in changes in the surface structure and/or trapped electron population near the surface of an insulator:

1. During or after mechanical deformation of inorganic ionic crystals or minerals such as metal oxides (or oxidized metal surfaces);

2. During the solidification of metals;

3. During changes in the crystalline structure, for example, phase changes, dehydration, or uptake of crystal water;

4. During some surface adsorption or desorption reactions;

5. During chemical reactions, e.g., the oxidation of metals or the chemisorption of oxygen and water on metal surfaces, and finally (this is obviously the only mechanism of dosimetric interest);

6. In inorganic crystals, glasses, semiconductors, and in several organic materials after exposure to ionizing radiation.

Depending on the depth of the electron traps from which exoelectron emission occurs, it can take place at temperatures from $-240°C$ (KBr) to above $600°C$. If the emission occurs around room temperature, a spontaneous "after-emission" is observed. It decays exponentially with time, if only one well-defined type of trap is involved. If higher temperatures are required to release the electrons, heating will usually result in several

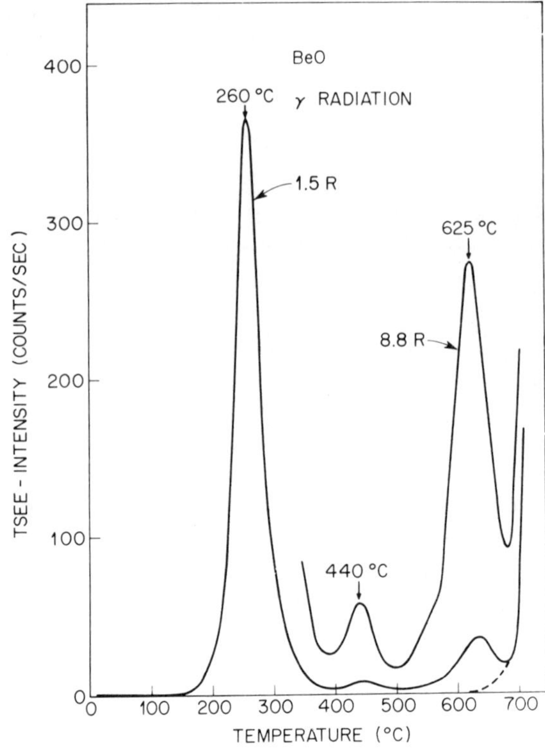

FIGURE 3-1. TSEE curve of a technical grade BeO powder, mixed with graphite powder and plated on a graphite sample holder if heated after exposure to gamma radiation inside a gas-flow G-M counter at a constant rate of 1°C/sec; dotted line indicates thermionic electron emission.

emission peaks at characteristic temperatures. When the emission rate is plotted as a function of temperature or time during heating, typical TSEE curves result (Figure 3-1). Exoelectrons can also be released from traps by optical stimulation at the proper wavelengths.

Average and mean energies between about 0.1 and 0.7 eV and maximum energies between about 2 and 8 eV have been reported, depending somewhat on the emission temperature (for a bibliography, see Becker, 1972). There is a widespread feeling that interesting information on the emission mechanism, the location of the traps, etc. can be gained from precise measurements of the exoelectron energy distribution; several laboratories (mostly in Germany) are presently employing more sophisticated methods of low energy electron spectroscopy in such studies (Braunschweig, 1972 and Brunsmann and Scharmann, 1973).

In one study exoelectron energies up to 90 eV

have been measured after bombardment of SiO_2 with monoenergetic low-energy electrons. It is difficult to explain such a high-energy on the assumption that the high-energy tail of the Maxwellian energy distribution of electrons in the conduction band can overcome the work function and leave the lattice. It has, therefore, been suggested that an electrical dipole layer is formed in the surface region, and electrons that are released from traps in deeper material layers are accelerated toward the surface.

Besides this dipole-layer and the Maxwell-tail theories, it was proposed (Bohun, 1970) that the energy that is released by the recombination of a hole and an electron is transferred to another electron localized in a nearby color center. The liberated electron can then leave the crystal. This effect can be interpreted as an internal Auger effect or as a "resonance" mechanism. Other theories of EE have also been suggested.

Despite these explanations, it is still not completely clear where the energy comes from which enables electrons to overcome the electron affinity (work function) and leave the crystal with an energy up to several eV and a rather high efficiency. "Holes" in the work function at growth steps, edges, etc. of the crystal (it can be shown with special imaging devices that the EE from a crystal surface is not homogeneous, but occurs mainly at such irregularities), potentials between surface and deeper crystal layers, energized sorption states, Auger processes, and possibly direct emission processes from traps without involvement of the conduction band probably all contribute to a varying degree to the emission process.

It should be mentioned that some EE phenomena are especially difficult to explain. One is the increasing EE which has been observed in some samples during consecutive heatings of the material without exposing it to radiation or any other EE inducing agent (Crase et al., 1971). This emission may reach its maximum after several heating and cooling-down cycles and has been called "self-excitation." Another strange effect is an EE which can occur during the cooling instead of the heating of a sample in the counter. Such "cooling-down peaks" may or may not coincide with the TSEE peaks which result during heating.

3-1-3. Kinetic Theory of EE

The lack of an adequate theory of exoelectron emission is to some degree due to difficulties in the interpretation of the experimental data. Exoelectrons originate in a very thin surface layer: The estimated thickness of the emitting layer is in the order of 10^{-5} to 10^{-4} cm for optical stimulation, more probably — in particular if the EE is not associated with TL — considerably thinner (according to Holzapfel, 1969, only electrons originating in a layer not thicker than 10^{-7} to 10^{-6} cm contribute directly to the observable EE). Therefore, EE can be expected to be quite sensitive to variations in the defect-structure of the surface in a given material and subject to "dirt effects" caused by humidity, oxygen, etc. Many early investigators were discouraged by badly reproducible results.

This situation has changed with the use of improved instrumentation and with the prevention of surface charges which may build up during emission. For example, mixing the emitter with a conductor such as graphite powder greatly increases the reproducibility of readings (Kramer, 1966). With more sophisticated devices, exoelectrons from a TSEE emitter can be observed with a standard deviation of less than 5% in the total counts in hundreds of repeated irradiation and heating cycles (Becker et al., 1971).

In developing a kinetic theory of EE, a distinction has to be made between actual surface traps, such as adsorbed species, and "volume" traps which are present throughout the material. Both can contribute to the emission spectrum. It was shown, for example, that a peak at $\sim 120°C$ in LiF is due to surface centers created by sorption effects, while a peak at $\sim 320°C$ is related to volume-type F and M centers (Gordan and Scharmann, 1968).

Emission from surface and volume traps can be expected to follow different kinetics. Emission occurring from surface states probably does not require conduction band involvement. For volume traps, it is assumed that the electrons are first excited from the trap into the conduction band from which they statistically overcome the electron affinity (for a kinetic treatment of the emission from a surface layer of definite depth, see Holzapfel, 1972). Electrons might be emitted directly from surface traps.

Experimental proof of the close relation between EE and thermoluminescence has been obtained by measuring light and electron emission simultaneously from samples during heating. In some cases, the TSEE and TL maxima and the fading rates are almost identical. Close agreement between TSEE and TL is, however, rare. In a few materials, TL peaks may be observed without corresponding TSEE peaks. In many more cases, TSEE maxima are observed without corresponding TL. This could be due to one or more of several reasons (e.g., electrons are ejected from the surface during nonradiative transfer, the photon emission associated with the TSEE cannot be detected by the photomultiplier, or the TSEE occurs from surface traps without corresponding volume traps which are responsible for TL).

There is, however, no doubt that at least in simple ionic crystals with a sufficient band gap, where the Maxwell-tail emission from volume traps probably dominates the emission process, a relation between TSEE, TL, and thermal conductivity exists. It may be represented in an oversimplified form, as in Figure 3-2. If such a pre-irradiated crystal is subjected to thermal stimulation, three effects will, in principle, occur simultaneously:

1. Electrical conductivity proportional to the electron concentration in the conduction band results in a "conductivity glow curve;"

2. Electron transfer from the conduction band to a deep activator level may be associated with photon emission (TL); and

3. TSEE takes place.

On the basis of this model, the TL glow curve analysis of Randall and Wilkins (1945), which was discussed in some detail in Chapter 2.2, has been applied to TSEE. Some simplifying assumptions have to be made about the trapped electrons and the detrapping process which take place upon heating to overcome mathematical difficulties. In the thus simplified treatment, the probability per unit P of an electron escaping from a trap of depth E below the conduction band at a temperature T is of the form

$$P = a_0 \, e^{-E/kT} \qquad\qquad 3.1$$

where a_0 is the frequency factor.

If the number of occupied traps is designated by c, then its reduction $-dc/dT$ during linear heating at a rate of $q = dT/dt$ follows the equation

$$\frac{-dc}{dT} = \frac{a_0 c}{q} \, e^{-E/KT} \qquad\qquad 3.2$$

The integration of this equation is given by

$$\int_{c_0}^{c} \frac{dc}{c} = \frac{a_0}{q} \int_{o}^{T} e^{-E/kT} \, dt \qquad\qquad 3.3$$

Since $T \ll E/k$, but greater than zero in the range of interest, the right integral can be approximated by

$$\frac{a_0}{q} \int_{o}^{T} e^{-E/kT} \, dT \simeq S(1 - \Sigma) \qquad\qquad 3.4$$

where

$$S = \frac{k \, a_0 \, T^2}{qE} \, e^{-E/kT} \qquad\qquad 3.5$$

and Σ represents the first several terms of the diverging series

$$\sum_{n=1}^{\infty} (-1)^{n+1} (n+1)! \, (kT/E)^n \qquad\qquad 3.6$$

For a given value of kT/E, the difference between alternate terms decreases until $(n+1)!$ begins to dominate the series. The summation Σ represents a cutoff of the series as long as this difference is small. In that case, Σ becomes a number much less

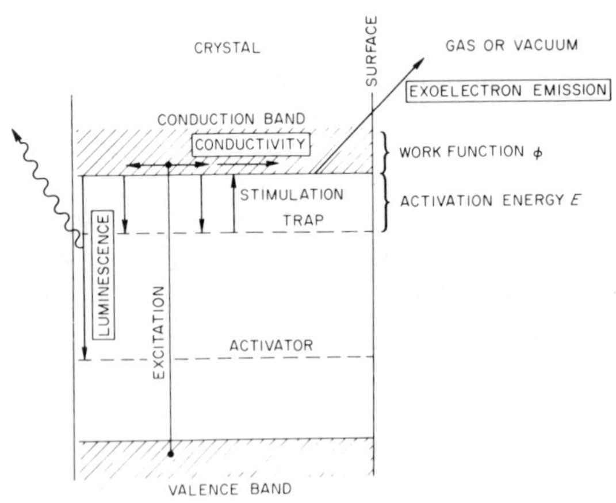

FIGURE 3-2. Simplified schematic diagram of the basic detrapping processes occurring in an irradiated ionic crystal during heating.

than unity. Therefore, integration of Equation 4 results in

$$C \equiv \frac{C(T)}{C_0} e S(1-\Sigma), \qquad 3.7$$

where C is the fraction of traps C_0 which dissociate between T and $T + dT$. In Figure 3-3, $-dc/dT$ is plotted as a function of T. A maximum in the TSEE curve corresponds to the most rapid electron detrapping rate $d^2 c/d^2 T = 0$, or $S(T_{max}) = 1$ and $c(T_{max}) = e^{-1}$. From this, the following relation between heating rate q and temperature T_{max} of the TSEE peak maximum is obtained:

$$q = \frac{a_0 k T^2_{max}}{E} \exp \left(\frac{-E}{k T_{max}} \right) \qquad 3.8$$

This equation implies a shift to a higher T_{max} with an increase in q similar to the well-known Randall-Wilkins peak shift in TL (see Chapter 2.2) which has indeed been observed in TSEE. If T_{max} is known, two different values of q and of E can be determined from

$$E = k \frac{T_{max_1} \cdot T_{max_2}}{T_{max_2} - T_{max_1}} \ln \frac{q_2 T^2 max_1}{q_1 T^2 max_2} \qquad 3.9$$

These equations have been used extensively to obtain term data (activation energy E and frequency factor a_0) for the various emission peaks of many emitting materials. Attempts have been made to use these data to identify traps, but it should be emphasized that this simple model does not satisfactorily explain all TSEE phenomena. In particular, some difficulties have arisen in the interpretation of the emission kinetics from crystals with a small band gap such as on ZnO.

Another method, more suitable because it uses the full information from experimental data, has been developed by Balarin and Zetzsche (1962) for determination of trapping depths. This method is particularly useful if an analog computer is employed for the evaluation of the data. If $\ln(\ln C)$ is plotted as a function of $1/T$, the slope of the resulting straight lines (Figure 3-4) indicates the activation energy because

$$\frac{E}{k} = \frac{d[\ln(-\ln C)]}{d(1/T)} \qquad 3.10$$

Another model for TSEE (Kriegseis and Scharmann, 1970) deals with two idealized cases. In one case, it is assumed that the emitting sample remains electrically neutral. In the other case, the sample is assumed to be electrically isolated and therefore becomes charged during emission. The most exact solutions of the kinetic equations that govern thermally stimulated current, luminescence, and EE have been presented by Kelly

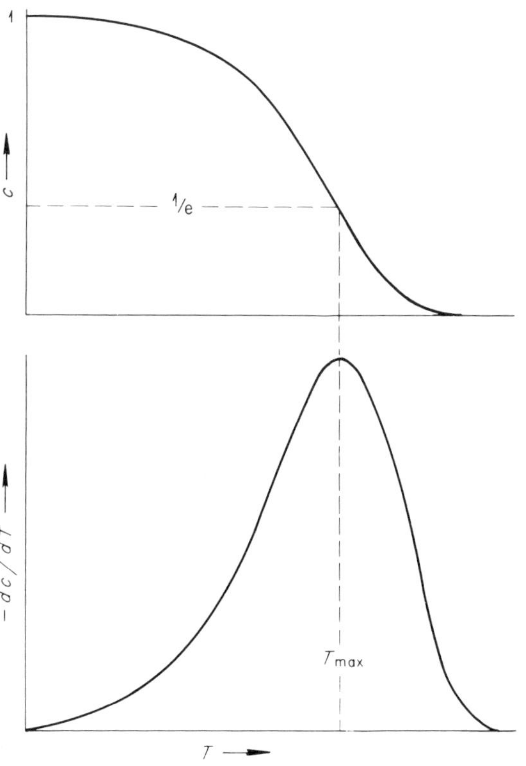

FIGURE 3-3. Electron trap population C and its reduction during linear increase of T.

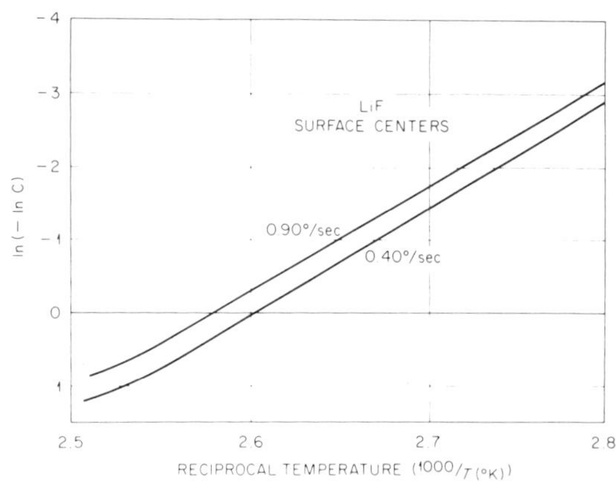

FIGURE 3-4. "Balarin-Zetzsche Curves" of the surface centers Σ of pure LiF. (After Holzapfel, 1970.)

(1972). In the case of an electrical neutral thin surface layer from which emission occurs, no TSEE takes place because each emitted electron is replaced by an electron from the surrounding vacuum; if, however, the sample charges up during emission, TSEE will occur.

3-1-4. Optical Stimulation

Depending on the depth of traps, exoelectron emission may also be optically stimulated by exposing the surface of the sample to light. OSEE is also called "induced photoelectric effect" and should not be confused with the fact that light — either ionizing, short wavelength UV, or extremely intense laser beams — can also populate traps, giving rise to spontaneous or stimulated EE. OSEE has not been studied nearly as much as TSEE, mainly because TSEE is considered easier to measure and "glow curves" are simpler to interpret.

A direct determination of the optical activation energy is, however, possible by the measurement of the wavelength dependence of the OSEE. An emission edge will be observed at a certain frequency corresponding to the transfer of electrons from traps to the conduction band (Figure 3-5). The existence of different trapping levels results in a step function of the EE rate plotted as a function of the photon energy if a comparable number of photons are absorbed in each trap and the quantum efficiency is assumed to be 100%.

The optical and thermal activation energies are not identical because of the different displacement energies of the ions in the crystal lattice in fields of different frequency. There is a strong effect of external electrical fields (polarization of the sample) on OSEE (Sujak and Gajda, 1968). Optical stimulation not only leads to the transfer of electrons from shallow traps into the conduction band, but may also result in the transfer of electrons from deep traps into (empty) shallower traps ("reactivation"). Another related effect is called "self-activation." It has been explained by a diffusion of electrons from deeper layers into the emitting surface (Holzapfel, 1969).

Alkali halides have been the subject of relatively intense OSEE studies by several groups (Bohun, 1970). One can distinguish between a "selective" OSEE, where the OSEE, if plotted as a function of wavelength, closely follows the corresponding optical absorption band, and a "non-

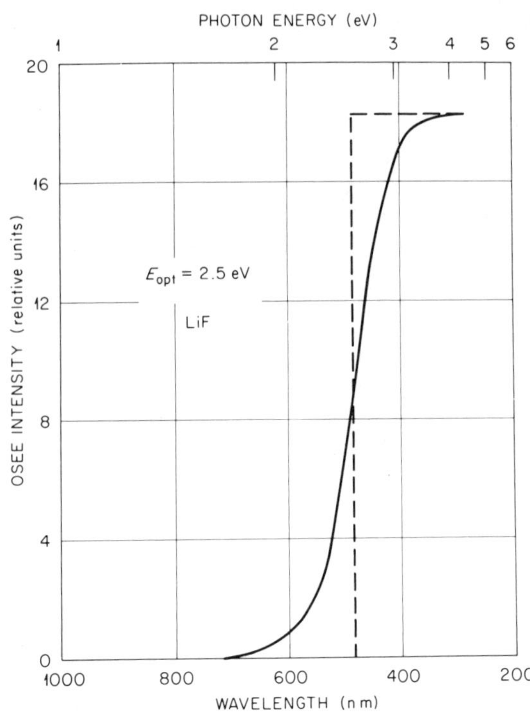

FIGURE 3-5. OSEE from a pure LiF sample as a function of the photon energy (wavelength), resulting in an optical activation energy of 2.5 eV. (After Holzapfel and Kramer, 1969.)

selective" OSEE, where such a relation does not exist. The former case was investigated in alkali halides (Petrescu, 1968). Several other emission mechanisms have also been suggested (Bohun, 1970). It should, however, be stressed that none of the various mechanisms and kinetic theories are suitable for the explanation of all EE effects, partly because of poor understanding of the physical processes involved and partly because of differences between the different types of traps involved.

3-2. Materials and Factors Affecting Their EE

Considerable efforts have been expended on studying the TSEE curves of a great many different materials (including rather exotic substances such as egg shells, hemoglobin, and moon dust) and on attempts to correlate the various peaks, as well as the factors affecting their location and intensity, with the production or destruction of well-defined electron traps. Those efforts have only been partially successful so far, but continuing progress is being made.

No attempt will be made here to review all the

scattered data in the literature, many of which have been obtained with ill-defined materials under less than reproducible experimental conditions (for example, the heating rate that affects peak locations frequently is not reported). As many different factors — in particular impurities and the thermal and radiation history — affect the TSEE curve and intensity of a given material, poor reproducibility of reported results is more the rule than the exception.

3-2-1. Halides and Sulfates

The most intensely studied alkali halide is lithium fluoride (Figure 3-6). There is one pronounced peak around 120 to 150°C which appears to be due to surface traps and a high-temperature peak around 270 to 350°C due to F and M centers (Gordan et al., 1969; Holzapfel, 1970; and Becker et al., 1970). If a lower heating rate is used for obtaining better peak resolution, additional peaks around 170 to 220°C are found in Mg,Ti activated TL-grade LiF (Attix, 1970 and Nash et al., 1971).

Several factors may affect the TSEE from LiF. For example, there is a LET dependence of the peak ratios and an effect of preheating which has been studied in some detail in LiF:Mg,Ti (see, for example, Kopp, 1971 and Burke and Beck, 1970). The TSEE curve also changes during storage of the irradiated sample (Nash et al., 1971). In a very pure LiF sample, a peak at 200°C was observed only during the first heating cycle. This has been attributed to desorption processes. Optical stimulation of EE in LiF is, of course, also possible

(Kaul and Sutter, 1970). In comparing the TSEE of the fluorides, chlorides, bromides, and iodides of lithium, sodium, and potassium, it was found that the temperature of the main peak maximum is approximately proportional to the square of the lattice constant (Kramer, 1963). Various other authors working with alkali halides from different sources with a different impurity content, thermal history, grain size, surface structure, etc. reported other results. As an example, TSEE in a commercial grade NaCl is compared in Figure 3-7 with the TL in this material. A compilation of peak temperatures and the dissociation of color centers, atomic activator centers, and holes ascribed to them in alkali halides is given in Table 3-1.

Calcium fluoride is the best-studied alkaline earth halide. "Chemically pure" CaF_2 showed three TSEE peaks at $\sim75°$, $155°$, and (most pronounced) $240°C$, corresponding to thermal excitation energies of 0.71, 0.91, and 1.03 eV, respectively (Sujak and Gieroszynska, 1967). Only one strong TSEE peak at $\sim290°C$ was observed in a Mn-activated CaF_2, as used for TLD by some authors, but a more complex TSEE curve by others. TL and TSEE do not completely agree in their characteristics (probably Y, Ho, Sm, and Tm impurities are acting as electron traps in this material — Puite and Arends, 1971; for the prompt and delayed TSEE from natural fluorite, see Nash et al., 1971). In BaF_2, maxima around $125°$, $180°$, and $326°C$ have been reported; in SrF_2 they occur around $118°$, $253°$, and $383°C$ (Sujak and Gasior,

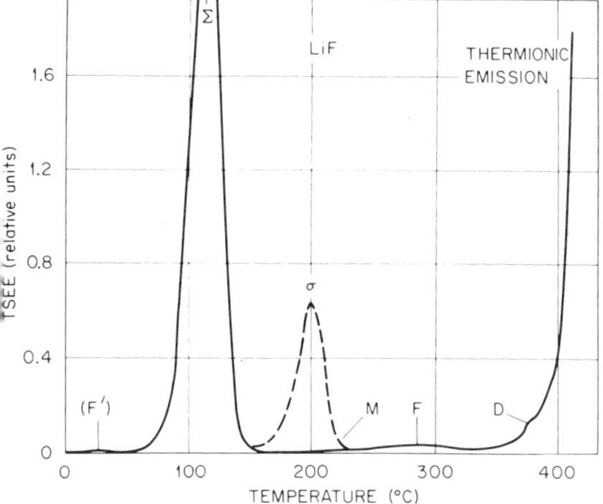

FIGURE 3-6. TSEE curve of very pure lithium fluoride; the dotted peak is only observed during the first heating cycle. (After Holzapfel, 1970.)

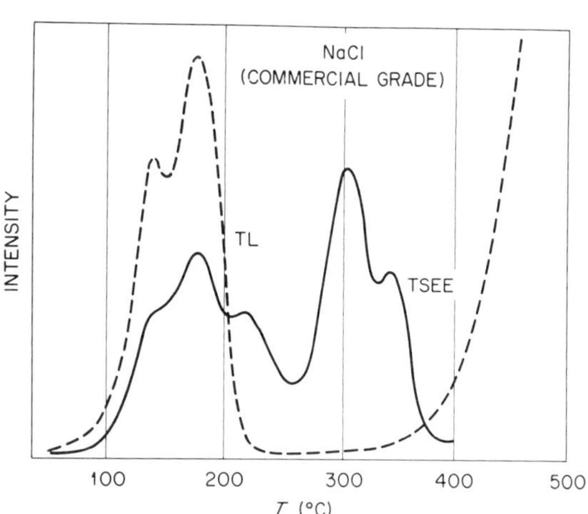

FIGURE 3-7. TSEE and TL curves of a commercial grade NaCl. (After Lewis, 1966.)

Proposed Correlation Between Observed TSEE Peaks (°K) and Centers in Unactivated and Activated Alkali Halide Crystals

Substance	V_i	F_i	V_k	In°	Ti°	Ag°	V_F	F
NaCl			170*					545
NaCl-Ti			180		220			
NaCl-Ag			160			230		
NaCl-Ca							400	
KCl	130	220*						540
KCl-Ti	130				310			
KCl-Ag	135	220*				345		
KCl-In	135		205	230				
KBr		150	170				235*	500

*Identification not certain
(After Bichivin and Käämbre, 1970.)

1968). In $BaCl_2 \cdot 2 H_2O$, TSEE can be stimulated by the loss of water, but there is also a real TSEE.

TSEE in $CaSO_4$ has been the subject of numerous investigations; various peaks between 100 and 430°C have been reported for pure and activated samples. An analytical grade material exhibited a response, as shown in Figure 3-8. Attempts to "activate" $CaSO_4$ with Mn, Sm, Pb, etc. (Novotny et al., 1970; Rotondi, 1970; and Nash et al., 1971) did not lead to conclusive, reproducible improvements.

Maximum sensitivity is, however, obtained by dehydration of the $CaSO_4$ (heating for about 100 min to 900 to 960°C). This effect is slowly reversible in humid atmosphere, accounting for frequently observed abnormally high fading in alkaline earth sulfates. Protection against humidity reduces the apparent fading rate (Figure 3-9) in such materials (Kramer, 1968 and Becker and Chantanakom, 1970). An effect that appears to be rather common in TSEE materials, namely, that the height ratio of the TSEE peaks depends on the LET of the radiation, was first observed in $CaSO_4$ (Becker and Chantanakom, 1968). It follows that the fading kinetics also is different, depending on the LET of the radiation.

In $BaSO_4$, exposure to O_2 and H_2O has a strong effect on the TSEE curve (Kriegseis and Scharmann, 1969). There are several peaks between 120 and 190°C. Strontium sulfate has a main peak at $\sim$250°C. There have been a number of TSEE studies with ZnS and, to a lesser degree, with CaS and SrS with or without the addition of activators such as Pb and Cu (see, for example, Levshin and Pipins, 1963). When heated in a reducing atmosphere, various TSEE peaks up to 350°C were observed in SrS, up to 250°C in ZnS, and $\sim$220°C in CaS:Sm. It seems that the TSEE characteristics of sulfides particularly depend on impurities and the method of preparation. For recent studies on the exoelectron emission from K_2SO_4, see Brunsmann and Scharmann (1973).

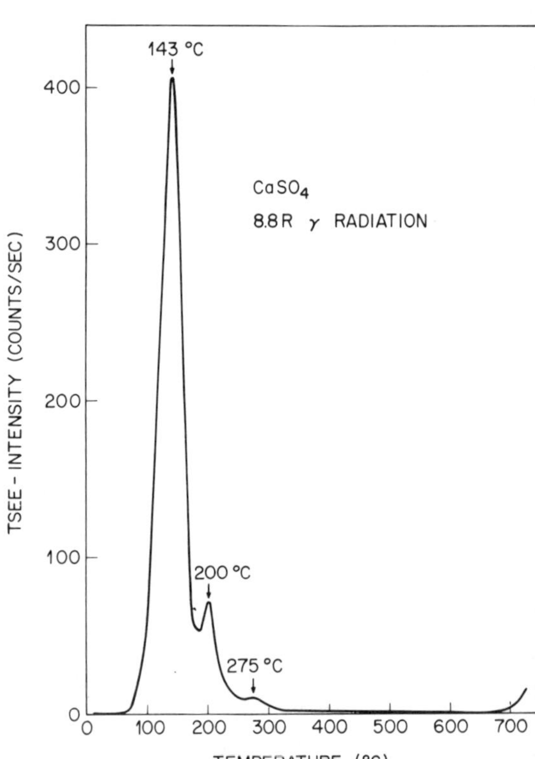

FIGURE 3-8. TSEE of an analytical grade $CaSO_4$ after 1 hr preheating at 940°C.

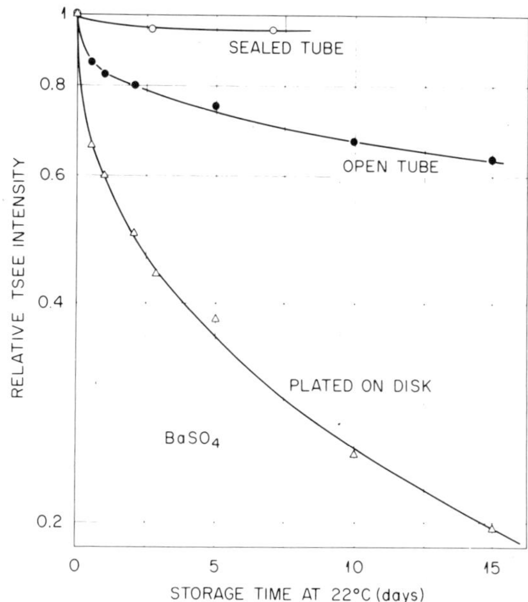

FIGURE 3-9. Fading of the radiation-induced TSEE in BaSO₄ at room temperature if the material is kept in a sealed tube, an open tube, and plated on the surface of an unprotected flat disk. (After Kramer, 1968.)

3-2-2. Metal Oxides and Complex Compounds

Because it is hard, thermally and chemically resistant, has a low atomic number, and some other desirable TSEE characteristics, such as high sensitivity and convenient peak location, BeO has been studied intensely in recent years, mainly from the dosimetric point of view. The high toxicity of finely dispersed unfired BeO powders apparently disappears as a result of high-temperature treatment.

In the initial studies, unfired or high-fired BeO powders have been mixed with graphite powders and plated on graphite planchets (Kramer, 1966; Holzapfel, 1968; Becker, 1968; Becker et al., 1970b; and Kriks, 1971). Both peak location and sensitivity of BeO from different sources were found to depend strongly on the material's thermal history, grain size, and impurity content. Some sintered, rather impure samples exhibited the highest sensitivity (Becker et al., 1970b and Figure 3-10). Later, it was discovered that certain ceramic beryllium oxides that are manufactured in different sizes and shapes under the trade name of Thermalox 995® (Brush Beryllium Co., Elmore, Ohio) are particularly sensitive.

These BeO ceramics have since been the subject of numerous studies at ORNL (Becker et al., 1970a and b; Crase et al., 1971; Becker et al., 1971; Gammage et al., 1971; Nagpal and Gammage, 1972; Moreno et al., 1972; and Becker 1972b) as well as in some European laboratories (Rotondi, 1971; Rasp and Siegel, 1972; Boros and

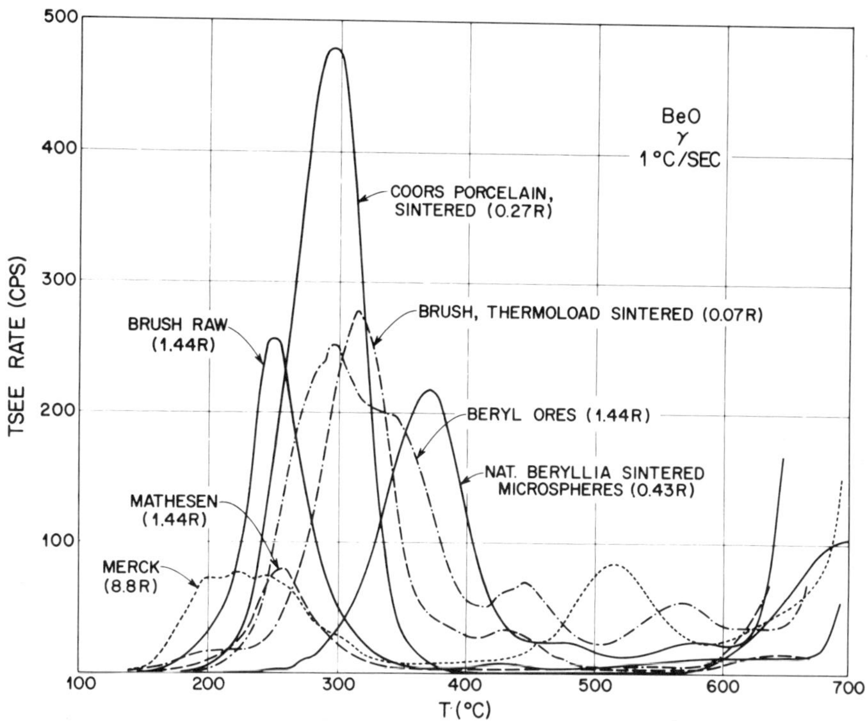

FIGURE 3-10. Comparison of TSEE curves of BeO powders or microspheres from different sources. (After Becker et al., 1970b.)

Szy, 1972; Ambrosetti and Oberhofer, 1972; and others). Some of the results are

a. If samples were coated with thin layers of Au, Pd, or Pt and heated, an increase of the main peak at ~325°C was observed. It disappeared during extended heating of the samples to temperatures exceeding ~600°C due to a coagulation of the metals into microcrystals along growth steps, resulting in a breakdown of the surface conductivity (Becker et al., 1970a).

b. In general, the TSEE peak at ~325°C appears to be closely associated with the surface conductivity, which is affected by the presence of SiO_2 in the emitting surface. The peak can be destroyed by removal of the SiO_2, and restored by reimpregnating the surface with SiO_2 (Gammage et al., 1971). Among other "activators" of interest are Li and Mn (Gammage et al., 1971 and Regulla et al., 1971), but the adsorption of various other gases, in particular of oxygen, also strongly affects the emission characteristics (Nagpal and Gammage, 1973 and Euler et al., 1973).

c. Several hours of preheating to 1,400°C increased the sensitivity substantially (Figure 3-11).

d. Unlike beta, gamma, and X-radiation, particles such as deuterons or alpha particles create semipermanent changes in the TSEE peak ratios. In Thermalox 995[®], particle irradiation in combination with a special heat treatment largely destroys the peak at ~325°C and strongly enhances a peak at ~450°C (Crase et al., 1971).

Exoelectron emission from BeO can also be stimulated optically (Ford et al., 1970). Sintered mixtures of BeO powder, ZnO, and SiO_2 have been studied in some detail by Brunsmann et al. (1972). Peaks around 340°C were observed; some of the compounds were apparently more sensitive than their constituents. Of the various modifications of Al_2O_3, without or with activators such as Cr, only TSEE from high-fired Al_2O_3 may become of dosimetric interest (Chryssou and Holzapfel, 1971). In CaO, a main peak at 180°C has been described (Bohun, 1962). In NiO, peaks at 160 and 260°C have been observed, in CuO at 140 and 200°C, and in ZnO, between 70 and 300°C.

In ZnO, there is a substantial difference in the TSEE (as well as the OSEE) curves between samples prepared in an excess of oxygen and those heated in H_2 (Figure 3-12); apparently the main maximum in the "oxidized" sample is due to physically or chemically adsorbed oxygen (Holzapfel and Nink, 1970). Perhaps negative ions are emitted from ZnO, indicating desorption processes. Adsorption and desorption of oxygen seem to play a major role as a surface trap in many metal oxides.

TSEE from other metal oxides has been used as a diagnostic tool for studies of the oxidation kinetics of metal and semiconductor surfaces. It is generally believed that pure semiconductor sur-

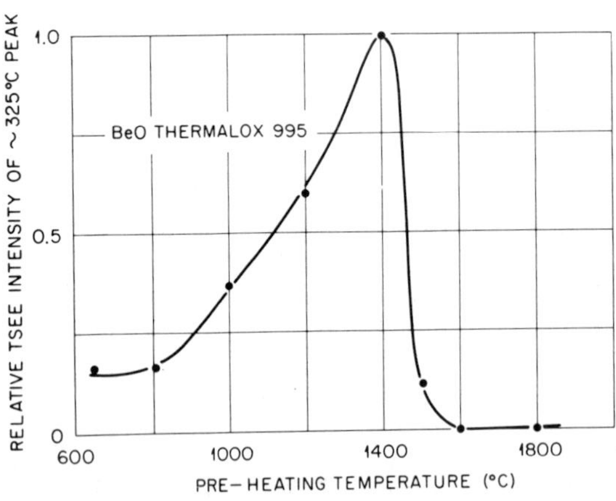

FIGURE 3-11.   Relative TSEE intensity of the ~325°C main peak in ceramic BeO (Thermalox 995) as a function of the preheating temperature, 4 hr heating time. (After Gammage et al., 1972.)

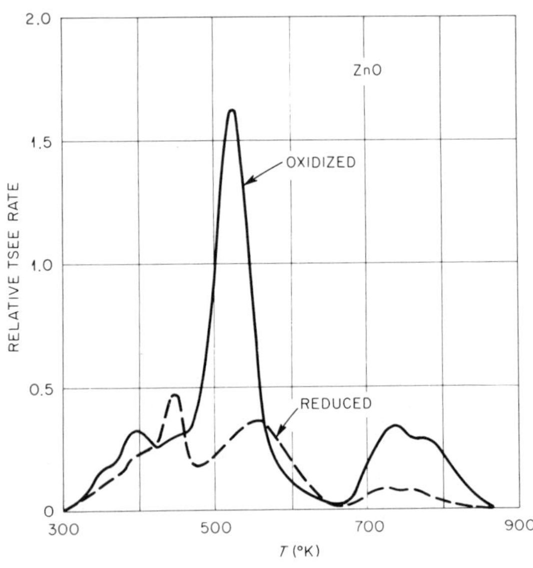

FIGURE 3-12.   TSEE from ZnO prepared with an excess of oxygen, and after annealing in H_2. (After Holzapfel and Nink, 1970.)

faces, like those of metals, do not emit exoelectrons, but that the observed TSEE from Si and Ge is also caused by thin oxide layers or adsorbed oxygen. For a good summary of these studies, see Minz et al. (1969) and Scharmann et al. (1970).

Only scattered information is available on the EE from more complex compounds. TSEE has been observed in glasses such as quartz glass, silver activated phosphate glass, and lithium borate (Gourgé, 1958; Becker, 1968; and Nash et al., 1971). EE that can be stimulated by an electrical field was found in certain ferromagnetic salts such as Seignette salt and barium titanate (Sujak and Kusz, 1965).

TSEE and OSEE also occur from many natural minerals. According to one recent study (Lausch and El Naggar, 1972), ceramic pellets of barium titanate activated with 0.03 M Sb, which are prepared by sintering mixtures of $BaCO_3$ or BaO with TiO_2 at 1,320°C, exhibit good sensitivity in a double-peak around 500°C. For example, limestone exhibits main TSEE peaks at $\sim$120 and 320°C (Lewis, 1966); clay (*bolus alba*) shows a peak at $\sim$230°C during heating and a "cooling-down" peak at $\sim$105°C (Kramer, 1970); a number of minerals such as granite, dunite, andesite, and granodiorite also exhibit TSEE (Becker, 1970 and Figure 3-13).

Some minerals such as granodiorite even show TSEE without previous irradiation, perhaps due to a self-dosing because of their content of radioactive elements. TSEE was also observed, without or with additional treatment such as preheating in air or etching, in lunar dust and rocks (Gammage and Becker, 1971). Even human bones (Brown and Dalton, 1972) and teeth may exhibit TSEE.

Some organic salts also exhibit TSEE. In quinine sulfate, dehydration is associated with intense electron emission. Even irradiated hemoglobin, either pure or associated with aminothiols, shows TSEE (Vladimirov and Vladimirov, 1968). Naturally, TSEE studies in organic compounds are limited by the low melting and/or disintegration points of most materials. Little OSEE work has been done with these substances which (due to their high hydrogen content) could be of interest in fast-neutron dosimetry. Some results with polymers (Arabin et al., 1972) do not appear to be promising.

From these data, it is obvious that a TSEE curve is by no means "characteristic" for a given chemical compound. The lack of agreement in the results of different investigators can frequently be explained with trivial experimental difficulties (it is, for example, not easy to determine the actual temperature of an emitting surface within a few °C), but in many cases there remain substantial differences which are of a more complex nature.

3-2-3. Effect of Preheating

The effect of preheating seems to be closely related to changes in surface structure, particle size, and electrical conductivity of the surface. The role of these parameters is little understood. Apparently EE occurs mainly, or exclusively, from disturbed regions in a crystalline surface such as edges, growth steps, and grain boundaries. Lack of electrical conductivity may inhibit EE because of the residual positive charge build-up at the surface during the emission process. If the total TSEE of an insulating surface is plotted as a function of the radiation dose, this inhibition effect is frequently reflected in a less than proportional increase of TSEE with dose (see, for example, the TSEE peak at $\sim$325°C in ceramic BeO:Si, Figure 3-14). This nonlinearity is reduced by decreasing the heating rate.

The charge build-up can also be avoided by the addition of sufficient amounts ($\gtrsim$10 to 20%) of conductors such as graphite powder to the powdered emitting substance (Kramer, 1966). Thin metal layers can have a similar effect, but also may reduce the EE due to their electron absorbing properties (Becker et al., 1970a). In many experimental studies about the effect of heat treatment, activators, etc. on the EE characteristics of various materials, it is difficult to decide whether the observed changes are actually due to those parameters, or simulated by changes

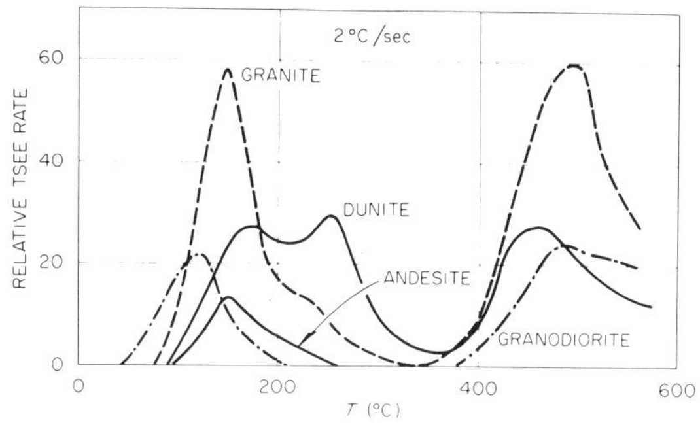

FIGURE 3-13. TSEE from different rock powders, mixed with graphite, after exposure to gamma radiation. (After Becker, 1970.)

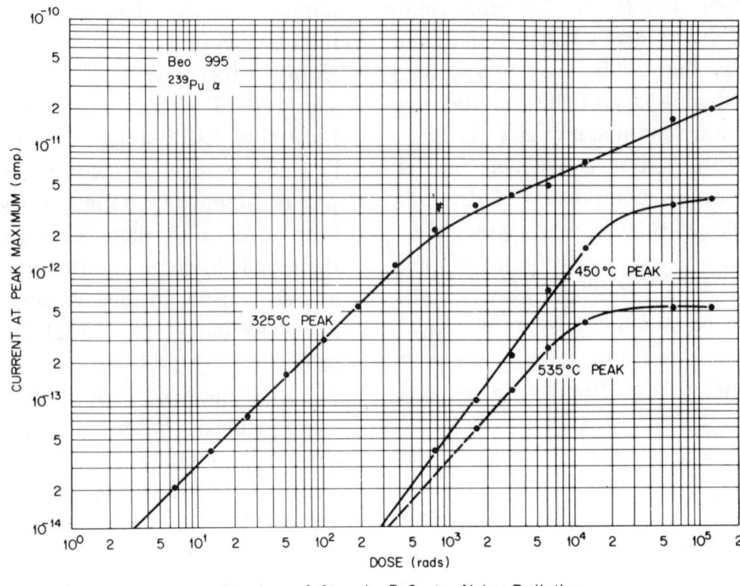

FIGURE 3-14.   TSEE from different peaks of ceramic BeO (Thermalox 995) as a function of gamma radiation dose. (After Crase et al., 1971.)

in the sample's conductivity, its surface structure, or the work function. Other experimental factors such as the electric field in the counter also affect the results.

Frequently, a substance heated for constant durations to increasing temperatures exhibits increased sensitivity up to some critical temperature, followed by a decrease in sensitivity (Figure 3-11). In many cases these sensitivity changes can be explained in terms of well-defined chemical reactions, such as dehydration, formation of new compounds, or decomposition. Some of those processes are obviously reversible. For example, $CaSO_4$ which has been sensitized by heating for 1 hr to 940°C (Hanle et al., 1966) slowly absorbs crystal water from a humid atmosphere, but may be resensitized by heating. A similar effect has been observed at higher temperatures ($\sim$1,000 to 1,400°C) in BeO. In some cases, the changes in the TSEE sensitivity correspond to similar changes in the TL.

In lunar dust from Apollo 12, oxidation and the removal of trapped solar wind occurred during heating, resulting in qualitative and quantitative changes of the TSEE curve (Gammage and Becker, 1971). In the case of mixtures of different materials that undergo a solid-state chemical reaction at a certain temperature, optimized TSEE may indicate the maximum reaction rate. In other materials, however, the effect of heat treatment is not as easily explainable (heating of LiF, NaCl, and CaF_2 results in an increase in sensitivity). In many cases the existing peaks are not simply enhanced (or reduced) by the heat treatment, but qualitative changes occur as well.

3-2-4. Effect of Pre-irradiation

Exposure of exoelectron emitters to doses of ionizing radiation not exceeding 10^2 to 10^3 rad induces a signal which can be totally annealed by optical or thermal stimulation, without permanently affecting its emitting properties. Much higher doses, in particular of densely ionizing radiation, affect their behavior semipermanently (the radiation damage can be thermally annealed), or permanently due to the creation of new traps (Gourgé, 1958 and Hanle et al., 1961). As an example, the effect of deuteron pre-irradiation on the gamma radiation response of ceramic BeO is illustrated in Figure 3-15.

Pre-irradiation of the ceramic BeO with low LET radiation up to several Mrad does not affect its TSEE. In other materials such as LiF, NaF, and CaF_2, however, existing TSEE maxima can be increased by preexposure to x-rays. With increasing doses, new TSEE maxima may appear. F centers frequently act as electron donors in TSEE (Bohun, 1963). In the case of the well-studied

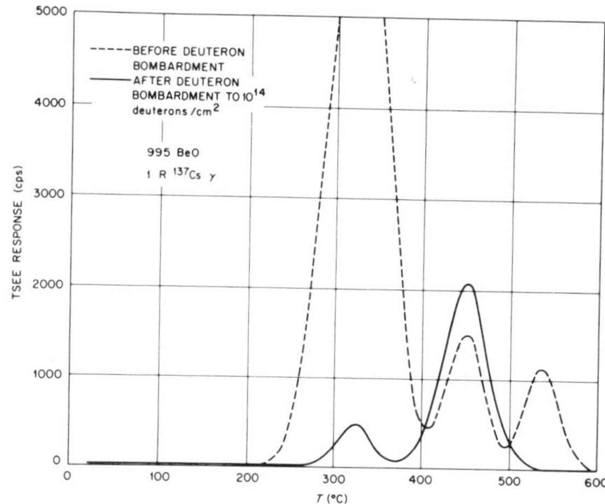

FIGURE 3-15. TSEE of ceramic BeO (Thermalox 995) before and after pre-irradiation to $\sim 10^{14}$ deuterons/cm², annealing at 700°C, and exposure to 1 R gamma radiation. (After Crase et al., 1971.)

alkali halides, there is a close relation between the disappearance of the F and M bands, and the TSEE and TL emission maxima. Pre-irradiation with x-rays also changes the TSEE curve of MgO and CaO.

There are several other effects of high doses of low LET radiation on EE. Gas discharges at the surface of exoelectron emitters modify their TSEE intensity (Figure 3-16); exposure to saturation dose, followed by partial annealing, may increase the TSEE intensity from a low-temperature peak in BeO by a factor of 4 to 7 (Kramer, 1968). Apparently complete filling of deeper traps prevents them from competing with shallower traps.

3-2-5. Impurities and Other Factors

The effect of impurities on TSEE is complex. One could speculate that activators that strongly affect TL (Mn, Ag, Cu, and rare earths) should not act as electron donors of sufficient depth and, therefore, would be ineffective. Indeed, some early studies indicated that the addition of Mn to $CaSO_4$ increases its TL significantly without affecting the TSEE. It was later demonstrated that this behavior is more the exception than the rule. For example, in ZnO, the addition of Ga_2O_3 increases the sensitivity, whereas the addition of Li_2O decreases it (Menold, 1960). In ZnS even very small concentrations of Cu have a pronounced effect on sensitivity; the earlier work on $CaSO_4$:Mn has not been confirmed by others who found an increase of the TSEE in $CaSO_4$ by the addition of Mn.

Only in a few cases, however, can the primary

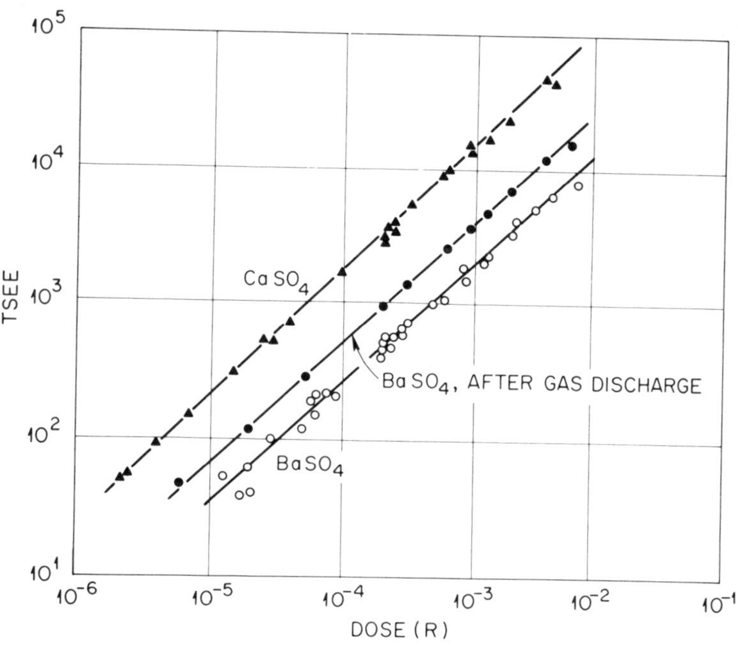

FIGURE 3-16. Response of a TSEE dosimeter based on $BaSO_4$ (closed tube, inside walls covered with $BaSO_4$:C) before and after a gas discharge in the tube, and of a $CaSO_4$ dosimeter of the same type to low gamma radiation doses. (After Kramer, 1968.)

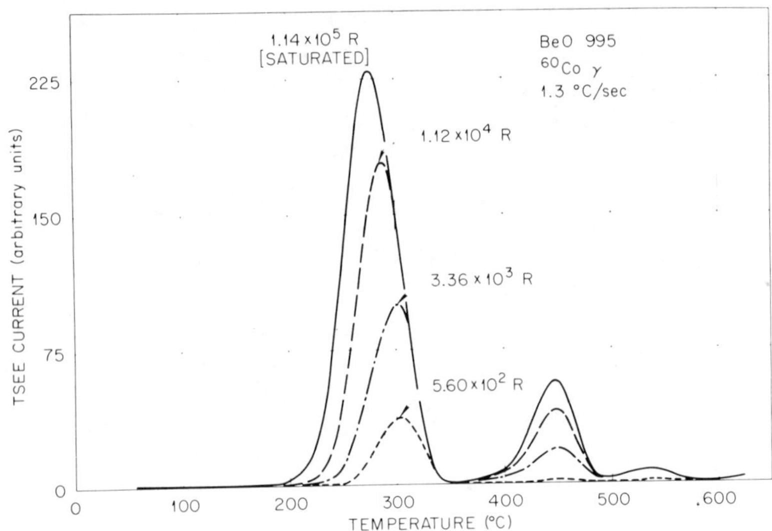

FIGURE 3-17.   TSEE response of ceramic BeO (Thermalox 995) after exposure to different gamma radiation dose-levels. (After Crase et al., 1971.)

effect of "activation" be clearly distinguished from secondary effects on TSEE such as changes in surface structure and electrical conductivity. In ceramic BeO that has been studied in great detail at ORNL, it seems possible to distinguish between "activators" which act through the creation of electron traps directly contributing to exoelectron emission (such as B, Ca, and Mn) and others with an indirect effect, namely, the promotion of emission from inherent traps by modification of such factors as surface conductivity, work function, or irregularities of the crystal surface (Moreno et al., 1972).

Other parameters affecting TSEE curves are

1. Heating rate. Increase in the heating rate produces narrower, higher peaks and higher peak temperatures;

2. Dose. The peak height ratio often depends on the radiation dose-level, some peaks growing faster and/or saturating earlier than others. Furthermore, the peak temperature may decrease with increasing dose (Figure 3-17);

3. LET. In many materials, peak height ratio as well as peak location depends on the LET of the radiation;

4. Storage time and temperature. Low-temperature traps fade more rapidly than high-temperature traps, thus creating a change in the peak height ratio which becomes more pronounced if the storage temperature approaches the peak temperature (Figure 3-18). The temperature during radiation exposure apparently has an influence already at lower temperatures (Nagpal and Gammage, 1972);

5. Light exposure. Visible light anneals surface traps by optical stimulation, and UV

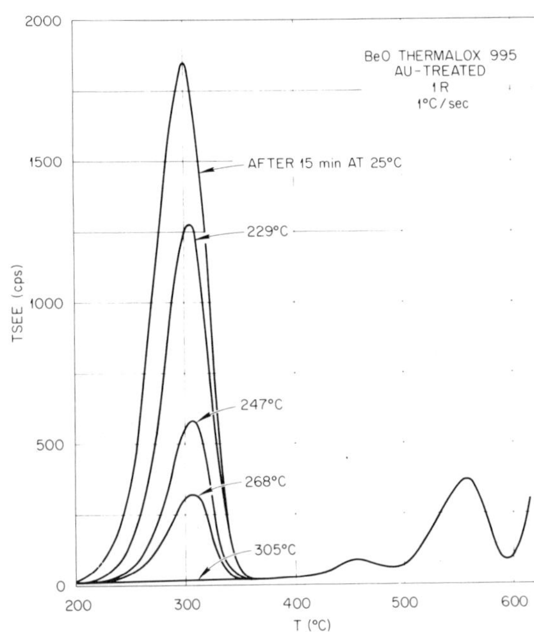

FIGURE 3-18.   Fading of the main peak in ceramic BeO during storage for 15 min at various temperatures. (After Becker et al., 1971.)

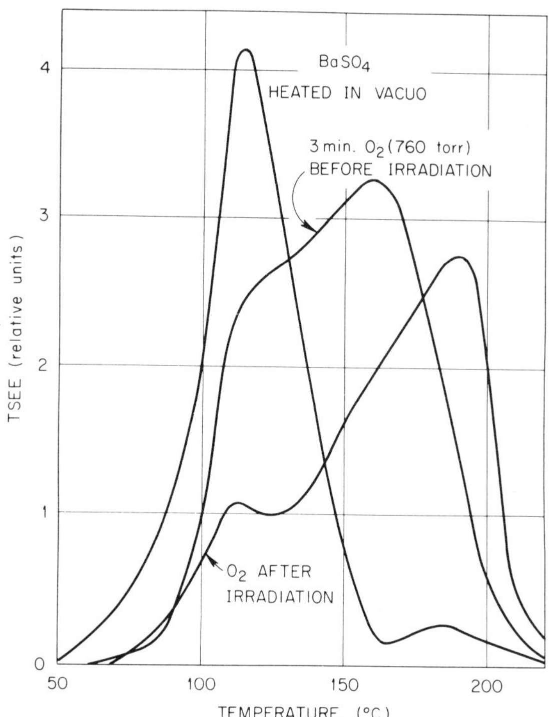

FIGURE 3-19. Effect of exposure to O_2 prior to, and after gamma irradiation on the TSEE curve of $BaSO_4$, whose surface had been previously cleaned by heating in vacuo. (After Kriegseis and Scharmann, 1969.)

induces as well as anneals signals. Exposure to sunlight may lead to both erasure of existing signals or induction of new ones, depending on the detector's irradiation history;

6. Humidity and gases. It can be shown that sorption of atmospheric constituents such as O_2 or H_2O both prior to and after irradiation (Figure 3-19) have pronounced effect on TSEE characteristics (see, for example, Kriegseis and Scharmann, 1969 and Nagpal and Gammage, 1972); and

7. Strange "memory" effects that could not be explained on the basis of gas absorption or humidity uptake have been observed in ceramic BeO under certain conditions (Becker, 1972b).

3-3. Instrumentation

With one exception — a simple TSEE reader for one type of TSEE dosimeter* — no commercial TSEE or OSEE readout instruments are available yet. Good instrumentation is, however, very

*Made by Atomika, Munich, Germany

important for obtaining reliable results. Several early investigators employed "point counters." These are a type of G-M counter which employs, instead of a wire, a symmetrical platinum sphere of ∿0.2 mm diameter at the end of a thin Pt wire and is operated in plain air (Sujak, 1962) or filled with an air/alcohol vapor mixture. Its properties can be improved with the use of self-quenching gases such as methane. A wealth of information on the design and operational characteristics of different types of such point counters for exoelectron measurements can be found in the literature (see, for example, Stepniowski, 1966; Frank et al., 1966; and Momose and Tamai, 1968).

Most modern investigators prefer gas-flow G-M or proportional counters with built-in heaters, operated with a mixture of 90 to 99% helium or argon with 10 to 1% ethylene, butane, or iso-butane (Ku and Pimbley, 1961; Petrescu, 1965; Kopp, 1970; Burke and Beck, 1970; Lillicrap, 1970; Niewiadomski, 1971; Becker, 1972b; and others) as counting gases. G-M counters for TSEE work have also been operated over the surface of an organic liquid whose vapors are quenching (Sujak et al., 1967).

The properties of gas-flow G-M counters with a straight or loop counting wire appear to be generally superior to those of point counters. Several types of commercial gas-flow G-M counters that had originally been built for the measurement of soft beta emitters have been modified for exoelectron measurements by installing a heater similar to the TL heaters and a thermocouple (Becker and Robinson, 1968). The excellent plateau of such counters varies only slightly with increasing temperature up to ∿400°C.

If higher temperatures are desirable, some additional measures become necessary. The heater has to be designed in such a way that no disturbing thermionic electron emission occurs from heated components (in this case, no background increase due to thermal electron emission will be observed up to approximately 670 to 750°C). Furthermore, it may be advisable to have the heated parts somewhat separated from the counting volume, possibly with a cooling system to prevent excessive heat transfer to the counting gas. Although a G-M counter is less temperature dependent than a proportional counter, there is a noticeable shift in the plateau if the temperature exceeds a few

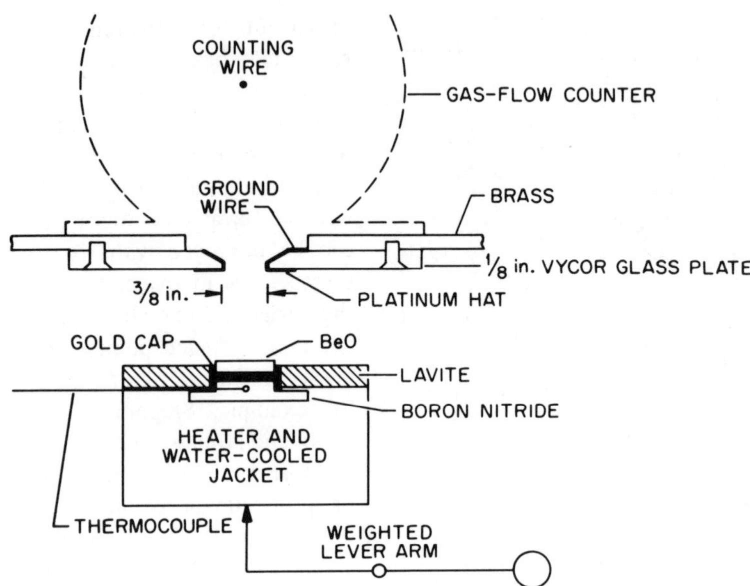

FIGURE 3-20. Diagram of TSEE counter for higher temperatures (<750°C) with water cooling and rapidly interchangeable samples. (After Gammage et al., unpublished.)

hundred °C. A cross section of such a high-temperature counter with water-cooling that is used at ORNL is given in Figure 3-20.

The distance between emitting surface and counting volume should be well reproducible. In some counters, an accelerating potential applied on a grid over the sample helps to inject the exoelectrons into the counting volume; such "suction potentials" may not only increase the total EE from a given sample, but also lead to a small shift in the peak location. A variation of such an exoelectron ejection aid is employed in the ORNL counter (Figure 3-20). A "pushing" potential of about −60 V to 100 V is applied to a metal cap at the bottom of the (electrically insulating) ceramic BeO sample, resulting in (a) a threefold increase in sensitivity, (b) a reduction of the standard deviation for multiple readings of the same detector from 4.7 to 3%, (c) smoothening of the curve appearance (Figure 3-21), and (d) disappearance of various disturbing "dirt effects" (Gammage and Cheka, to be published).

The counters are usually operated in the G-M region because of the larger signal. However, deadtime losses at high emission rates limit the usability of the G-M counter. With a deadtime of 200 μsec, for example, corrections for count losses become necessary around 10^5 cpm (in comparison, the deadtime of a proportional counter is $\sim$0.1 μsec and the counter electronics become the limiting element).

In several TSEE readout instruments use has been made, therefore, of high-amplification proportional counters, with the counting wire as a loop (Attix, 1970), a straight wire (Moreno et al., 1972), or as a point counter (Fintelmann, 1968). The counting gas could be methane or a mixture of methane with argon or helium. There is only a small difference between the plateaus of external gamma radiation and internal beta radiation, but a substantial difference for the exoelectron plateau in proportional as well as G-M counters (Kriks, 1971 and Figure 3-22).

The heater plate consists of a material which combines good thermal and poor electrical conductivity. The sample holders should be grounded (earthed) electrically and be good thermal conductors. A common problem in TSEE studies is the determination of the actual temperature of the emitting surface because this surface represents the interface between the cool counting gas and the sample which is heated from below. With sophisticated differential heating technique employing two identical counters, of which one is used to automatically correct the temperature gradient of the other (Figure 3-23), it has been possible to obtain accuracies within a few degrees (Holzapfel, 1970).

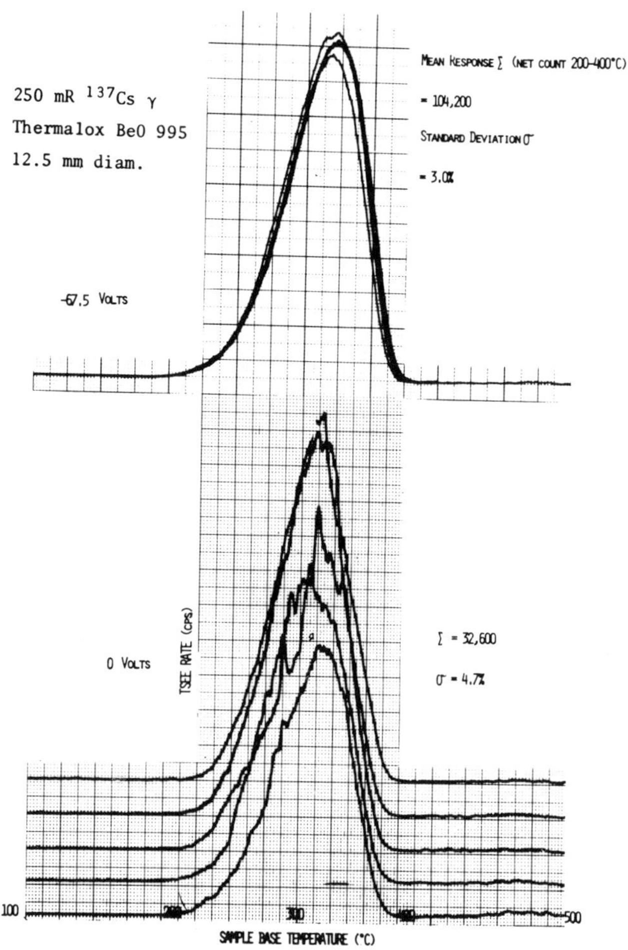

FIGURE 3-21. Effect of a "pushing potential" of −67 V on the back of a ceramic BeO sample on TSEE peak height, shape, and reproducibility. (After Gammage and Cheka, unpublished.)

As in TLD studies, it is advisable to provide for linear heating with a choice of variable heating rates. Several more or less sophisticated linear heating devices have been used for TSEE readers. An ORNL design (Thorngate, 1969) permits precise linear heating rates between 0.5 and 10°C/sec. The block diagram of the TSEE reading unit and the linear heater is given in Figure 3-24.

TSEE readers can be designed to permit the simultaneous measurement of TL (Puite and Arends, 1971 and others). Such readers help minimize errors if there is an interest in comparing TL and TSEE curves. A variety of special G-M and proportional counters have been developed for exoelectron dosimetry. Various suggested TSEE reader designs permit the rapid change of samples (Engelland, 1969), optical as well as thermal stimulation (Kriks, 1971), and easy changeability of samples as well as good reproducibility (repeated reading of the same sample can be done with a reproducibility of a few percent — Figure 3-20).

In some of the earliest observations of EE, ionization chambers and electrometers were used for quantitative exoelectron measurements. This method still has advantages for measurements of intense TSEE up to the saturation level (Kramer, 1968). A simple electrometric device with a gold collection plate over the heated sample and a vibrating reed electrometer (Figure 3-25) has been used for TSEE current measurements in the 10^{-14} to 10^{-10} A range, thus covering with a combination of a G-M counter and an ionization chamber almost ten decades in TSEE intensity (Crase et al., 1971 and 1972). A similar reader has also been used by LaRiviere and Tochilin (1972).

All the devices described have in common that the samples are handled in a normal atmosphere

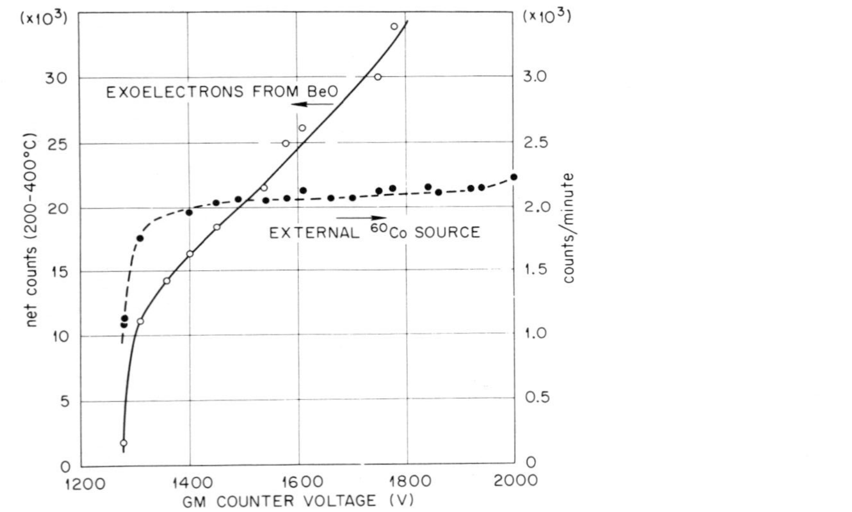

FIGURE 3-22. Plateaus for external gamma radiation and exoelectrons from BeO in a G-M counter. (After Becker, 1972b)

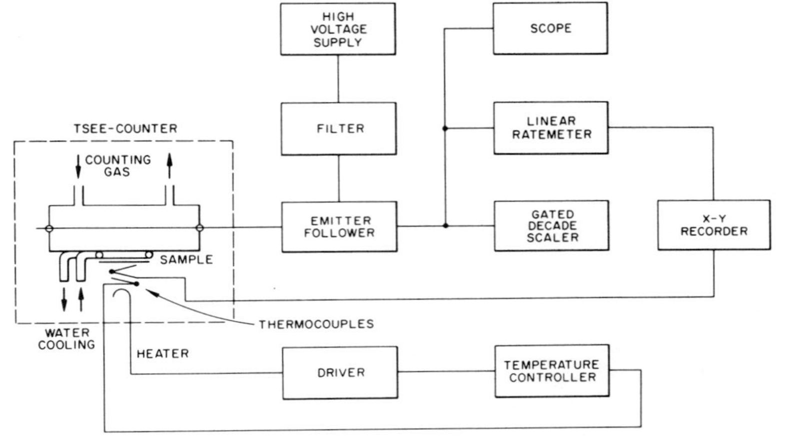

FIGURE 3-23.   Diagram of TSEE reader with reference heater to correct for the temperature gradient in the sample, for the determination of the precise temperature of the emitting surface. (After Holzapfel, 1970.)

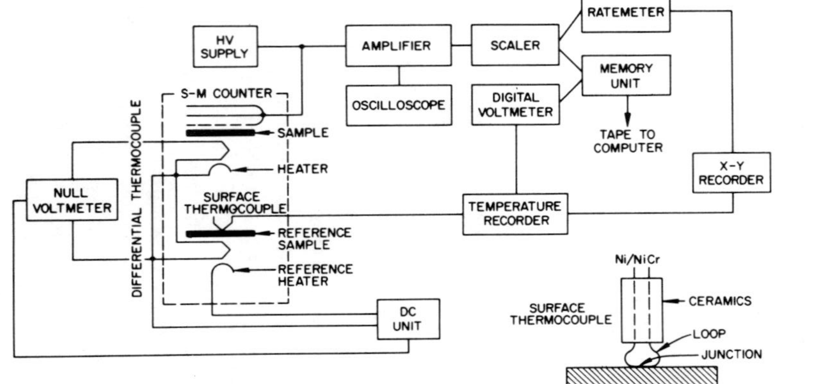

FIGURE 3-24a.

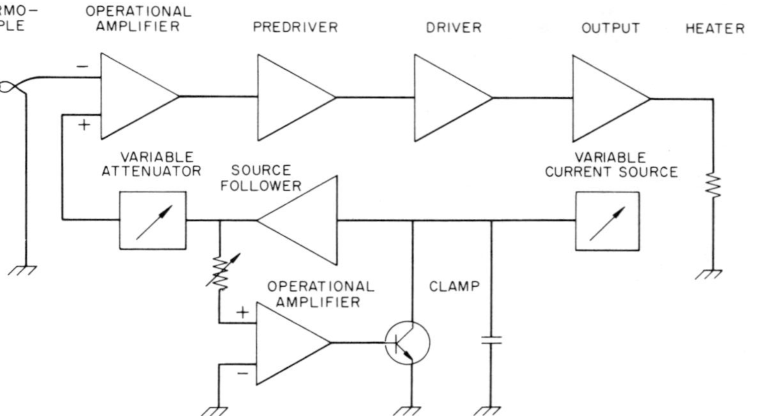

FIGURE 3-24b.

FIGURE 3-24a. Diagram of the ORNL TSEE counter (after Oberhofer and Robinson, 1970a), and b. of the ORNL controller for the linear heater with variable heating rate. (After Thorngate, 1969b.)

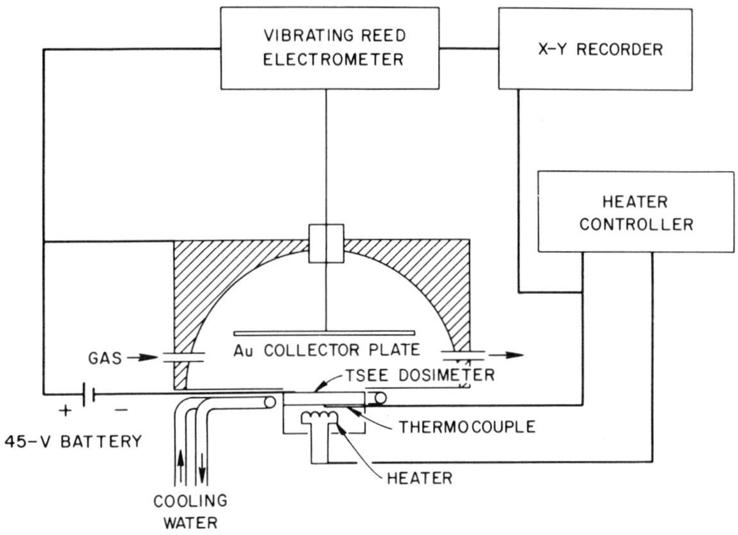

FIGURE 3-25. Diagram of ORNL ionization chamber for the measurement of high TSEE rates. (After Becker et al., 1971.)

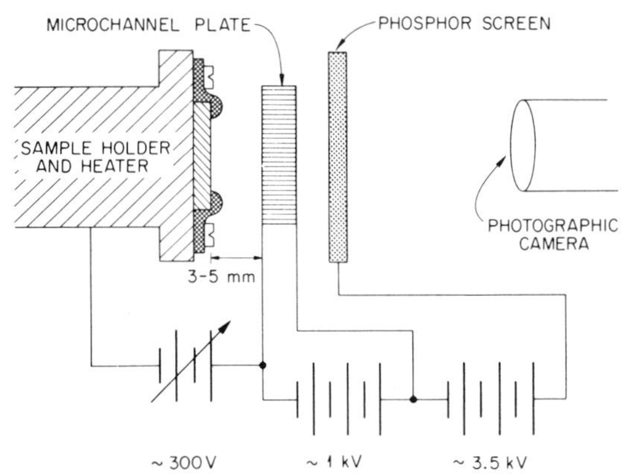

FIGURE 3-26. Diagram of instrument for visualization of the spatial distribution of exoelectron emission. (After Bräunlich, 1971.)

and heated in the counting gas. This may lead to an intolerable interference with the surface characteristics of the sample if "clean," freshly created surfaces or instable compounds are to be investigated. In such cases, exoelectron studies have to be carried out in ultrahigh vacuum and an electron multiplier is employed for detecting the accelerated exoelectrons. A scintillator screen/photomultiplier combination has also been considered.

Electron multipliers have been used widely, but sample-changing remains a rather time-consuming process (see, for example, Onsgaard and Hougs, 1973). An interesting variation of the simple multiplier, namely, a microchannel plate consisting of a bundle of closely packed channeltron multipliers (Figure 3-26), may be used for obtaining information on the spatial distribution of the TSEE from single crystals (Bräunlich, 1971 and Bräunlich and Rosenblum, 1972). A resolution of 0.05 mm results in interesting photographs of the emitting surface, for instance, if radiation fields with steep gradients have to be analyzed (microdosimetry). Localization of the exact point of EE is also possible with OSEE, using a microbeam of stimulating light through a microscopic system (Figure 3-27). A resolution of $\sim$0.01 mm can be obtained by this method.

As in TLD studies, besides the integral count between two temperatures, a TSEE "glow curve" is usually recorded with a conventional X-Y recorder to detect instrument malfunctions, spurious signals (the peak of a tribo-signal is frequently different from the radiation-induced signal), etc. It may, however, have certain advantages to use a multichannel analyzer and optical display of the TSEE curve instead (Moreno et al., 1972 and Moreno, 1973).

3-4. Dosimeter Design and Applications

Obviously, EE is an interesting diagnostic tool for the selective investigation of trapping centers at (or close to) the surface of a material, and of changes occurring in them, with little or no interference from the bulk of the material. TSEE

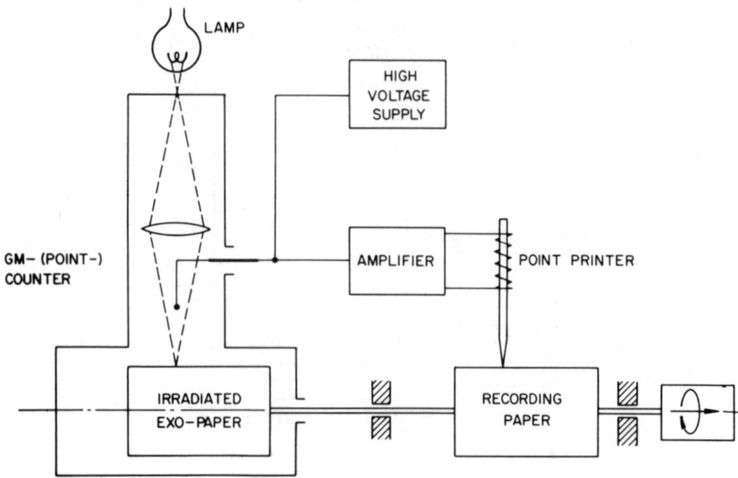

FIGURE 3-27. Simple mechanical device for translating the spatial distribution of dose in an "exoelectron radiograph" via optical stimulation into a visible, permanent record. (After Kramer, 1970.)

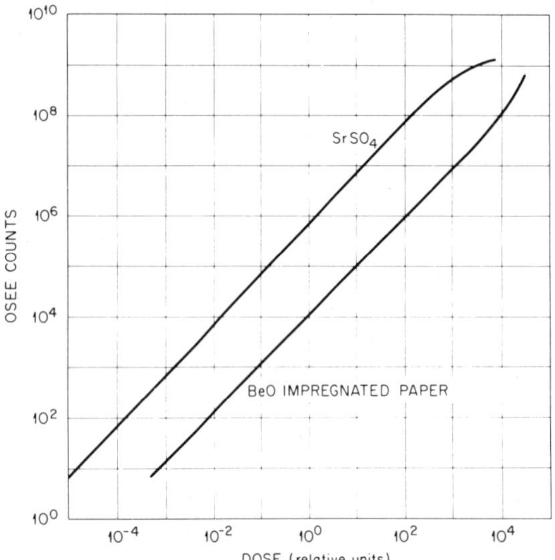

FIGURE 3-28. OSEE from paper coated with a mixture of BeO or $SrSO_4$ and graphite, as a function of radiation dose. (After Kramer, 1971.)

is also an extremely sensitive analytical tool. It has potentials as an imaging technique ("exoelectron photography") and in various areas of biophysical research (for a summary of nondosimetric applications, see Becker, 1972). The use of exoelectron emission in the early diagnosis of human caries, of structural metal fatiguing, and for the dating of artifacts is being considered. The most important application so far is, however, in dosimetry.

A proportionality between radiation dose and exoelectron emission in calcium sulfate was first pointed out in 1957 (Kramer, 1957 and Gourgé, 1958), but further studies followed only slowly (Kramer, 1962 and Hanle et al., 1966). Seriously limiting the practical usability of such early "dosimeters" were the nonlinearity of the response, saturation occurring at low doses, and a rapid fading.

Later it was discovered that the addition of an electrical conductor such as graphite to the emitter greatly enhanced the reproducibility and linearity of response (Kramer, 1966 and Kriks, 1970). During the last few years, a rapidly increasing number of laboratories have become interested in exoelectron emission because it was realized that several properties make exoelectron dosimeters an interesting alternative to the widely used TL dosimeters:

1. Extreme sensitivity of some detectors such as preheated $CaSO_4$ makes it possible to measure doses in the μR range without sophisticated instrumentation or rapid fading in the detectors.

2. Absence of disturbing infrared emission does not limit the temperature of usable emission peaks to the rather narrow range of about 140 to $260°C$ as in TLD, where it results in a limited choice of phosphors and rapid fading at elevated

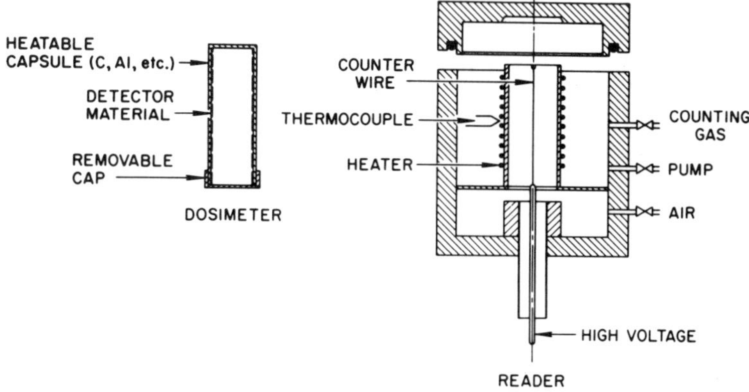

FIGURE 3-29. Closed capsule TSEE dosimeter with the exoelectron emitter coated at the inside (left) and schematic diagram of reader for this dosimeter type. (After Kramer, 1968.)

temperatures. Instead, use can be made of very stable, deep trapping levels in a wide variety of materials.

3. The thin sensitive layer makes exoelectron emitters essentially two-dimensional detectors. This offers great advantages if short-range types of radiation such as recoil protons in fast-neutron dosimetry or low energy beta radiation (^{3}H) are to be measured.

4. The materials are rather inexpensive and flexible in their composition and dosimetric properties (including energy response) within wide limits.

Only a few examples of exoelectron dosimeters, representing the approaches in different countries, will be reviewed here. In Germany, it was found that some types of commercial papers can be used as dosimeters if they contain exoelectron emitting additives (Kramer, 1970, 1971, 1972 and Kawanishi, 1972). Stocks of ordinary paper have been used for depth-dose measurements at low Z/high Z interfaces. More sensitive and more linear in their response, however, were papers that have been prepared either by incorporating the detector material and graphite during the manufacturing process or by coating the paper with a mixture of the detector and graphite. Using such paper dosimeters, a dose-range of eight orders of magnitude can be covered with $SrSO_4$, or slightly less with BeO (Figure 3-28).

The reading principle is simple: Part of the coated paper is exposed to a narrow light beam whose intensity may be varied, e.g., by neutral density filters. The optically stimulated exoelectrons are counted in a G-M counter. Only a small fraction of the total exoelectrons is released. The main advantages of this method are the following: Because only a very small fraction of the total area is required for the reading, many readouts can be conducted within the same dosimeter for establishing the proper reading range, for increasing the precision by multiple readings, or for performing a "filter analysis" of a mixed radiation field similar to that in a film badge.

An interesting variety of this evaluation technique consists of measuring the amount of light which is required to release a preset number of exoelectrons (Kramer, 1972). Simple and inexpensive manufacturing of paper dosimeters makes them ideal one-way dosimeters. Of course, the dosimeters have to be protected from light and some other potentially disturbing factors by sealing them into a protective layer.

Several different dosimeter designs for thermal stimulation have been suggested (Kramer, 1968 and Müller and Stolz, 1972). One consists of a capsule whose inside is coated with the sensitive material (Figure 3-29). It can be read by placing the open end over the counting wire, filling the system with counting gas, and heating of the capsule walls to release the exoelectrons. Using $CaSO_4$, doses corresponding to a few minutes of background radiation have been measured (Figure 3-16).

Another design is based on small, sealed, heatable G-M counters, whose inner walls are coated with the detector material. This principle has been employed in the first mass-produced TSEE personnel dosimeters in the Soviet Union (Beskorskii et al., 1967 and 1971). The detector unit (Figure 3-30) consists of a steel cylinder, 100 mm in diameter. The inner surface is coated with 0.1 mg/cm^2 of a specially prepared Mn-activated $CaSO_4$ and the counter is filled with a mixture of 90% argon and 10% isopentane at 100 torr. Only pulses in the temperature region of 220 and 300°C are counted during heating at a rate of 1°C/sec.

As can be seen in Figure 3-31, the response is not strictly linear. The standard deviation of the response of the mass-produced dosimeters is 3 to 5%. Background noise corresponds to 0.1 mR, and less than 6% fading has been found after three weeks of storage of 65°C. However, there is a strong energy dependence of the dosimeter, which requires a 2 mm lead compensation filter. This makes the dosimeter rather bulky, heavy, and rules out its use for the measurement of lower photon energies.

A number of laboratories still use a type of dosimeter consisting of a mixture of the finely powdered TSEE material and graphite powder, with an acetone added, plated in the central recess of a small graphite disk and dried. This procedure results in a thin, smooth layer. However, such detectors tend to vary in their response, and the layer easily peels off as a result of repeated reuse or mechanical shock. Small tablet-shaped disks of ceramic BeO have a much better thermal, chemical, and mechanical stability, a low atomic number, a convenient TSEE peak location, and may be purchased inexpensively in large, reproducible quantities.

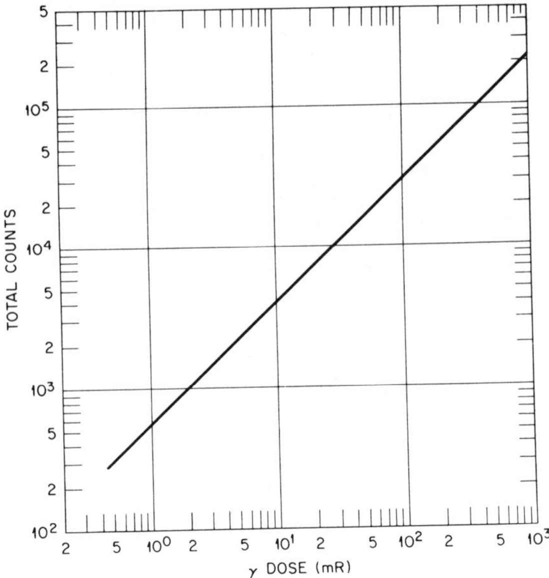

FIGURE 3-31.   Gamma radiation response of sealed G-M tube TSEE dosimeters as in Figure 3-27. (After Beskorskii et al., 1971.)

The reproducibility of multiple readings of the same detector over extended periods of time, as well as of groups of identically prepared detectors, may be within 2 to 5% (Figure 3-21), but larger and somewhat erratic deviations are also not uncommon (Becker et al., 1971; Gammage and Cheka, personal communication; and Braunschweig, 1972). It requires some time for an open detector to equilibrate with the environment. Strange "memory" effects (dependence of sensitivity and the time between the first and consecutive readings) have been observed in some ceramic BeO detectors. The response becomes sublinear at higher dose levels and/or heating rates. Using BeO

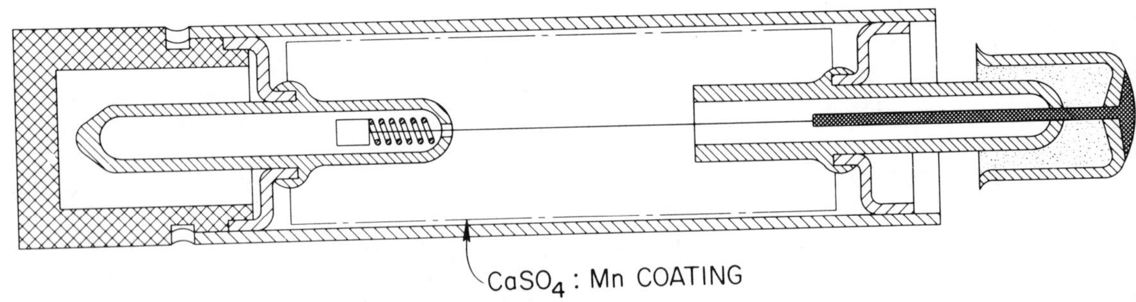

FIGURE 3-30.   Diagram of small sealed G-M counter, whose inside walls are coated with $CaSO_4$:Mn, as a TSEE dosimeter. (After Beskorskii et al., 1971.)

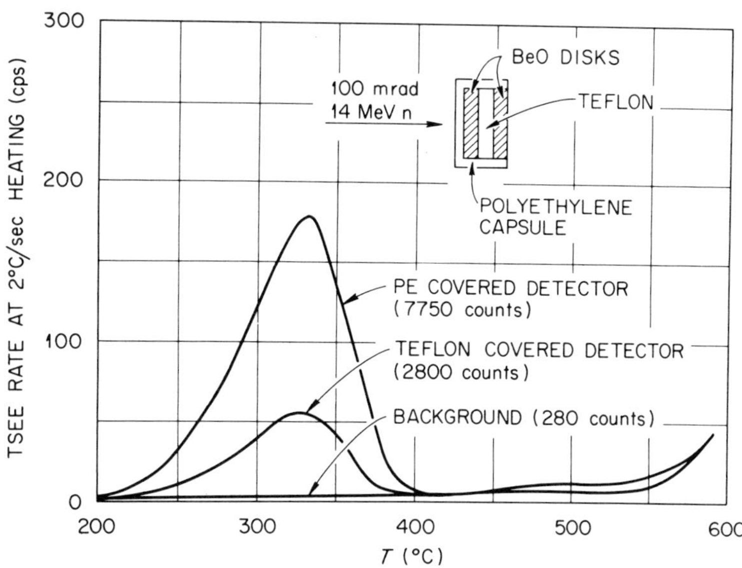

FIGURE 3-32. TSEE curves from ceramic BeO detectors, covered with polyethylene and with Teflon, in a mixed neutron-gamma radiation field. (After Crase and Becker, 1970.)

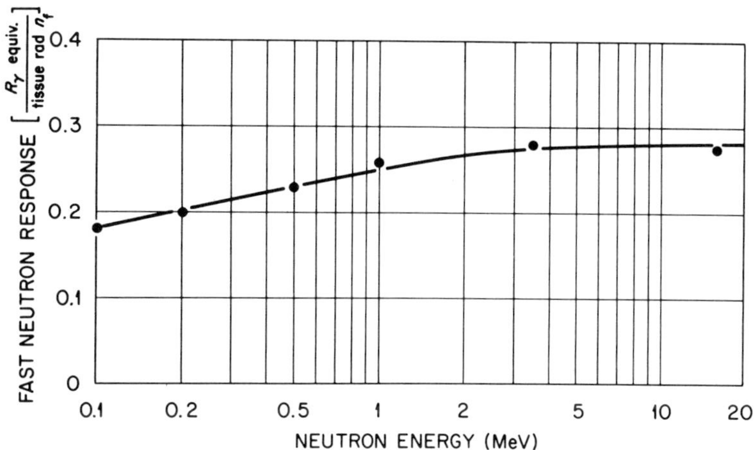

FIGURE 3-33. Energy dependence of the fast-neutron response of polyethylene-covered ceramic BeO. (After Becker and Crase, 1970.)

disks 12 mm in diameter, sensitivities of more than 1,000 counts/mR have been obtained.

The inherent sensitivity of BeO to thermal neutrons is small. However, if covered by suitable converter materials that undergo (n,α) reactions such as B or Li compounds, the detectors may be used for highly sensitive thermal neutron measurements. Most inorganic TSEE dosimeters have a low fast-neutron sensitivity due to the absence of hydrogen. Fast neutrons can be measured by observing the effect of recoil photons from an external hydrogeneous radiator (Becker and Crase, 1970).

A pair of detectors is used, one being covered with a hydrogen-free substance to measure the gamma radiation dose and the other covered with a hydrogen-containing foil. The difference between both (Figure 3-32) is an indication of the fast-neutron dose. The fast-neutron response varies only relatively little, between 18 and 28% of the

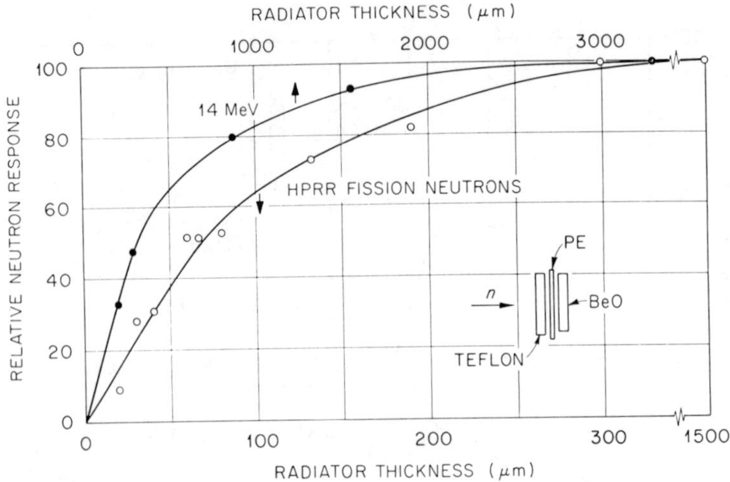

FIGURE 3-34.   Relative fast-neutron response of a ceramic BeO detector as a function of the thickness of the polyethylene radiator layer. (After Becker and Abd-El Razek, 1973.)

gamma radiation sensitivity on a rad per rad basis with neutron energies between at least 0.1 and $\sim$20 MeV (Figure 3-33), indicating that the detector "sees" only the constant-LET stem of the recoil proton tracks.

In an intercomparison of neutron dosimetry system with the unmoderated and moderated fission spectrum of the ORNL Health Physics Research Reactor (Lee et al., unpublished), TSEE detector pairs based on this principle gave excellent results. Of course, the response of such detectors depends on the direction of neutron incidence, the response decreasing rapidly if the angle of incidence approaches or exceeds 90°. This angular dependence becomes more pronounced with decreasing neutron energy. Combinations of several detectors can be used for determining the direction of neutron incidence (Becker and Gammage, unpublished).

Information on the effective fast-neutron energy and energy distribution has been obtained by measuring the response of TSEE detectors either as a function of the thickness of the hydrogeneous radiator layer, or as a function of an absorber layer thickness between detector and a (thick) radiator (Figure 3-34). The fast-neutron response of TSEE detectors has also been studied by some other groups (Laitano and Rotondi, 1972 and Rasp and Siegel, 1972). It would be desirable to increase the neutron to gamma radiation response ratio of the detectors to improve their performance in mixed radiation fields with a

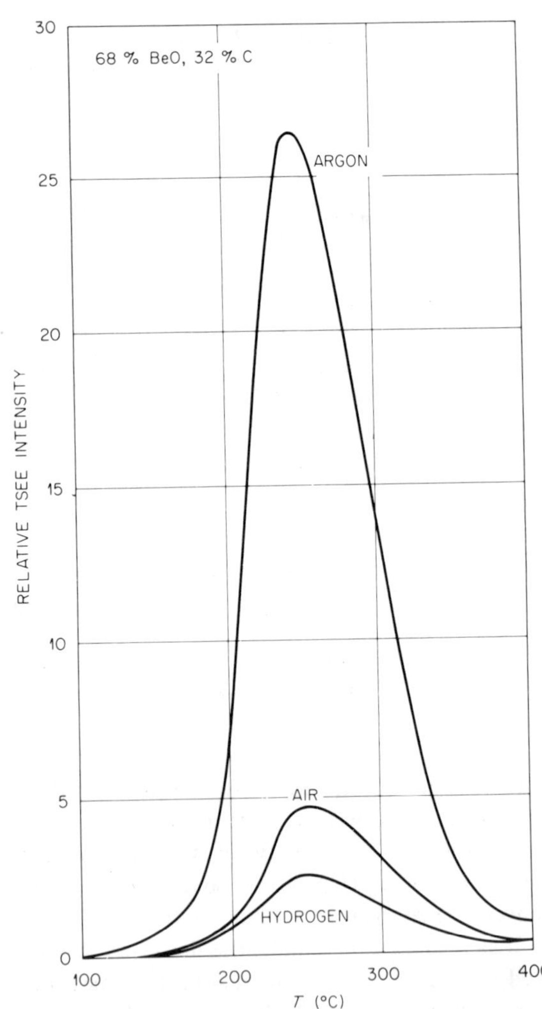

FIGURE 3-35.   TSEE response of a BeO/graphite mixture exposed to the same dose of 50 kV x-rays in hydrogen, air, and argon. (After Kramer, 1971.)

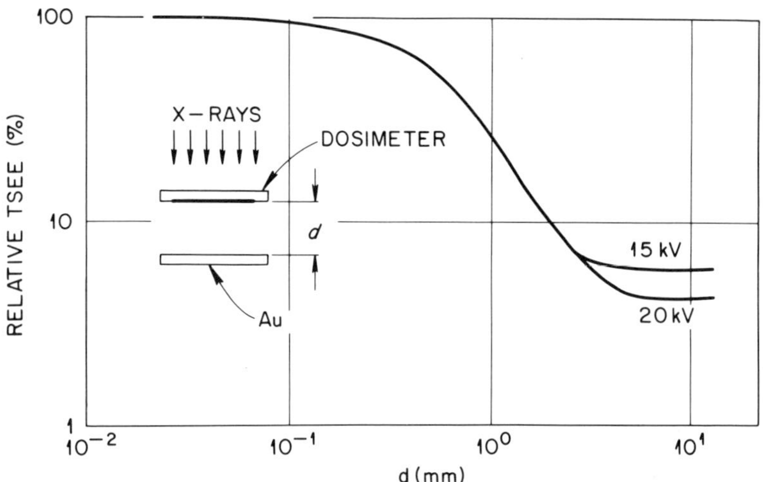

FIGURE 3-36. Measurement of the range of secondary electrons that are backscattered from a gold foil during exposure to soft x-rays. (After Kramer, 1971.)

relatively low neutron contribution to total dose. Some detectors exhibit a better response to high LET protons than BeO (Becker, 1968), and some MgO samples apparently are almost LET-independent in their sensitivity (Ritz and Attix, personal communication).

As one would expect in a detector system in which the sensitive layer is essentially two-dimensional, the atomic number Z of the detector's immediate environment has a strong effect on its photon energy response. Indeed, the sensitivity of a BeO/graphite detector for 50 kV rays was found to vary substantially, depending on the Z of the gas in which it was exposed (Figure 3-35). Of course, this effect also occurs if the detector surface is covered with solid materials of varying Z (Wilson, 1971; Rotondi, 1971; and Samuelsson and Ennow, 1972).

By a short heating to temperatures between 500 and 700°C, the TSEE dosimeters can be completely annealed. They may, in principle, be used hundreds of times without permanent change in the dosimetric characteristics, but have to be protected from disturbing environmental influences such as intense light, particularly direct sunlight, and surface contaminations. Scratching with hard objects induces a disturbing tribo-signal.

Among the other dosimetric applications of TSEE dosimeters are

a. Integrating measurement of low doses over extended periods of time;

b. Measurement of short-range radiation, for example, from tritium in solid surfaces, gases or liquids; of the range of secondary electrons which are released by low energy rays (Figure 3-36); of radon in air; or of incorporated radionuclides (Boros and Szy, 1972); and

c. Evaluation of radiation fields with a very high spatial resolution, for instance, the delta ray distribution around a thin x-ray beam in various gases, or steep gradients at interfaces between low Z and high Z materials (Kramer, 1971; Ennow, 1972; and Samuelsson and Ennow, 1972), or the measurement of fast-neutron dose distributions at interfaces with a different hydrogen content, for example, bone/tissue (Becker, unpublished).

Presently, exoelectron emission is a still relatively little known and poorly understood physical effect, but this can be expected to change. Commercial instruments for thermal and/or optical stimulation will probably become available and EE may find more and wider applications. Although some feasibility tests at ORNL (Gammage and Cheka, unpublished) showed encouraging results in personnel and environmental monitoring, TSEE is unlikely to replace TLD and other "established" techniques in these fields in the near future. It seems likely, however, that there will be an increasing number of applications in research and for problems which are difficult to solve by other methods. There are still problems, mainly concerning stability and reproducibility of the detectors, but most researchers in this young field feel cautiously optimistic about its prospects.

REFERENCES

Ambrosetti, A. and Oberhofer, M., Some Reproducibility Checks with Brush Ceramic BeO Thermalox 995 as a TSEE Dosimeter, Proc. Seminar on Possible Applications of Exoelectrons in Dosimetry, EUR-4904, 1972.

Arabin, G., Kriegseis, W., and Scharmann, A., Untersuchungen zur Exoelektronenemission von Polymeren, *Z. Naturforsch.,* 27a, 1350, 1972.

Attix, F. H., A Proportional Counter for TSEE, Report of NRL Progress, May 1970, 1.

Balarin, M. and Zetzsche, A., Bestimmung der Aktivierungsenergie für die Beweglichkeit von Gitterdefekten durch semilineares Aufheizen, *Phys. Status Solidi,* 2, 1670, 1962.

Becker, K., Some Studies on Radiation Dosimetry by TSEE, Proc. Sec. Int. Conf. Luminescence Dosimetry, CONF-680920, 1968, 200.

Becker, K., Principles of thermally stimulated exoelectron dosimetry, *IAEA At. Energ. Rev.,* 8, 173, 1970.

Becker, K., Stimulated exoelectron emission from the surface of insulating solids, *CRC Crit. Rev. Solid State Sci.,* 3(1), 39, 1972a.

Becker, K., Applied Dosimetry Research Annual Report, in ORNL-4811, Oak Ridge National Laboratory, Oak Ridge, Tenn., 1972b.

Becker, K. and Chantanakom, N., On the LET dependence of thermally stimulated exoelectron emission, *Nucl. Instr. Meth.,* 66, 353, 1968.

Becker, K. and Robinson, E. M., Integrating dosimetry by thermally stimulated exoelectron (after-) emission, *Health Phys.,* 15, 463, 1968.

Becker, K. and Chantanakom, N., On the thermal fading characteristics of some TSEE dosimetry materials, *Atompraxis,* 16, 1, 1970.

Becker, K. and Crase, K. W., A sensitive integrating fast neutron dosimeter based on TSEE, *Nucl. Instr. Meth.,* 82, 297, 1970.

Becker, K., Cheka, J. S., and Gammage, R. B., Progress in exoelectron dosimetry, *PTB-Mitt.,* 80, No. 5, 1970a.

Becker, K., Cheka, J. S., and Oberhofer, M., Thermally stimulated exoelectron emission, thermoluminescence, and impurities in LiF and BeO, *Health Phys.,* 19, 391, 1970b.

Becker, K., Cheka, J. S., Crase, K. W., and Gammage, R. B., New Exoelectron Dosimeters, IAEA-SM-143/37, Adv. Phys. and Biol. Rad. Detectors, IAEA, Vienna, 1971, 35.

Becker, K., Gammage, R. B., Crase, K. W., and Cheka, J. S., Recent Advances in TSEE Dosimetry, Proc. Third Int. Symp. on Microdosimetry, Stresa 1971, EUR-4810, Euratom, Luxembourg, 1972, 675.

Becker, K. and Abd-el Razek, M., Fast neutron response of TSEE, *Nucl. Instr. Meth.,* to be published.

Beskorskii, A. I., Schergalina, G. A., and Goichberg, E. J., Using TSEE for ionizing radiation dosimetry, *Prib. Tekj. Eksp.,* 1, 36, 1967.

Beskorskii, A. I., Schergalina, G. A., Kusmin, V. V., Minajeva, E. E., Misev, I. P., and Sokolov, A. D., A High-Sensitivity Thermally Stimulated Exoelectron Emission Element for Gamma Dosimeters, IAEA-SM-143/69, Adv. Phys. Biol. Rad. Detectors, Vienna, 1971, 37.

Bichivin, V. and Käämbre, H., Manifestation of hold processes in thermostimulated exoelectron emission, *PTB-Mitt.,* 80, No. 5, 1970.

Bohun, A., Thermoemission und Photoemission von Natriumchlorid, *Czech. J. Phys.,* 4, 91, 1954.

Bohun, A., Thermostimulierte Exoemission und Lumineszenz von MgO und CaO, *Czech. J. Phys.,* 12, 328, 1962.

Bohun, A., Exoelektronenemission von Ionenkristallen, *Phys. Status Solidi,* 3, 779, 1963.

Bohun, A., The physics of exoelectron emission on ionic crystals, *PTB-Mitt.,* 80, No. 5, 1970.

Boros, L. and Szy, D., Experiences with Thermally Stimulated Exoelectron Emission Dosimeters in Environmental and Incorporational Dosimetry, Paper, Second European Congress on Radiation Protection, and *Izotoptechnika,* 6, 1972.

Bräunlich, P., Exoelectron imaging of dielectric materials, *J. Appl. Phys.,* 42, 465, 1971.

Bräunlich, P. and Rosenblum, B., The Exoelectron Microscope — A New Tool in Science, Paper, Rolla Conf. on Surface Properties and Surface States of Electronic Materials, Rolla, Miss., 1972.

Braunschweig, Seminar on the Possibilities of Application of Exoelectrons in Dosimetry, Commission of the European Communities, EUR-4904, 1972.

Brown, L. D. and Dalton, P., The Application of TSEE Measurements to the Study of Supralinearity in BeO, Proc. Seminar Anwendungsmöglichk, Exoelektronen für d. Dosimetrie, EUR-4904, Luxembourg, 1972, 79.

Brunsmann, U., Euler, M., Kriegseis, W., and Scharmann, A., A New Material for Exoelectron Dosimetry, Proc. Symp. on Dosimetry Techniques, IAEA, Vienna, 1972.

Brunsmann, U. and Scharmann, A., Exoelectron energy measurements on K_2SO_4 after x-ray irradiation and electron bombardment, *Phys. Status Solidi,* 15, 525, 1973.

Burke, G. de P. and Beck, W. L., Dosimetry Study of TSEE with LiF, HASL-236, USAEC Health and Safety Laboratory, New York, 1970.

Chryssou, E. and Holzapfel, G., Exoelectron emission from the system γ-a Al_2O_3, *Phys. Status Solidi,* 8, 47, 1971.

Crase, K. W., Becker, K., and Gammage, R. B., Parameters Affecting the Radiation-Induced Thermally Stimulated Exoelectron Emission from Ceramic Beryllium Oxide, ORNL-TM-3572, Oak Ridge National Laboratory, Oak Ridge, Tenn., 1971.

Crase, K. W., Gammage, R. B., and Becker, K., High dose-level response on ceramic BeO, *Health Phys.,* 22, 402, 1972.

Engelland, W., Ein Methandurchflußzähler zur schnellen Auswertung von Exoelektronen-Dosimetern, *PTB-Mitt.,* 1/69, 15, 1969.

Ennow, K., TSEE Measurements of Dose Gradients at Interfaces, Paper IAEA/SM-160/35, Proc. Symp. Dosimetry Techniques Applied to Agriculture, Industry, Biology, and Medicine, IAEA, Vienna, 1972.

Euler, M., Kriegseis, W., and Scharmann, A., The influence of oxygen adsorption centres upon the exoelectron emission of BeO, *Phys. Status Solidi,* 15, 431, 1973.

Fintelmann, D., Proc. 4th Czech. Conf. on Electronics and Vacuum Phys., Prague, 1968, 395.

Ford, L. H., Holzapfel, G., and Kaul, W., Photostimulated exoelectron emission of BeO, *Z. Angew. Phys.,* 30, 1970.

Frank, M., Knoll, P., and Müller, F., Eignung der TSEE von LiF zur Dosimetrie, *Isotopenpraxis,* 2, 369, 1966.

Gammage, R. B. and Becker, K., Exoelectron Emission and Surface Characteristics of Lunar Materials, *Proc. Sec. Lunar Science Conf.,* Vol. 3, MIT Press, Cambridge, Mass., 1971, 2057.

Gammage, R. B., Crase, K. W., Becker, K., and Moreno y Moreno, A., Chemically and Radiation-Induced Changes in the TSEE Characteristics of Ceramic BeO, *Proc. Third Int. Conf. Luminescence Dosimetry, Risö-Rep. 249,* Danish AEC, Risö, Roskilde, 1971, 573.

Gobrecht, H. and Barthow, G., Uber die Elektronenemission bei der Kristallisation, *Z. Phys.,* 146, 1, 1956.

Gordan, P. and Scharmann, A., Zur Elektronennachemission in LiF-Kristallen, *Z. Phys.,* 217, 309, 1968.

Gordan, P., Scharmann, A., and Seibert, J., The discrimination of charged particles during exoelectron emission, *Phys. Status Solidi,* 33, 97, 1969.

Gourgé, G., Untersuchungen der Exoelektronenemission und Lumineszenz Anorganischer Kristalle, *Z. Phys.,* 153, 186, 1958.

Gourgé, G. and Hanle, W., Neue Ergebnisse über Exoelektronenemission an Nichtmetallen, *Acta Phys. Austr.,* 10, 427, 1957.

Hanle, W., Kanzler, G., and Scharmann, A., Zur Elektronennachemission bei Ionenkristallen, *Z. Phys.,* 162, 483, 1961.

Hanle, W., Scharmann, A., Seibert, G., and Seibert, J., Über die Eignung der Elektronen-Nachemission von $CaSO_4$ zur Dosimetrie, *Nukleonik,* 8, 129, 1966.

Holzapfel, G., Ph.D. Dissertation No. D 83, Technical University, Berlin, 1968.

Holzapfel, G., Diffusion controlled exoelectron emission phenomena in BeO, *Phys. Status Solidi,* 33, 235, 1969.

Holzapfel, G., TSEE of LiF, *Z. Angew. Phys.,* 29, 107, 1970.

Holzapfel, G. and Kramer, J., Untersuchung der Elektronenhaftstellen in LiF mittels Exoelektronen, Paper, Spring Meeting German Physical Soc., Munich, 1969.

Holzapfel, G. and Kaul, W., Eds., Proc. Third Int. Symp. on Exoelectrons, PTB Braunschweig 1970, *PTB-Mitt.,* 80, No. 5, 318, 1970.

Holzapfel, G. and Nink, R., Thermally and optically stimulated exoemission from ZnO, *Phys. Status Solidi,* 3, K181, 1970.

Holzapfel, G., Thermionic Emission from Electron Traps, *Vacuum,* in press, 1972.

Kawanishi, S., Application of exoelectron dosimeter, *Genshiryoku Kogyo,* 18(6), 39, 1972.

Kelly, P., Thermally stimulated exoelectron emission, *Phys. Rev.,* 13(5), 749, 1972.

Kopp, D. T., TSEE from LiF, Ph.D. Thesis, University of Kansas, Lawrence, 1971.

Kramer, J., *Der Metallische Zustand,* Vandenhoeck & Rupprecht, Göttingen, 1950.

Kramer, J., Oberflächenuntersuchung an Metallen und Nichtmetallen mit Exo-und Photoelektronen, *Z. Phys.,* 133, 629, 1952.

Kramer, J., Der Nachweis Ionisierender Strahlung mit Exoelektronen, *Z. Angew. Phys.,* 15, 20, 1962.

Kramer, J., Die Exoelektronenemission an Alkali-Halogeniden, *Z. Naturforsch.,* 18a, 1022, 1963.

Kramer, J., Exoelektronen-Dosimeter für Röntgen- und Gammastrahlen, *Z. Angew. Phys.,* 20, 411, 1966.

Kramer, J., Dosimetry with Exoelectrons, Proc. Sec. Int. Conf. Luminescence Dosimetry, CONF-680920, 1968, 180.

Kramer, J., New aspects in application of exoelectrons, *PTB-Mitt.,* 80, No. 5, 1970.

Kramer, J., Optical Stimulation of Exoelectrons, *Proc. Third Int. Conf. Luminescence Dosimetry, Risö-Rep. 249,* Danish AEC, Risö, Roskilde, 1971, 622.

Kramer, J., Einige Bemerkungen zur Dosimetrie mit Exoelektronen, Proc. Seminar in Possible Applications of Exoelectrons in Dosimetry, Braunschweig, 1972.

Kriegseis, W. and Scharmann, A., Beeinflussung der Thermisch Stimulierten Elektronennachemission von $BaSO_4$ Durch Sorbierte Fremdstoffe, *Z. Naturforsch.,* 24a, 862, 1969.

Kriegseis, W. and Scharmann, A., Energy distribution of thermally stimulated exo-electron emission, *Phys. Status Solidi,* 33, K41, 1969.

Kriegseis, W. and Scharmann, A., A model for the description of exoelectron emission, *PTB-Mitt.,* 80, No. 5, 1970.

Kriegseis, W., Scharmann, A., and Seibert, J., Dosimetrie mit Exoelektronen, *Kerntechnik,* 13, 11, 1971.

Kriks, H. J., Über die Anwendung der EE von BeO in der Dosimetrie Ionisierender Strahlen, *PTB-Mitt.,* No. 4, 251, 1970.

Kriks, H. J., Investigations on Exoelectrons with a Special Proportional Counter, IAEA-SM-143/5, Adv. Phys. Biol. Rad. Detectors, Vienna, 1971, 17.

Ku, T. C. and Pimbley, W. T., Effect of temperature on the emission of electrons from abraded surfaces of Be, Ca, Al, and Mg, *J. Appl. Phys.,* 32, 124, 1961.

Laitano, R. F. and Rotondi, E., Fast Neutron Dosimetry by Means of TSEE of Beryllium Oxide, Proc. Int. Symp. Neutron Dosimetry, Neuherberg, Germany, 1972.

LaRiviere, P. D. and Tochilin, E., An ionization chamber for the measurement of thermally stimulated exoelectrons, *Health Phys.,* 22, 198, 1972.

Lausch, W. and El Naggar, S., $BaTiO_3$-Keramik als neuer Exoelektronen-Emitter, *Radiochem. Radioanal. Lett.,* 12, 177, 1972.

Levshin, V. L. and Pipins, P. A., TSEE from crystal phosphors made from ZnS, *Sov. Phys. Solid State,* 5, 691, 1963.

Lewis, D. R., Geological applications of exoelectron phenomena, *Bull. Geol. Soc. Am.,* 77, 761, 1966.

Lillicrap, S. C., A dose reader using TSEE from LiF crystals, *Br. J. Radiol.,* 43, 277, 1970.

Menold, R., Die Exoelektronen-Emission (Kramer-Effect) von ZnO, *Z. Phys.,* 157, 499, 1960.

Minz, R. I., Kortov, V. S., and Krjuk, V. I., Eds., Investigation of the Surface of Structural Materials with the Exoelectron Emission Method, Ural Polytechnic. Inst. C. M. Kirov, Sverdlovsk, 1969.

Momose, Y. and Tamai, Y., Geiger-Müeller counter for the detection of exoelectrons, *Tohoku Daigaku Hisuiyoeki Kagaku Kenkyusho Hogoku,* 17, 435, 1968.

Moreno, A., Gammage, R. B., Cheka, J. S., Nagpal, J. S., and Becker, K., Further Studies on TSEE Activators in BeO, ORNL-TM-3668, Oak Ridge National Laboratory, Oak Ridge, Tenn., 1972.

Moreno, A., On the detection and recording of thermally stimulated exoelectrons, *Nucl. Instr. Meth.,* in press.

Müller, F. and Stolz, W., Dosimetrie ionisierender Strahlung durch Messung der Exoelektronenemission, *Isotopenpraxis,* 8, 161, 1972.

Nagpal, J. S. and Gammage, R. B., Effect of gases, temperature and other factors on the sensitivity of BeO, *Radiat. Eff.,* in press.

Nagpal, J. S. and Gammage, R. B., Adsorbed gases, atmospheric, and temperature effects on the TSEE from BeO, *Radiat. Eff.,* in press.

Nash, A. E., Ritz, V. H., and Attix, F. H., Storage Stability of TL and TSEE from Six Dosimetry Phosphors, *Proc. Third Int. Conf. Luminescence Dosimetry, Risö-Rep. 249,* Vol. 3, Danish AEC, Risö, Roskilde, 1971, 1123.

Nassenstein, H., Die Elektronenemission von Festkörperoberflächen nach mechanischer Bearbeitung und Bestrahlung, *Z. Naturforsch.,* 10a, 944, 1955.

Niewiadomski, T., TSEE Dosimetry Studies, *Proc. Third Int. Conf. Luminescence Dosimetry, Risö-Rep. 249,* Danish AEC, Risö, Roskilde, 1971, 612.

Novotny, J., Spurny, Z., and Binova, M., EE from activated $CaSO_4$ phosphors, *J. Phys. Chem. Solids,* 31, 1412, 1970.

Oberhofer, M. and Robinson, R. M., A TSEE G-M gas-flow counter, *Kerntechnik,* 12, 34, 1970.

Onsgaard, J. and Hougs, E., A channel electron multiplier as detector of thermally stimulated electrons, *Int. J. Appl. Radiat. Isotopes,* 24, 55, 1973.

Petrescu, A., On the L-bands and the first exciton bands in the photoemission spectra of alkali iodides, *Phys. Status Solidi,* 20, 333, 1968; On the photoemission spectrum of LiF:Mg, *Phys. Status Solidi,* 29, K1, 1968.

Petrescu, P., Electron emission spectroscopy of alkali halides, *Phys. Status Solidi,* 9, 539, 1965.

Puite, K. J. and Arends, J., Trapping Centers in CaF_2:Mn from TL and TSEE Measurements on Undoped and Mn Doped CaF_2 Samples, *Proc. Third Int. Conf. Luminescence Dosimetry, Risö-Rep. 249,* Danish AEC, Risö, Roskilde, 1971, 680.

Puite, K. J., Ph.D. Thesis, University of Groningen, Holland, 1971.

Randall, J. T. and Wilkins, M. H. F., *Proc. Roy. Soc. Ser A,* 184, 347, 1945.

Rasp, W. and Siegel, V., Neutronendosimetrie mit Hilfe von Exoelektronen, Proc. Int. Symp. Neutron Dosimetry, Neuherberg, Germany, 1972.

Regulla, D. F., Drexler, G., and Boris, L., Low-Z Activated Beryllium Oxide as a High Sensitive Radiation Detector in TSEE Dosimetry, *Proc. Third Int. Symp. Luminescence Dosimetry, Risö-Rep. 249,* Vol. 2, Danish AEC, Risö, Roskilde, 1971, 573.

Rotondi, E., The TSEE Response of Ceramic BeO Covered with Different Absorbers During Gamma and X-Irradiation, *Proc. Third Int. Conf. Luminescence Dosimetry, Risö-Rep. 249,* Danish AEC, Risö, Roskilde, 1971.

Rotondi, E. and Bordoni, R., Dosimetry Applications of TSEE, CNEN Report RT/FI-(70) 37, 1970 and EE from $CaSO_4$ Doped with Different Concentrations of Mn, Sm, and Pb, RT/FI (70) 4, 1970.

Samuelsson, Ch. and Ennow, K., Radiation dosimetry with exoelectron emitting materials, *Acta Radiol. Suppl.,* 313, 127, 1972.

Scharmann, A., Kriegseis, W., and Seibert, J., Surface effects of exoelectron emitting solids, *PTB-Mitt.,* 80, No. 5, 1970.

Stepniowski, I., On the properties of point counter with quenching vapour above free liquid surface (exoelectron detector), *Acta Phys. Pol.,* 30, 365, 1966.

Sujak, B., Luftspitzenzähler als Detektor der Exoelektronen, *Acta Phys. Pol.,* 22, 137, 1962.

Sujak, B. and Gieroszynska, K., On the influence of calcination and x-ray dosage on the exoemission of electrons and TL of CaF_2, *Acta Phys. Pol.,* 32, 541, 1967.

Sujak, B., Pietrzak, R., and Stepniowski, I., Open cylindrical counter with quenching vapour above the free surface of liquid (detector of exoelectrons), *Acta Phys. Pol.,* 32, 949, 1967.

Sujak, B. and Gasior, S., Exoelectron emission from BaF_2 and SrF_2 induced by x-rays, *Acta Phys. Pol.,* 33, 231, 1968.

Sujak, B. and Gajda, R., Influence of external electric field of photostimulated exoemission of electrons in vacuum, *Acta Phys. Pol.,* 23, 239, 1968.

Sujak, B. and Kusz, J., Surface electroluminescence of single crystals of seignette salt, *Acta Phys. Pol.,* 33, 845, 1968.

Thorngate, J. H., An Instrument that Provides Linear Temperature Ramps for Use with TSEE Experiments, ORNL-TM-2687, Oak Ridge National Laboratory, Oak Ridge, Tenn., 1969.

Vladimirov, V. G. and Valadimirov, G. G., Thermostimulated exoelectron emission of hemoglobin associated with certain aminothiols, *Radiobiology,* 168, 1968.

Wilson, J. F., The Effect of Radiation Energy, Dose Level, and the Atomic Number of the Immediate Environment on the TSEE from BeO and LiF, ORNL-TM-3325, Oak Ridge National Laboratory, Oak Ridge, Tenn., 1971.

4. RADIOPHOTOLUMINESCENCE

4.1. Materials and Mechanism

4.1.1. History and Definition

Alterations of the ultraviolet-excited luminescence spectrum of numerous inorganic compounds after exposure to beta or gamma radiation have been described as far back as 1912. Beginning in about 1920, Przibram and his students in Vienna (Przibram, 1924 to 1936) extensively studied this effect which mainly involved rare earth impurities in different minerals and synthetic inorganic solids. He also introduced the term "radiophotoluminescence" (RPL), which means (after a redefinition in 1925) that a material which is originally nonluminescent under visible or ultraviolet light is made responsive to such excitation by pretreatment with ionizing radiation. In 1951, Schulman et al. restricted the term to the creation of new quasistable luminescent centers, which are not (or only to a minor degree) destroyed by the excitation radiation.

During investigations of nonphotographic effects of radiation on solids under the aspects of radiography and dosimetry, it was discovered that dilute solidified "solutions" of silver salts in ionic crystals, such as alkali halides, Na_2SO_4, and $BaCl_2$, as well as in certain glasses (an alumino-phosphate glass was mainly used which had been prepared in the Bausch & Lomb Laboratories for a different purpose) exhibit a strong orange RPL during excitation with the 365 nm mercury line (Weyl et al., 1949; Schulman et al., 1950; and Figure 4-1). Extensive studies of this effect, in particular in a metaphosphate glass (composition by weight 44% $Al(PO_3)_3$, 23% $Ba(PO_3)_2$, 23% KPO_3, 8% $AgPO_3$), led to the development of the world's first mass-produced solid-state dosimetry system.

More than four million units have been made, mainly as an accidental dosimetry system for the 10 to 1,000 R exposure range by the U.S. Navy (Schulman et al., 1953 and Ballinger and Harris, 1959). In England and Sweden similar systems have been developed (Peirson, 1958). A glass with a reduced energy dependence became available in 1960 (Ginther and Schulman). Small glass needles (fluorods) became quite popular for high-dose personnel dosimetry and in radiobiological research (Schulman and Etzel, 1953 and Hardy and Werts, 1968).

The experience of some investigators with the early glasses, such as the fluorods with a high background luminescence and relatively poor

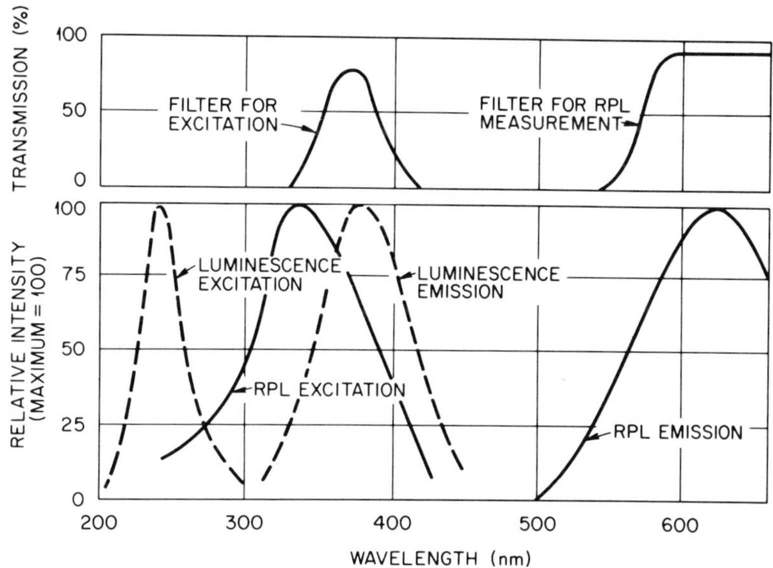

FIGURE 4-1.   Luminescence and radiophotoluminescence (RPL) excitation and emission bands in a typical Ag-activated dosimeter glass and filters used in reader (schematical).

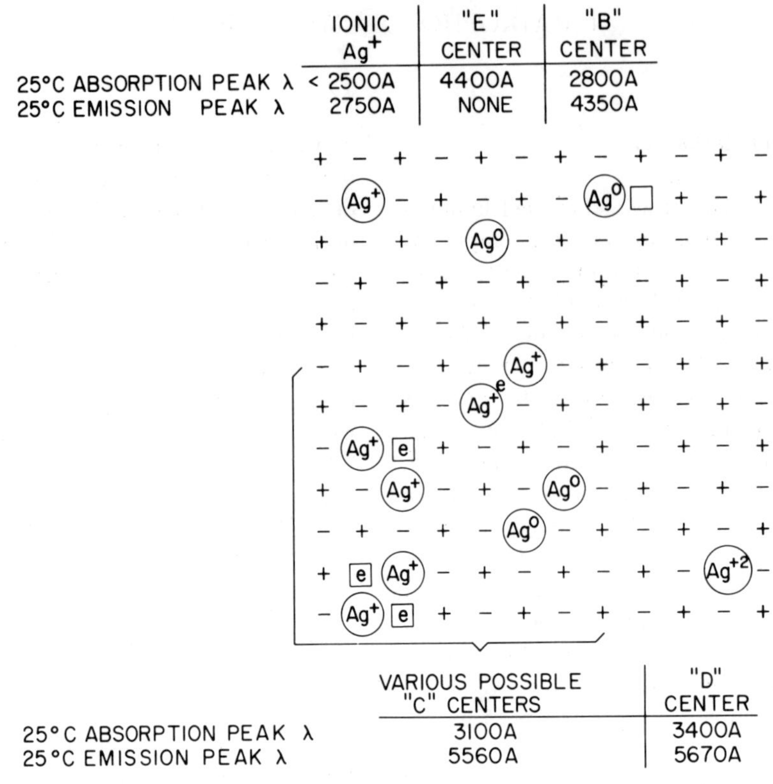

FIGURE 4-2. Models of various centers formed by irradiation of Ag^+-doped potassium chloride. (After Schulman, 1967.)

sensitivity, accuracy, and stability, and a developing widespread interest in some countries in TLD systems have somewhat masked the considerable progress in glass dosimetry in recent years. Glasses have been developed which have three times the sensitivity, much less energy dependence, and only one hundredth of the predose of the early glasses. In addition, some basic advantages of RPL systems — in particular, the great reliability of individual dose readings and the permanence of the radiation effect which permits an unlimited number of remeasurements or interim measurements during dose integration — would justify the more extensive use of these devices in personnel routine and emergency dosimetry as well as in research.

The second phase of RPL dosimetry began in 1961 with studies on improved glasses and readers at Toshiba, Japan (Yokota et al., 1961). Doses in the 10 mrad range became measurable with such systems. They stimulated further research and extensive application of glass dosimeters in many fields. Partly on the basis of Yokota's early work, other new glasses having a similar or even better background (predose), sensitivity, and energy response have since been made in several other countries such as France (Francois et al., 1965 and Carpentier et al., 1971), Germany (Becker, 1965 and 1968; Anon., 1969; Käs, 1972; Käs et al., 1972a) U.S. (Becker and Cheka, 1968), India (Nagpal, 1969), the Soviet Union, and in Japan (Yokota et al., 1969a and b). Progress in this field is still quite rapid. Bibliographies (Becker, 1966, 1967, 1968 and 1969 and Piesch, 1972) and review articles list about 500 publications on this subject, of which only some more important and/or more recent ones will be mentioned here.

4.1.2. Principle

The nature of centers and type of processes in dosimeter glasses are not yet completely understood (see, for example, Lell and Kreidl, 1967). Relatively well studied are the radiation-induced effects in Ag^+-doped alkali halides which may, to a certain degree, serve as a model substance for the processes involved in glasses. High energy radiation induces several types of centers in such crystals (Figure 4-2).

After irradiation, absorption bands in the near

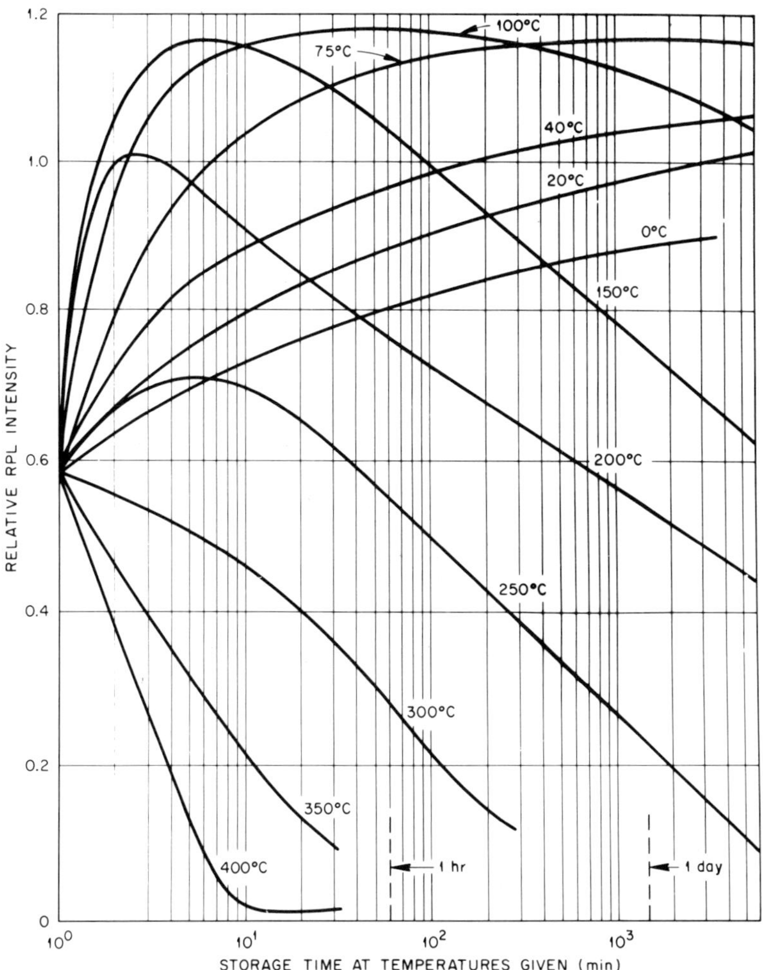

FIGURE 4-3. Radiophotoluminescence in the Yokota FD-2 dosimeter glass in the 10^{-2} to 10^{-3} rad exposure range as a function of storage temperature after exposure to a constant gamma radiation dose. (After Becker, 1965.)

ultraviolet are produced. Illumination with this light produces visible luminescence. The RPL was originally ascribed to neutral silver atoms (centers E in Figure 4-2) or Ag^+-modified color (F) centers (B). According to further studies, more complex centers such as (Ag^+ ion + hole + positive ion vacancy) and (Ag^+ ion pair + electron + negative ion vacancy) (C,D) are more probable. It is, however, doubtful to which degree these effects in simple crystals parallel the processes in glasses containing a high silver concentration.

Dosimeter glasses show a peculiar behavior after a short-time radiation exposure: Immediately after exposure there is an increase in RPL intensity (build-up), which is later superimposed by fading.

Besides depending on temperature (Figure 4-3), the kinetics also depends on the silver concentration in the glass (Figure 4-4), on the glass base composition (Figure 4-5) and on the LET of the radiation (Becker and Cheka, 1969). A mathematical treatment of the kinetics on the basis of a simplified band model (Figure 4-6) has been carried out (Vogel and Becker, 1965) assuming the following processes: Ionizing radiation of sufficient energy lifts electrons to the conduction band. Some of them are directly trapped by positively charged silver atoms or aggregates, thus forming a new type of center. Others are first trapped in centers which do not contribute to the radiophotoluminescence before being transferred

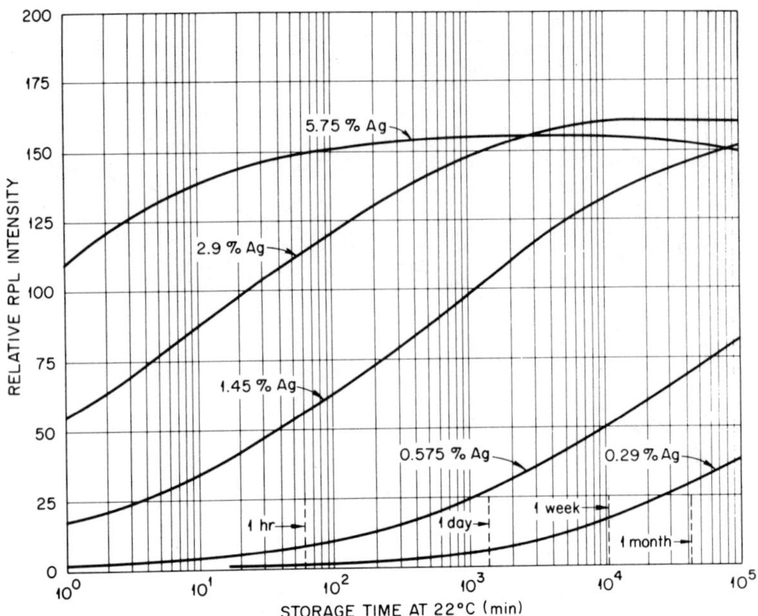

FIGURE 4-4.   RPL as a function of storage time at ambient temperature for a constant glass base composition, but varying silver content. (After Becker, 1965.)

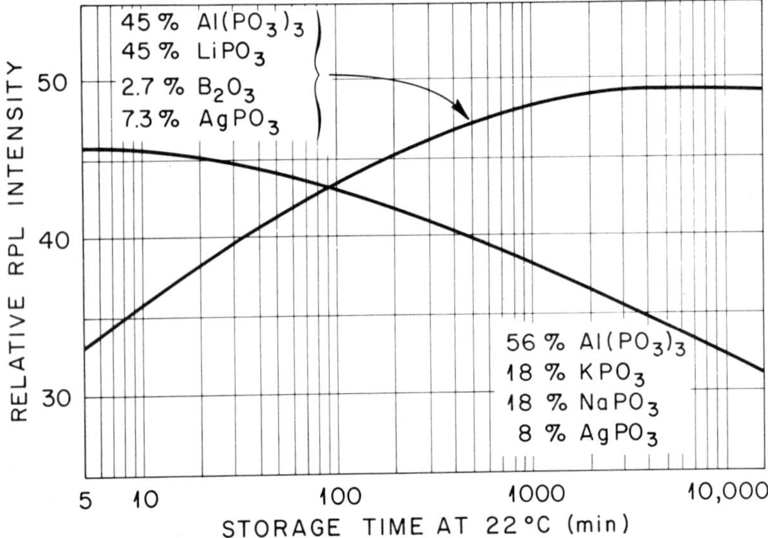

FIGURE 4-5.   RPL as a function of storage time at ambient temperature in two glasses with approximately the same silver content, but different base composition. (After Becker, 1967.)

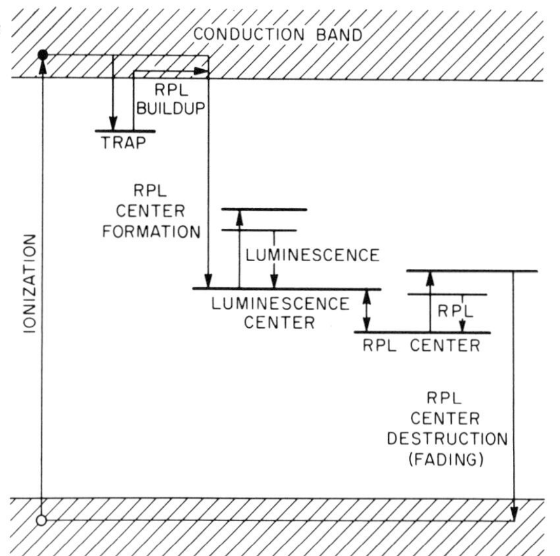

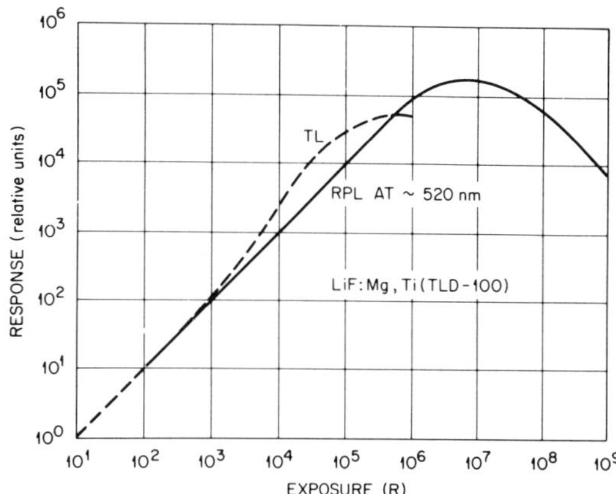

FIGURE 4-7. Radiophotoluminescence and thermoluminescence (dotted line) of TLD-grade LiF:Mg,Ti as a function of dose. (After Regulla, 1972.)

FIGURE 4-6. Schematic band diagram of the main processes occurring in an Ag-activated dosimeter glass: Electrons are lifted during irradiation into the conduction band and trapped either by luminescence centers, transforming them into radiophotoluminescence centers, or they are trapped in ineffective traps, from which they can be retrapped, as above. (After Vogel and Becker, 1965.)

thermally to the effective centers. This explains the build-up. The increased RPL build-up with decreased silver concentration can be explained by the changes in the relation of the quantities of traps to luminescence centers. Fading or ultraviolet bleaching may be explained by a recombination of electrons from the excited state of the RPL centers with electron vacancies in "hole" traps.

Findings concerning the structure of the centers are consistent with this theory. From electron spin resonance measurements (Yokota and Imagawa, 1967), it has been concluded that most important centers in dosimeter glasses at room temperature are $(Ag^+ + hole)$ having the capture level $dx^2 - y^2$ of $4d^9$ (Ag) and $2P\sigma$ oxygen, the $(Ag^+ + electron)$ with the 5S (Ag) orbital capture level, the latter being particularly important in glasses with a small $AgPO_3$ concentration and a $(Ag^+ + electron)$ pair coupled by exchange interaction. At higher doses exceeding several thousand R, higher aggregates of $(Ag^+ + electron)$ (triads, etc.) are increasingly added. The main reason for RPL build-up with time is the increase of $(Ag^+ + electron)$ triads and the decrease of holes in the PO_4 tetrahedrons. Several questions, such as a quantitative treatment of the fading processes and the polarization of the RPL light (Kishii and Sakurai 1965), remain, however, unanswered.

4.1.3. Composition and Basic Properties

So far, mostly silver-activated inorganic materials have been used, because silver seems to be superior to other potential activators such as cadmium and lanthanides, which also exhibit RPL in certain minerals and crystals. Other elements, however, may exist with similar or even better RPL properties if combined with an optimal "solvent." For example, Käs and Scharmann (1972) report a sensitizing effect of 0.001% of TiO_2.

Suitable inorganic crystals are difficult to grow with identical properties and the solubility of silver ions is often rather limited. RPL has, for example, been observed in silver chloride crystals (Vinduska, 1970). Only one system based on polycrystalline pellets of silver-activated sodium chloride was recommended for practical dosimetry (Druskina, 1965). Silver-activated ammonium salts and lithium hydride, which should be more sensitive to fast neutrons due to their hydrogen content, are very unstable. The LiF:Mg,Ti that is used to a large extent in thermoluminescence dosimetry has been shown to exhibit detectable RPL after exposure to gamma radiation exposures exceeding about 10 R if excited at 450 nm. The main emission maximum is around 520 nm (Regulla, 1972 and Claffy et al., 1971). There is no "supralinearity" of the RPL response comparable to that of the TL (Figure 4-7).

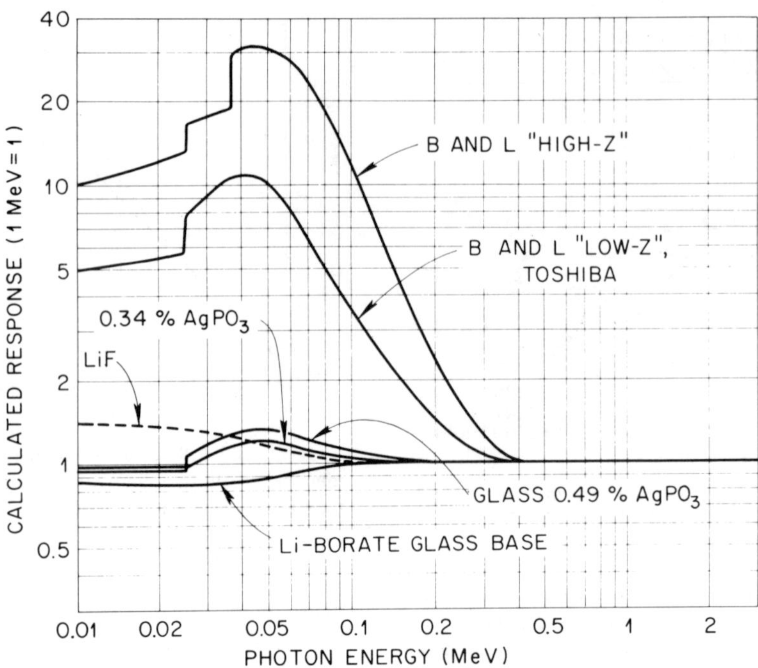

FIGURE 4-8. Calculated photon energy response of different commerical dosimeter glasses and an experimental glass based on lithium borate. (After Becker and Cheka, 1969.)

More studies have, however, been directed toward the modification or improvement of the metaphosphate glass base (the glass components and their ratios may be quite different without seriously affecting the RPL). For instance, experimental glasses having high concentration of P_2O_5, B_2O_3, SiO_2, BeO, and other oxides have been prepared. Most of these studies, of which only relatively few results have been published because of the commercial interests involved, led to glass bases consisting mainly of only a few metal metaphosphates such as $Al(PO_3)_3$, $Mg(PO_3)_2$, $Ba(PO_3)_2$, KPO_3, and $NaPO_3$, with additions such as P_2O_5, BeO, B_2O_3, or SiO_2 for improvement of the weathering stability or other reasons (see Table 4-1). For a detailed recent discussion of the technology of dosimeter glass production, see Käs et al. (1972b).

Lithium oxide/boron oxide glasses with a composition between $(Li_2O \cdot 2B_2O_3)$ and $(Li_2O \cdot 6B_2O_3)$ also are excellent solvents for silver salts (silver metaphosphate). The resulting glasses exhibit a strong RPL in the same spectral region as the metaphosphate glasses (Becker and Cheka, 1969) and have a considerably improved photon

energy response, particularly if their silver content is reduced (Figure 4-8). Other modern RPL glasses contain up to 75 mol % fluorides (mostly LiF) to reduce the energy dependence of the old phosphate glasses (Käs, 1972 and Käs et al., 1972a). Their energy dependence of response is not as low as of the borate glasses, but their weathering resistance is higher. Quite different inorganic and organic glass bases may also be suitable.

From the wide choice of possible compositions, the glass has to be "designed" to represent an optimized compromise between the following main requirements:

1. Low predose — The background luminescence of the unexposed glass was quite high in the Schulman-type glass (equivalent to about 10 to 40 R gamma exposure, depending on the fluorimetric reading technique). It has since been reduced, mostly by using the purest materials. All luminescent ions, particularly those having a strong emission in the orange-red, such as Mn^{++} (Peirson, 1958), as well as "quenching" ions such as Fe^{++}, have to be very carefully avoided. Improved formulas and special melting techniques (the

TABLE 4-1

Composition and Properties of Some Radiophotoluminescence Glass Dosimeters

Source	Name	Composition (% by weight)						Predose (R γ exposure equivalent)[2]	Maximum energy dependence[3]	Relative sensitivity (^{60}Co)	Reference
		Ag	Al	Li	P	O	Others				
Bausch & Lomb	High Z[1]	4.3	4.7		28.4	44.1	10.8 Ba, 7.7 K	∿10−40	30−32	1	Schulman et al., 1951
Rochester, N.Y.	Low Z[1]	4.3	4.7	1.9	33.7	52.3	3.1 Mg	∿10−40	7−10	1	Ginther and Schulman, 1960
Toshiba	FD-1	3.7	4.6	3.6	33.5	53.7	0.85 B	∿0.1−0.3	7−10	2.2	Yokota and Nakajima, 1965
Tokyo, Japan	FD-3	3.3	5.1	3.6	34.5	53.5		∿0.1−0.3	7	2.2	Yokota and Nakajima, 1965
	FD-5	0.52	6.1		33.1	51.3	8.9 Na	∿0.1	3.5	3.2	Yokota et al., 1971a
	FD-7	0.17	6.1		31.5	51.2	11.0 Na	∿0.05−0.1	2.8	3.6	Yokota et al., 1971
CEC, Montrouge,	Type 1	2.9	3.5	2.5	33.6	52.3	4.7 Na, 0.5 Be	∿0.6	6.5−7.3	1.9	Francois et al., 1965
France	Type 31	2.4	3.5	0.7	34.1	52.8	4.8 Na, 1.7 Be				Francois et al., 1969
Jenaer Glaswerke	RPL-I	4.6	3.1	4.45	34.1	52.8	0.9 Mg	∿0.3	7.5−11	2.1	Jahn, 1969
Schott & Gen.,	RPL-II	3.67	4.67	3.47	33.5	57.7	0.9 B	∿0.3	7		Jahn, 1969
Mainz, Germany	RPL-III	0.62	0.51	7.34	34.7	55.85	0.95 B	∿0.5	4−5	3.1	Becker, 1968
	RPL-V	0.58	6.1		32.3	50.2	6.45 K, 4.4 Na	∿0.3	6.8	2.4	Becker, 1968
	9929[4]	0.4	6.1		33.2	51.5	8.86 Na	∿0.25	3.5−4.7	2.4	Becker, 1968
	9930[4]	0.58	6.1		31.5	50.0	12.9 K	∿0.4	7.1	1.7	Becker, 1968
ORNL, Oak Ridge, Tenn.	[4]	0.13		4.51	0.13	67.6	24.1 B	∿0.4	<1.2	0.3	Becker and Cheka, 1969

[1] Not in production anymore.

[2] Depends on optical system of the reader, precleaning of the glass (removal of fluorescent surface contamination), and time elapsed since manufacturing or thermal annealing (measurable dose build-up occurs in low-predose glasses during extended storage).

[3] The lower value is usually the measured and the higher value the calculated ratio between maximum sensitivity at ∿30−40 keV and minimum sensitivity at 1.25 MeV.

[4] Experimental glasses, not commercially available.

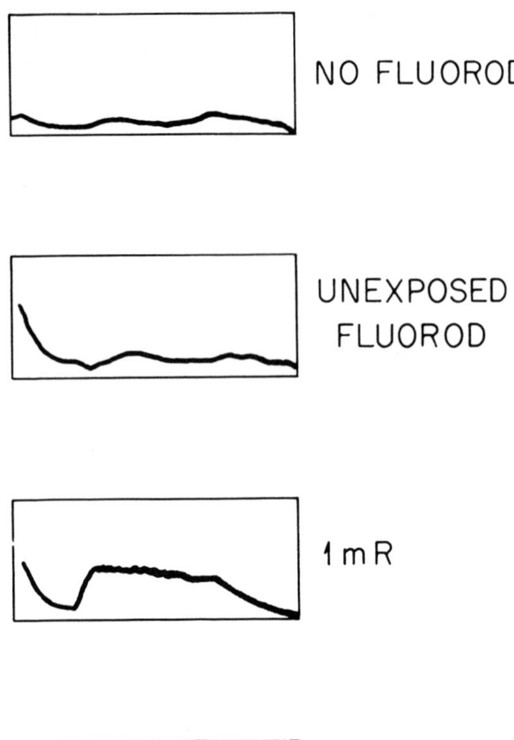

FIGURE 4-9A. Oscilloscope traces (20 mV/cm, 5 μsec delay time) of RPL signal in Schulman-type dosimeter glasses as a consequence of pulsed UV laser excitation. (After Kastner et al., 1967.)

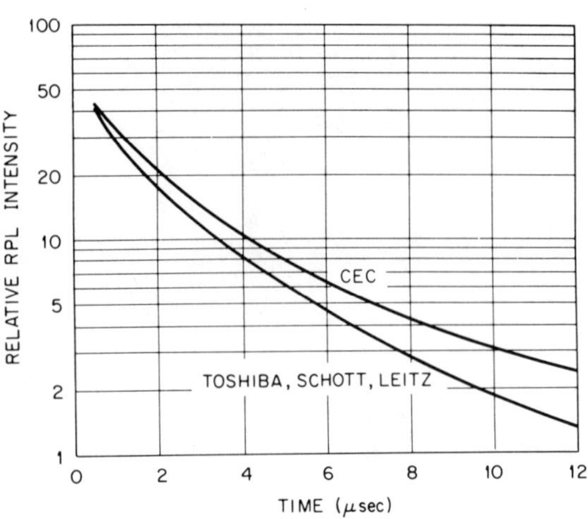

FIGURE 4-9B. Decay time of radiophotoluminescence after pulsed excitation in dosimeters glasses of different manufacturers, exposed to 2 R of gamma radiation. (After Hillenkamp and Regulla, 1971.)

crucible material, atmosphere in the oven, time and temperature of the melting, and heat after-treatment may be among the factors which affect the predose — see, for example, Käs et al., 1972b) helped also to make it possible to produce glasses in large amounts and with reproducible properties, which have predoses of less than 0.3 R if read in an optimized RPL reader.

In some glasses, the decay time for the radiation-induced luminescence is much slower (about 3 μsec) than for the "predose" (about 0.3 μsec — Kastner et al., 1968, and 1970). After pulsed UV laser excitation, delayed RPL measurement permits detection of gamma doses as low as 1 mrad even in the old high-predose fluorods (Figure 4-9). Unfortunately, these promising results have not been confirmed completely for other glass types (apparently the type of luminescent impurities and their fluorescence decay times

can differ more or less from the RPL, even if the RPL decay times in various types of glasses are rather similar (Figure 4-9B — also see Barthé et al., 1969). Attempts are being made to use the principle in practical glass dosimeter readers (Yokota et al., 1971b and Hillenkamp and Regulla, 1971).

2. **High sensitivity** — Silver concentration, glass base composition, and certain "quenching" cations have a strong influence on the glass sensitivity. It is not yet possible to explain or to predict the effect of the glass base on sensitivity. If, for instance, $Ba(PO_3)$ and KPO_3 in the Schulman-type dosimeter glass are replaced by $LiPO_3$, the sensitivity is increased by a factor of two. If, on the other hand, $LiPO_3$ in a special experimental glass is replaced by $Mg(PO_3)$, a sensitivity drop by a factor of 20 is observed (Becker, 1966).

The silver concentration is also important. Depending on the glass base, maximum RPL is usually obtained with glasses containing 1 to 2% silver (Figure 4-10). The reason the conventional dosimeter glasses contain more silver (2.5 to 4.5%) is that the build-up of RPL in glasses with low silver content is a slow process. However, glass bases have been found which show a fast build-up with a lower silver content and high final sensitivity.

An increase of the $LiPO_3$ content of glasses partially compensates for the slower build-up rate due to a reduced Ag concentration (Becker, 1965

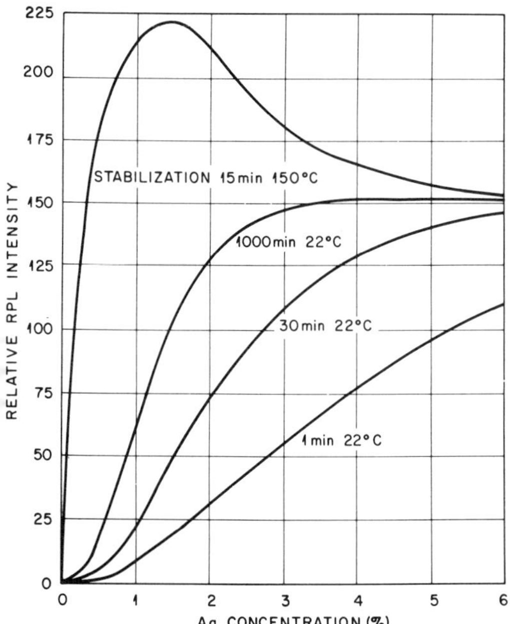

FIGURE 4-10. RPL intensity as a function of silver concentration for different storage times at ambient temperature, and after "stabilization" for 15 min at 150°C. (After Becker, 1966.)

and Glass RPL III in Table 4-1). An even faster build-up despite a silver content of only 0.4% has been obtained in a glass in which $LiPO_3$ was replaced by $NaPO_3$ (Becker, 1968 and Glass no. 9929 in Table 4-1). Essentially the same glass became commercially available later (Yokota et al., 1971a and Glass FD-5 in Table 4-1).

3. Low energy dependence – For the more important application of RPL glasses in the dosimetry of X- and gamma radiation, a minimum photon energy dependence is desirable. A reduction of the effective atomic number Z of the glasses is possible by replacing high Z cations in the glass base by low Z ones, and by reduction of the silver content. The maximum sensitivity is usually between 40 and 50 keV, while the minimum is between 0.3 and 2 MeV. The energy dependence of the glass can, therefore, be expressed as

$$\text{E.D.} = \frac{\text{relative sensitivity at } \sim 45 \text{ keV}}{\text{relative sensitivity at } \sim 1 \text{ MeV}} \qquad (4.1)$$

This value can be calculated under the assumptions of electron equilibrium, monoenergetic radiation, and narrow-beam geometry, neglecting the self-shielding of the glass as well as the fluorescence and electron escape:

$$\text{E.D.} = \frac{[(\mu_{en}/\rho)_{glass}/(\mu_{en}/\rho)_{air}] \, 50 \text{ keV}}{[(\mu_{en}/\rho)_{glass}/(\mu_{en}/\rho)_{air}] \, 1 \text{ MeV}}, \qquad (4.2)$$

(μ_{en}/ρ) = mass energy transfer coefficients for air and glass, calculated from

$$\frac{\mu_{en}}{\rho} = \frac{\Sigma(\mu_{en})_i}{\rho_i} \, P_i,$$

where P_i = percentage by weight of the element i.

If the fluorescence and electron escape from the glass are considered (Feige et al., 1965), the calculated values have to be reduced by about 10%. Experimental values obtained with X-radiation of 50 keV effective energy are about 60 to 80% of the calculated values, depending on the degree of filtration because even well-filtered X-radiation has a rather broad energy distribution. If monoenergetic fluorescence X-radiation is used, the agreement is usually within the limits of accuracy of calculation and measurement. Using the lightest cations available, the energy dependence may be reduced from about 30 (U.S. Navy glass) to about 10 in modern phosphate glasses. Because of the dominating effects of the Ag, however, further reduction is only possible if the silver concentration is also reduced. In experimental metaphosphate glasses, factors of about 4 have been obtained (Becker, 1965; Yokota and Nakajima, 1965; and Yokota et al., 1971a).

Lithium borate glasses that have, if unactivated, an energy dependence of less than 1 can be activated with 0.3 to 3% $AgPO_3$ (Becker and Cheka, 1969). The energy dependence of glasses containing less than about 0.5% $AgPO_3$ is comparable to that of LiF (Figure 4-8). Such glasses need, however, a stabilizing heat "development" prior to reading and have to be protected from excess humidity.

The energy dependence of dosimeter glasses may, of course, be compensated to a large degree for photon energies exceeding 35 to 80 keV, depending on the degree of sophistication of metal filters or filter combination. Below this energy, several glasses behind different filters have been used as a "poor man's spectrometer," similar to the filter analytical techniques in film dosimetry. These techniques are discussed in more detail in Section 4.3.1.

4. Stability – Phosphate glasses are less stable

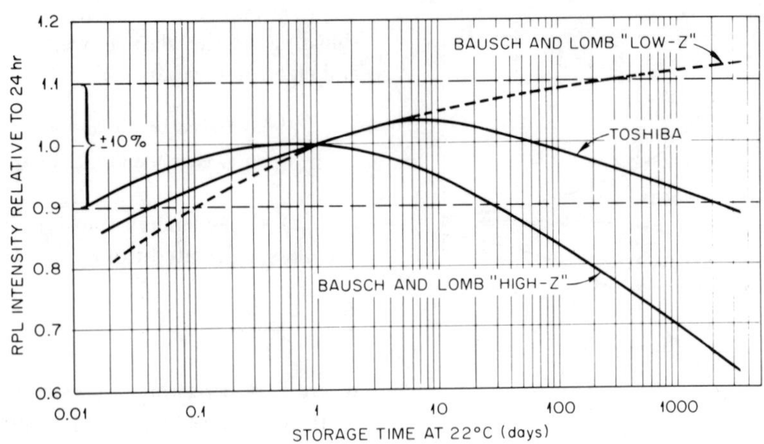

FIGURE 4-11A.   Build-up and fading of RPL in three types of glasses (Bausch & Lomb high Z and low Z, Toshiba FD-1) at room temperature ($\sim 22°$C) during ten years of storage, normalized for the RPL one day after exposure. (After Cheka, 1964, 1968, and unpublished.)

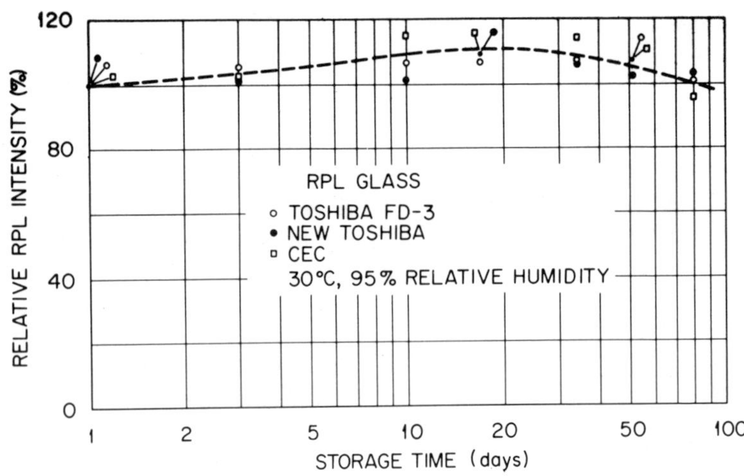

FIGURE 4-11B.   RPL in three modern dosimeter glasses as a function of storage time at 30°C and 95% relative humidity, normalized for the response one day after exposure.

against humidity and certain chemicals than conventional glasses. The surface of RPL glasses, particularly those containing much $LiPO_3$ or B_2O_3, may soften and become sticky when kept for extended periods at high humidity without protection. Simple washing prior to reading completely removes the sticky layer without affecting the stored information. Increasing the glass resistance against disturbing environmental effects usually is not very important because the glasses are encapsulated for use. For other purposes, such as in vivo dosimetry, however, it is desirable to improve resistivity by modification of the glass base (Yokota and Nakajima, 1965 and Käs, 1972) or by coating of the glass surface with inorganic (SiO_2) or organic materials (polymers).

5. Fast build-up and slow fading of RPL — Unfortunately, these requirements somewhat contradict each other because glasses with a fast build-up usually also exhibit a pronounced fading and vice versa (Becker, 1967). Glasses that exhibit an RPL which is constant within about $\pm$ 10%

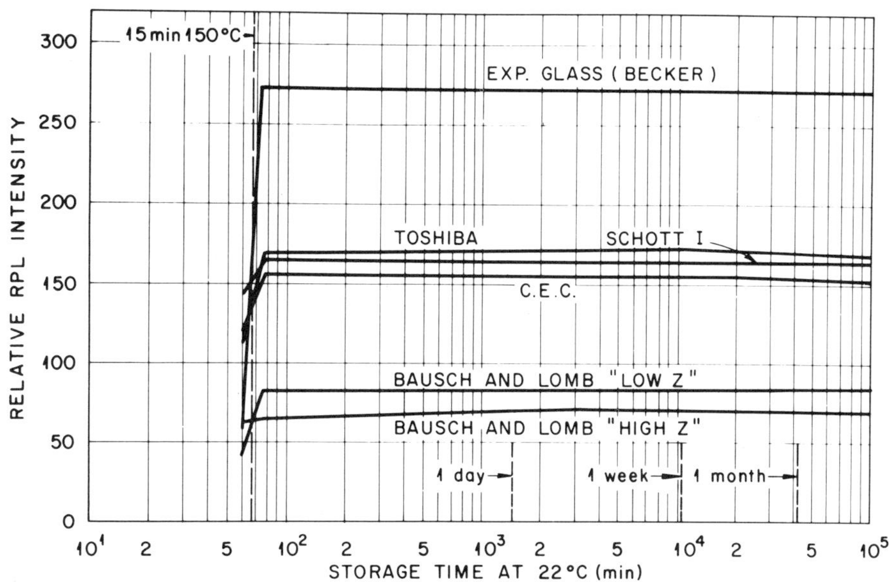

FIGURE 4-12. Storage stability of different RPL glasses of the same dimensions after exposure to the same gamma radiation dose, prior to and after heat stabilization for 15 min at 150°C. (After Becker, 1967.)

between 1 hr and several years after irradiation at normal temperature can be considered a reasonable compromise.

Extensive long-term studies on RPL fading have been carried out with glasses that have been exposed to various dose-levels and stored at ambient temperature ($22\pm2°C$) for 10 years at ORNL by Cheka (1964, 1968, and unpublished). In the Bausch and Lomb "high Z" glass, maximum RPL is already reached less than 1 day after irradiation, but consecutive fading is rather rapid ($\sim30\%$ during the first year). In the "low Z" as well as the Toshiba FD-2 glass, stabilization proceeds slower (maximum RPL after about 100 days), but the fading rate is also much reduced (Figure 4-11A). At least in the FD-2 glass, the RPL between 1 hr and 10 years after exposure does not vary more than within $\pm10\%$ due to build-up and fading, thus meeting the above criteria. In Figure 4-11B, the stabilities of three modern dosimeter glasses have been compared in a simulated tropical climate of 30°C and 95% relative humidity. The changes that occur in the RPL between one day and three months after exposure are within ca $\pm8\%$.

Obviously, stability in such glasses does not represent a problem in most practical applications including personnel dosimetry. Only under special conditions, such as accurate measurements immediately after exposure, a correction or "stabilization" treatment may be desirable. Measurements at increased temperatures also require special precautions. "Stabilization" (optimization of RPL) can be obtained in the Toshiba FD-2 glass by storing it for 24 hr at 75°C, for 15 to 20 min at 100°C, or for 5 min at 150°C (Figure 4-3).

Obviously, the precise time-temperature combination for stabilization depends on the inherent RPL kinetics of the glass, but a treatment of 10 to 15 min at 150°C represents a good compromise for different glasses. Some results of such a treatment are given in Figure 4-12. There is no observable effect of the dose-level on the RPL kinetics up to exposures of about 1,000 R. (For some recent fading studies in C.E.C. glasses, see Barthé et al., 1972.)

Build-up and fading kinetics depend not only on the previously discussed parameters of storage temperature, glass base composition, silver concentration, and the (indirect) effect of dose level at gamma doses exceeding several thousand rad (see section 4.2.1), but also on LET. Some glasses have been found to show a substantially different build-up speed, as well as a different thermal stability of their RPL after exposure to thermal neutrons instead of gamma radiation (also see section 4.2.2).

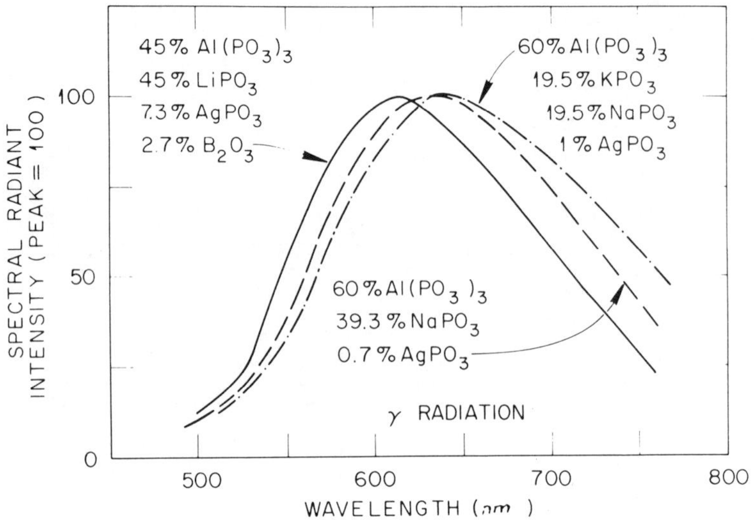

FIGURE 4-13. "Absolute" RPL light spectra of various dosimeter glasses. (After Becker and McQuilling, 1967.)

As can also be seen in Figure 4-3, a 10 to 20_ min 400°C treatment is sufficient to anneal the RPL completely. If the glass is kept at high temperatures for an extended period of time, a small predose increase may be observed. Also, after exposure of the glass to extremely high doses, there may be some permanent predose increase because of the formation of new luminescence centers. Even after numerous annealings there is no change in sensitivity (Becker, 1965). Glasses may, therefore, be annealed many times for reuse. Dirt should be removed carefully before heating the glasses because surface contaminations may become "burned-in" into the glass surface and cause an apparent predose increase. A different behavior is observed due to the interference of discoloration kinetics at gamma dose-levels exceeding ∿1,000 rad.

Other factors, such as the costs of the high-purity constituents and the manufacturing process, the desired size and shape of the glasses (square blocks of different sizes, plates, needles, cylindrical units, and spheres have been prepared for different purposes) also may be important factors for the development of a dosimeter glass.

Accurate knowledge of RPL spectra is important for choosing an optimal combination of optical filters and RPL detectors that give best sensitivity and high background (predose) discrimination. Spectra measurements without correction of the RPL light detector response characteristics are only of little value. Schulman et al. (1951) first measured RPL "absolute" spectra (corrected for the photomultiplier response) of metaphosphate glasses with 365 nm UV excitation in the 420 to 640 nm range. They found a 630 nm peak for glasses containing 2 to 8% AgPO₃ and a second broad maximum around 500 nm for 12 to 16% AgPO₃ glasses. In the U.S. Navy DT-60 glass (8% AgPO₃) there is a small, poorly resolved peak around 480 nm and a main peak at 640 nm. A rather symmetric spectrum peaking at 610 nm (50% of maximum RPL at 545 and 685 nm) was reported by Yokota and Nakajima (1965) for their glass.

More systematic studies by Becker and McQuilling (1967) resulted in the absolute RPL spectra in Figure 4-13 for different glasses, with maxima between 610 and 640 nm and a slow decrease into the red. At least for higher doses, there is no effect of the dose on the shape of the spectrum. Also during build-up, no significant change in the spectrum takes place. In the 0.5 to 5% silver range no effect of silver concentration is observed and the glass base also has not much influence on the spectrum unless it is altered rather dramatically. In lithium borate glasses, the RPL maximum is located between 575 and 590 nm (Becker and Cheka, 1969). Differences have also been observed in the RPL spectra of glasses after exposure to neutron instead of gamma radiation (Becker and Cheka, 1968 and Regulla, 1968).

Several authors studied the effect of temperature during irradiation on the glass sensitivity. Between -10 and +50°C an increase of about 0.3%/°C was found (Schulman et al., 1953). Others report an almost linear increase of 0.2 to 0.3%/°C between 0 and 50°C. In most of these experiments the effect is partially simulated by the temperature influence on build-up kinetics. Experiments involving a stabilization treatment of the glasses have shown only a 0.1 to 0.2%/°C increase of sensitivity between -20 and +80°C (Becker, unpublished).

Schulman et al. (1953) first observed an influence of the reading temperature on the RPL intensity. They determined the coefficient to be -0.7%/°C. Other authors reported somewhat higher values between -0.86 and -1.1%/°C, but -0.9%/°C seems to be a good average of these values. It will, however, be superimposed by the temperature dependence of the instrument and the standards if the whole reader is used at different temperatures. The temperature dependence is much smaller than that of the glass alone with the U.S. Navy CP-85 A/PD reader. This result has been confirmed by several investigators who did not find any temperature dependence if the measurement was done in a Toshiba reader against a Sm-activated standard.

The effect of light on glasses depends mainly on the pre-irradiation of the glass and the wavelength of the light. Like increased temperature, infrared light increases the RPL build-up and fading rate. Visible light has only very little or no effect. Near ultraviolet in the region of the RPL excitation causes a slow destruction of RPL centers ("bleaching"). This effect is too small to disturb the usual fluorimetric reading, which requires only a few seconds. It may, however, cause a considerable loss in RPL intensity if the glass is exposed to bright sunlight for extended periods of time (Schulman et al., 1953; Blaylock and Witherspoon, 1964; Kiefer et al., 1965; and others). It has been found (Becker, 1965) that the bleaching at constant UV intensity is quite rapid at first and becomes smaller as a linear function of log time. During the first hour of intense exposure with 365 nm UV, the RPL intensity dropped by almost 15%; 5 hr is required to obtain a 30% fading.

Short wavelength ultraviolet light can also create new RPL centers (Becker, 1964). Because of its strong absorption in the glass (the UV absorption edge of modern glasses is around 300 nm; 254 nm UV were used in the experiments), the effect is localized to the glass surface and the kinetics of the UV-induced RPL is different. If glasses are exposed to direct sunlight, all effects that have been mentioned may superimpose each other to a varying degree. Indirect, natural, or artificial room illumination causes no observable effect on the glasses.

No effect of dose-rate on the response of RPL glasses has been observed in careful studies by several investigators, at least in the range of 10^{-4} R/min (Yokota et al., 1962) to $> 10^{10}$ R/sec (Tochilin and Goldstein, 1966). A single report of a 10% drop of sensitivity between 0.17 and 5 x 10^3 R/min (Kondo 1961) seems to be in error.

There is only a small directional dependence of the photon radiation response of glass blocks, plates, and rods, even for low photon energies (less than ±10% for photon energies above 30 to 40 keV). A serious directional dependence may be introduced, however, by asymmetrical energy compensation filters or a badge design which results in a strong attenuation of the radiation under certain angles of radiation incidence. This directional dependence can be avoided by a symmetrical design, in particular by the use of conical holes that overlap in such a way that the projection in the glass volume stays constant under different angles of incidence (Becker, 1963 and Maushart and Piesch, 1967).

4.2. Dosimetric Properties

4.2.1. Dose Range and Accuracy

The RPL intensity is a linear function of dose up to at least several thousand rad of gamma radiation or a corresponding X-radiation, beta, or thermal neutron dose. The exact dose-level on which the response becomes gradually sublinear depends on the glass dimensions and the readout geometry. If the path length of the UV in a small glass needle is only 1 mm, for example, the response may be linear up to 10^5 rad (Barr et al., 1961).

For the usual size glass squares and readers, the RPL reaches a maximum at about 3 to 5 x 10^4 rad of gamma radiation and decreases with further irradiation (Cheka, 1964; Becker, 1965c; and Figure 4-14). If the RPL is, however, not measured in the orange-red spectral region but at a shorter

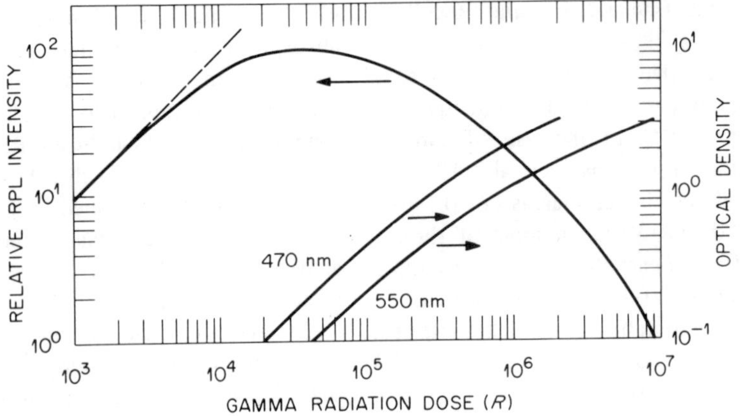

FIGURE 4-14. RPL intensity and optical density at various wavelength in Toshiba FD-2 glass block. (After Becker, 1965c.)

wavelength such as 420 nm, the RPL increases monotonously with dose up to $\sim 10^6$ rad (Becker, 1965c). It is also possible to calibrate the dosimeters for the region of decreasing RPL with increasing dose above the point of maximum RPL. Measurements up to $\sim 10^9$ rad thus become feasible (Freytag, 1971).

The cause of nonlinearity is the absorption band with a maximum at 315 to 320 nm that extends into the visible region. Absorption measurements at 320 nm may be used for dosimetry above a few hundred rad; after exposure to several thousand rad, the glass will be visibly discolored. It becomes slightly yellow at 10^4 rad, yellowish-brown at 10^5 rad, deep brown at 10^6 rad, and almost black at 10^7 rad. The absorption at 320 nm is an almost linear function of the silver concentration (Becker, 1965c). If the phosphate glass base without silver activator is exposed to high radiation doses, an absorption band in the red is observed.

The discoloration was first used for high-dose measurements by Schulman et al. (1955) and was later studied by Rabin and Price (1955) and Davison et al. (1956). Unlike the RPL build-up after exposure, the absorption band at 320 nm starts to fade immediately after exposure. The fading rate depends somewhat on the dose-level and strongly on the silver concentration (at very low silver concentration, a build-up after exposure can be observed). Of course, it also depends on storage temperature; the discoloration may be completely annealed by a high-temperature treatment (Figure 4-15).

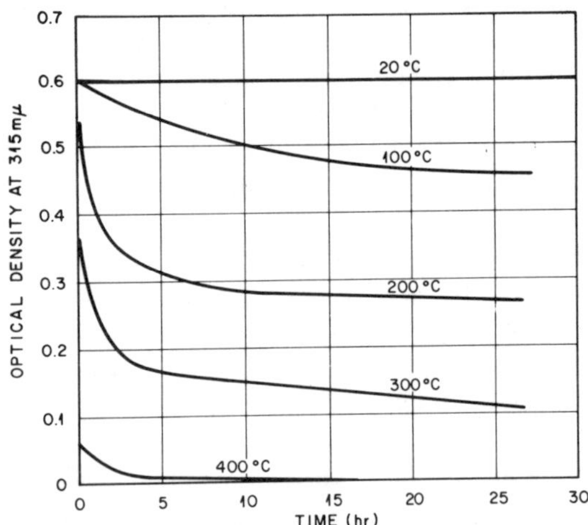

FIGURE 4-15. Fading of the 320 nm absorption band in a Toshiba dosimeter glass at different storage temperatures. (After Becker, 1965c.)

Corrections for the fading that occurs during and after exposure may be made by the use of kinetic equations (Rabin and Price, 1955). The radiation-induced absorption band at 320 nm is particularly stable in lithium borate glasses with a low silver content (Cheka and Becker, 1969). Only a few percent fading occurs in such glasses during one week at ambient temperature; by choosing the proper wavelength, measurements at temperatures in the 200°C range became possible.

Because the radiation-induced absorption bands are less stable than the RPL, the kinetics of RPL build-up and fading becomes different from the above-discussed low-level behavior if the glass has been exposed to doses exceeding several thousand rad. With the decrease in the absorption of the excitation-UV during storage, the build-up of the RPL becomes more pronounced (Becker, 1965c). If the glass is heated, more absorption than RPL centers are destroyed and the linear response range is extended to higher doses (Lee et al., 1961). RPL glasses also exhibit thermoluminescence when heated (Regulla, 1968). There are two broad TL glow peaks at about 200 and 280°C. This effect can, in principle, also be employed for high-dose measurements.

For most applications the lower limit of sensitivity is more interesting than the high-dose range. It is determined mainly by the following factors:

1. Predose of the glass. If a glass such as the older U.S. glasses exhibits a background luminescence corresponding to a gamma dose of 30 to 40 rad, only doses exceeding several rad may be measured with a reasonable degree of accuracy; on the other hand, modern glass dosimeters usually have a very uniform predose corresponding to only a few hundred mrad.

2. Consistency of predose and sensitivity. Because it is obviously not desirable to calibrate each individual dosimeter glass, a sufficient consistency of the dosimetric properties within large badges of dosimeters is important. Deviations in the dimensions of the glasses, inhomogeneities, luminescent surface contaminations, and fluctuations in the positioning of the glass in the reader and the inherent reader precision all contribute to the overall standard deviation in predose and sensitivity. There have been considerable problems with early glass types, but in modern glasses the standard deviation is as low as 0.1 to 0.6% (Menkes, 1966a and b). Careful cleaning and accurate positioning of the glasses in the reader are, of course, important for obtaining good accuracy, in particular in the low-dose range.

3. Glass size. In principle, accuracy and sensitivity should be independent of the glass size. However, under most experimental conditions it is easier to obtain precise results if the volume of the glass is not too small.

4. Number of readouts. The accuracy of a measurement is substantially increased by either repeated reading of the same detector, or by exposing more than one detector (Figure 4-16).

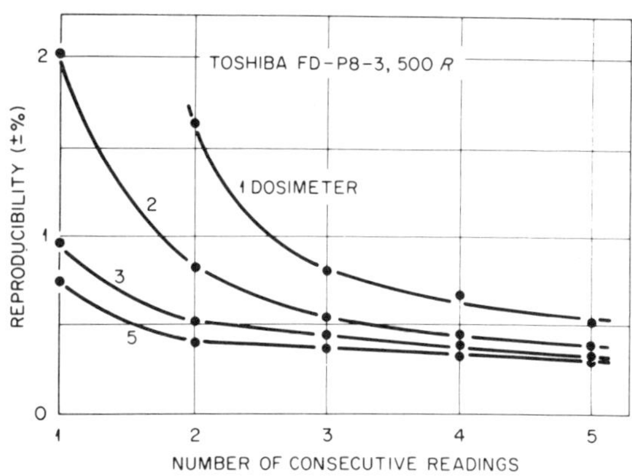

FIGURE 4-16. Influence of either multiple reading of one, or of more than one simultaneously exposed Toshiba glass blocks on the standard deviation of the dose determination under laboratory conditions.

Some accuracies that have been determined with a modern system are compiled in Table 4-2. Depending on the degree of care taken and the dose-level, the standard deviations are usually between 0.5 and 3% as compared with the DT-60 glasses of the U.S. Navy with σ = 6 to 20%. For obtaining a high reproducibility, the glasses have to be cleaned from luminescent impurities by a sequence of washings (preferably in an ultrasonic bath) with liquid detergents, diluted acid (chromic acid and nitric acids are recommended because of their oxidizing effect), and organic solvents such as pure acetone or alcohol (Becker, 1964; Kathren, 1966; and Yokota and Yuhei, 1968).

If the once-cleaned glasses are always well protected from contamination inside their encapsulation and not exposed to dust or touched with the fingers during evaluation, or if doses in excess of several rad are to be measured, such a cleaning procedure may not be necessary. It is, however, advisable to clean the glasses prior to their annealing at higher temperature and prior to their first use because they may have accumulated dirt as well as background dosage during extended storage.

If those precautions are taken, modern glasses offer a degree of accuracy and stability which, in combination with the unlimited remeasurability of the results, make them eminently suitable for precision measurements, intercalibrations by mail, etc. They have already been used for international comparisons of medical radiation sources by the International Atomic Energy Agency (Dorofeev and Somasundaram, 1971) and their use as a

TABLE 4-2

Standard Deviation of Gamma Radiation Dose Measurements with RPL Glasses

Glass type and size (mm)	Reader	Dose (rad γ)	Standard deviation (%)	Reference
Toshiba FD-P8-1 (8 x 8 x 4.7)	Toshiba FGD-3B	1–1000	<3	Becker, 1964
		Predose	<4	Kathren, 1966
	FGD-6	0.01	4	Yokota and Nakajima, 1965
		0.015	3.5	Yokota and Nakajima, 1965
		0.025	3	Yokota and Nakajima, 1965
		0.05	1.5	Yokota and Nakajima, 1965
FD-P6-1 (6 x 6 x 3.3)	FGD-3B (modified)	Predose to 35	$\sim$0.35	Menkes, 1966
	FGD-3B	Predose to 20	$\sim$1.6	Menkes, 1966
FD-1 fluorods (1 x 6)	FGD-3B	50	6	Yokota and Nakajima, 1965
		1000	3.9	Yokota and Nakajima, 1965
		Predose to 1150	$\sim$2.5	Menkes, 1966
FD-1, Bausch & Lomb high Z and low Z fluorods (1 x 6)	Bausch & Lomb and Turner Assoc.	10–5000	2-5	Cheka, 1964
FD-1	Toshiba		3	Routti and Van de Voorde, 1972

secondary standard in developing countries is being considered.

4.2.2. LET, Neutron, and Charged Particle Response

As in almost all other solid-state detectors, an increase of the LET of the ionizing radiation results in a decrease in response. For alpha particles with an energy of several MeV, only 16 to 20% of the gamma radiation sensitivity per rad is observed in RPL glasses (Kondo, 1961 and Tochilin et al., 1963), and with protons of various energies, the results given in Figure 4-17 have been obtained. There are probably several reasons for this behavior, including thermal annealing effects along the track, but the primary reason appears to be saturation ("overkill" of the available RPL centers) in a narrow channel along particle track due to short-range, low energy delta electrons (Katz et al., 1972).

There are qualitative and quantitative differences in the LET response of RPL glasses. Differences in the RPL spectra (Figure 4-18) as well as in the optical absorption spectra in some glasses following exposure to thermal neutrons instead of

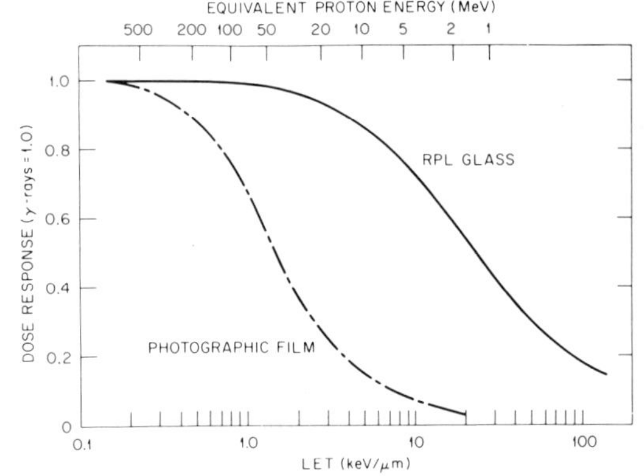

FIGURE 4-17. Response per tissue rad of a Toshiba RPL glass and a photographic film as a function of the linear energy transfer of the radiation, normalized for the response to gamma radiation. (After Tochilin, unpublished.)

gamma or X-radiation strongly suggest the formation of different RPL and absorption centers (Becker and Cheka, 1968). Deviations in the RPL build-up and fading as a function of LET have also been observed (Figure 4-19). The neutron response of RPL glasses has been the subject of numerous

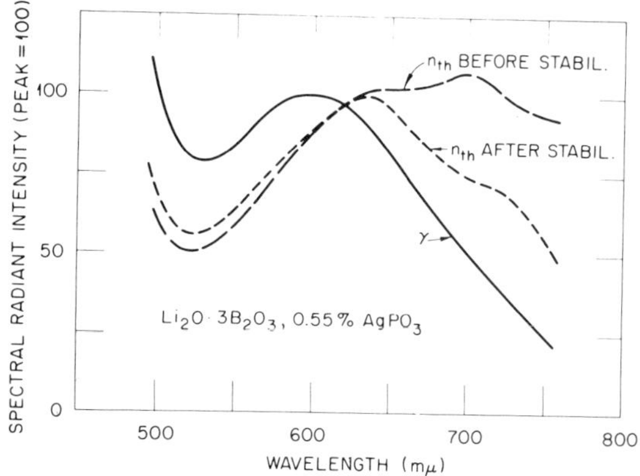

FIGURE 4-18. Radiophotoluminescence spectra induced in an experimental lithium borate glass by gamma radiation and by thermal neutrons before and after stabilizing heat treatment, normalized for emission maximum. (After Becker and Cheka, 1968.)

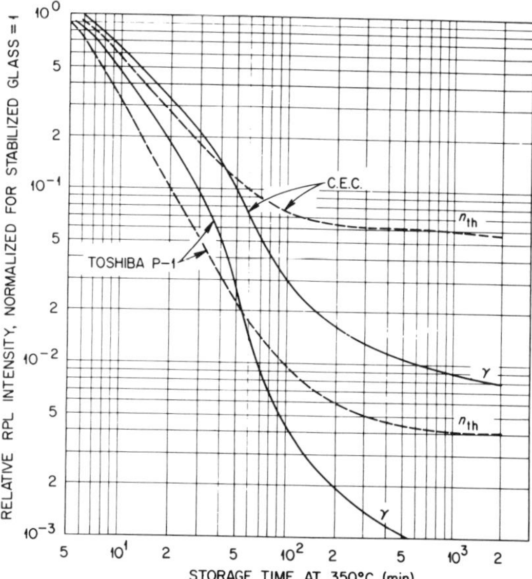

FIGURE 4-19. Relative gamma and thermal neutron induced RPL intensity in two commercial glass types as a function of storage time at 350°C, normalized for reading after stabilization. (After Becker and Cheka, 1968.)

studies (for compilations, see Becker and Tuyn, 1965 and Piesch, 1972). In glasses that contain larger amounts of thermal neutron absorbers, such as lithium and boron, the thermal neutron sensitivity is high, while the inherent fast-neutron sensitivity is rather small in all glasses.

The thermal neutron response of a glass depends chiefly on

a. its content of thermal neutron absorbers (Ag, B, Li, etc.);

b. the glass dimensions, which affect the reabsorption of beta and gamma radiation from neutron-activated silver; and

c. the glass surroundings, particularly shields containing neutron absorbers or "converters."

The results of various measurements of the thermal neutron sensitivity (expressed as the dosimeter reading in rad gamma equivalent over the first collision neutron dose in the glas in rad; 1 rad thermal neutrons equals 2.9×10^9 n/cm^2; to obtain the first collision dose in tissue the values should be multiplied by a factor of 11) can be summarized as follows: For larger blocks of the high Z glass inside a photon energy compensation shield (1.1 to 1.25 mm lead), sensitivities of 0.24 to 0.38 (24 to 38% of the gamma radiation response) have been found (Ballinger and Harris,

1959 and Peirson, 1958). For unshielded glass rods of the same glass, the sensitivity is between 0.4 and 0.1 (Kondo, 1960 and Bernard et al., 1961). The main neutron absorber in this glass is silver, resulting in a relatively strong dependence of the response on the glass geometry.

The lithium-containing low Z fluorods exhibited a thermal neutron sensitivity of $\sim$8 (Bernard et al., 1961); for the original Toshiba glass blocks, sensitivities between 10 and 15 have been reported for either unshielded glasses or blocks shielded with such metals as brass, tin, and bismuth (Yokota et al., 1961; Miyanaga et al., 1963; Piesch, 1964; and Becker and Tuyn, 1965). A shield consisting of 0.3 to 1 mm of cadmium reduces the sensitivity of blocks and needles to 3.5 (Yokota et al., 1961 and Piesch, 1964). With lithium shields* of sufficient thickness, the thermal neutron response can be drastically reduced (Bernard et al., 1961).

For measurements of thermal neutron fluxes in mixed radiation fields, either glasses with the same base composition, but a varying silver content (Kondo, 1960b) or, better, glasses with a constant

---

*Lithium shields are superior to other thermal neutron absorbers such as ^{10}B and ^{113}Cd because the neutron absorption is not associated with the emission of penetrating radiation. The Li density in LiF exceeds that of Li metal. A capsule of LiF with 3 mm wall thickness attenuates the thermal neutron flux by a factor of 10^6. Such shields can easily be prepared (Svikis, 1963).

silver content, but varying concentrations of Li and/or B (Becker and Tuyn, 1965) may be used.

The components of mixed radiation fields can be measured separately by the use of dosimeter pairs which are encapsulated in various thermal neutron absorbers (Yokota et al., 1961; Miyanaga et al., 1962; Piesch, 1964; Becker and Tuyn, 1965; and others). Mostly dosimeter pairs with and without lithium shields, or placed in Sn and Cd filters of equal thickness, have been employed.

In another method, glass pairs with identical gamma sensitivities, but different thermal neutron sensitivities, are used. This can be done either by replacing natural lithium in a lithium phosphate glass by $^7LiPO_3$ (Kerr and Cheka, 1967) or by reducing the thermal neutron sensitivity of the glass by modifications in its composition. For example, a glass consisting of 60% $Al(PO_3)_3$, 19.5% KPO_3, 19.5% $NaPO_3$, and 1% $AgPO_3$ corresponds in its gamma response characteristics to the Toshiba FD-1 glass, but has only 1% of its thermal neutron sensitivity due to its reduced silver content and the absence of Li and B (Becker, 1968).

The experimental results are in good agreement with the response values R calculated by the formula

$$R = 0.18\,D + R_{Ag} \tag{4.3}$$

where

$$
\begin{aligned}
D \quad &= \quad \text{energy absorbed in the glass for the}\\
&\qquad (n, \alpha)\text{-reactions of Li and B,}\\
R_{Ag} &= \quad \text{the component of the response due}\\
&\qquad \text{to the silver activation, and}\\
0.18 &= \quad \text{the relative efficiency of the alpha}\\
&\qquad \text{particles.}
\end{aligned}
$$

D can be calculated by the formula

$$D = V \times \phi \times g^{-1} \times 1.6 \times 10^{-8}\,(N_{^{10}B}\,\sigma_{^{10}B}\,E_{^{10}B} + N_{^6Li}\,\sigma_{^6Li}\,E_{^6Li}) \times \frac{1 - \exp\,-\sum_i N_i \sigma_i L}{\sum_i N_i \sigma_i L} \tag{4.4}$$

where

V = glass volume in cm^3

N_i = number of ^{10}B or ^6Li atoms per cm^3

σ_i = cross section of the (n,α)-reaction in ^{10}B and ^6Li per cm^2

E = energy released by the reactions ($E_{^{10}B}$ = 2.5 MeV, $E_{^6Li}$ = 4.78 MeV)

g = weight of the glass in g

L = glass thickness in cm.

The silver effect has been calculated from the formula

$$R_{Ag} = 2.15\ \text{rad}/10^{10}\ \text{n/cm}^2 \times \frac{1 - \exp\,-\sum_i N_i \sigma_i L}{\sum N_i \sigma_i L} \tag{4.5}$$

The value of 2.15 rad/10^{10} n/cm^2 was obtained experimentally (Kondo, 1961 and Amato and Malsky, 1961). Using Equation 4.4, the effects of glass thickness, ^{10}B and/or ^6Li enrichment, and other parameters have also been calculated. It was found that self-shielding inside glass blocks easily leads to saturation in the thermal neutron response. For this reason, the incorporation of other strong thermal neutron absorbers such as gadolinium into the glass block does not significantly increase the thermal neutron sensitivity (Becker and Tuyn, 1965).

The fast-neutron response of RPL glasses is rather small and difficult to measure precisely. The available results indicate an increase with increasing neutron energy (Figure 6-35b). For neutrons between 0.5 and 1.5 MeV, values of 0.003 to 0.007 (0.3 to 0.7% of the gamma response in rad) have been measured (Peirson, 1958 and Bernard et al., 1961). For 14 MeV neutrons the response of bare glass blocks and needles is around 0.07 to 0.1; encapsulation of glass rods in a hydrogenous radiator (Perspex$^{®}$) resulted in an increase to 0.21 (Kondo, 1961 and Piesch, 1964).

Incorporation of hydrogen into the glass would increase its fast-neutron sensitivity. Unfortunately, this is difficult because glasses with a high water content or containing ammonium salts are rather unstable. Other approaches such as the use of silver-activated hydrogen compounds (ammonium salts, lithium hydride) or incorporation of finely powdered glass into organic polymers have also met with limited success, partly because of stability and background (predose) problems and partly because of the LET response of such systems.

Another method for fast-neutron measurements is based on the use of moderators which thermalize the neutrons for detection, either by optimized moderators in front of (or around) the

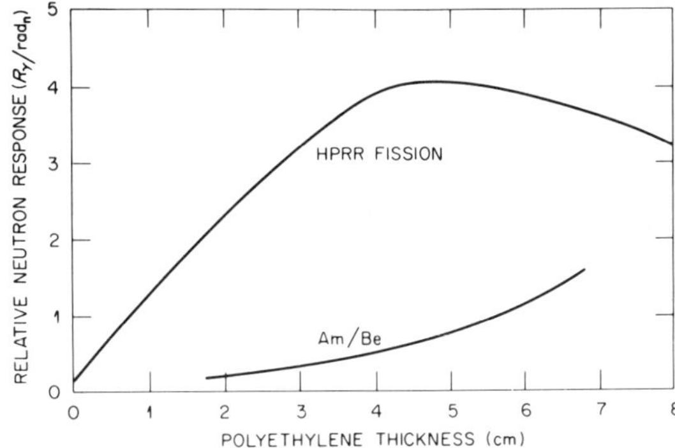

FIGURE 4-20.   Neutron sensitivity, given as the reading of the glass block in R gamma equivalence, divided by the fast-neutron tissue dose in rad, as a function of the thickness of polyethylene discs on both sides of the glass, for the HPRR reactor fast fission spectrum, and neutrons from an Am/Be source. (After Becker and Tuyn, 1965.)

glass or by using the human body as a backscattering and moderating medium (albedo method). If an unshielded Yokota type glass is exposed at the front of a thorax phantom to an unmoderated fission spectrum from a fast reactor, the neutron and gamma components are indicated almost rad equivalent; in a free-air exposure, only 1 cm of polyethylene around the glass is required to obtain a rad equivalent response for fission neutrons (Becker and Tuyn, 1965). The response of such systems is, however, strongly dependent on the neutron energy (Figure 4-20) as well as the distance of the detector from the moderating medium (body).

Almost rem equivalent measurements between thermal and about 10 MeV neutrons are possible by surrounding the glass with a large spherical moderator. The response characteristics that can be obtained with a uniform moderator (20 to 25 cm diameter of pressed wood or polyethylene) may be improved further with moderator-absorber-moderator sequences similar to the Anderson-Brown neutron rem counter shield. Such devices are, of course, rather bulky and heavy, and are only of practical interest for neutron area monitoring, for example, around accelerators or reactors.

There are two other approaches to neutron dosimetry with RPL glasses which are not based on measuring RPL. One is fission-fragment track etching (see Chapter 5). Heavy charged particle tracks such as fission-fragments, but not alpha or proton tracks, can be made visible in the micro-

scope by preferential etching of the dosimeter glass with hydrofluoric acid, sodium hydroxide solutions, etc. (Becker 1966b). The registration efficiency for fission-fragments is over 90%; the response is about 1.2×10^{-5} tracks per neutron per barn for fissile material layers whose thickness exceeds the range of the fission-fragments.

As the RPL measurement is not affected by the etching of the glass surface, a combination of a fissile foil and a glass could, in principle, be used for measurements in mixed radiation fields. However, the penetrating radiation that is emitted by most fissile materials leads to a rapid predose increase, thus limiting this method to short contact times between glass and fissile material. Of course, this does not apply to high energy neutron fission reactions in such elements as bismuth and gold. Non-RPL glasses that contain 40% ThO_2 and exhibit a high sensitivity to neutrons above 2 MeV have been prepared (Yokota et al., 1969b).

The second method is based on the detection of the neutron-induced radioactivity in the glass. In the common Toshiba glasses, there are three activation reactions of dosimetric interest: $^{31}P(n,p)^{31}Si$ (halflife 2.6 hr) for the detection of fast neutrons above 2.5 MeV; $^{31}P(n,\gamma)^{32}P$ (halflife 14.3 days) for the detection of thermal and epithermal neutrons, both in the range of about 1 to 10^3 rad; and $^{109}Ag(n,\gamma)^{110}Ag$ (halflife 253 days) for the detection of thermal and epithermal neutrons above $\sim$10 rad.

If the glass is used in a protective and energy-compensating encapsulation, the activation of this encapsulation (Sn, Cd, Cu, etc.) also can aid the evaluation. Two measurements of the induced activity in the glass at different times after exposure (determination of the ^{31}Si after <15 hr and of the ^{32}P after 24 hr) help to distinguish between the effect of thermal and fast neutrons. Use can be made of the counting of the induced activity of the Cerenkov light production inside the glass block. It can easily be measured in a liquid-scintillation counter for tritium (Piesch, 1966).

In Figure 4-21, the sensitivity obtained immediately after exposure of a glass block to different neutron energies is presented. Obviously, the use of the activation technique is restricted to such cases of experimental or accidental exposures where the evaluation is possible during a relatively short time after exposure. It should also be noted that the reaction $^{23}Na(n,\gamma)^{24}Na$ complicates the

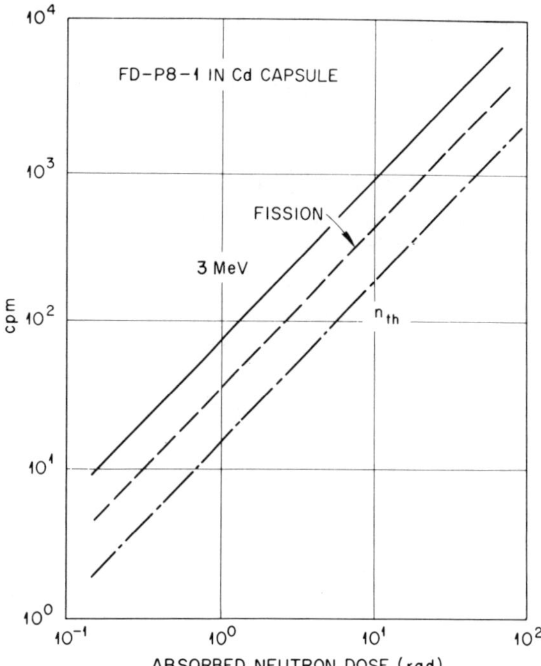

FIGURE 4-21. Neutron-induced radioactivity in a Toshiba dosimeter glass block as measured by detection of the Cerenkov radiation in a liquid scintillation counter, for different neutron energies. (After Piesch, 1966.)

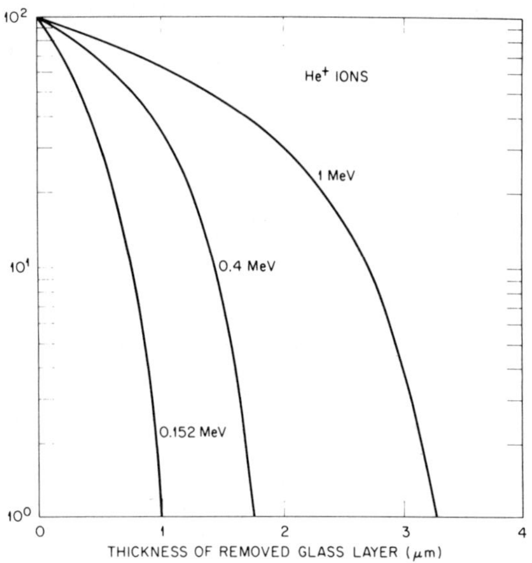

FIGURE 4-22. Residual radiophotoluminescence of a dosimeter glass block irradiated with helium ions of different energies as a function of the thickness of the surface layer removed by fractionated etching, normalized for the reading of the unetched glass. (After Becker, 1968B.)

evaluation of sodium-containing glasses. More suitable activation detectors for neutron accident dosimetry are special arsenic phosphate glasses (Piesch, 1971).

The effect of external exposure of dosimeter glasses to charged particles and electrons depends not only on dose and LET, but also on their penetration into the glass, the effective volume dose being small if only a thin surface layer of the glass is affected by the radiation. For electrons and protons exceeding 50 to 100 MeV, one would expect that the sensitivity of RPL glasses equals that of gamma radiation on a rad per rad basis. This has been confirmed in studies involving electrons in the 50 to 950 MeV range, protons between 150 and 730 MeV, and 910 MeV alpha particles (Tochilin et al., 1963). For measurements at lower electron energies in the 10 to 50 MeV range, see Tosi et al. (1969) and Westerholm and Hettinger (1971).

In cases where radiation of low penetrating power affected the uppermost layers of the glass only, measurements of the range and depth-dose distribution become possible by a "peeling" method: The initial RPL of the glass is measured first, then remeasured after removal of a thin layer of known thickness by chemical etching, etc., until all RPL has been removed. The residual RPL can be plotted as a function of the thickness of the removed glass layers (Figure 4-22). The range of beta radiation from ^{3}H, ^{63}Ni, and ^{35}S, of alpha radiation from uranium, neptunium, and plutonium, and of protons, deuterons, and helium ions of different energies has thus been measured (Becker, 1968b).

If less spatial resolution is required, similar studies can be carried out with glass plates of different thickness which are exposed by submersion into the solution of a beta emitter (Figure 4-23A). A more advanced modification of this technique makes use of trapezoidal-shaped glass in a special holder with internal reflection of the excitation UV (Figure 4-23B), permitting the selective measurement of the RPL in a thin surface layer.

4.3. Dosimeter and Reader Design and Applications

4.3.1. Dosimeters

The problems associated with a photon energy response of RPL glasses, as described in Section 4.1.3., may be overcome to a large extent by the

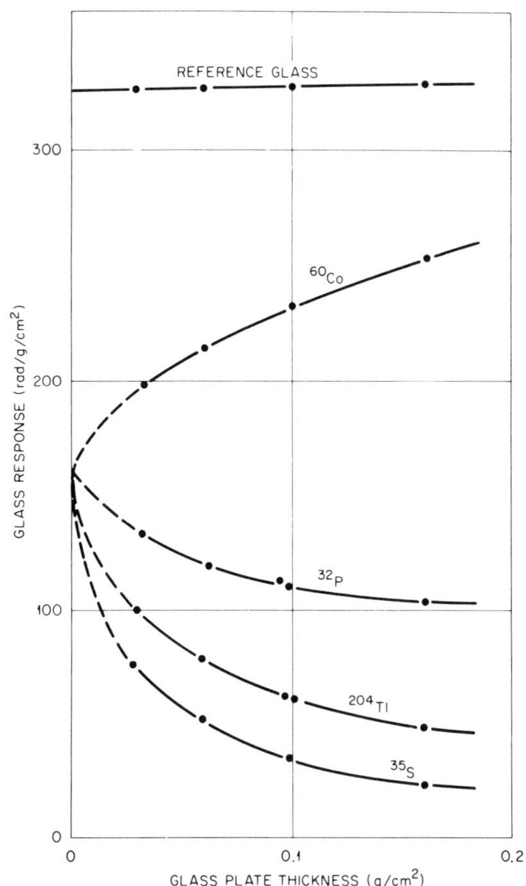

FIGURE 4-23A. Response of thin plates of RPL glass of varying thickness to ^{60}Co gamma radiation and beta radiation from different nuclides. (After Yokota et al., 1968.)

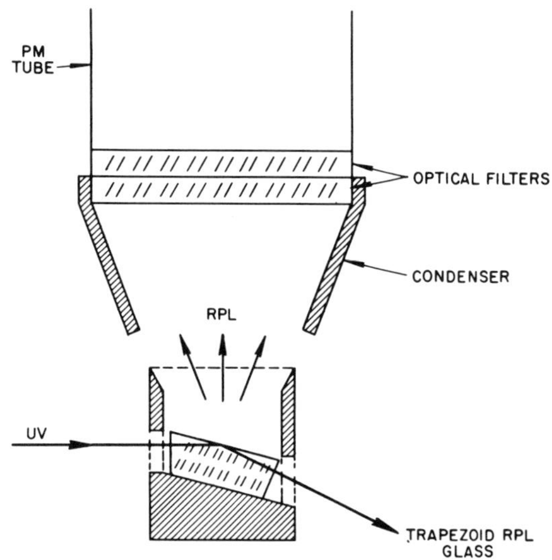

FIGURE 4-23B. Arrangement for readout of trapezoid glass for increased sensitivity of RPL measurements close to the glass surface. (After Yokota et al., 1968 and 1972.)

use of low Z constituents. The remaining energy response is usually modified by encapsulation in metal filters, either to approximate the photon energy response of water or soft tissue ("energy independence"), or that of one or more critical organs. A large volume of information is available on these techniques (see the reviews by Becker, 1967 and Piesch, 1972), but only a brief summary of the results can be given here. In comparing the results of different authors, one should be aware of the fact that not only differences in glass composition and geometry, but also the arrangement of the filter(s) and the spectral distribution of the x-rays used in the measurements (weak filtration tends to simulate a flat response) affect the results.

There are various possibilities of combining an RPL glass with one or more metal filters, including partial covering (perforated filters), composite filters (the sequence of metals may affect the result), and glass-filter-glass sequences. The simplest method is a single metal of uniform thickness around the glass. With 1 to 1.5 mm of lead it is possible to compensate the energy dependence of the old high Z glasses between approximately 120 keV and a few MeV within ± 10 to 20% (Schulman et al., 1953).

Below 120 keV the sensitivity drops rapidly, with a narrow sensitivity peak at the K_a edge of lead. Other investigators reported a flat response above 150 keV with 0.75 mm tantalum. For low Z glasses, thinner filters and/or materials with a lower Z are required. For the Bausch and Lomb low Z and the Toshiba FD-2 glass, a reasonably flat response above 80 to 90 keV is obtained by 2 mm copper, 0.7 mm cadmium, or 0.5 mm lead (Becker, 1963).

Composite filters containing several metals either mixed (alloy, hot-pressed powder mixtures) or in a stack-type sequence with the heaviest metal on the outside may give an even better energy response. One combination for low Z glasses consists of 0.5 mm tantalum, 0.75 mm tin, and an inner layer of 0.37 mm Teflon® and gives a flat response for high Z glass above 115 keV (Auxier et al., 1960). With 1 mm brass and 0.3 mm of lead, the response is quite flat down to about 80 keV with low Z glass (Yokota and Nakajima, 1965).

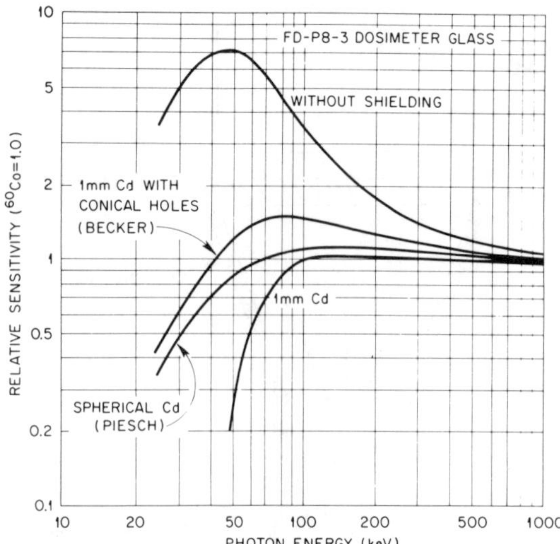

FIGURE 4-24. Photon energy dependence of response of Toshiba FD-P8-3 dosimeter glass blocks without shielding, with a simple 1 mm Cd shield, and with different types of perforated Cd filters. (After Piesch, 1972.)

Another even more effective method is partial filtration. Holes in the lead filters were used in the early 1950's for improving the energy dependence of the U.S. Navy DT-60 glass; a flat response above 80 keV was reported (Schulman et al., 1953). Similar arrangements have been recommended by other authors (Peirson, 1958; Riegert and Spinks, 1958; and Hardt and Lutz, 1963). Better results with the perforated filter technique may, of course, be obtained for low Z glasses. Figure 4-24 demonstrates the effect of 1 mm cadmium shielding with and without perforations. A spherical encapsulation is more complicated than a square one, but further improves the characteristics (response between 45 keV and 1.2 MeV within ± 18% for all possible angles of radiation incidence). The perforations may be filled with plastics or low Z metals without greatly modifying the response.

For the measurement of photon energies of less than ∿40 keV, such as diagnostic x-rays, one has to employ either lithium borate glasses with a low silver content or two glass blocks with different filtration. Both glasses are read and the results multiplied with appropriate weighting factors. For example, the reading for a Lucite®-covered low Z glass has been multiplied by 0.15 and the response of a glass covered with 2 mm tin added, resulting in a reading which is energy-independent down to 35 keV within ± 20% (Miyanaga and Yamamoto, 1963). This method has later been improved (Yokota and Nakajima, 1965) and modified to permit readings in the 15 to 40 keV energy range (Figure 4-25).

Whenever information on the effective photon energy is desired, for instance, for a better determination of organ doses in personnel dosimetry, the ratio of glass readings behind different filters is used similarly to the filter-analytical methods common in film dosimetry (Maushart and

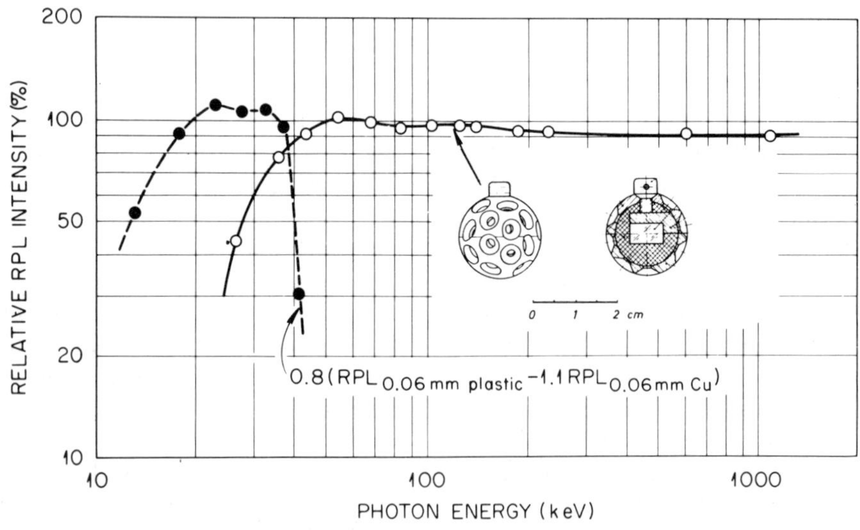

FIGURE 4-25. Photon energy dependence of response of Toshiba FD-P8-3 glass in spherical tin capsule with conical holes and additional pairs of glasses covered with copper and plastic to facilitate dose determinations at low photon energies. (After Maushart and Piesch, 1965.)

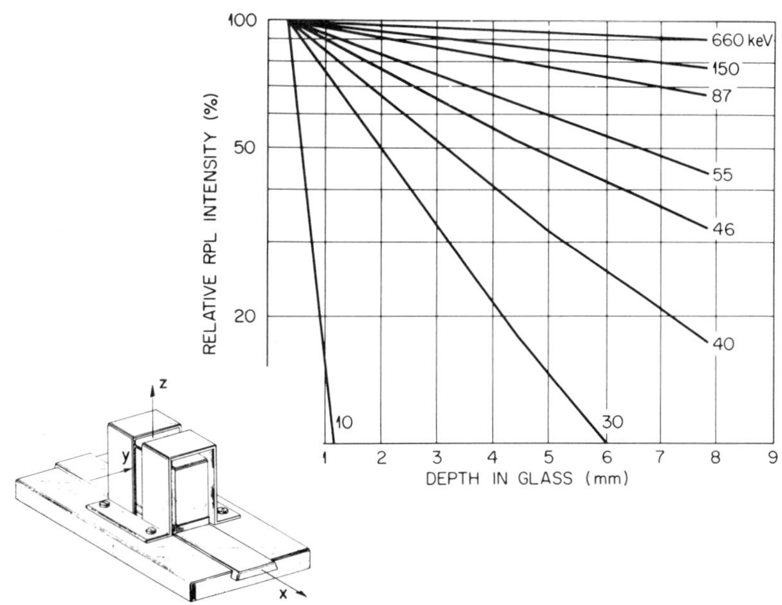

FIGURE 4-26.   Relative RPL depth-dose distribution in Toshiba FD-P8-3 glass blocks after exposure to x-rays of different energies and scanning device for measuring the depth-dose distribution. (After Kiefer and Piesch, 1969b.)

Piesch, 1967). The evaluation of such combinations may, in principle, be done automatically by simultaneous reading of more than one glass and computation of the effective energy and correction factors (Becker, 1966c), or measurement of different areas of the same, partially shielded glass.

In a new automatic reader (Dade et al., 1971), one photomultiplier "sees" the total area of a glass plate behind a partial shield, while another measures the unshielded area only. The difference in the RPL intensities is electronically translated into effective energy and dose information and printed out. For phantom or in vivo "spectrometry," sets of glass needles that had been encapsulated in metal tubes of different wall thickness have also been used (Malsky et al., 1961).

A somewhat different approach for obtaining information on the effective photon energy from a single glass square is based on the "scanning" of the RPL in the glass with a specially modified reader. This results in information on the depth-dose distribution in the glass, which depends on the photon energy in case of monodirectional radiation incidence (Piesch, 1972; Yokota and Muto, 1971; and Toivonen, 1971). The principle is illustrated in Figure 4-26. In case of a dosimeter that is worn on the front of the trunk, scanning also helps to decide whether a person has been exposed from the front or the back (Figure 4-27). Unfortunately, the practical value of this method is limited by the fact that personnel exposures are rarely monodirectional.

Another concept for glass dosimeter filter design has been promoted recently by the Karlsruhe group (for a summary, see Piesch, 1972). It is not based on an attempt to measure "energy-independent" in free-air exposures to various photon energies and directions of incidence, but to design filters in such a way that the glass dose indication takes the effect of the

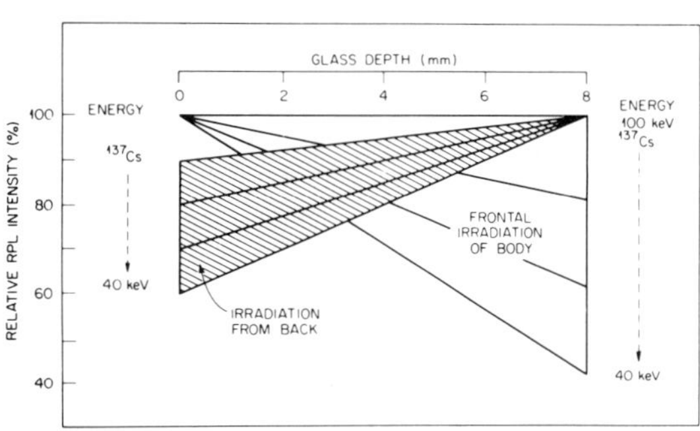

FIGURE 4-27.   Schematical presentation of the RPL depth-dose distribution in RPL glass blocks attached to the front of a thorax phantom after exposure from the front and from the back. (After Piesch, 1968.)

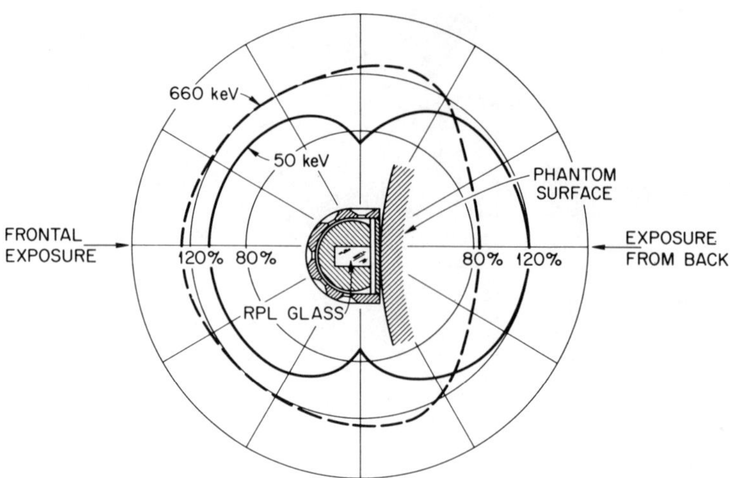

FIGURE 4-28. Response of asymmetrically shielded RPL glass block at the front of a human phantom to 50 keV x-rays and 660 keV ^{137}Cs gamma radiation, relative to the absorbed dose in the testes. (Piesch, 1968.)

human body into account. An asymmetrical, hemispherical shield around a glass block can be dimensioned in such a way that it responds adequately to exposures from the side and the back if attached to the chest of a standard man over a wide photon energy range. The perforated 2 mm tin shield in Figure 4-28 approximates the dose to the ovaries, testes, gut mucosa, and bone marrow (but not the lens of the eye) within ±25% for all angles of incidence and photon energies between 70 keV and 1.2 MeV.

The actual badge design is not only determined by the requirement of filter optimalization, but by several other factors. Some types of dosimeters have been devised for the special requirements of military and civil defense dosimetry, while others are designed for evaluation in a special type of reader. In general, the encapsulation should

    a.   protect the glass from mechanical damage, dirt, water, direct sunlight, and other disturbing environmental effects;

    b.   guarantee the reliable identification of badge and wearer;

    c.   be tamperproof, but fast and easy to open for readout; and

    e.   be small, lightweight, and inexpensive.

There may be additional specifications with regard to temperature and pressure resistance which further complicate the task of dosimeter design.

The first practical dosimeter for large-scale use, DT-60/PD, was developed in the early 1950's by the Polaroid Corp. (Schulman et al., 1953). It was later modified several times. One version, the DT-60 B/PD, contained a 17 x 17 x 4.7 mm glass block in a plastic holder with built-in 1 mm lead compensation filters having a central, brass-filled hole. The dosimeter was circular, 40 mm in diameter, 17 mm thick, and weighed 53 g (Figure 4-29). More than four million of this dosimeter have been produced.

Other modifications of this early badge design have been described by Peirson (1958) and Hardt and Lutz (1963). Later the U.S. Army considered a combination of a small cylindrical RPL glass as a gamma radiation detector and a silicon diode (Chapter 6.1) for fast-neutron measurements as an "administrative" high dose-level dosimeter. These badges are, however, nowadays mostly of historic interest because they are based on obsolete glass types.

More recently, several other badge types for military and civil defense purposes have been proposed. In Germany a flat, lightweight identification badge that contains one or two small glass plates (15 x 6 x 1.5 mm) was designed by the Total Corp. (Hardt, 1965). It can only be opened with a special electromagnetic device. Another German design contains a cylindrical glass in a cylindrical metal capsule to be read in a special portable reader (Buttler et al., 1968). Some of these glass badges are pictured in Figure 4-29.

FIGURE 4-29. Some typical glass dosimeter designs: The DT-60B/PD of the U.S. Navy open (upper left) and closed (upper middle); a civil defense identification badge type dosimeter by Total, Germany (upper right); a dosimeter for nuclear installations made by Total, Germany (center); a dosimeter with two differently shielded glass blocks made by Toshiba, Japan (lower left open, lower middle closed); and opened and closed spherical glass dosimeter holder which is used in the Karlsruhe and Jülich Nuclear Research Centers in Germany.

In routine dosimetry, fluorods have been used for over a decade as an additional high-dose accident dosimeter in the ORNL badge. In recent years they have been replaced by single Toshiba glass blocks as a long-term integrating device. RPL glasses are also in use as a back-up for other detectors in France (Francois et al., 1965). So far, only in Karlsruhe and Geesthacht, Germany are glasses routinely used as the only personnel monitor for the employees of a large nuclear research center. Special designs include finger-ring dosimeters. Inert materials have to be used for the badge (in some of the early studies, ammonia and formaldehyde from the bakelite and fluorescent adhesives produced substantial changes in the glass readings).

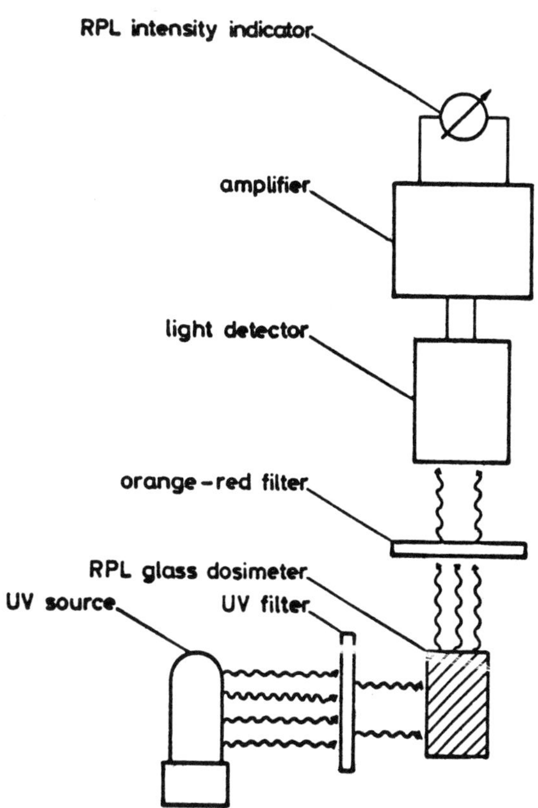

FIGURE 4-30. Schematical principle of a RPL glass reader.

4.3.2. Readers

For a crude, semiquantitative estimation of high radiation exposures, the RPL of glasses under an UV lamp may be visually compared to that of a series of fluorescent standards. With a dark-adapted eye, doses above about 10 rad can be seen. It is also possible to integrate the RPL output of UV-exposed glasses photographically (Hopkins and Baxter, 1971). In general, such methods are not recommended because of very poor sensitivity and accuracy.

In principle, a RPL reader consists of an ultraviolet light source with a strong output in the excitation region, a filter which is transparent for the excitation UV only, and a detector for the RPL light. An additional orange-red filter between glass and light detector helps to discriminate against scattered UV (Figure 4-30). The UV source is usually a high-pressure mercury lamp with a strong emission at 365 nm, and the light detector a red-sensitive photomultiplier.

Because of changes in the output of the UV source, the sensitivity of the multiplier, temperature dependence of the circuit, etc., fluorescent standards have to be used before and between measurements for the calibration of the instrument. As RPL in irradiated glasses may be subject to changes over extended periods of time, manganese- or samarium-activated glasses are frequently employed as reference standards. The luminescence spectrum and the decay time of such glasses may, however, be quite different from the RPL. This can cause differences in the results if the standards are compared in different types of fluorimeters.

The first RPL readers were developed by the U.S. Naval Research Laboratory for laboratory [model CP-95(XN-3)PD] and field use (CP-95/PD). The field reader was produced in large quantities and, after extensive tests, improved several times (Schulman et al., 1953). In England, a reader for the same dosimeter type was developed which had a Cs-Sb cathode as RPL detector (Peirson, 1958). The U.S. Navy also developed a battery-operated reader (model B-DT-60; Schaffert, 1957).

Improved small battery-operated readers with a xenon-filled flash tube for the RPL excitation became available in 1960. Similar readers were made commercially in Germany and France in the mid-1960's (see, for example, Buttler, 1968) for emergency dosimetry in nuclear installations, civil defense, and in "tactical" military dosimetry. Prototypes of semiautomatic (Baciotti et al., 1967) and automatic (Yokota, personal communication) readers have also been developed.

Several earlier commercial readers for the evaluation of fluorod needles (Bausch & Lomb, Rochester and G. K. Turner Assoc., Palo Alto, Calif.) are not in production any more. The most advanced commercial instrument for laboratory and research use is currently the one made by the Tokyo Shibaura Electric Co. (Toshiba) in Tokyo. The first of a series of models, FGD-3 B, became available in 1961. It featured a fan-cooled high-pressure mercury lamp and a special red-sensitive PM-tube with an Ag-Bi-Cs photocathode.

The most recent model FGD-6 (Figure 4-31) is completely transistorized. Measurements can be made either by a deflection or a zero method, and a small pressure of filtered air is maintained inside the reader to prevent dust contaminations (Yokota and Nakajima, 1967). A similar, more compact, but less versatile reader for a special type of glass

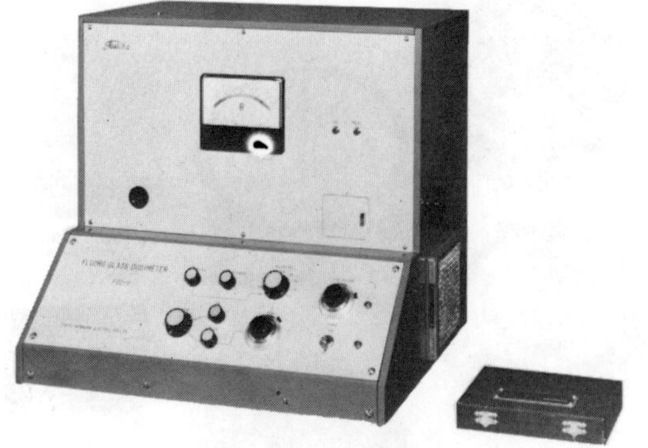

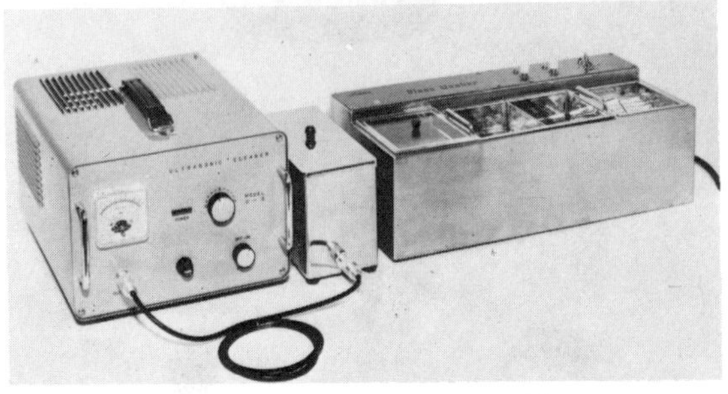

FIGURE 4-31. Advanced laboratory and research RPL glass reader and ultrasonic cleaning device for glasses. (Courtesy of Toshiba, Tokyo.)

dosimeters, which may also be used by untrained personnel under field conditions, is made in Germany (Figure 4-32).

The present trend in RPL reader development is toward increasingly sophisticated designs which permit, for example, scanning of the dose distribution in individual glasses (Toivonen, 1971 and Piesch, 1972), pulsed excitation and delayed RPL measurement to reduce predose and surface dirt effects (Yokota et al., 1971) or automatic reading of large numbers of dosimeters. A prototype automatic reader that has been developed by Frieseke and Hoepfner, Erlangen, Germany is designed to evaluate simultaneously the unfiltered and the partially filtered areas of a circular glass plate (for details of the dosimeter design, see Schmidt and Pfister, 1970) and calculate an energy-independent dose as well as the effective photon energy (Figure 4-33).

4.3.3. Applications

The most important potential application of glass dosimetry is personnel dosimetry. The main advantages of RPL glasses as compared with photographic film in personnel monitoring may be summarized as follows:

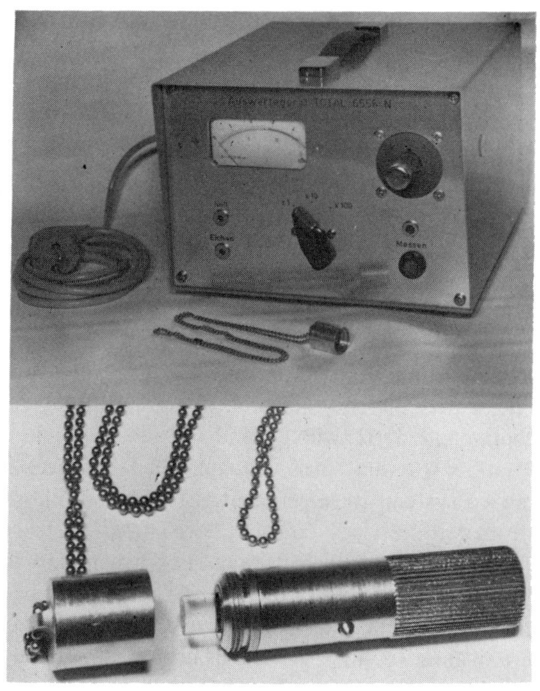

FIGURE 4-32. Compact RPL reader for cylindrical glasses in special capsules (after Buttler et al., 1966) and close-up capsule and magnetic capsule opener. (Courtesy of Total, Ladenburg, Germany.)

a. much higher accuracy in the important range between 50 mrad and several thousand rad for X- and gamma radiation;

b. less energy dependence in modern glasses, and consequently easier adjustability to the photon response requirements;

c. greater dynamic range, with one detector covering the whole range of interest and increasing accuracy of the readings with increasing dose;

d. minimal fading during at least ten years of storage at ambient temperature and sufficient stability even under unusual climatic conditions such as in tropical countries;

e. excellent storage stability, insensitivity to light, solvents, laboratory gases, etc.;

f. reusability after thermal annealing, as well as dose integration over extended periods of time with an unlimited number of interval measurements;

g. simple, fast evaluation which can also be done by personnel with little training, and battery-operated readers for on-the-spot field measurements.

FIGURE 4-33. Automatic glass dosimeter reader providing printout of the dose in steps of 50 mrad, of the effective photon energy, and the dosimeter number. (Courtesy of Frieseke and Hoepfner, Erlangen, Germany.)

In the majority of cases these advantages outweigh some advantages of the film. The reusability also compensates for the higher initial costs. According to a German study (Kiefer et al., 1965), a personnel dosimetry program with monthly readings of RPL glasses costs $2 per annum, while an equivalent film badge service is four times as expensive. Of course, the results of such comparisons depend on many factors, such as local labor costs, mailing expenses, size of the service, and degree of automation (mailing and record-keeping may be the most expensive part of a dosimetry service).

Compared with ionization-chamber pocket dosimeters, the main advantage of the RPL glass is the greater dynamic range, dose-rate independence up to very high dose-rates, better directional response, insensitivity to mechanical shocks, lack of self-discharge, and its smaller volume, weight, and price. A more serious "competitor" for glass dosimeters is an advanced TLD system.

The basic disadvantage of thermoluminescence dosimeters is that the TLD signal is largely destroyed by the readout procedure and remeasurements of doubtful or particularly important dose readings are difficult or impossible unless backup detectors are used. The RPL dosimeter, on

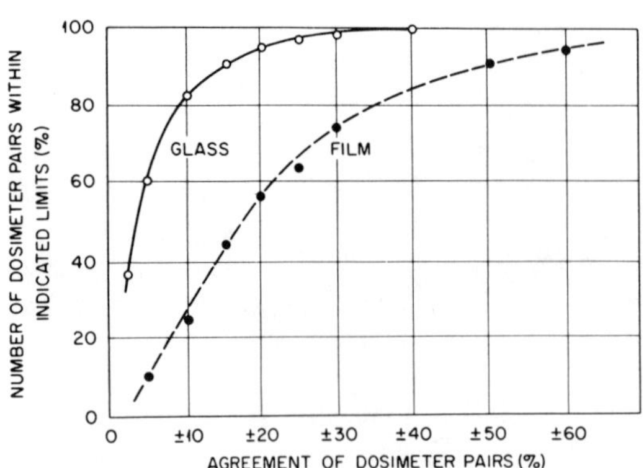

FIGURE 4-34. Agreement of dose readings in glasses and photographic film dosimeter pairs worn by radiation workers in a nuclear installation. (After Piesch, 1967.)

the other hand, represents a "document" of the radiation exposure which can be reread many times in the same or other institutions, and may be introduced as legal evidence in court. Some TLD materials are, however, clearly superior to RPL with regard to sensitivity and/or flat photon energy response or sensitivity to beta radiation below 1 MeV.

No experimental comparisons of the accuracy of TLD and glass dosimeters have yet been reported. Glass and film dosimeters have been compared under laboratory conditions which simulated the practical conditions, including different doses and photon energies between 45 keV and 1.25 MeV, different angles of radiation incidence, storage for several weeks, different temperatures and humidities during storage between irradiation and evaluation (Becker, 1968a). Under these conditions, 88% of the glass dose values in the 0.1 to 1,000 rad dose-range were found to be within ±30% of the true exposure, the maximum differences being −50 and +66%. Above 0.4 rad, 95% of the dose values were found within the ±30% limits; above 5 rad the average error was ±4.6%. Under the same conditions less than 10% of the films indicated the doses to within ±30%; no evaluation was possible above 100 rad.

In another study (Piesch, 1967), radiation workers carried pairs of film badges, ionization pocket chambers, and glass dosimeters simultaneously. As can be seen in Figure 4-34, more than 50% of the glass pairs worn by the same person agreed within ±5%, but only 10% of the film dosimeters were within these limits. For doses above 200 mrad, the maximum discrepancies between film and pocket dosimeter readings amounted to less than ±50% compared to −100 and +130% differences between film and pocket dosimeter results. Similar tests in other laboratories confirmed these results (Yokota et al., 1970 and Ewen and Erlenbach, 1973). It can be said that RPL dosimeters are clearly superior to film dosimeters in most personnel monitoring situations.

No general advice can be given with regard to the organizational aspects of a RPL monitoring program. For dosimeters that may be in use under extreme climatic conditions for long periods of time, a periodic check of predose and sensitivity is recommended. It will depend on the particular situation whether small, portable readers (decentralized system) or larger, centralized, more or less automated services are preferred. In nuclear installations, an RPL badge could be assigned to a person and read once a month or a quarter, to be annealed or replaced by a new glass only after a certain dose (perhaps 1 rad) has been accumulated (small additional dose increments become more difficult to measure if the "background" increases).

Glass dosimeters have been employed in many other areas too. Only a few examples are given here. One area of application of glass needles (fluorods) has been medical and radiobiological research (Fowler, 1963). Between 1959 and 1965, Amato, Roswit, and Malsky (for a summary and references, see Roswit et al., 1965) studied extensively the various aspects of phantom and in vivo dosimetry using different types of glass needles and spheres, shields, and readers. For example, they placed gold-encapsulated glass needles with catheters into the human bladder and with rubber and plastic tubes at different locations into the bronchial tree. Dosimeters were also introduced in tandem into the pharynx and esophagus down to the stomach, and even into the brain. Measurements of the dose and energy distribution in phantoms and dog cadavers exposed to 1,000 kVp X-radiation have been made by Maillie et al. (1965). In these experiments the fluorods had 0.05 and 1 mm lead shields. Other authors reported similar studies (for a bibliography, see Becker, 1966 and 1969).

Despite the energy dependence of the glass, several investigators report that in gamma radiation depth-dose studies with a narrow-beam geometry, the response even of high Z glasses closely approximates that of Fricke dosimeters and ionization chambers. Only when low energy photons contribute substantially to the total dose does good shielding become important. Phantom and in vivo beta radiation and high energy electron measurements have also been carried out. Of course, the depth-dose distribution in the glass may have to be considered in such measurements. For glass needles irradiated in a solution of ^{32}P, for example, only 26% of the gamma radiation response has been observed (Yokota et al., 1962). It increased to 62% for ^{90}Sr/^{90}Y (Kastner et al., 1965).

Further examples for the application of glass dosimeters are

a. Radioecology programs (Kaye, 1963 and Chapuis et al., 1972);

b. Mapping of high-intensity gamma radiation fields around fuel elements, reactors, ^{60}Co sources, etc. (Schulman et al., 1953; Rabin and Price, 1955; Gibson, 1968; and others);

c. Background dose measurements in the environment of nuclear reactors (at least one year of integration is required for obtaining an accuracy of about ±20%; on the other hand, the excellent stability of the glasses makes them eminently feasible for very long monitoring periods — Kiefer and Piesch, 1969b);

d. Thermal neutron dosimetry in and around research reactors;

e. Space radiation dosimetry, e.g., during the Gemini IV flight and in biosatellite experiments (Von Borstel and Smith, 1972); and

f. Measurement of the radiation exposure of radiologists performing contrast-media studies of the gastrointestinal tract, venography, and urethrocystography (Rytilae and Perttala, 1970), and in radiotherapy (Labäu et al., 1972).

REFERENCES

Anon. (E. Leitz GmbH, Wetzlar, Germany), Silver-Activated Phosphate Glass, British Patent 1,169,312, 1969.

Amato, C. G. and Malsky, J. S., Radiophotoluminescent γ-ray dosimetry of mixed neutron γ-ray fields, *Radiology,* 76, 290, 1961.

Auxier, J. A., Bernard, C. H., and Thornton, W. T., Silver Metaphosphate Glass for Gamma-Ray Measurements in Coexistent Neutron and Gamma Radiation Fields, *Proc. Symp. Select. Topics Radiation Dosimetry,* IAEA, Vienna, 1960.

Baciotti, J. P., Blanc, D., and Teyssier, J. L., Un prototype de lecteur de lumidosimetres, *Nucl. Instr. Meth.,* 50, 93, 1967.

Ballinger, E. R. and Harris, P. S., Field Study of AgPO$_3$ Glass Personnel Dosimeter, LA-2998, Report 4526, Los Alamos, N.M., 1959.

Barr, N. F., Stark, M., Hands, J., and McLaughlin, J. E., Dosimetry with small silver-activated glass rods, *Health Phys.,* 7, 48, 1961.

Barthé, J., Blanc, D., Commanay, L., Francois, H., and Teyssier, J. L., Mesure du temps de décroissance de la luminescence de lumidosimètres après excitation par un eclair ultraviolet, *Compte Rendu,* 268, 201, 1969.

Barthé, J., Blanc, D., and Commanay, L., Fading of Radiophotoluminescence in Silver-Activated Glass Dosimeters, Paper IAEA/SM-160/44, Proc. Symp. Dosimetry Techniques Applied to Agriculture, Industry, Biology, and Medicine, IAEA, Vienna, 1972.

Becker, K., Phosphatglasdosimeter für die Routineüberwachung in kerntechnischen Anlagen, *Nukleonik* 5, 154, 1963.

Becker, K., Die Meßbarkeit kleiner Gamma-Dosen mit Film und Glasdosimetern, *Kerntechnik,* 6, 199, 1964a.

Becker, K., UV-induzierte Radiophotolumineszenz in silveraktivierten Metaphosphatgläsern, *Z. Naturforsch.,* 19a, 1233, 1964b.

Becker, K., Sources of Error in Phosphate Glass Dosimetry: Luminescence and Absorption Alterations Between Exposure and Evaluation, Proc. Symp. Personnel Dosimetry for Radiation Accidents, IAEA, Vienna, 1965a, 169.

Becker, K., A new dosimeter glass with improved properties, *Nucl. Instr. Meth.,* 36, 323, 1965b.

Becker, K., High gamma-dose response of recent Ag-activated phosphate glasses, *Health Phys.,* 11, 523, 1965c.

Becker, K. and Tuyn, J. W. N., Thermal neutron response and intermediate and fast neutron personnel dosimetry with silver-activated phosphate glasses of different composition, *Health Phys.,* 11, 1225, 1965.

Becker, K., Radiophotoluminescence dosimetry — a bibliography, *Health Phys.,* 12, 1376, 1966a.

Becker, K., Nuclear track registration in dosimeter glasses for neutron dosimetry in mixed radiation fields, *Health Phys.*, 12, 769, 1966b.

Becker, K., Photographic, glass, or thermoluminescence dosimetry, *Health Phys.*, 12, 995, 1966c.

Becker, K., Radiophotoluminescence dosimetry, *IAEA At. Energ. Rev.*, 5, 43, 1967.

Becker, K. and McQuilling, D. W., Radiophotoluminescence Spectra of Silver-Activated Phosphate Glass, USNRDL-TR-67-66, U.S. Naval Rad. Defense Lab., San Francisco, 1967.

Becker, K., Some Characteristics of RPL Dosimetry Glasses, Paper SM-78-31, Proc. Symp. Solid-State and Chem. Radiat. Dosimetry, IAEA, Vienna, 1967a, 131.

Becker, K., Recent progress in radiophotoluminescence dosimetry, *Health Phys.*, 14, 17, 1968a.

Becker, K., Range and Depth Dose Distribution of Low Energy Charged Particles in Dosimeter Glasses, *Proc. 1st Int. Congr. Radiation Protect., Rome 1966,* Pergamon Press, Oxford, 1968b, 135.

Becker, K. and Cheka, J. S., Low-Z Radiophotoluminescent Glasses Based on Lithium Borate, Proc. Sec. Int. Conf. Luminescence Dosimetry, Gatlinburg, CONF-680920, 1968.

Becker, K., RPL dosimetry bibliography II, *Health Phys.*, 17, 631, 1969.

Becker, K. and Cheka, J. S., Silver-activated lithium borate glass as radiophotoluminescence dosimeter with low energy dependence, *Health Phys.*, 16, 125, 1969.

Bernard, C. H., Thornton, W. T., and Auxier, J. A., Silver metaphosphate glass for x-ray measurements in coexistent neutron and gamma radiation fields, *Health Phys.*, 4, 236, 1961.

Blaylock, G. G. and Witherspoon, J. P., Environmental factors affecting glass rod sensitivity, *Health Phys.*, 11, 549, 1965.

Buttler, W., Glass Dosimeter and Portable Reader for Accidental Dosimetry, Proc. Int. Symp. Luminescence Dosimetry, CONF-650637, AEC-Symp. Ser. No. 8, 1967.

Buttler, W., Maushart, R., and Piesch, E., *A* Phosphate Glass Dosimeter Adapted to the Requirements of Health Physics Applications, *Proc. 1st Int. Congr. IRPA,* Vol. 1, Pergamon Press, Oxford, 1968, 131.

Carpentier, S., Chapuis, A., Delarue, R., and Francois, H., Glass Having a Base of Metaphosphates for Dosimeters Which Are Responsive to X-Rays and Thermal Neutrons, British Patent 1,226,626, 1971.

Chapuis, A., Chartier, M., Francois, H., and Soudain, G., RPL Dosimetric Glasses Suitable for Radioecology, Paper IAEA/SM-160/61, Proc. Symp. Dosimetry Techniques Applied to Agriculture, Industry, Biology, and Medicine, IAEA, Vienna, 1972.

Cheka, J. S., Stability of radiophotoluminescence in metaphosphate glass, *Health Phys.*, 10, 303, 1964.

Cheka, J. S., Long-term stability of radiophotoluminescence in metaphosphate glass, *Health Phys.*, 15, 363, 1968.

Cheka, J. S. and Becker, K., High-level glass dosimeters with low dependence on energy, *Nucl. Appl.*, 6, 163, 1969.

Claffy, E. W., Gorbics, S. G., and Attix, F. H., Radiation-Induced Optical Absorption and Photoluminescence of LiF Powder for High-Level Dosimetry, *Proc. Third Int. Conf. Luminescence Dosimetry, Risö-Rep. 249,* Vol. 2, Danish AEC, Risö, Roskilde, Denmark, 1971, 756.

Dade, M., Hoegl, A., and Maushart, R., A RPL Dosimetry System With Fully Automated Data Evaluation, *Proc. Third Int. Conf. Luminescence Dosimetry, Risö-Rep. 249,* Vol. 2, Danish AEC, Risö, Roskilde, Denmark, 1972, 693.

Davison, S., Goldblith, S. A., and Proctor, B. E., Glass dosimetry, *Nucleonics,* 14(1), 34, 1956.

Dorofeev, G. A. and Somasundaram, S., IAEA International Glass Dosimetry Intercomparison Experiment 1970, Proc. Symp. Adv. Phys. Biol. Radiat. Detectors, IAEA, Vienna, 1971, 59.

Druskina, L. S., Radiophotoluminescence of polycrystalline pellets of silver-activated sodium chloride, *Izv. Akad. Nauk SSSR, Ser. Fiz.,* 29, 434, 1965.

Ewen, K. and Erlenbach, H. R., Long-term comparison of glass dosimeters, and pocket electrometers under practical conditions, *Health Phys.*, 24, 679, 1973.

Feige, Y., Sever, J., and Alterovitz, S., Calculation of γ-Ray Energy Response For a Flat Silver Metaphosphate Glass Dosimeter, Israel AEC Report No. IA-1003, 1965.

Francois, H., Bourbigot, Y., Grand-Clement, H. A., Portal, G., and Soudain, G., Dosimétrie Personnelle à Fonctions Multiples Adaptée à la Dosimétrie des Fortes Irradiations Accidentelles, Proc. Symp. Personnel Dosimetry for Radiation Accidents, IAEA, Vienna, 1965, 319.

Francois, H., Radiophotoluminescent Detectors, Particle Track Detectors and Liquid-Filled Ionization Chambers, in *Radiation Protection Monitoring,* IAEA STI/PUB/199, IAEA, Vienna, 1969, 343.

Freytag, E., Measurement of high doses with glass dosimeters, *Health Phys.,* 20, 93, 1971.

Gibson, N. N., Silver Phosphate Glass Dosimetry and its Use at the U.S. Army Nuclear Defense Laboratory, AD-681891, Edgeweed Arsenal, Md., 1968.

Ginther, R. J. and Schulman, J. H., New glass dosimetry is less energy dependent, *Nucleonics,* 18(4), 92, 1960.

Hardt, H. J. and Lutz, E., Verbesserung der Energieunabhängigkeit von Zählrohren, *Nukleonik,* 5, 39, 1963.

Hardt, H. J., The Development of an Individual Dosimeter for Measurement of High-Level Radiation Doses, Proc. Symp. Personnel Dosimetry for Radiation Accidents, IAEA, Vienna, 1965, 127.

Hardy, K. A. and Werts, R., Effects of Laboratory Reprocessing in Silver Phosphate Glass Microdosimeters, Report AD-681497, School of Aerospace Medicine, Brooks AFB, Texas, 1968.

Hillenkamp, F. and Regulla, D. F., Laser Pulse Excitation of Radiation Induced Photoluminescence in Silver-Activated Phosphate Glasses, *Proc. Third Int. Conf. Luminescence Dosimetry,* Risö-Rep. 249, Vol. 2, Danish AEC, Risö, Roskilde, 1971, 718.

Hopkins, B. J. H. and Baxter, R. C., Photographic technique for dosimetry using silver-activated phosphate glass plates, *Health Phys.*, 20, 645, 1971.

Käs, H. H., Radiophotolumineszenz-Dosimeter mit geringer Gamma-Energieabhängigkeit, *Atomkernenergie*, 20, 75, 1972.

Käs, H. H. and Scharmann, A., Sensibilisierung von Radiophotolumineszenzgläsern, *Z. Naturforsch.*, 27a, 1523, 1972.

Käs, H. H., Becker, M., and Scharmann, A., Dosimetrieeigenschaften von Radiophotolumineszenzgläsern auf Fluoridbasis, *Atomkernenergie*, 20, 151, 1972a.

Käs, H. H., May, A., and Scharmann, A., Einfluß der Schmelz- und Tempertechnologie auf die Dosimetereigenschaften silveraktivierter Phosphatgläser, *Glastechn. Ber.*, 45, 182, 1972b.

Kastner, J., Roberts, D. R., and Prepejchal, W., Internal ^{90}Sr beta-ray dosimetry with fluorods, *Am. J. Roentgenol.*, 94, 984, 1965.

Kastner, J., Langs, R. K., Cameron, B. A., Paesler, M., and Anderson, G., Electro-Optical Techniques for Ultrasensitive RPL Dosimetry, Proc. Sec. Int. Conf. Luminescence Dosimetry, CONF-680920, 1968, 670.

Kastner, J., Strash, A. M., Langs, R. K., Lutz, W. J., and Dunlop, J. T., RPL Dosimetry Employing an UV Laser, in ANL-7760, Part I, Argonne National Lab., Argonne, Ill., 1970, 13.

Katz, R., Sharma, S. C., and Homayooufar, M., Detection of energetic heavy ions, *Nucl. Instr. Meth.*, 100, 13, 1972.

Kathren, R. L., Improved cleaning technique for silver metaphosphate glass dosimeters, *Health Phys.*, 12, 1524, 1966.

Kaye, S. V., Use of glass rod dosimetry in the ORNL ecology program, *Health Phys.*, 8, 888, 1963.

Kerr, G. D. and Cheka, J. S., A lithium-7 phosphate glass detector for exposure measurements in mixed neutron gamma-ray radiation fields, *Health Phys.*, 16, 231, 1969.

Kiefer, H., Maushart, R., and Piesch, E., Erfahrungen mit Phosphatglasdosimetern zur Personendosimetrie, *Atompraxis*, 11, 88, 1965.

Kiefer, H. and Piesch, E., Die Ermittlung der Strahlenqualität und der Dosis von Röntgenstrahlung über eine Tiefendosismessung in silberaktivierten Phosphatgläsern, *Atompraxis*, 15, 19, 1969a.

Kiefer, H. and Piesch, E., Ergebnisse der mit Orts- und Personendosimetern bestimmten Jahresdosis im Kernforschungszentrum Karlsruhe, *Strahlentherapie*, 138, 696, 1969b.

Kishii, T. and Sakurai, J., Polarization of radiophotoluminescence of fluorodosimeter glass, *J. Phys. Soc. Jap.*, 20, 1271, 1965.

Kondo, S., Simultaneous Measurement of Thermal Neutron Fluxes and Gamma Contamination Dose by Silver-Activated Phosphate Glass, *Symp. Select. Topics Radiation Dosimetry*, IAEA, Vienna, 1960a.

Kondo, S., Neutron response of silver-activated phosphate glass, *Health Phys.*, 4, 21, 1960b.

Kondo, S., Responses of silver-activated phosphate glass to α-, β- and γ-rays and neutrons, *Health Phys.*, 7, 25, 1961.

Labäu, V., Matache, G., Nicolae, M., and Spiridon, M., Application of the RPL Glass Detectors for Absorbed Dose Measurement in Radiotherapy, Paper IAEA/SM/160/13, Proc. Symp. Dosimetry Techniques Applied to Agriculture, Industry, Biology, and Medicine, IAEA, Vienna, 1972.

Lee, P. K., Ballinger, E. R., and Schweitzer, W. H., A Heat Treatment Which Extends the Usable Range of AgPO$_3$ Glass Dosimeters, LA-2575, Los Alamos, N.M., 1961.

Lell, E. and Kreidl, H. J., Radiation effects in complex glasses, in *Interaction of Radiation with Solids*, Plenum Press, New York, 1967, 199.

Malsky, S. J. and Roswit, B., Measurement of radiation dosage, *J.A.M.A.*, 187, 839, 1964.

Maushart, R. and Piesch, E., Phosphate glasses as routine personnel dosimeters, in *Luminescence Dosimetry*, AEC Symp. Ser. 8, U.S. Atomic Energy Comm., 300, 1967.

Menkes, C. K., A glass dosimeter washing technique for improved reproducibility, *Health Phys.*, 12, 429, 1966a.

Menkes, C. K., Modification of the Toshiba FGD-3B glass dosimeter reader to improve accuracy and reproducibility, *Health Phys.*, 12, 852, 1966b.

Miyanaga, I. and Yamamoto, H., Studies on silver-activated metaphosphate glass as a personnel monitoring dosimeter, *Health Phys.*, 9, 965, 1963.

Nagpal, J. S., Development of Silver-Activated Phosphate Glass for Dosimetric Purposes, Paper SM-114-7, *Proc. Symp. Radiat. Protect. Monitoring*, IAEA, Vienna, 1969, 299.

Peirson, D. H., Phosphate Glass Dosimeter, AERE EL/R 2590, UKAEA Research Establishment Harwell, Didcot, England, 1958.

Piesch, E., Die Verwendung von silberaktivierten Metaphosphatgläsern zur Bestimmung einer Personen- und Ortsdosis von Gamma- und Neutronenstrahlung, *Atompraxis*, 10, 268, 1964.

Piesch, E., Neutron Dose Measurements by Means of Cerenkov Effect in Gamma and Activation Detector, in *Proc. Symp. Neutron Monitoring*, IAEA, Vienna, 1966.

Piesch, E., Intercomparison of Results for Film, Glass and Ionization Chamber Dosimeters Worn together in Routine Personnel Monitoring, Proc. Symp. Radiation Dose Measurements, Stockholm, OECD/ENEA, Paris, 1967, 151.

Piesch, E., Routine Dosimetry with Phosphate Glasses, Proc. Sec. Int. Conf. Luminescence Dosimetry, Gatlinburg, CONF-680920, 1968.

Piesch, E., Development of New Neutron Detectors for Accidental Dosimetry, Proc. Symp. Advances Radiation Detectors, IAEA, Vienna, 1971.

Piesch, E., Development in RPL dosimetry, in *Topics in Radiation Dosimetry*, Vol. 1, Attix, F. H., Ed., Academic Press, New York, 1972.

Przibram, K., *Z. Phys.*, 20, 196, 1924; *Z. Phys.*, 41, 833, 1927; *Z. Phys.*, 44, 542, 1972; *Z. Phys.*, 68, 403, 1931; *Z. Phys.*, 102, 331, 1936.

Rabin, H. and Price, W. E., Mapping radiation fields with silver-activated phosphate glass, *Nucleonics*, 13(3), 33, 1955.

Regulla, D. F., Differences of Gamma and Neutron Induced RPL Spectra of Phosphate Glass Dosimeters, Proc. Sec. Int. Conf. Luminescence Dosimetry, Gatlinburg, CONF-680920, 1968, 518.

Regulla, D. F., Lithium fluoride dosimetry based on radiophotoluminescence, *Health Phys.*, 22, 491, 1972.

Riegert, A. L. and Spinks, J. W. T., Energy dependence of silver phosphate glass dosimeters, *Int. J. Appl. Radiat. Isotopes*, 3, 125, 1958.

Roswit, B., Malsky, S. J., and Amato, C. G., *In vivo* radiation dosimetry for clinical and experimental radiation therapy, in *Progress in Clinical Cancer*, Grune & Stratton, New York, 1965, 96.

Routti, J. T. and van de Voorde, M. H., Intercomparison of high-dose dosimeters in accelerator radiation fields, *Nucl. Instr. Meth.*, 99, 563, 1972.

Rytilae, A. and Perttala, Y., Average doses to the radiologist in contrast-media studies of the gastrointestinal tract, lower extremity venography, and urethrocystography, *Health Phys.*, 18, 123, 1970.

Schaffert, J. C., Portable self-powered reader for DT-60 glass dosimeter, *Nucleonics,* 15(12), 60, 1957.

Schmidt, Th. and Pfister, H., Eigenschaften eines neuen Katastrophenfall-Glasdosimeters, *Atompraxis,* 16, 264, 1970.

Schulman, J. H., Evans, L. W., and Ginther, R., X-Ray Sensitive Screen, U.S. Patent 2,524,839, 1950.

Schulman, J. H., Ginther, R. J., Klick, C. C., Alger, R. S., and Levy, R. A., Dosimetry of x-rays by radiophotoluminescence, *J. Appl. Phys.*, 22, 1479, 1951.

Schulman, J. H. and Etzel, H. W., Small volume dosimeter for x-rays and γ-rays, *Science,* 118, 184, 1953.

Schulman, J. H., Shurcliff, W., Ginther, R. J., and Attix, F. H., Radiophotoluminescence dosimetry system of the U.S Navy, *Nucleonics,* 11(10), 52, 1953.

Schulman, J. H., Klick, C. C., and Rabin, H., Measuring high doses by absorption changes in glass, *Nucleonics,* 13(2), 30, 1955.

Schulman, J. H., Principles of Solid-State Luminescence Dosimetry, Paper SM-78-1, Proc. Symp. Solid-State and Chemical Radiation Dosimetry, IAEA, Vienna, 1967, 3.

Svikis, V. D., Dense LiF for gamma-free neutron shielding, *Nucl. Instr. Meth.*, 25, 93, 1963.

Tochilin, E., Lyman, J. T., Attix, F. H., and West, E. J., The dose response of glass, thermoluminescent and film dosimeters to high energy charged particles, *Radiat. Res.*, 19, 200, 1963.

Tochilin, E. and Goldstein, N., Dose rate and spectral measurements from pulsed x-ray generators, *Health Phys.*, 12, 1705, 1966.

Toivonen, M., Some Ways of Applying the Capabilities of Various Luminescence Methods in Personnel Monitoring, *Proc. Third Int. Conf. Luminescence Dosimetry, Risö-Rep. 249*, Vol. 2, Danish AEC, Risö, Roskilde, 1971, 742.

Tosi, G., Turchi, L., and Maffi, A., Dosimetry of high energy (12 to 32 MeV) electron beams using photoluminescent glass, *Minerva Fisiconucl.*, 13, 135, 1969.

Vinduska, O., Photoluminescence of AgCl crystals after x-ray irradiation, *Czech. J. Phys.*, 20, 755, 1970.

Vogel, H. and Becker, K., Zur Theorie der Radiophotolumineszenz in Gläsern I, *Nukleonik*, 7, 18, 1965.

Von Borstel, R. C. and Smith, R. H., Biology Division Habrobracon Experiment P-1079, ORNL-TM-3628, 1972.

Westerholm, L. and Hettinger, G., The Response of Radiophotoluminescent Glass to ^{60}Co Gamma and 10-30 MeV Electron Radiation, *Proc. Third Int. Conf. Luminescence Dosimetry, Risö-Rep. 249,* Vol. 2, Danish AEC, Risö, Roskilde, 1971, 727.

Weyl, W. A., Schulman, J. H., Ginther, R. J., and Evans, L. W., On the fluorescence of atomic silver in glasses and crystals, *J. Electrochem. Soc.,* 95, 70, 1949.

Wingate, C., Tochilin, E., and Goldstein, N., Response of LiF to Neutrons and Charged Particles, Proc. Int. Conf. Luminescence Dosimetry, Stanford, CONF-650637, AEC-Symp. Ser. No. 8, 1965, 421.

Yokota, R., Nakajima, S., and Sakai, E., High-sensitivity silver-activated phosphate glass for the simultaneous measurement of thermal neutrons, γ- and/or β-rays, *Health Phys.*, 5, 219, 1961.

Yokota, R., Nakajima, S., and Osawa, H., Fluoroglass dosimeter, *Toshiba Rev.,* 1962.

Yokota, R. and Nakajima, S., Improved fluoroglass dosimeter as a personnel monitoring dosimeter and microdosimeter, *Health Phys.,* 11, 241, 1965.

Yokota, R. and Nakajima, S., Recent Improvements in Radiophotoluminescence Dosimetry, Proc. Int. Symp. Luminescence Dosimetry, Stanford, CONF-650637, AEC Symp. Ser. No. 8, 1967.

Yokota, R. and Imagawa, H., Radiophotoluminescent centers in silver-activated phosphate glass, *J. Phys. Soc. Jap.,* 23, 1038, 1967.

Yokota, R., Muto, Y., and Miyake, T., β-Dosimetry with Silver-Activated Phosphate Glass, Proc. Sec. Int. Conf. Luminescence Dosimetry, CONF-680920, 1968.

Yokota, R. and Yuhei, M., Improved dosimeter glass washing technique, *J. Nucl. Sci. Technol.,* 5, 35, 1968.

Yokota, R., Kinoshita, M., Maruyama, M., and Nishiwaki, Y., Improved RPL Dosimeter and New Nuclear Track Detector, Paper SM-114-5, *Proc. Symp. Radiation Protection Monitoring,* IAEA, Vienna, 1969a, 287.

Yokota, R., Kinoshita, M., Maruyama, M., and Nishiwaki, Y., Recent Improvement of RPL Dosimeter and New Nuclear Track Detector, Proc. Regional Seminar for Asia and the Far East on Radiation Protection Monitoring, IAEA, Vienna, 1969b.

Yokota, R., Muto, Y., and Miyake, T., Performance of glass and film dosimeters in personnel monitoring, *Health Phys.*, 19, 316, 1970.

Yokota, R. and Muto, Y., Simplified method of finding quantum radiation energy and its incident direction using a single dosimeter glass, *Health Phys.*, 21, 597, 1971.

Yokota, R., Muto, Y., Naoi, J., and Yamagi, I., Silver-activated phosphate dosimeter glasses with low energy dependence and higher sensitivity, *Health Phys.*, 20, 662, 1971a.

Yokota, R., Muto, Y., Koshiro, Y., and Sugawara, H., New Type of High-Sensitive and Soil-Insensitive RPL Dosimetry, *Proc. Third Int. Conf. Luminescence Dosimetry, Risö-Rep. 249,* Danish AEC, Risö, Denmark, 1971b.

Yokota, R., Muto, Y., and Miyake, T., β-Dosimetry with trapezoidal shaped silver-activated phosphate glass, *Health Phys.*, 22, 516, 1972.

5. TRACK ETCHING

5-1. Mechanism

Because of various disadvantages of photographic nuclear track emulsions, much effort has been spent on the search for alternative methods of nuclear track registration in condensed media. The presently most widely used is track etching, which is based on the discovery that microscopic etch pits are formed in the surface of solids which have been irradiated with heavy particles and etched (Young, 1958). This important observation remained unexploited for several years, until the study of charged particle registration by preferential etching was taken up again in the early 1960's (Price and Walker, 1962; Fleischer and Price, 1963a and b; and Fleischer et al., 1965a).

It was shown that tracks could be made visible in a variety of dielectric materials such as minerals (mica), natural or artificial inorganic glasses, and especially in organic polymers (this category of track detectors proved to be considerably more sensitive than the inorganic detectors). The visual appearance of the etched track depends on many factors, including the particle and detector characteristics, the etching procedure, and the microscopic technique. Some typical fission-fragment micrographs in different materials are shown in Figure 5-1A. In some materials, "tracks" become elliptical or circular etch pits during extended etching (Figure 5-1B). The best images of the track structure may be obtained with replica techniques, and scanning electron microscopy of the replica. As an example, the etched track of a cosmic radiation zinc ion in the polycarbonate helmet of an astronaut is given in Figure 5-1C.

Since 1962, nuclear track registration in solids has burgeoned with fundamental and applied research having implications for numerous fields, including nuclear physics and chemistry, biology, lunar research, space radiation physics, geochronology, and radiation dosimetry (papers on the subject are presently published at a rate of about five per month). Among the many applications of the track etching technique are dating of meteorites, minerals, archeological artifacts, and art objects; quantitative analysis of alpha-emitting or fissile elements; studies of unusual nuclear reactions; research on heavy ions in cosmic radiation; studies of the radiation history, age, and erosion rate of lunar minerals; the search for new super-heavy elements and exotic particles such as magnetic monopoles; uranium exploration; the production of filters with uniform pore sizes; counting and size analysis of submicroscopic particles; and the tracing of water (for additional information, see reviews by Becker, 1968 and 1972; Fleischer et al., 1969a; Blanc, 1970; Monnin and Isabelle, 1970; Sakanoue, 1970a and b; Allkofer, 1971; Fleischer et al., 1972a and Avan et al., 1972). This chapter will, however, be restricted to a brief description of the principles and basic techniques of track etching and some of its more important applications in radiation dosimetry and health physics in general.

Although there is still much to be done on rather basic aspects of understanding the detection principles, optimizing the detector sensitivity and stability, and on etching and evaluation methods, track etching has already greatly advanced several areas of radiation dosimetry, especially neutron dosimetry. For example, combinations of suitable fissile elements and fission-fragment detectors (preferably polycarbonate foils to be evaluated by spark counting) represent one of the most sensitive, accurate, and simple methods for integrating fast-neutron dosimetry which are currently available.

The usability of fissile materials in personnel dosimetry may be limited due to their costs, radiotoxicity, and/or spontaneous fission rate. However, detectors based on this principle are quite well established in over one hundred laboratories as a standard method for neutron or high energy ionizing particle flux and spectrum measurements, in accident dosimetry, research, and in and around reactors and accelerators (there is also at least one commercial service offering track etching neutron dosimetry, namely, EG & G, Goleta, Calif.).

The direct interaction of fast neutrons with sensitive polymer foils offers an interesting alternative to the use of fissile materials, at least for the higher dose-levels. Foils based on direct interaction are by far the least expensive and simplest fast-neutron detectors. Special applications of alpha particle track detectors include personnel dosimeters for uranium miners, advanced techniques

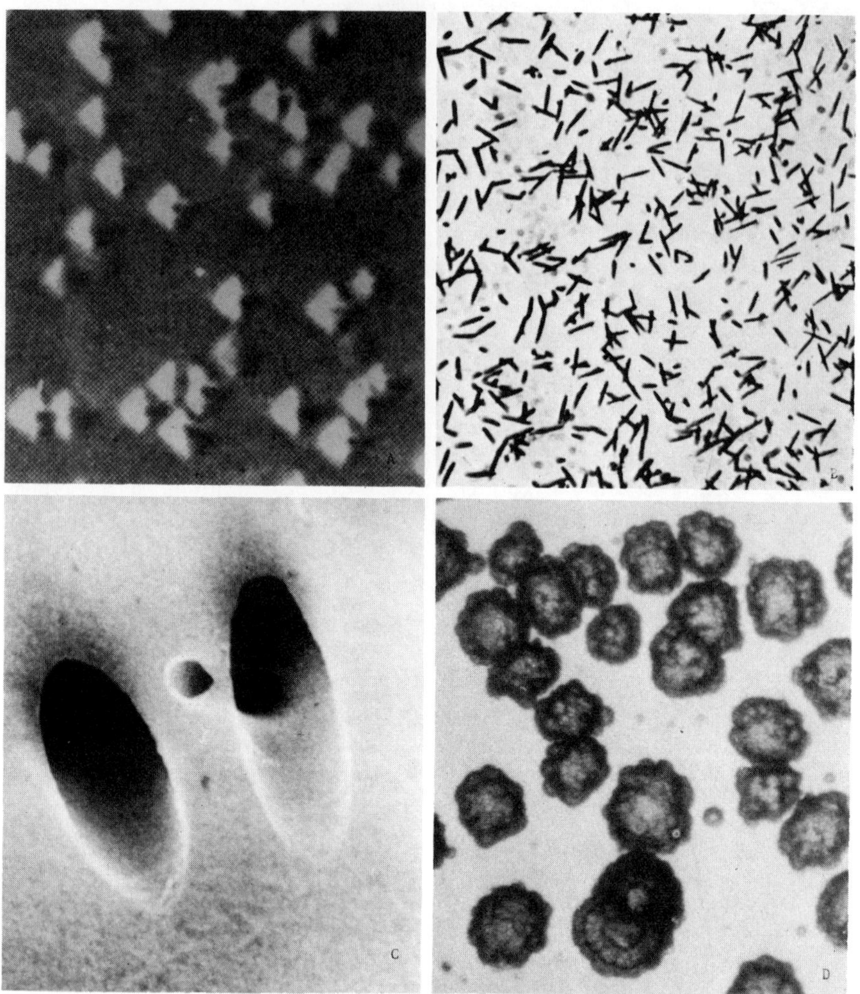

FIGURE 5-1A.   Visual appearance of fission-fragment tracks under different conditions (from upper left to lower right):A. square etch pits in a lithium fluoride crystal, one of the first particle etch micrographs made in 1958 by D. A. Young; B. tracks in polycarbonate after usual etching as seen in an optical microscope; C. the same tracks in a scanning electron microscope at a higher magnification (the smaller track in center representing a shorter etching time); and D. tracks "amplified" by etching in a high-frequency electrical field.

of alpha autoradiography, and ultrasensitive thermal neutron detectors.

It is safe to predict that the potential applications of nonphotographic nuclear track registration in solids for dosimetric purposes are not yet exhausted. More sensitive organic detectors or improved etching techniques may permit the registration of recoil protons over a larger range of energy, which would be of great advantage in fast-neutron dosimetry. Advanced evaluation techniques may simplify the automatic counting of recoil particle tracks. Perhaps most importantly, new radiation damage amplification principles not based on etching but on chemical processes, such as "free-radical photography," luminescence, "glass chambers" (Doke, 1972), preferential crystallization or vitrification, or "decoration" techniques may lead to further improved track detectors.

Among these techniques, track registration in large single silver chloride crystals deserves particular attention. Such crystals, whose particle

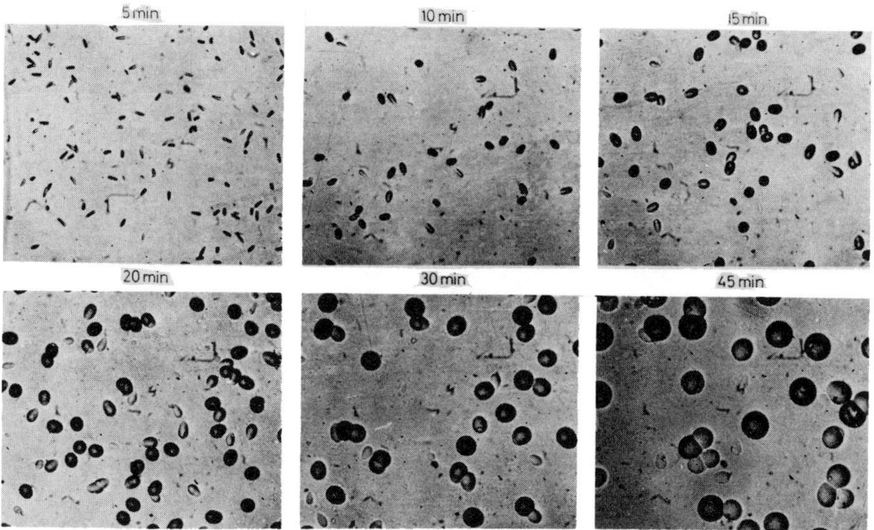

FIGURE 5-1B. Appearance of fission-fragment tracks in a phosphate glass as a function of etching time in 28% NaOH at 60°C.

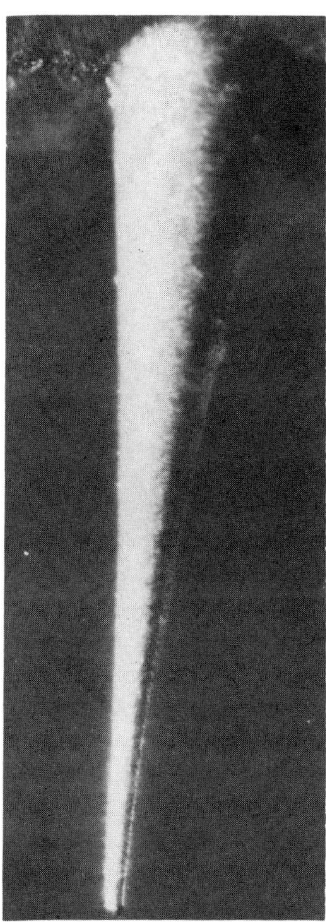

FIGURE 5-1C. Scanning electron micrograph of replica of cosmic radiation zinc ion track (length 700 μm) from astronaut Lovell's helmet after Apollo mission. (After Fleischer et al., 1970.)

detection threshold can be varied within wide limits by doping with varying amounts of cadmium and lead and/or the intensity of simultaneous exposure to yellow light, are insensitive in the dark and can, therefore, easily be switched "on" and "off." The "latent" tracks are made visible by silver precipitation in an electrical field (Schopper and Haase, 1971; Childs, 1972; and Schott et al., 1972). The use of such detectors is being considered for space radiation dosimetry (Henig et al., 1972).

Another system of interest consists of foils made from leuco crystal violet in tetrabromethane with polysterene as a binder (Fotland, 1968). It is almost colorless before irradiation, but becomes violet if exposed to gamma radiation in the 10^4 to 10^7 rad range. If "developed" by post-irradiation exposure to light at a characteristic wavelength, its sensitivity is greatly enhanced, and alpha or low energy electron tracks become visible in these foils. Unfortunately, they are also sensitive to UV and not very stable thermally.

The known dielectric track etching detectors fall into two distinct categories, namely, inorganic crystals and glasses (and a few natural products) and synthetic organic polymers. Only particles that lose energy at a rate greater than approximately 15 MeV-cm^2/mg can be detected in inorganic materials (for some recent work on track detection in glasses and quartz, see Sowinski et al., 1972; and Fleischer and Hart, 1972) and amber (Uzgiris and Fleischer, 1971). All synthetic organic polymers that have been tested record particles of less than approximately 4 MeV-cm^2/mg, and some cellulose nitrates as low as 1 MeV-cm^2/mg. These two groups of materials are also quite different in their other physical and chemical properties; track formation in them cannot be explained by a single mechanism.

For the track formation phenomenon in minerals and in organic glasses, an "ion-explosion spike" model (Fleischer et al., 1965b) has been advanced which suggests that a cylindrical region of positive charge is produced along the path of heavy charged particles due to ejection of electrons, and that the mutual repulsion of the remaining heavy positive ions results in atomic displacements in the crystal lattice. This process causes a cylindrical region of imperfections which is more easily attacked by an etching reagent than the surrounding, undamaged material, and ultimately leads to microsopically visible etch pits or tracks. It appears that the damage along the track of a heavy ion in an inorganic insulator is not continuous but that highly damaged regions alternate with "islets" of little damage (Maurette, 1970).

The formation of etchable tracks in organic polymers is closely related to the ionization produced by delta electrons in a channel around the track. It has been calculated that local doses exceeding several Mrad can be caused by low energy delta rays in the immediate environment of a particle track (Monnin, 1968 and 1970; Katz and Kobetich, 1968; Fain et al., 1971; Katz et al., 1972; Fain et al., 1972; Paretzke, personal communication; and others). It is known that high doses of gamma or electron radiation (Goland and Mateosian, 1973) enhance the bulk etching rate of plastics (Figure 5-2), and even the "track" of thin laser beams in plastics can be enlarged by preferential etching (Uzgiris and Fleischer, 1973).

The most plausible theory for latent track formation in polymers is, therefore, based on a radiochemical damage mechanism: Etchable tracks are formed by radiolytic scission of long polymer chains into shorter fragments and the production of reactive, low molecular weight radiolytic products which are more easily dissolved by etchants than the surrounding, undamaged bulk plastic.

Of the produced delta rays, those with higher

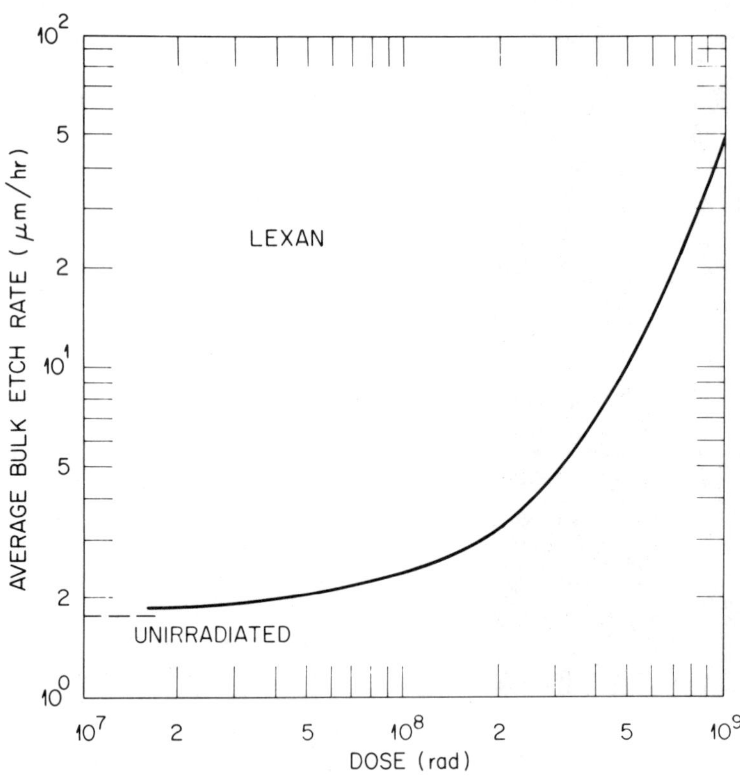

FIGURE 5-2. Average bulk etching rate of polycarbonate (Lexan®) as a function of ^{60}Co gamma radiation dose. (After Frank and Benton, 1970b.)

energies have a considerable range and deposit their energies over a large volume of the bulk material. As the "damage threshold" for preferential etchability is related to a critical dose of $\sim10^6$ to 10^7 rad in the region immediately surrounding the track (the diameter of the region in which above-threshold doses occur has been calculated to be around 20 to 100 Å for alpha particles up to several MeV in energy), only secondaries with energies less than a few hundred eV (the range of a 100 eV electron in polymers can be estimated to be $\sim$0.2 to 0.3 $\mu g/cm^2$) can deposit substantial amounts of their energies within the etchable portion of the path of the ionizing particle.

Electron energies of 9 to 15 eV are sufficient to cause ionization in polymers. Electrons around $\sim$6 to 10 eV cause excitation and those with an energy $\lesssim$6 eV cause vibration. The contribution of these three effects to the total energy which is imparted to the system within the "direct energy deposition period" (10^{-13} sec) has been calculated and is given in Figure 5-3. About 30% of the energy is deposited by the ion very close to its

trajectory (Fain et al., 1972). Ionization and excitation lead to free-radical and ion-molecule reactions and the subsequent production of excited and chemically altered species (for the mechanism of the chemical damage along particle tracks in various polymers, see Boyett et al., 1970; Veprik et al., 1970b; Paretzke, 1972a and Veprik et al., 1972). Strictly speaking, there is no well-defined damage threshold below which no preferential etching takes place. Under practical conditions with given etching and observation conditions, however, only particle impacts that create a sufficient amount of damage in a certain minimum volume of the plastic lead to visible tracks or etch pits (Figure 5-4) when the "preferential" etch rate, V_T, along the track exceeds the bulk etch rate, V_G, sufficiently. This results in a depression in the detector surface, the cone angle Φ of which depends on the ratio of V_T/V_G. The shape of the tracks in a given detector has to be determined empirically. In some materials, mostly plastics, a high preferential etching ratio leads to needle-like or cone-shaped tracks such as in Figure 5-1C.

In other materials, such as inorganic glasses, circular or oval lens-shaped etch pits dominate (Figure 5-16); in crystals such as LiF or mica, long etching of fission-fragment tracks results in regularly shaped square or rhombic etch pits (Figure 5-1A). Based on the knowledge of the etching rates V_T and V_G, purely geometrical considerations in many cases permit a good description of the track shape and growth kinetics (Somogyi, 1967; Tuyn, 1968; Haseganu, 1972; and Somogyi and Szalay, 1972). In the long narrow tracks other processes such as the diffusion of the etchant affect the kinetics and the situation becomes more complex (Henke and Benton, 1971). Even the dimensions of narrow tracks are usually remarkably well defined.

The "threshold" concentrations of the altered chemical species that are required for the formation of a visible track not only depend on the molecular and physical structure of the material, but also on its pre- and post-irradiation treatment, etching conditions, etc. Little is known about the relation between the physical and chemical properties of a material and its capability to register tracks, and it is difficult to predict the sensitivity of a new material. However, systems that are highly unstable chemically, such as cellu-

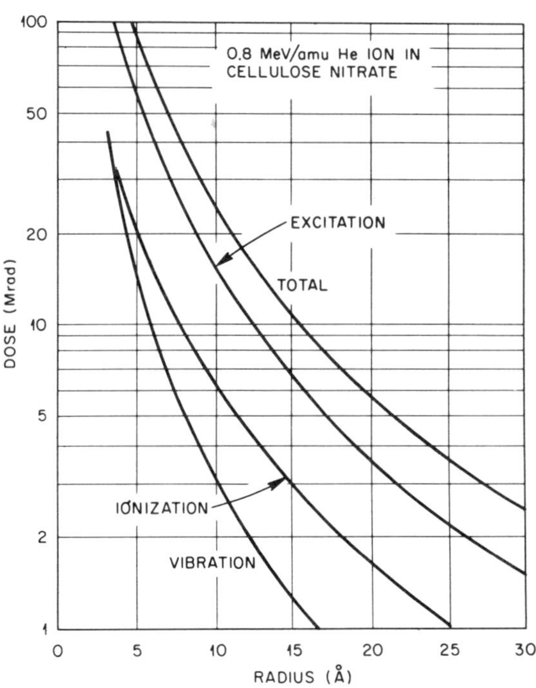

FIGURE 5-3. Calculated contribution of excitation, ionization, and vibration to total energy which was initially (during first $\sim10^{-13}$ sec) imparted along a helium ion track in cellulose nitrate, as a function of distance from the track. (After Fain et al., 1971.)

179

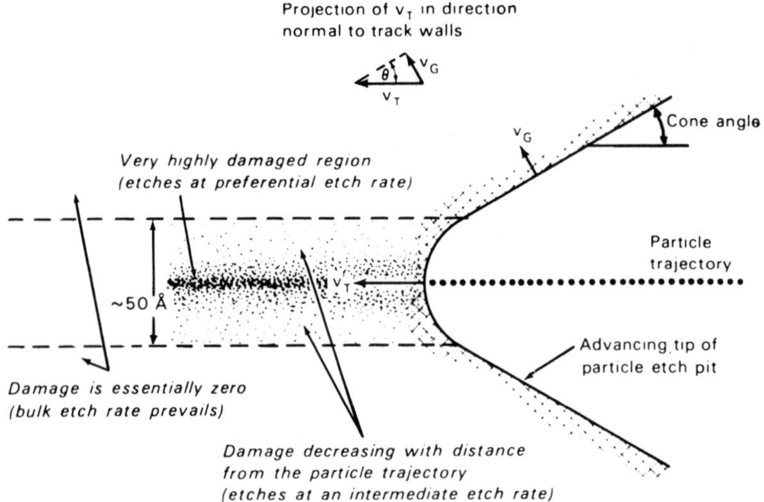

FIGURE 5-4.   Schematic diagram of submicroscopic damage trail along particle track, and proceeding preferential etching. (After Henke and Benton, 1971.)

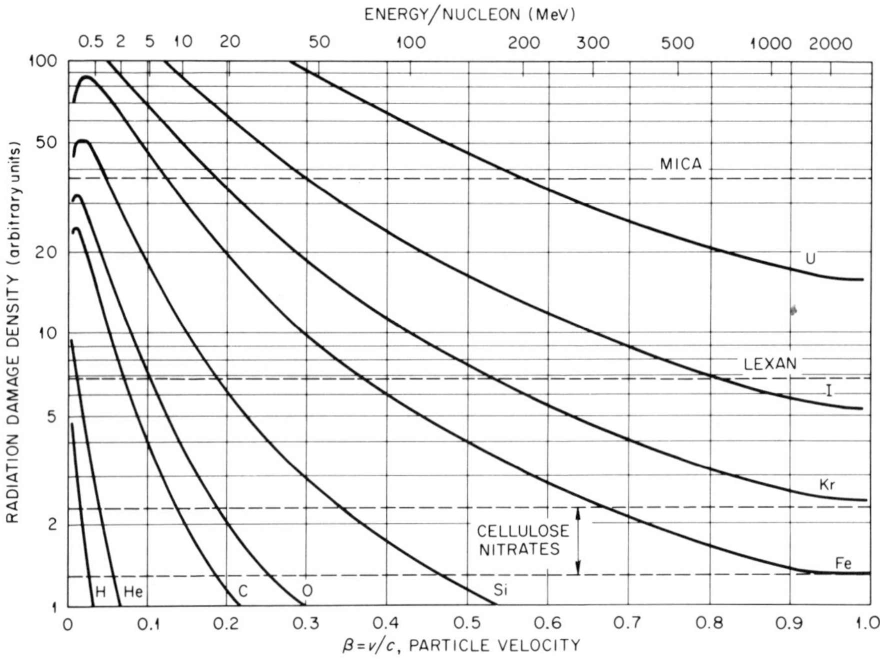

FIGURE 5-5.   Detection threshold of three widely used track detectors for charged particles of different atomic number and energy. (After Fleischer et al., 1969a.)

lose nitrate, tend to be more sensitive than radiation-resistant polymers (Veprik et al., 1972).

In Figure 5-5, the detection thresholds for three common track detectors, muscovite mica, polycarbonate (Lexan[1] or Makrofol[2]), and cellulose nitrates of different sensitivities, are indicated for ions as a function of their velocity (energy/nucleon). Obviously, fission fragments can be

[1] General Electric Co., Pittsburg, Pennsylvania.
[2] Farbenfabriken Bayer A.G., Leverkusen, Germany (made in the U.S. in license by Kimberly & Clark Co. under the name of Kimfol[®]).

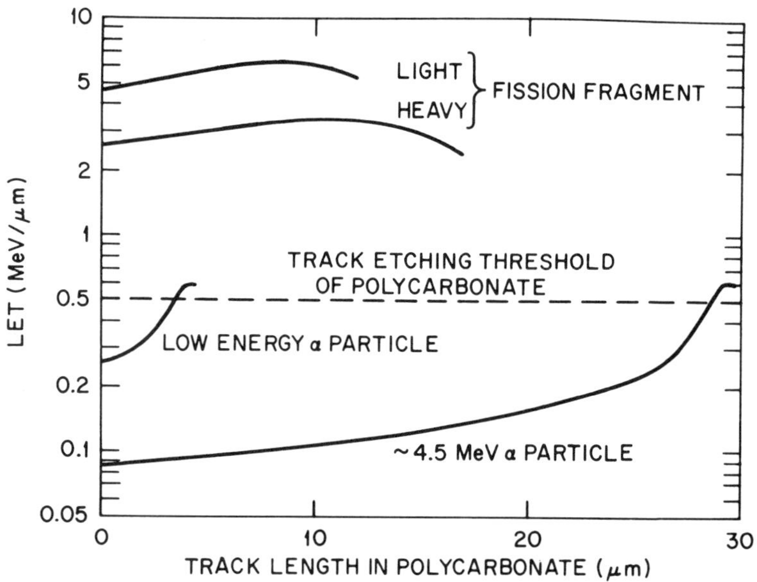

FIGURE 5-6. Energy deposition along the track of fission fragments, low energy alpha particles, and high energy alpha particles in relation to the track etching threshold in polycarbonate (Makrofol). (After Prêtre, 1972.)

TABLE 5–1

Critical Detection Angle and Efficiency of Various Fission-fragment Detectors

Material	Minimum angle of incidence for detection	Detection efficiency for isotropic (2π) incidence (%)
Soda-lime glass	35° 30′	42.3
Obsidian glass	26° 00′	56.7
Tektite glass	25° 45′	56.8
Quartz	7° 15′	87.5
Mica	4° 30′	92.0
Makrofol®	3° 00′	95.0
Lexan®	2° 30′	95.7

(After Khan and Durrani, 1972b.)

recorded along their whole range even in relatively insensitive polymers such as polycarbonate (Figure 5-6), while an alpha particle reaches the registration threshold only close to the end of its track and has, therefore, to be slowed down to become visible by surface etching. With the most sensitive cellulose nitrate foils, α particles up to 4 to 5 MeV in energy can be recorded (Veprik et al., 1970b; Nicolae, 1970; Anno, 1970; and others).

The sensitivity of a track detector can also be expressed as the critical angle of incidence to its surface under which a fission fragment may still be detected. In Table 5-1, some such critical angles and the corresponding detection efficiencies for the detector to isotropically incident ("2π geometry") fission fragments have been compiled. Obviously, the resulting etch pits become smaller as the particle incidence approaches the critical detection angle, and are more likely to become invisible under given optical observation conditions during extended etching because they are too shallow ("overetched"). This effect is illustrated in Figure 5-7 for fission fragments in a natural glass (tektite) with a critical angle of ~26°. The steeper the angle or particle incidence, the higher is the

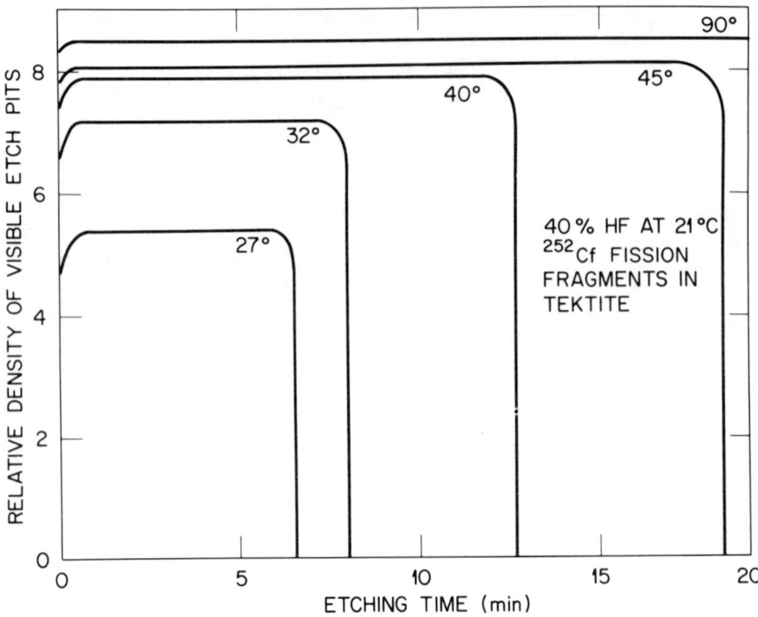

FIGURE 5-7. Effect of prolonged etching on the number of visible fission-fragment etch pits for different angles of particle incidence in a tektite glass, at optical magnification 450x. (After Khan and Durrani, 1972a.)

detection efficiency and the probability that the etch pits will remain visible after extended etching.

Some early claims that proton tracks have been made visible by etching of sensitive polymers were erroneous (after intense proton bombardment of polymers the observed etch pits may have been due to proton-induced recoil nuclei). There is, however, more recent evidence that protons can indeed be detected in a specially etched cellulose acetate (Varnagy et al., 1970) and cellulose nitrate (Carpenter and LaFleur, 1972). Obviously, one has to work very close to the sensitivity limit of the best detectors to register proton tracks. Therefore, the use of recoil proton tracks in fast-neutron dosimetry appears to be impractical in track etching for the time being.

After etching, the continued visibility of a track is limited only by the melting, softening, or chemical decomposition of the detector material. Before etching, however, the "latent" damage zone can be partially or completely annealed at substantially lower temperatures. This process normally occurs without permanent changes in the detector's recording capabilities. It should be noted that in some sensitive polymers (in particular in cellulosics) extended heating results in permanent changes of sensitivity and etching

rate (Figure 5-8) due to changes in the structure of the material.

If the melting point is as high as it is in many minerals and inorganic glasses, temperatures up to several hundred degrees may not cause any fading even over extended periods of time. In natural crystalline quartz, only 20% of latent fission-fragment tracks fade during 150 days at 710°C (Kosanke, 1972). It can be calculated by assuming a simple relationship between fading rate and temperature, as illustrated in Figure 5-9, that they would be stable at room temperature much longer than the age of the solar system (Fleischer and Hart, 1970 and Fleischer et al., 1971). Quartz, mica, or high-melting glasses may, therefore, be used for track detection at temperatures exceeding several hundred °C, for example, in smoke stacks or inside a reactor. Some conventional glasses, however, such as microscope slides exhibit substantial fading after storage for several days at 60 to 100°C (Stolz and Dörschel, 1969 and Reimer et al., 1972).

No fading of the latent tracks has been observed in polymers at temperatures below about 50°C. Even in 10 μm polycarbonate foils stored for three months at 30° and 95% relative humidity before etching and spark counting of the fission-fragment tracks, no fading could be detected

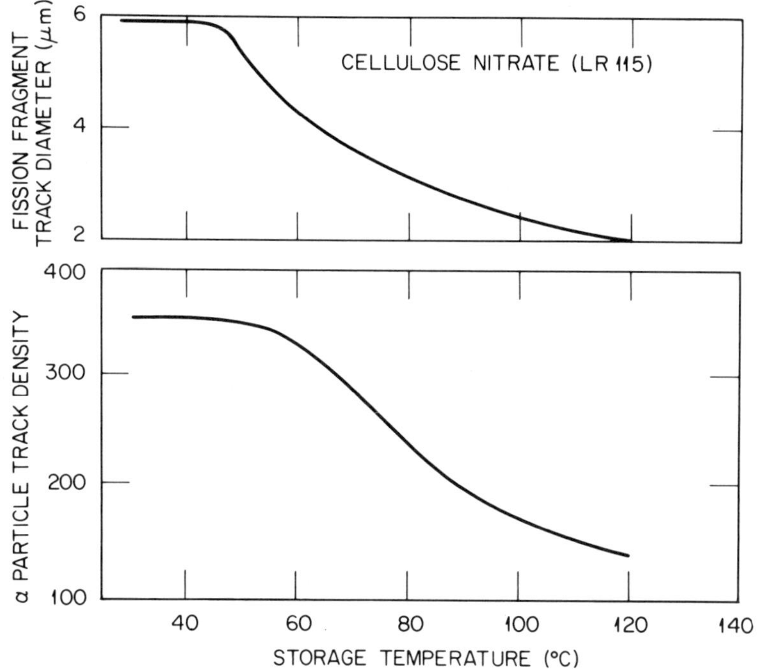

FIGURE 5-8.   Effect of extended (4 to 15 hr) preheating on (a) the diameter of fission-fragment tracks and (b) the relative alpha particle density that is observed in LR 115[®] cellulose nitrate (Kodak-Pathé, France), irradiated and uniformly etched after the thermal treatment. (After Marchetti et al., 1972.)

(Figure 5-10). Only if the softening temperature of a plastic is approached, rapid annealing begins to take place. For example, the accumulated "background" alpha radiation effect due to atmospheric radon in a cellulose nitrate foil can be annealed prior to use by overnight storage at ~80°C. In some of the less fading-resistant polymers the process of fading may actually compete with the progress of etching if the etching temperature exceeds 60 to 70°C, thus reducing the apparent sensitivity (Barbier and Renard, 1972). In such cases, extended etching at lower temperatures is advisable.

In other foils such as polycarbonate (Makrofol), cellulose triacetate (Triafol T[®]),* or cellulose butyroacetate (Triafol B[®]*), higher temperatures are requried for annealing (Becker, 1969a and Heinzelmann and Schüren, 1972). As can be seen in Figures 5-11A and B, fission fragment tracks in Makrofol E and Triafol B are still stable at 60°C, but the fading rate rapidly accelerates around 100°C. Extended etching can make some

"invisible" tracks reappear. Of course, not only the number of visible etch pits or tracks, but also their dimensions (size distribution and length), undergo pronounced changes during fading; such changes may be used as an indicator of the detector's thermal history (Somogyi, 1972a and Wagner, 1972).

The fading kinetics in polymers is affected by several parameters other than temperature. For instance, exposure of a cellulose nitrate foil to intense ultraviolet light induced fading of latent alpha particle tracks (Hasegawa et al., 1968). High LET tracks, such as those of fission fragments, tend to fade less rapidly than alpha particle tracks in a given material (this process can to some extent be used for discrimination against low LET particle tracks).

The atmosphere in which a foil is stored between exposure and etching also can have a strong effect on the etching parameters. As can be seen in Figure 5-11C, little change is observed in Lexan[®] during storage in air and inert gases, but

*Farbenfabriken Bayer A.G., Leverkusen, Germany.

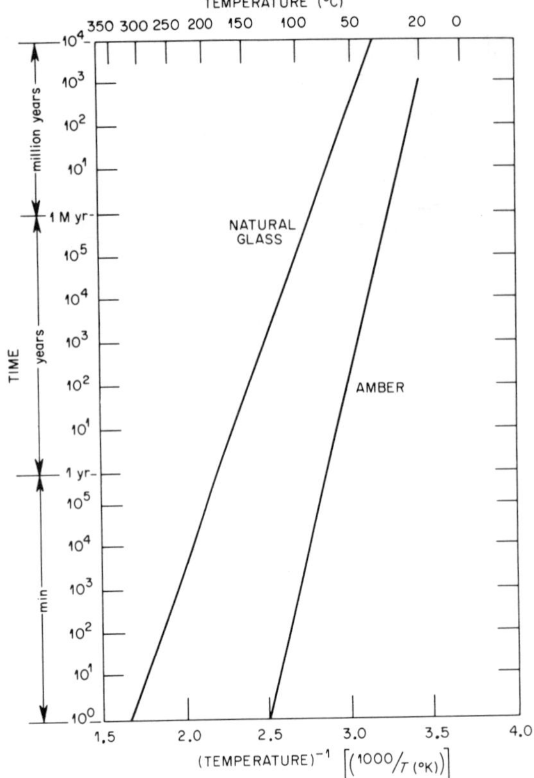

FIGURE 5-9. Latent fission-fragment track stability in a basaltic glass from the Mid-Atlantic Ridge (After Fleischer et al., 1971) and in amber (After Uzgiris and Fleischer, 1971), as determined by measuring at various higher temperatures the time required for total annealing of the latent tracks.

the track length substantially changes in oxygen and in vacuum. There is also evidence that the fading in some polymers is slightly accelerated at high relative humidities and that "old" tracks behave differently from "young" tracks in minerals and glasses. Apparently, physical as well as chemical processes contribute to the thermal fading of latent tracks.

5-2. Detectors and Etchants

Optimized etching conditions have to be established empirically for each detector material. In principle, any chemical that attacks at a sufficient speed can be used as an etchant. For example, the use of different alkali hydroxides yielded differences in both the etching rate and sensitivity of various polymers (Blanford et al., 1970; Dutrannois, 1971; and Enge et al., 1972a), but little differences in the final sensitivity. As an example, the etching rates of fission-fragment tracks in polycarbonate are given in Figure 5-12. More for convenience than for well-established technical reasons, some simple etchants, such as ~30% KOH or NaOH and concentrated or diluted HF, have dominated the etching work for many years (a particularly widely used "standard" etchant is 28% KOH; as KOH is hygroscopic and takes up CO_2 from the atmosphere forming K_2CO_3, it should be prepared from freshly opened KOH bottles, and its concen-

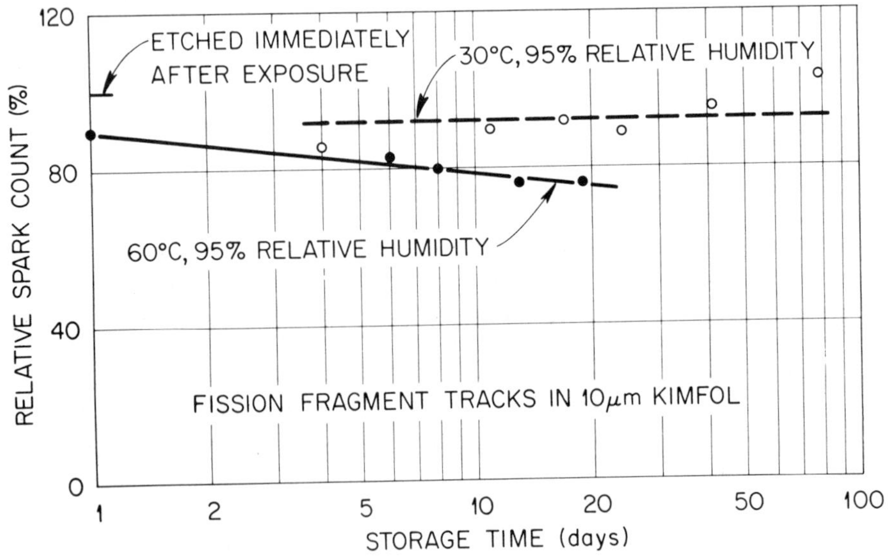

FIGURE 5-10. Relative number of fission-fragment spark counts in 10 μm polycarbonate (Kimfol®) as a function of postexposure, preetching storage time of the foil under adverse climate conditions. (After Becker et al., 1972.)

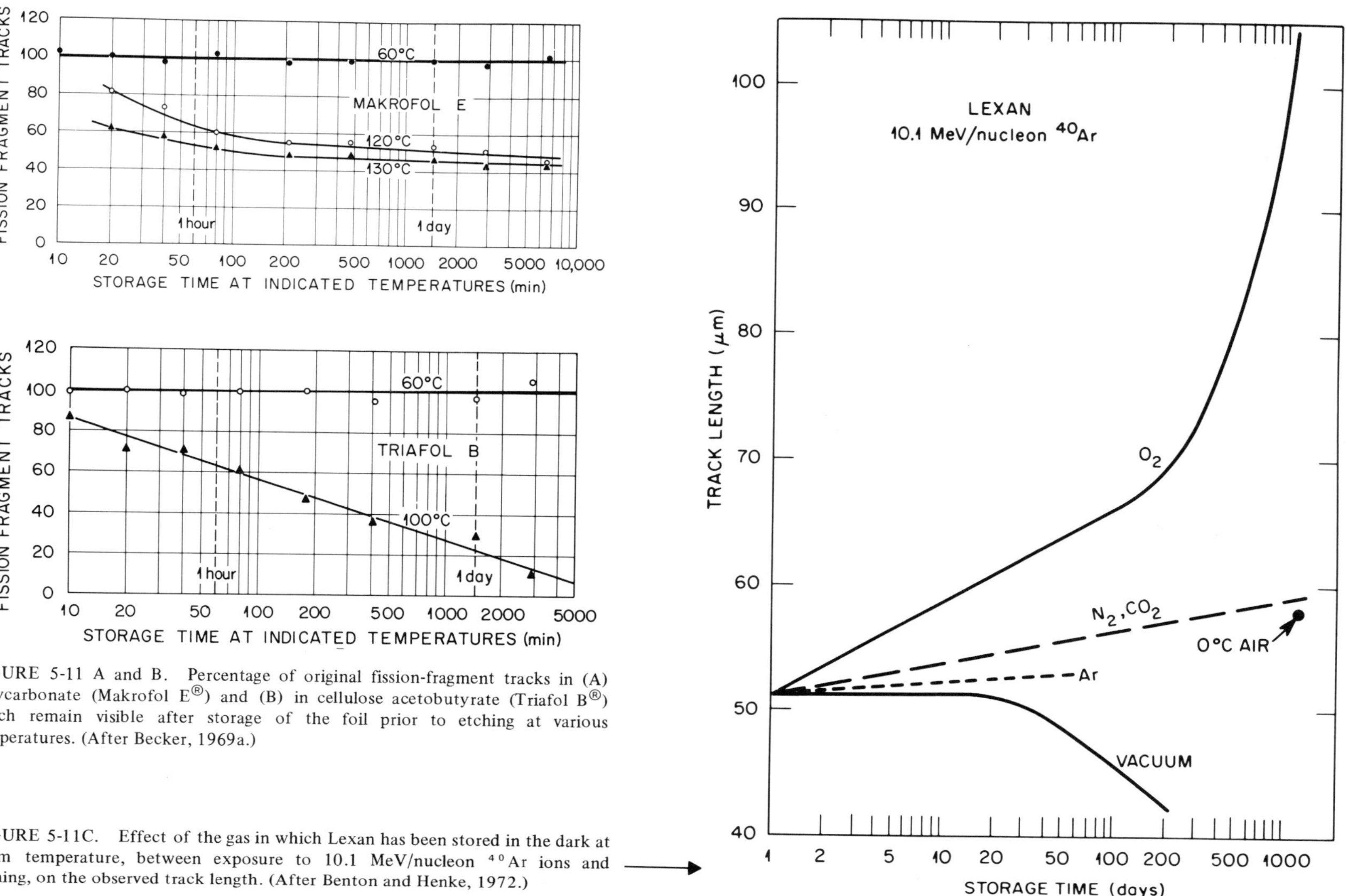

FIGURE 5-11 A and B. Percentage of original fission-fragment tracks in (A) polycarbonate (Makrofol E®) and (B) in cellulose acetobutyrate (Triafol B®) which remain visible after storage of the foil prior to etching at various temperatures. (After Becker, 1969a.)

FIGURE 5-11C. Effect of the gas in which Lexan has been stored in the dark at room temperature, between exposure to 10.1 MeV/nucleon ⁴⁰Ar ions and etching, on the observed track length. (After Benton and Henke, 1972.)

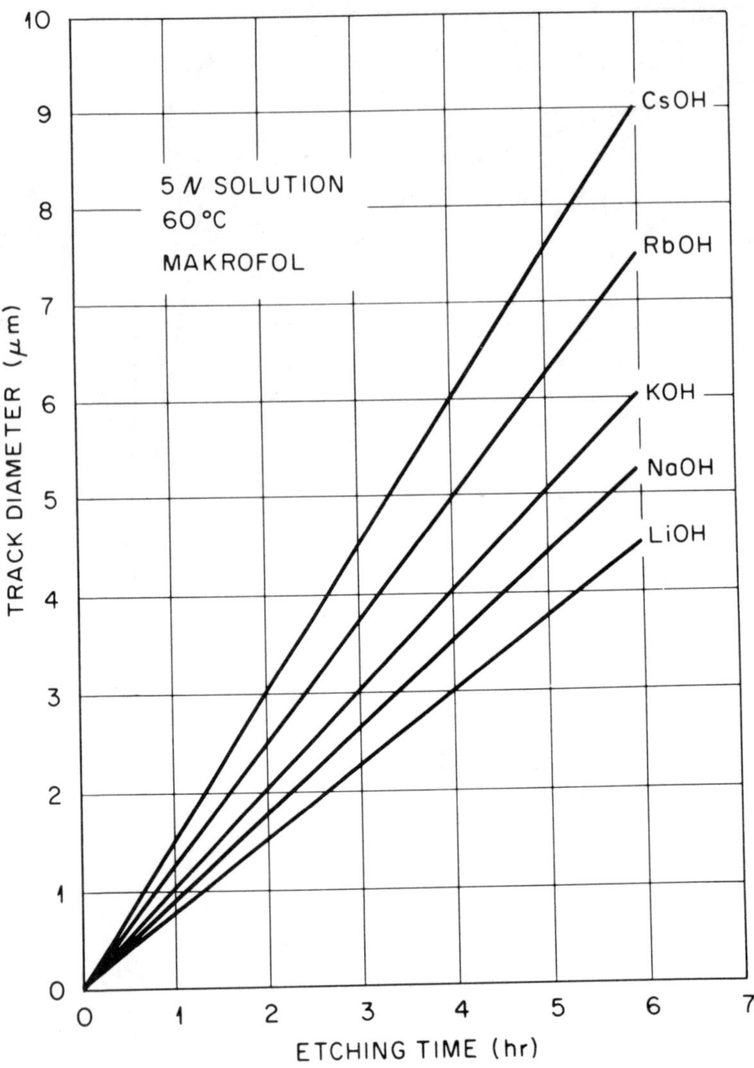

FIGURE 5-12.   Diameter of etched fission-fragment tracks in polycarbonate (Makrofol®) as a function of etching time in 5 N alkali hydroxide solutions at 60°C. (After Dutrannois, 1971.)

tration checked with a hydrometer — its density should be 1.270 g/cm³ at 15°C).

Only more recently have more sophisticated procedures, such as multistep treatments or more complex etchants including oxidizing reagents, been used to a larger extent, leading to much improved results in numerous polymers in which track etching had previously been difficult or impossible. It may even be possible to distinguish between the effect of different damage mechanisms along the track by variation of the etching conditions (Haseganu, 1971). Some etching procedures that have been suggested for fission-fragment detection in various materials have been compiled in Table 5-2. The data in this table are, however, rather arbitrary; actual "optimum" etching conditions may differ from those indicated.

In a given detector material the etching speed depends mainly on the following parameters: the type and velocity of the charged particle; the chemical composition of the etchant; the temperature, concentration, and viscosity of the etchant; the chemical and physical pre- or post-irradiation treatment of the detector material; and the physical or chemical con-

TABLE 5–2

Recommended Etching Conditions for Fission-fragment Track Etching in Some Dielectrics

Material	Etchant composition, etching time, and temperature (°C)
A. Inorganics	
Mica (lepidolite)	15% HF, 20 sec, 50°
	48% HF, 3–70 sec, 23°
Mica (muscovite)	20% HF, 2 h, 23°
	20% HF, 12 min, 52°
	15% HF, 20 min, 50°
	48% HF, 10–40 min, 23°
Mica (phlogopite)	20% HF, 5 min, 23°
	15% HF, 1 min, 50°
	48% HF, 1–5 min, 23°
Quartz (SiO_2)	48% HF, 24 hr, 23°
Borate glass	H_2O, 1 min, 23°
Obsidian (vulcanic glass)	48% HF, 30 sec, 23°
Phosphate glass	48% HF, 5–20 min, 23°
Silica glass (fused quartz)	48% HF, 1 min, 23°
Soda-lime glass (microscope slide, window	48% HF, 5 sec, 23°
glass)	5% HF, 2 min, 23°
B. Organics	
Cellulose acetate (Kodacel[®], Triafol T[®])	28% KOH, 30 min, 60°
	$NaCO_3$, 30 min, 100°
Cellulose acetate butyrate (Triafol B[®])	6.25 NaOH, 12 min, 70°
	($KMnO_4$, NaClO)
Cellulose nitrate (DaiCell[®], Nixon-Baldwin)	6.25 N NaOH, 2 min, 70°
	6.25 N NaOH, 4 min, 55°
	6.25 N NaOH, 2–4 hr, 23°
	28% KOH, 30 min, 23°
	($KMnO_4$, NaClO, $K_2Cr_2O_7$)
Cellulose propionate (Cellidor[®])	28% KOH, 100 min, 60°
Cellulose triacetate (Kodacel TA401[®],	6.25 N NaOH + 15% NaClO (2:1 to 1:3),
Triafol TN[®])	40°
	28% KOH, 60 min, 60°
Cormophenol (Ambrolithe[®], Phenoplaste[®])	NaOH, 1 hr, 40°
Ionomeric polyethylene (Surlyn[®])	10 g $K_2Cr_2O_7$ + 35 cm^3 30% H_2SO_4, 1 hr,
	50°
Polyimide (H-film)	$KMnO_4$ (25% aq), 1.5 hr, 100°
Polycarbonate (Lexan[®], Makrofol[®], Kimfol[®],	6.25 N NaOH, 20 min, 50°
Merlon[®])	33 cm^3 30% H_2SO_4 + $K_2Cr_2O_7$ (10 g), 2 hr,
	85°
	(6.25 N NaOH), 2 hr, 23°
	2.5 N NaOH + 4% Benax, 20 min, 70°
Polyethylene	10 g $K_2Cr_2O_7$ + 35 cm^3 30% H_2SO_4,
	30 min, 85°
	10 g $K_2Cr_2O_7$ + 5 cm^3 30% H_2SO_4 + 29 g
	H_2O, 90°
Polyethylene terephthalate (Mylar[®],	6.25 N NaOH, 10 min, 70°
Chronar[®], Melinex[®])	$KMnO_4$ (25%), 1 hr, 55°
Polymethylmethacrylate (Plexiglas[®], Lucite[®])	$KMnO_4$ (sat), 50 min, 85°
Polyoxymethylene (Delrin[®])	5% $KMnO_4$, 10 hr, 60°
Polyphenoxide	$KMnO_4$, (25% aq), 4 min, 100°
Polystyrene	$KMnO_4$ (sat), 2.5 hr, 85°
	10 g $K_2Cr_2O_7$ + 35 cm^3, 30% H_2SO_4, 3 hr,
	85°

(After Fleischer and Hart, 1970 and Blanc, 1970.)

Recommended Etching Conditions for Fission-fragment Track Etching in Some Dielectrics

Material	Etchant composition, etching time, and temperature (°C)
Polyvinylacetochloride	$KMnO_4$ (25% aq), 30 min, 100°
Polyvinylchloride	$KMnO_4$ (sat), 2.5 hr, 85°
	$KMnO_4$ (25% aq), 2 hr, 55°
Polyvinyl toluene	$KMnO_4$ (sat aq), 30 min, 100°
Silicone-polycarbonate copolymer	6.25 NaOH, 20 min, 50°
Siloxane-cellulose copolymer	8 N NaOH + Benax, 3 hr, 85°

dition of the foil during irradiation. For example, "hardening" of polymer foils by aging or heating prior to etching may decrease the etching speed (Figure 5-8), while "softening" its structure with damaging agents such as H_2O_2 (Becker, 1968b), NH_3 (Heinzelmann and Haschke, 1971a), ozone (Somogyi, 1972b), SO_2, N_2O, NO, and J_2 (Chave and Monnin, 1970), or ultraviolet light (Crawford et al., 1968) in some cases dramatically increases the etching speed.

Even the humidity content of the foil when etched may affect the etching rate (Becker, 1968b and Enge et al., 1970). As such effects of external factors on the track etching parameters have considerable influence on the reliability of quantitative measurements, they have recently been studied in detail by various investigators (see, for example, Girshin et al., 1970; Somogyi and Srivastava, 1970; and Nicolae, 1972).

The effect of those agents on sensitivity is much less pronounced than that on the etching speed, and increases in the sensitivity of some of the less sensitive polymers have been obtained by exposing them to UV of the proper wavelength (Henke et al., 1970) or by soaking them in ethanol/ammonia prior to etching (Heinzelmann and Haschke, 1971b). In the case of cellulose nitrate foils, the degree of nitration, molecular weight, and the type and concentration of additives all have been shown to slightly affect sensitivity and etching speed (Benton, 1968 and Veprik et al., 1970a). A cellulose nitrate made by the Dai Nippon Co. in Japan (trade name DaiCell®) appears to be one of the most sensitive materials so far.

An interesting method for enhancing the track registration sensitivity and etching speed in cellulose acetobutyrate foils consists of the application of high voltage, square, or sinusoidal waveforms during the etching process (Tommasino, 1970). The foil to be etched is placed between two containers filled with the etchant in such a way that it electrically separates the containers; Pt electrodes in the containers are connected with a 2,000 V, 1 kHz H.V. function generator. By this "electrochemical" etching, tracks can easily be enlarged to visibility with the naked eye (Sohrabi and Becker, 1971).

There is a definite effect of the presence of oxygen during irradiation on the sensitivity of at least some organic foils such as cellulose acetate (Becker, 1968b; Boyett et al., 1970; and Girshin et al., 1970). In the presence of oxygen, dry foils are substantially more sensitive than in high vacuum or in an inert gas. The etching kinetics is also affected (if exposed under anoxic conditions, the etching proceeds more slowly). This oxygen effect is not observed if the foil is humid during exposure, indicating that charged particles may "carry their own oxgyen supply" by the radiolysis of water.

The effect of the etchant's concentration on the etching rate (as measured, for example, by the increase in track diameter) is complex and depends on both etchant and detector. Etching rate and registration sensitivity in cellulose nitrate do not change significantly with increasing concentration for hydroxide solutions >6 N, while in Lexan there is a substantial increase even at 12 N, as shown in Figure 5-13A. Etching temperature and interruptions of the etching procedure also affect this concentration dependence. As exemplified in Figure 5-13B, the bulk etching rate in cellulose nitrate "saturates" in ~ 3 N NaOH during uninterrupted etching, but increases rapidly with concentration if the etching procedure is frequently interrupted in 10 to 30 sec intervals.

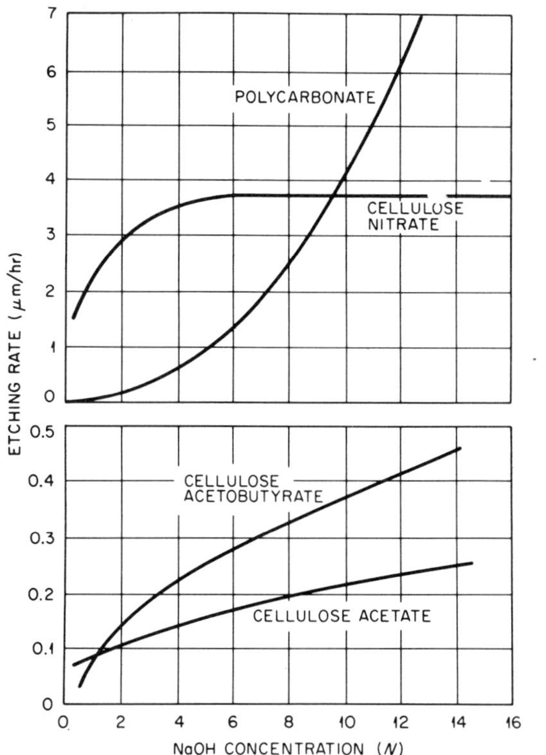

FIGURE 5-13A. Rate of increase in the diameter of tracks in various detector materials as a function of the NaOH concentration. (After Blanford et al., 1970.)

Stirring has a similar influence. This behavior (which is not necessarily the same inside narrow tracks) is explained by assuming a colloid layer of semidamaged cellulose nitrate molecules on the surface (Enge et al., 1972a).

The sensitivity of cellulose nitrate, acetate, etc. can also vary considerably as a function of the etchant concentration (Somogyi et al., 1970; Enge et al., 1972a; and Baroni et al., 1972a). Even the etching temperature can modify the sensitivity. Etching of α particle tracks in Makrofol in 30% KOH for 30 min at 90°C produced more tracks than "optimal" etching for 2 to 3 hr at 60°C (Heinzelmann and Haschke, 1971). Normally the (bulk) etching rate of a track detector in a given etchant is rather simply related to the etching temperature, as indicated in Figure 5-14.

It is a basic disadvantage of track etching that only tracks at the surface that are accessible to the etchant can be made visible. One rather complicated method for the "development" of volume tracks has been suggested so far (Benton, 1971): The foil is bombarded with heavy ions normal to the detector surface. Etching results in parallel

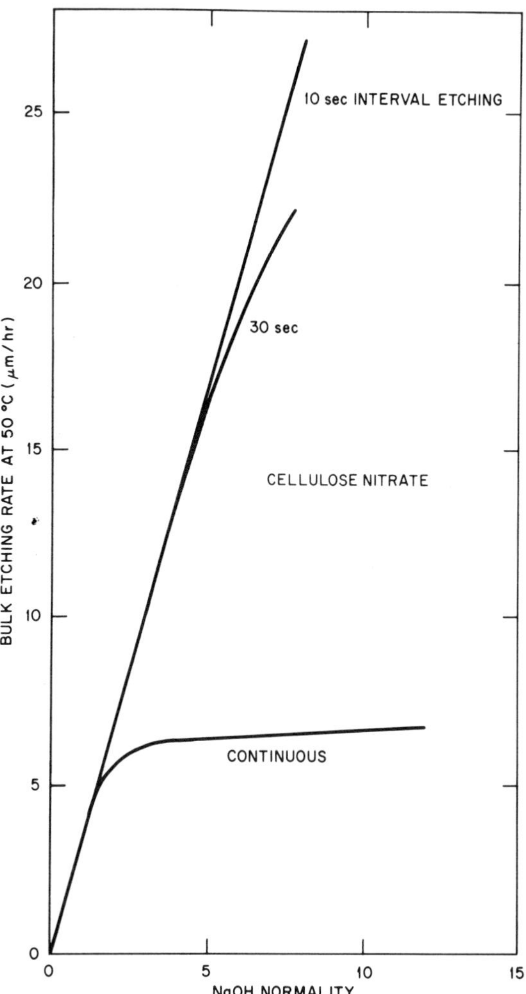

FIGURE 5-13B. Bulk etching rate of cellulose nitrate in NaOH at 50°C as a function of the etchant concentration for uninterrupted etching and for interval etching. (After Enge et al., 1972a.)

large channels through which the etchant can penetrate and attack smaller tracks inside the polymer.

Most of the materials listed in Table 5-2 are commercially available in sheets or foils of reasonably uniform thickness and chemical properties. Of the inorganic detector materials, artificial mica is more uniform in its uranium content and has less background "tracks" than natural mica. Special glasses doped with U and Th (Yokota et al., 1968; Nomura et al., 1971; and Schreurs et al., 1971) are also quite uniform in their properties.

Even if the etching is done under carefully controlled conditions, for instance, in a thermostatically controlled bath with ultrasonic agitation,

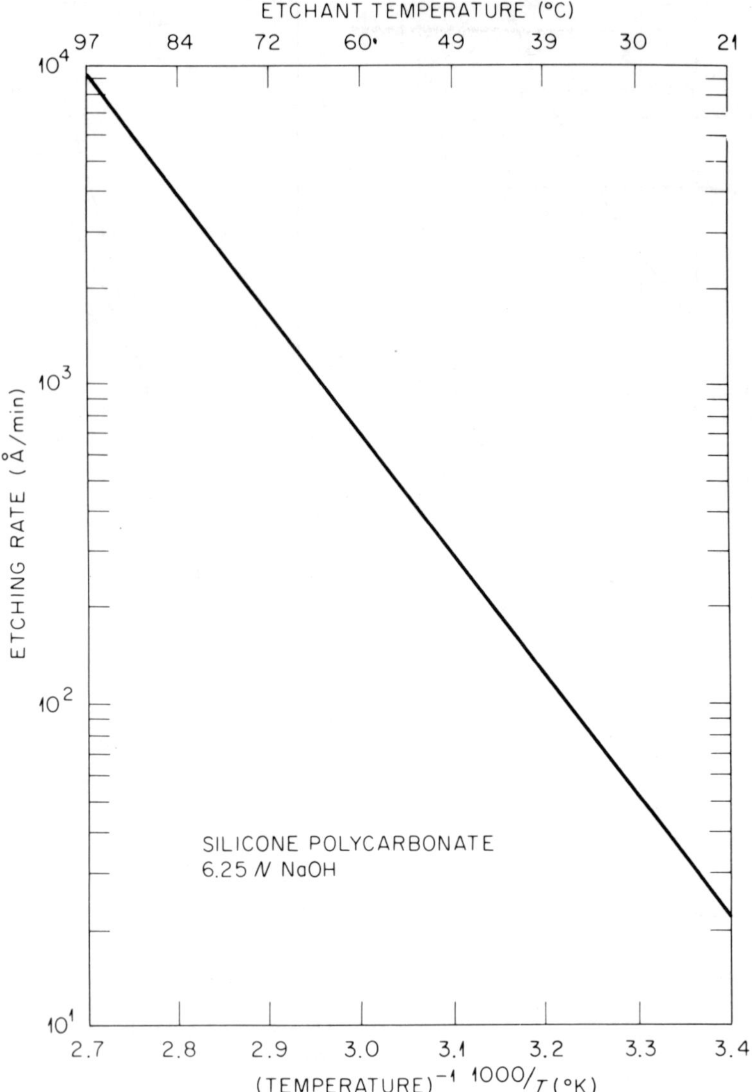

FIGURE 5-14. Variation of the general etching rate in a (stretchy) silicone polycarbonate with temperature of the etchant (After Fleischer et al., 1972b.)

identical tracks may, however, not reach the level of visibility exactly simultaneously, due to small local differences in the foil's sensitivity. Lack of reproducibility of polymers from batch to batch may require individual calibration in case of measurements requiring high precision, such as particle and energy determination.

Foil thickness is not an important factor if visual track counting is employed. When the evaluation process is based on the perforation of thin foils, uniform film thickness may become crucial for reproducible results. Detector films such as Makrofol (Kimfol) for heavy charged particle (fission-fragment) detection are commercially available in several thicknesses, but more sensitive cellulose-nitrate foils for alpha particle detection have to be prepared in the laboratory unless one "peels" the 10 μm red-dyed cellulose nitrate layer off from the 0.1 mm polyester base in the Kodak-Pathe LR 115[®] material (Audran and Renard, 1972). The procedure most often used in making cellulose nitrate films involves casting concentrated polymer solutions onto flat, level surfaces such as a pool of mercury and allowing slow evaporation of solvents.

For the convenient evaluation technique of

"spark-counting" of alpha particle tracks, foils with reproducible properties have been prepared as follows (Johnson and Becker, 1970): The cellulose nitrate is usually obtained (for instance, from the Hercules Powder Co., Wilmington, Del.) in the form of a powder containing 30% ethyl alcohol (films poured from highly nitrated, high viscosity cellulose nitrate have, in general, superior etching and spark-counting characteristics compared with other grades such as Parlodion®). To 24.3 g of this powder (containing 17 g cellulose nitrate) solvents, cosolvents, and plasticizers are added sequentially: 124.7 g ethyl acetate, 4 g isopropyl alcohol, 5 g butyl alcohol, 8 g Cellusolve acetate (trade name for ethylene glycol monoethyl ether acetate), and 4 g dioctyl phthalate.

After aging about four days at room temperature, a 10 to 20 g aliquot (depending on the desired film thickness of about 10 to 20 μm) of this stock solution is diluted to 240 ml with ethyl acetate. Then 40 ml aliquots of the diluted stock solution are transferred to glass dishes 15 cm in diameter, which are positioned on a carefully preleveled surface (or partially filled with mercury). The dishes are $\sim$80 to 90% covered by a glass plate to maintain a slow evaporation rate and to help prevent dust contamination. One to two days is required for drying of the foils at room temperature. They are then removed by water flotation and blotted to dryness between layers of filter paper. Prior to use, the films should be thermally annealed at 80°C overnight to remove residual solvents and background tracks.

If track densities are determined by visual counting, the etching conditions do not have to be controlled very precisely (variation of the etchant composition, temperature, and etching time within 5% or 5°C may still yield sufficiently reproducible results). More precise measurements of track dimensions or some nonvisual evaluation techniques require substantially better reproducibility of the etching conditions. In particular, changes in the concentration of the etchant due to evaporation have to be minimized; corrections may be required for changes in the "activity" of the etchant. It is known that the track etching procedure at least in polycarbonate is improved by a certain concentration of etch products in the sodium hydroxide solution, which can be determined by optical absorption measurements (Peterson, 1970 and Paretzke, 1972b).

There appears to be a "latent" period at the beginning of the etching process (diffusion of the etchant into the narrow damaged channel, etc.), during which little track growth can be observed. Also, etching times are not strictly cumulative, and differences in the appearance and behavior of tracks may occur between continuously etched polymers and those etched in intervals (see, for example, Baroni et al., 1972b and Enge et al., 1972b).

Precise measurements of the track dimensions in organic and inorganic detectors have been used widely in physics and space research and for particle type and energy identification (Fleischer et al., 1969a; Price et al., 1969; Benton and Henke, 1969; Enge et al., 1972b; Stern and Price, 1972; Schaefer et al., 1972; and others). The high resolution of this method permits distinguishing neighboring isotopes as heavy as ^{10}B and ^{11}B. For the energy spectrometry of fission fragments, special glasses with a strong energy dependence of the etch pit size have been developed (Figure 5-15A – also see Khan et al., 1972).

Determination of the Z and energy of heavy ions is usually only of limited interest in radiation dosimetry. A notable exception is the measurement of the cosmic radiation dose to astronauts and other organisms in space, including the crew and passengers of very high-flying aircraft (SST). The heavy particle dose to the Apollo astronauts has been determined by track etching of their Lexan helmets (Fleischer et al., 1970 and Comstock et al., 1971), or by the evaluation of Lexan foils they have been carrying (Schaefer et al., 1972 and Benton and Henke, 1971 to 1972). The high LET nuclear recoil particle exposure in Biosatellite III has also been measured with plastic detectors (Benton et al., 1972). Others (see, for example, Allkofer et al., 1971 and Haseganu and Nicolae, 1971) determined the heavy particle flux as a function of altitude in the atmosphere (Figure 5-15B).

If a sensitive detector has been exposed to fission fragments, the tracks become visible after approximately the same etching time. Further etching increases the track diameter (for an extended period, the track diameter is a linear function of etching time – Figure 5-12), but does not noticeably affect the number of visible tracks unless they become overetched. If, however, a sensitive detector is exposed to a wide energy spectrum of low LET particles (for example, alpha

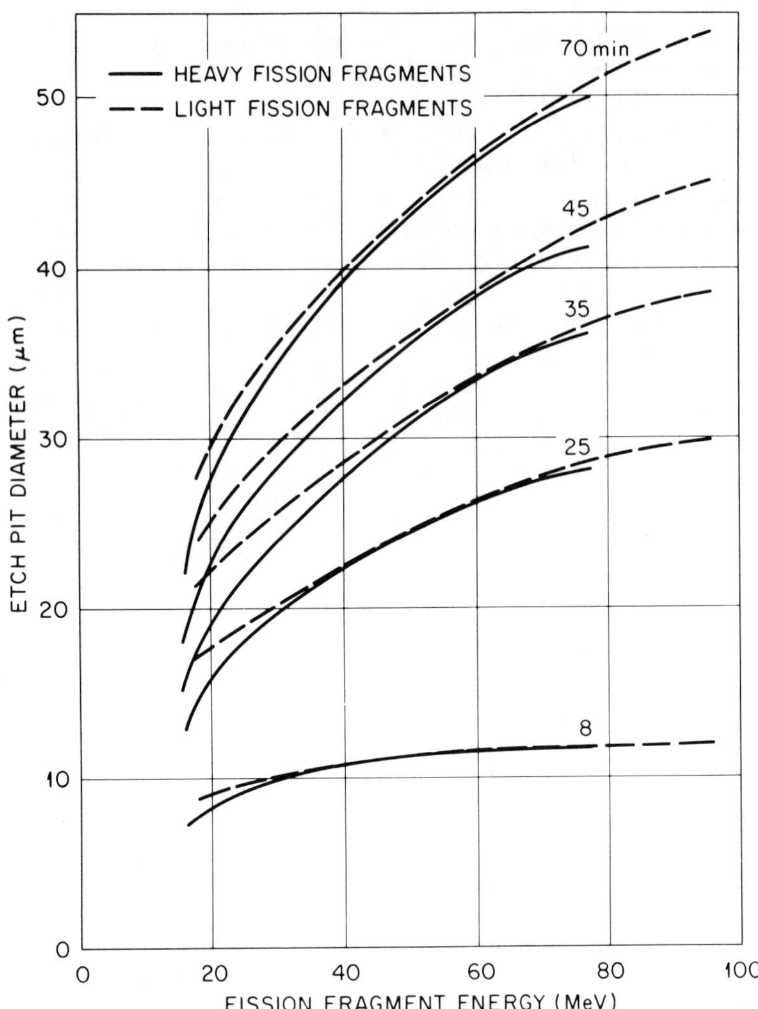

FIGURE 5-15A.   Mean etch pit diameter as a function of the energy of light and heavy fission fragments for different etching times in a glass consisting of 79% P_2O_5, 7% Al_2O_3, 9% ZnO, and 5% B_2O_5. (After Fiedler and Schreck-Köllner, 1972.)

particles from a thick uranium foil with energies from close to 0 to 4.4 MeV) the number of visible tracks becomes a more complex function of etching time. There are three main reasons for this:

1. Low energy, high LET particle tracks at the surface become visible earlier than the smaller tracks or etch pits caused by particles which are closer to the detection threshold of the detector material (Figure 5-16).

2. Some low LET particles reach the LET detection threshold only after slowing down in a certain depth of the material. Such subsurface tracks become susceptible to preferential etching after bulk etching has removed the surface layer to that depth.

3. After extended etching, small surface tracks are likely to become overetched.

The result is a curve such as the one given in Figure 5-17, whose shape depends on the type of detector foil, the particle energy distribution, and the etching conditions. If this wide spectrum of latent tracks is present throughout the foil (for example, in a thick polymer with fast-neutron-induced "internal" recoil nuclei tracks), the track density usually reaches a peak after a certain etching time, representing an equilibrium between the appearance of new and the disappearance of overetched tracks.

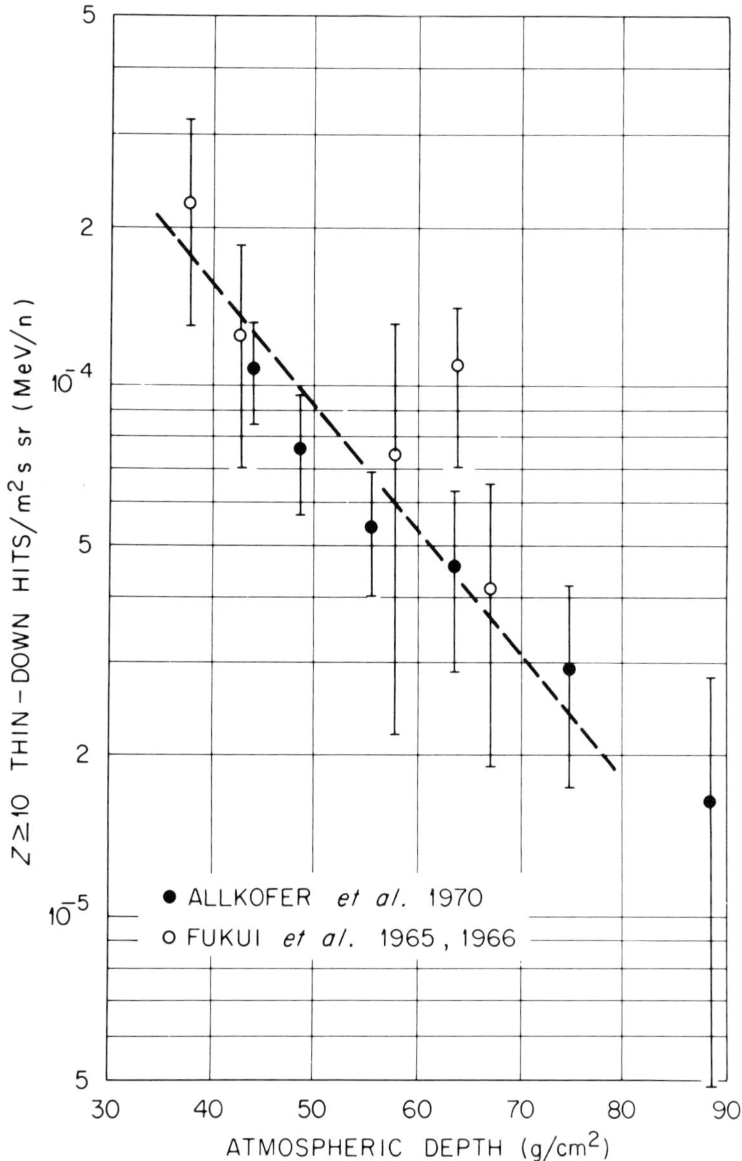

FIGURE 5-15B. Flux of heavy (Z ⩾ 10) ions from the cosmic radiation as a function of depth in the atmosphere as measured with track etching detectors. (After Allkofer et al., 1971.)

5-3. Evaluation Techniques

Most dosimetric applications of solid-state particle detectors require the determination of the number of etch pits per unit area of detector surface. Visual track counting in a microscope is the simplest way of doing this, but it usually is a slow, tedious, and not very accurate process which may be subject to systematic errors. This is true in particular when the track densities are very low requiring the scanning of large areas, very high (overlapping of tracks), or poor visibility of small tracks makes it difficult to discriminate against artifacts such as scratches, dust particles, and "fog" at the foil's surface.

In some cases the fog can be reduced by a special pretreatment such as exposure to high doses of gamma radiation or "flame polishing." In polycarbonate foils that have been in contact with an intense alpha emitter over an extended period of time, a strong increase in background fog due to

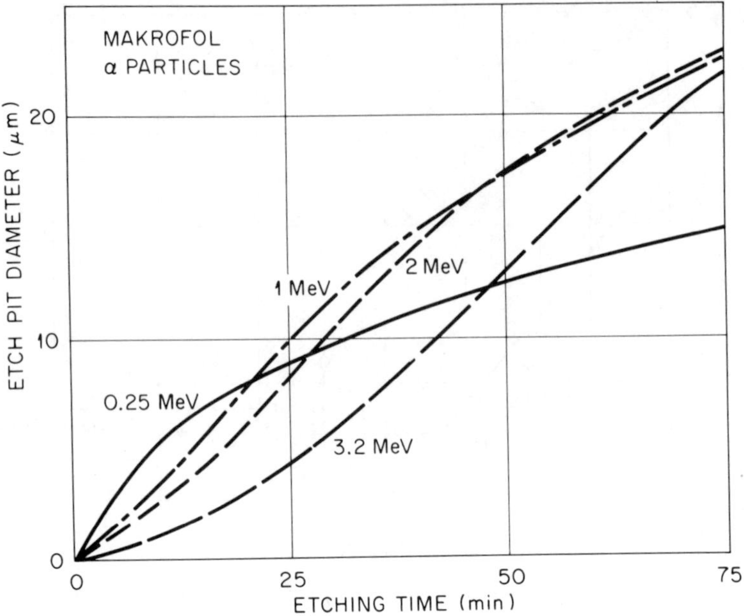

FIGURE 5-16. Alpha particle etch pit diameter in a polycarbonate foil as a function of etching time for different alpha particle energies. (After Tuyn, 1969.)

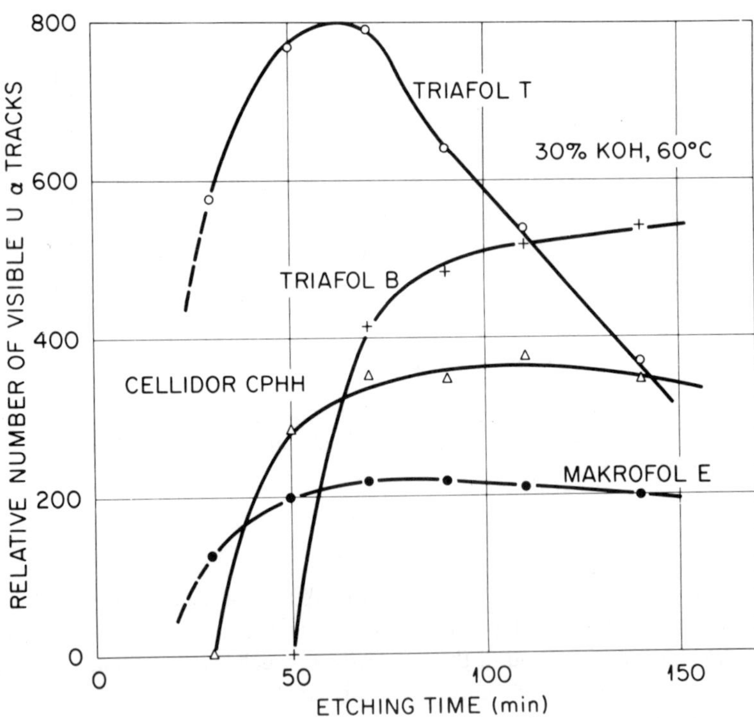

FIGURE 5-17. Relative number of visible alpha particle tracks in various materials after exposure to a thick uranium foil, as a function of etching time in 30% KOH at 60°C. (After Becker, 1969a.)

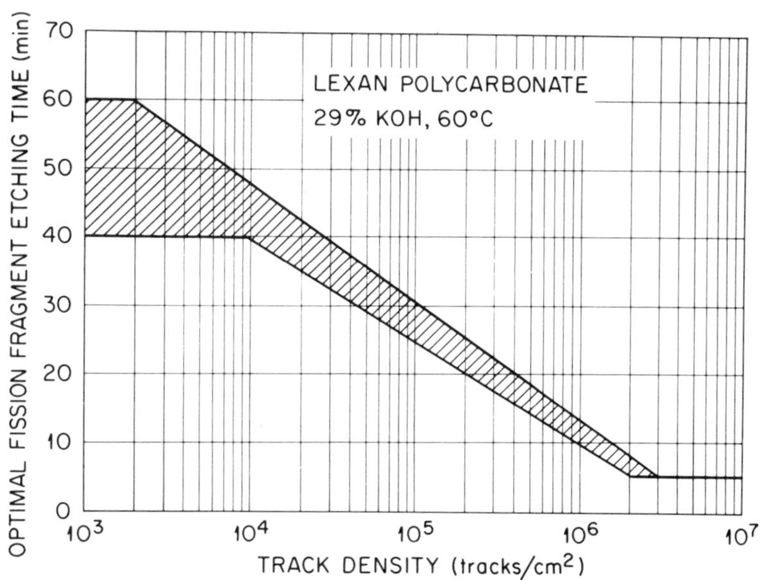

FIGURE 5-18. Optimal fission-fragment etching time of a polycarbonate for visual microscopic counting, as a function of track density. (After Prêtre et al., 1968.)

alpha recoils can obscure the fission-fragment tracks and modify the etching kinetics. In such cases, reduced etching time improves the track visibility (Stone, 1969).

Visual track counting can be made a little easier by some simple aids including projection, adjustment of the etch time to the track density to permit the use of an optimized optical magnification (Figure 5-18), or special techniques of contrast amplification. One such technique, Nomarski's method of interference contrast amplification (Piesch, 1970a), is strictly optical. In another, a special foil consisting of a thin ($\sim$ 10 μm) particle sensitive layer of intensely red-dyed cellulose nitrate is coated on a less sensitive base (Barbier, 1971). Etching creates easily visible clear holes against a red background (the contrast can be further increased by the use of green optical filters). Still others recommend scanning electron microscopy, or counting of fission-fragment tracks in reflecting light after silver-coating the etched surface (Seitz et al., 1973).

Another procedure consists of filling the tracks after etching with a dye such as red ink for contrast amplification (Khan, 1971 and Guenther, 1971). In another method, the etchant penetrates the sensitive polymer layer through the track and reacts with certain chemicals in another, underlying material layer resulting in local absorption changes and thus an "amplification" of the track (Sherwood, 1970). Despite these improvements, visual track counting remains a less than ideal method, particularly when speed, high accuracy, and/or the evaluation of large numbers of foils, or large areas are required. Therefore, automatic counting procedures are highly desirable and a wide variety of them have been devised. In Table 5-3, some basic possibilities are compiled.

Fission-fragment tracks or etch pits usually appear clearly and well defined on a background which is free of disturbances. They can also be easily enlarged by extended etching. Both factors make it much easier to automate the track counting by microscopic scanning in solid-state detectors than in photographic track detectors. Several types of scanning devices such as the "Quantimet®" have been designed and used in track studies (Bitter et al., 1967; Becker, 1969b; Oosterkamp et al., 1970; Guenther, 1971; Cohn and Gold, 1972; Abmayr et al., 1972; Carpenter, 1972; and others). Most of those rather complex electro-optical devices with or without the need for a computer should, however, be regarded largely as obsolete since simpler methods, in particular the "spark-counting" technique which will be described later, have become available.

Under dark field illumination, tracks or etch pits appear bright against a dark background. The

TABLE 5-3

Some Methods for Automatic Track or Etch Pit Counting in Solids

Microscopic techniques	Macroscopic techniques	
	For low track densities	For high track densities
1. Electrooptical scanning devices	1. Dye, gas, or reactive chemical penetration through perforations in membranes	1. Light scattering at etched surface
2. Microdensitometry	2. Light penetration through one-side etched, other-side aluminized thin foils	2. UV penetration through perforated foils

3. Measurement of scattered light in dark-field microscope
  a. Pulse-counting type
  b. Integral intensity
	3. Electric discharges (sparks) through perforations
	  a. Scanning electrode
	  b. Fixed electrodes
		3. Light absorption in thin dyed foils

	4. α particle or electron penetration through perforated foils	4. Dielectric constant or resistance across perforated foil 5. Fraunhofer diffraction of laser light

integral amount of light that is scattered into the microscope objective by the individual tracks can easily be measured by a photomultiplier placed at the location of the eyepiece. If the number of etch pits is too small for good averaging statistics, the integral measurement can be replaced by a pulse-type measurement in which the sample is moved perpendicularly to a slit in the optics (Becker, 1966).

A number of methods do not require the use of a microscope. For low densities, several techniques have been described which are based on the use of thin polymer foils, with the track resulting in a small hole in the foil after etching. Solutions of dyes can be forced through such holes and appear as enlarged spots on a white filter paper (Cross and Tommasino, 1967 and Khan, 1971). The penetration of ammonia through the perforations is recorded with Ozalid® paper (Blok et al., 1969). If a detector has been etched long enough, projection through a microfilm reader, a photographic enlarger, or a slide projector permits rapid scanning of large areas.

If the track densities are high, the change in the appearance of the etched glass or plastic surface can be seen directly. The proportionality between the track density and the amount of light scattered from such surfaces has been used as an indicator of the track density (Becker, 1966; Schultz, 1968; Abmayr et al., 1972; and Platzer et al., 1972). A more convenient method of measuring the amount of scattered light is optical densitometry (Tuyn, 1967; Prêtre et al., 1968; Khovanivich, 1970; Kashukeev et al., 1971; and Khan, 1971).

The transparent holes in a thin dyed detector foil (Barbier, 1971) can also be subjected to optical densitometry, especially if the wavelength chosen for the densitometry is adjusted to the optical absorption peak of the dye. Of course, as in all other techniques based on the measurement of macroscopic integral effects instead of individual track recognition, the optical density depends not only on the track density but also on the etching time. In another nonmicroscopic technique, Fraunhofer-diffraction, using a 1 mW He-Ne gas laser as light source, has been applied successfully to the measurement of the average alpha-particle etch pit diameter in cellulose acetate (Varnagy et al., 1973).

Several other physical parameters, such as the dielectric constant of a foil between electrodes, will depend on the number of holes in the foil. The potential between two electrolytes separated by a perforated foil also is related to the degree of perforation. The penetration of alpha particles through a perforated foil, which can easily be measured with a scintillation detector on its other side, can be a sensitive indicator of the total hole area (Dörschel, 1969 and Khan and Durrani, 1972c). A similar procedure uses low energy electron penetration measurements (Shoup and Haywood, unpublished).

By far, the most elegant method for the automatic counting of tracks in polymers is based on the counting of electrical discharges occurring through the etched perforation in thin foils. Originally, this was done by placing the foil on one electrode and moving the other electrode, a knife-edge, over it (Lark, 1969 and Cross and Tommasino, 1968). The method has later been improved by avoiding moving parts (Cross and Tommasino, 1970 and Cross, 1970). Presently, it represents probably the least complicated, fastest, and most accurate way for determining low and medium track densities in a relatively large area (for higher track densities, optical scattering and absorption methods are still the methods of choice).

The standard procedure in "spark counting" of the small ($\sim$ 2 to 10 μm diameter) perforations representing fission-fragment tracks in, for example, a 10 μm polycarbonate (Kimfol®)* foil, is as follows: After rinsing and drying, the etched foil is placed on an electrode [rounded edges are recommended for the electrode (Prêtre and Heusi, 1972) but otherwise size and shape have little influence on the results]. Clamping of the thin foils in retainer rings or attaching them to plastic rings by rubber cement simplifies their handling and marking. The film is then covered with a piece of one-sided aluminized Mylar®,** the aluminized side of which is facing the etched film, making contact with an outer, grounded electrode.

When a positive voltage of about 500 V is applied to the thin Al layer, sparks occur through the perforations in the etched film. Each spark causes the evaporation of aluminum in an area several orders of magnitude larger than the original hole in the detector film. The typical diameter of the evaporated Al spot is about 0.2 mm. Figure 5-19 shows, as an example, the perforation at the location of a 3 MeV alpha particle track in a commercial cellulose nitrate film (Kodak LR 115®) and the evaporated aluminum area on the Mylar® foil after sparking (note the separate sparking despite overlap of the evaporated zones).

No multiple sparking can occur through individual holes if all parameters are properly adjusted; a plainly visible "replica" of each hole remains in the aluminum layer. The replica provides an image of the track distribution for possible radiographic applications, as well as a permanent record of the results which can easily be filed. Several such replicas may be made from the same detector film.

The counter is coupled to a scaler through a simple, G-M-type quenching circuit so that each spark is recorded in the scaler. The spark counter also is a simple device which requires only a RC pulse-shaping circuit, a HV supply, and a scaler for auxiliary apparatus. A simple RC quenching circuit operates the scaler reliably and accurately. The whole portable spark counting equipment is shown in Figure 5-20.

If the spark is measured as a function of sparking voltage, a "plateau" is obtained. As can be seen in Figure 5-21, a much longer plateau is obtained if the aluminized Mylar® is the positive electrode. The optimum conditions in a given spark counter depend on several parameters, including foil thickness, thickness of the Al layer (a thick layer requiring more energy for evaporation), voltage, nature and pressure of the gas in which the sparking takes place, the resistor and the condenser of the counter, etching conditions, and type and thickness of the foil and particle to be registered.* Optimum conditions should be empirically established by some simple experiments involving variation of etching time, voltage, etc. (even the pressure which is applied during counting and humidity of the fore may have some influence on the results — Distenfeld et al., 1972, Nishiwaki et al., 1973).*** In Figure 5-22 the

---

*Peter J. Schweitzer Div., Kimberly-Clark Corp., Lee, Mass.; the material is identical with a Makrofol® foil made by Farbenfabriken Bayer, Leverkusen, Germany.
**McCordi Metallizing Corp., Mamaroneck, N.Y.
***According to a recent survey of six laboratories that have used the spark counter extensively etching times for the 10 μm Kimfol varies between 40 min at 70°C and 3 hr at 40°C in 28% KOH for fission fragments. The most popular capacitance is 100 pF; a pressure of 100 to 500 g/cm^2 is applied during counting. A foam or rubber padding makes it easier to apply uniform pressure. Presparking voltage varies between 500 to 1,500 V, but record counts are always made around 500 V.

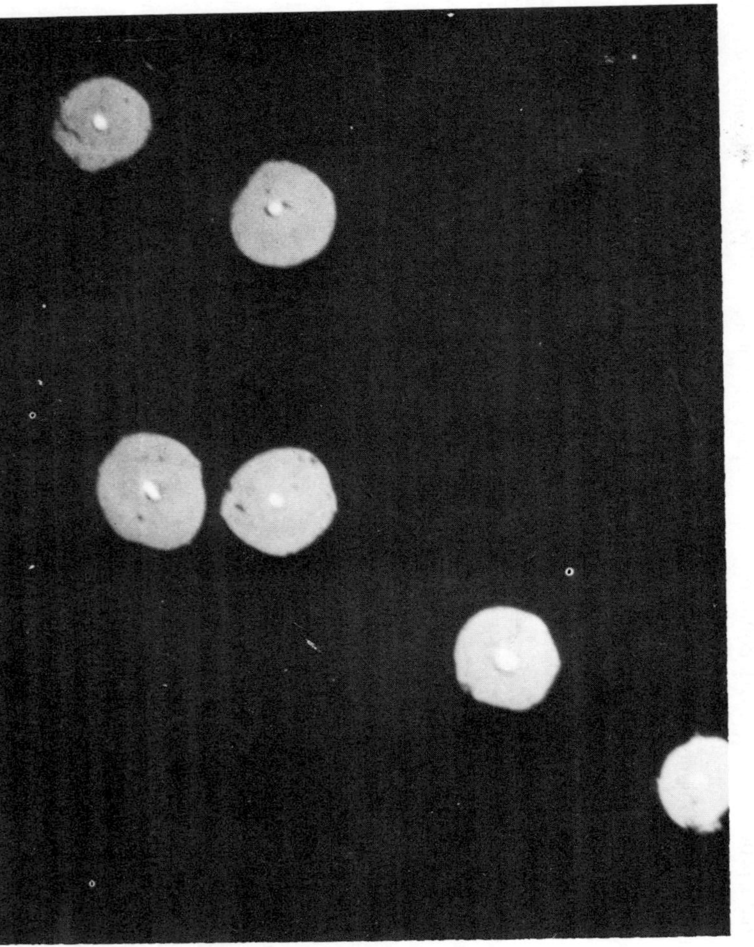

FIGURE 5-19. Alpha particle tracks from a ^{239}Pu source in red-dyed Kodak-Pathe LR 115® cellulose nitrate foil after 2 hr etching at 40°C in 28% KOH (small central perforations) and evaporated spots in aluminized Mylar after sparking in air at 500 V. (After Becker, unpublished.)

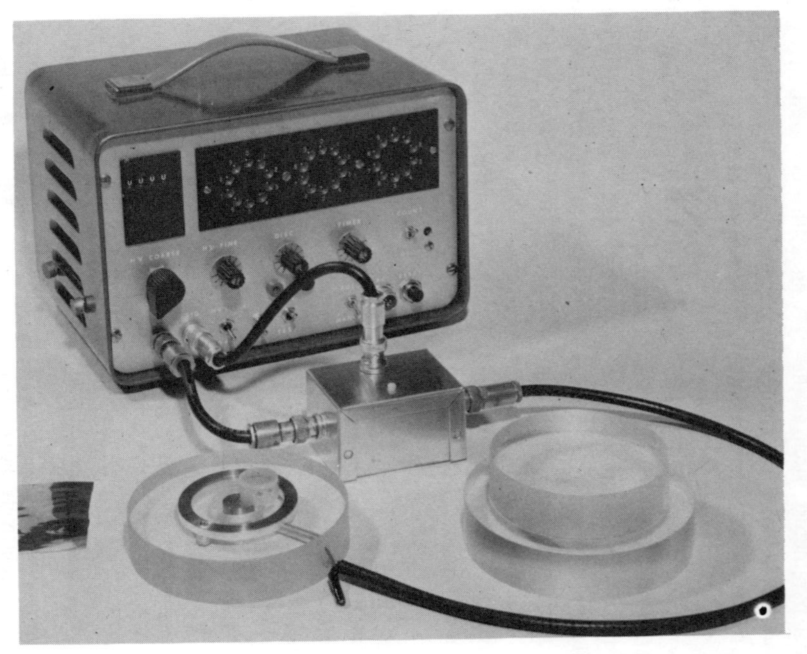

FIGURE 5-20. Portable automatic alpha particle and fission-fragment counter based on sparking. (After Johnson and Becker, 1970.)

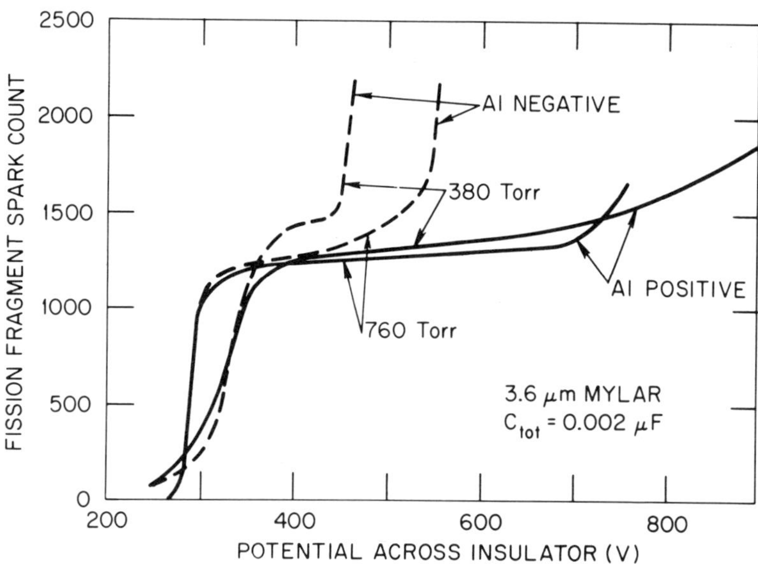

FIGURE 5-21. Fission-fragment spark count in a 3.6 μm Mylar foil as a function of the sparking potential at normal and reduced air pressure, with different polarity of the thin Al layer. (After Cross and Tommasino, 1972.)

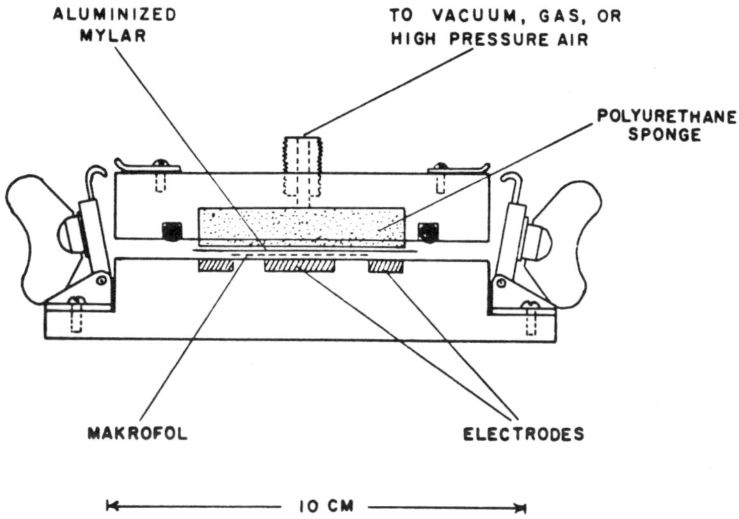

FIGURE 5-22. Automatic track counter to be used for sparking in different gas atmospheres and/or pressures. (After Cross and Tommasino, 1972.)

design is given for a simple device in which the sparking can be performed at a controlled atmosphere and pressure.

The sparking operation also completes "development" of the holes by spark-punching incompletely etched tracks, enlarging their diameter, and cleansing them from residual etchant and debris. Therefore, foils have to be sparked at least twice, the first time at a higher voltage to assure completion of the development process. Sparking at an increased air pressure of 6 atm and a high voltage of 1,200 to 1,500 V gives optimum results for presparking (Cross and Tommasino, 1972). The second (and subsequent) counts for data collection are taken at a reduced voltage since some multiple sparking begins to

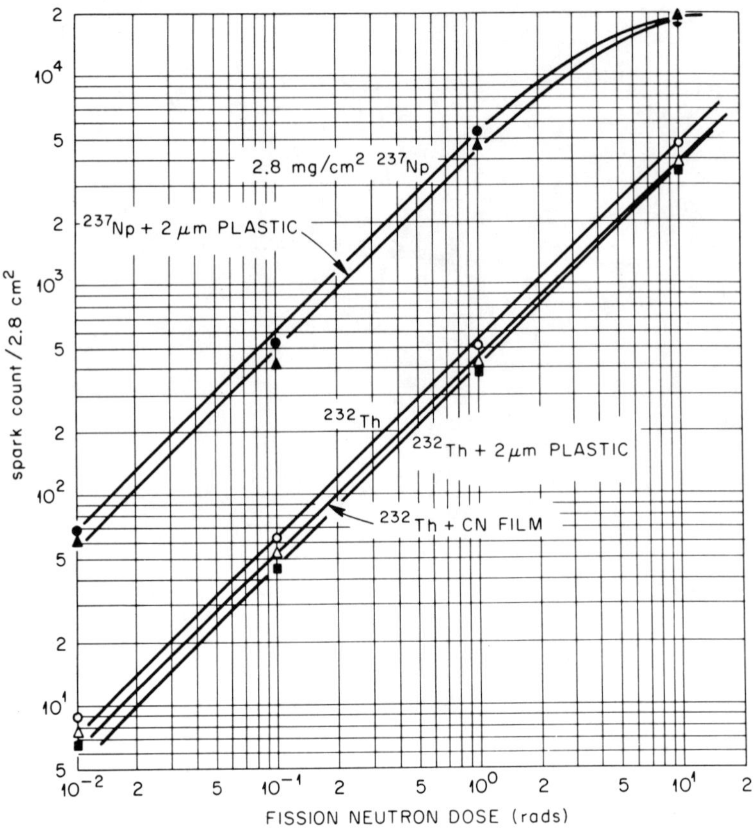

FIGURE 5-23.   Spark count after optimized etching as a function of fast fission neutron dose from the Health Physics Research Reactor, for different fissionable materials. (After Sohrabi and Becker, 1971.)

occur at normal air pressure and about 600 to 800 V. The evaporated aluminum is partially coated out on the film and the positive electrode, making it necessary to clean both occasionally by wiping the electrode with cotton, and rinsing the film in a NaOH or KOH solution after multiple sparking.

If cellulose nitrate or polycarbonate is "over-etched" or thinner than $\sim$3 to 4 μm, electrical breakdown occurs even in unperforated areas. Only with a special technique employing high voltage pulses, is it possible to count polycarbonate foils as thin as 3.5 μm (Geisler and Philips, 1972). Mylar films also appear to permit reduced thickness. In fission-fragment spark counting with polycarbonate films, only a very small "background" count ($<$0.01 count/cm²) from unexposed, etched films is observed.

The spark counting technique was originally developed for the counting of heavily ionizing fission fragments. It is now also applied to the counting of alpha particle tracks in cellulose nitrate

foils with an almost 100% registration efficiency (Johnson et al., 1970). It should, however, be noted that alpha track spark counting is not a routine and easy matter because the detector foils are less likely to be uniform and reproducible, the etching parameters have to be more carefully controlled, and environmental radiation or artifacts in the foil may cause a higher background count.

The useful dynamic range is limited at the high-density end to between 500 and 3,000 tracks/cm² in the spark counting technique. This is mainly due to eventual loss of most of the Al on the aluminized Mylar, resulting in electrically isolated unsparked "islands" in the Al layer (Oltman et al., 1970). In an optimized system in air at normal pressure, the linearity of the system is usually satisfactory to approximately 3,000 tracks/cm², after which the precision decreases (Figure 5-23). The range can be extended to substantially higher track densities by two methods: One uses a moving electrode by placing the aluminized foil on a cylinder which slowly rolls over the detector foil. Another more elegant method makes use of the fact that the sparking in helium gas of 3,200 torr pressure at 230 V produces much smaller holes without occurrence of multiple sparking. Track densities up to 5,000 tracks/cm^2 can thus be spark counted with only 10% maximum counting loss (Cross and Tommasino, 1972).

The reproducibility of repeated track counting of a foil in routine work is approximately $\sigma = 3\%$, but more precise measurements are obtainable by experienced operators if a number of precautions are taken. A standard deviation of only 0.5% has been reported for track densities up to 1,000 tracks/cm^2 if foils are presparked at a high air pressure at a high voltage, and if the actual sparking is carried out in helium, with the Al layer as a positive electrode at a low voltage (Cross and Tommasino, 1972). The detection efficiency for fission fragments is close to 100% if their angle of incidence is around 90°; for isotropic incidence (thick radiation foil in contact with detectors), it can drop to about 70% because not all tracks result in a perforation of the foil.

According to a recent survey (Griffith, personal communication), more than half of the more than 100 laboratories around the world working on track etching in mid-1972 are primarily interested in track density determinations and more than half employ the microscope for visual counting, followed by spark counting (36 groups) and T.V. computer systems (20 groups) as favorite means of evaluation. Basic research and neutron dosimetry are the most common applications, followed by geological studies, cosmic ray research, cross section measurements, and alpha autoradiography. Fission fragments and alpha particles are the most commonly detected particles; polycarbonates, mica, and cellulose nitrate are the most popular detector materials.

5-4. Dosimetric Applications of Fission-fragment Registration

So far, the most widespread dosimetric application of track etching has been its use in neutron dosimetry by fission-fragment registration. In 1963, Walker et al. first drew attention to the possibility of using (n,f) reactions in (or in the environment of) fission-fragment detectors for thermal- and fast-neutron measurements. The basic

technique is extremely simple: An ordinary microscopic slide contains enough uranium to permit neutron fluence measurement in reactors by simply counting the fission-fragment tracks after a few seconds of etching of the exposed slide in HF. The use of fission-fragment registration in mica, different glasses, or plastic foils for neutron flux measurements spread rapidly (Debeauvais et al., 1964 and Prevo et al., 1964).

It was observed that the track density for the neutron energy of interest can be equated to the neutron fluence by a constant, weighted by the fission cross section of the fissile material (Becker, 1965). This relationship between track density and fluence arises from the fact that the detection efficiency of organic detectors for fission fragments, as well as the range and energy distribution of the fission products in the common fissile materials, is relatively constant. Therefore, if the fission-foil thickness exceeds the maximum range of the fission fragments (about 10 mg/cm^2) and does not depress the neutron flux by absorption, the sensitivity of the dosimeter is always approximately 1.2×10^{-5} (tracks/cm^2)/(neutron/cm^2-barn). A more accurate value of 1.16×10^{-5} tracks/(neutrons/barn) was determined later (Prêtre et al., 1968).

Because of this proportionality between neutron sensitivity and fission cross section, the known cross section may be used for neutron energy dependence calculations of various detector combinations. Sets of fissionable materials with different fission thresholds in a ^{10}B absorber sphere, in combination with fission-track detectors, have been used for obtaining additional information on fast neutron spectra, for example, of different types of reactors (Becker, 1965). This technique was soon adopted by other laboratories (see, for example, Kerr and Strickler, 1966) and has since been employed widely as a standard method for neutron flux determinations.

The advantages of the fission-fragment registration method of fast-neutron dosimetry over the older technique of recoil-proton registration in photographic nuclear track emulsions are obvious:

1. Solid-state track detectors are normally insensitive even to high doses of beta, gamma, and X- radiation (no changes in the etching kinetics have been observed for doses up to 10^6 to 10^7 rad);

2. There are no serious deterioration,

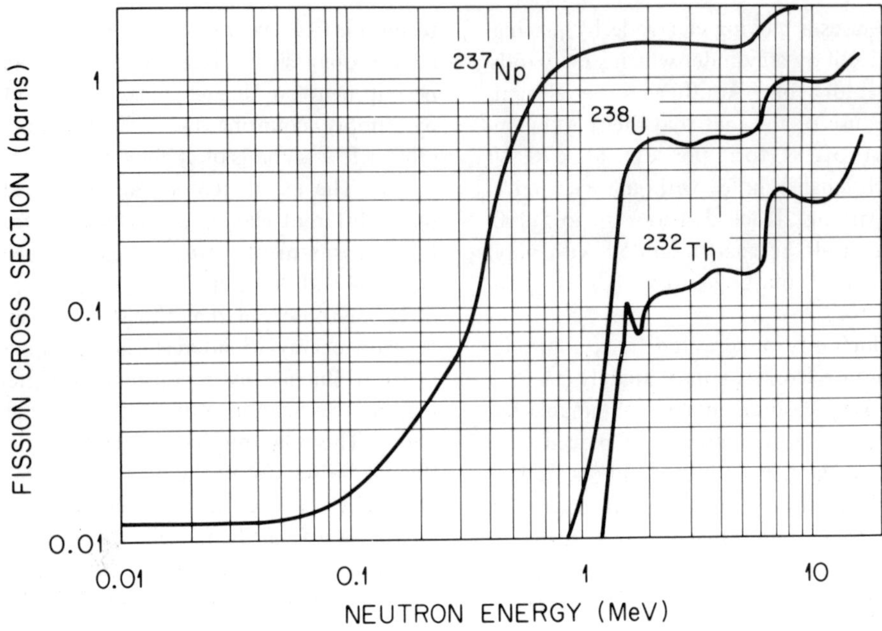

FIGURE 5-24. Fission cross section of different fast-neutron threshold detectors as a function of neutron energy.

storage, and fading problems involved unless the temperature far exceeds room temperature; the detectors are much less sensitive to light, pressure, and other environmental factors;

3. Both "development" (etching) and evaluation are much easier than that of photographic films, and can easily be automated;

4. Due to possibility of easily reading larger areas with automatic counting methods, there is a much higher effective sensitivity;

5. A wider dose-range can be covered by simple means such as dilution of the fissile material; and

6. The detectors can easily be switched "on" and "off" by coupling or decoupling during or between irradiations of the fissile foil and the detector foil.

There are, however, some difficulties for dosimetric applications. One is related to the neutron energy dependence of the detector sensitivity. Although the fission cross section of ^{238}U agrees above $\sim$2 MeV quite well (within ±30%) with the ICRP recommended neutron rem curve (Becker, 1967), in the interesting region below $\sim$2 MeV a rapid drop in sensitivity occurs. The same is true for thorium, but ^{237}Np has a lower neutron energy "threshold" at about 0.6

MeV (Figure 5-24) and its higher cross section insures a higher sensitivity (e.g., at 3 MeV, the ^{237}Np fission cross section is 12 times higher than in ^{232}Th). On the other hand, ^{237}Np is more expensive and radiotoxic than natural or depleted uranium or thorium. The average range of fission-fragments, which is about 4.5μm in ^{232}Np and $\sim$8 μm in ^{232}Th, gives foils with such thickness maximum sensitivity. Some reduction of the thickness will, however, reduce the sensitivity less than proportional due to the reduced self-absorption.

There are other properties that limit the choice of fissile materials. For example, the high radiotoxicity of plutonium makes it impractical in personnel monitoring. The high spontaneous fission rate in ^{238}U causes a considerable "background" increase, which would seriously interfere with low-level measurements over long periods of time, and enriched ^{235}U is expensive and subject to administrative restrictions. From the spontaneous fission point of view, ^{232}Th with a spontaneous fission halflife of $>10^{21}$ years (corresponding to not more than $\sim$6 x 10^{-8} fission/g/sec) is superior to ^{237}Np ($>10^{18}$ years), ^{232}U ($\sim$2 x 10^{17} years), and ^{238}U ($\sim$6.5 x 10^{15} years). However, the spontaneous fission rate in

^{237}Np can be considered tolerable for most practical purposes.

Over extended periods gamma and beta radiation from a uranium foil may give a substantial dose to its environment such as the skin close to a personnel dosimeter or other gamma-sensitive detectors nearby. Larger amounts of ^{232}Th are difficult to shield in a small detector unit. The minimum thickness of ^{232}Th that results in almost full sensitivity is 5.5 mg/cm^2. A detector with 9 cm^2 effective area and 4 μm ^{232}Th thickness delivers about 2 mrad during one working week (40 hr) to its immediate environment. This dose could be reduced substantially by separating ^{232}Th from its daughter products; however, much of the gamma activity would reappear within a few years. It exhibits a sensitivity to the neutron spectrum around a synchrotron of 29 mrem/track x cm^2 (Distenfeld et al., 1972).

Also, preparing uniform thin layers of fissile materials is not easy. Some authors simply evaporate a solution, for example, of uranyl acetate, on a carrier foil, but the uniformity and mechanical and chemical stability of such layers are less than desirable (Koshijima, 1971). Electrochemical plating of layers exceeding $\sim$0.5 mg/cm^2 is difficult. Vacuum deposition creates a layer which rubs off easily. A more promising method is burning viscous organic solutions of neptunium nitrate into the carrier after "painting" it or dipping it into the solution.

Some other methods such as the preparation of Np alloys (Cross, 1970), the coating of a ceramic carrier with a Np-containing glass (Yokota, personal communication), or suspending the fissionable material in cellulose nitrate foils (Song and Lee, 1970) are considered less desirable because they result in a "dilution" of the fissile material in the sensitive layer with a corresponding reduction in sensitivity. The contamination risk can be reduced by coating the fissile material with a thin polymer layer or sealing the fissile material in a $\sim$2 μm polycarbonate (Kimfol) foil. The latter method results in a good protection and a tolerable reduction in the sensitivity of 20 ± 5% (Sohrabi and Becker, 1972).

Dependence of the detector sensitivity on the direction of neutron incidence usually is not much of a problem. There is some disagreement in the results reported in the literature, but the error that is introduced due to directional dependence is small. According to some calculations (Heinzelmann, 1968), fast neutrons incident at a grazing instead of perpendicular angle result in a sensitivity drop of $\sim$12%; values of 17 to 19% have been measured for neutrons in the 4.5 to 15 MeV range (Stolz and Dörschel, 1969). According to other calculations and experiments (Koroleva and Kraitor, 1971), the directional effect in the 0.5 to 10 MeV neutron energy range depends on the sensitivity of the fission-fragment detector. The directional dependence is less pronounced with polymers than with inorganic detectors because of the differences in the critical particle detection angle. No directional dependence has been found for thermal neutrons.

Of course, a substantial directional response is introduced by neutron absorption, backscattering, and energy changes in the human body if such a detector is attached to its surface, as in personnel dosimetry. A typical result, obtained with a thorax phantom, ^{237}Np and ^{232}Th radiators, and fission neutrons, is given in Figure 5-25. Due to the different low energy cut-off energies for the

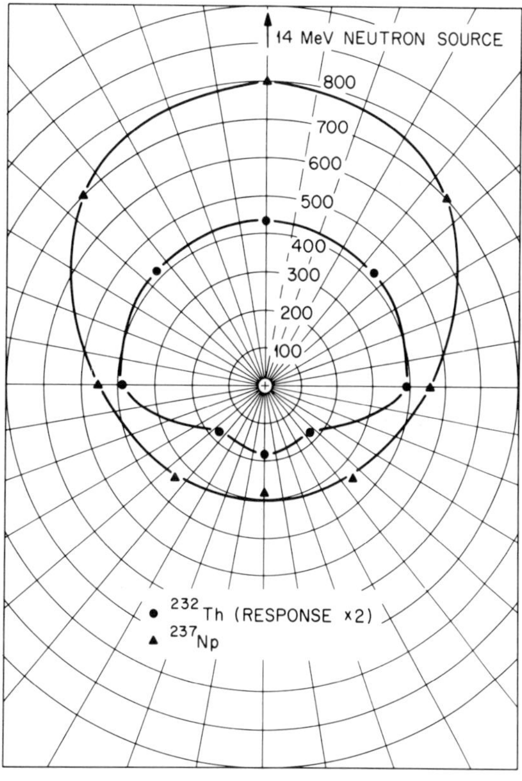

FIGURE 5-25. Directional response of fission-fragment detectors (in spark counts for 2 cm^2 of sensitive area per rad of neutrons) consisting of a thick ^{232}Th foil, and of 2.8 mg/cm^2 ^{237}Np and attached to a water-filled thorax phantom, for monodirectional fission neutrons from an unshielded, fast reactor. (After Sohrabi and Becker, 1971.)

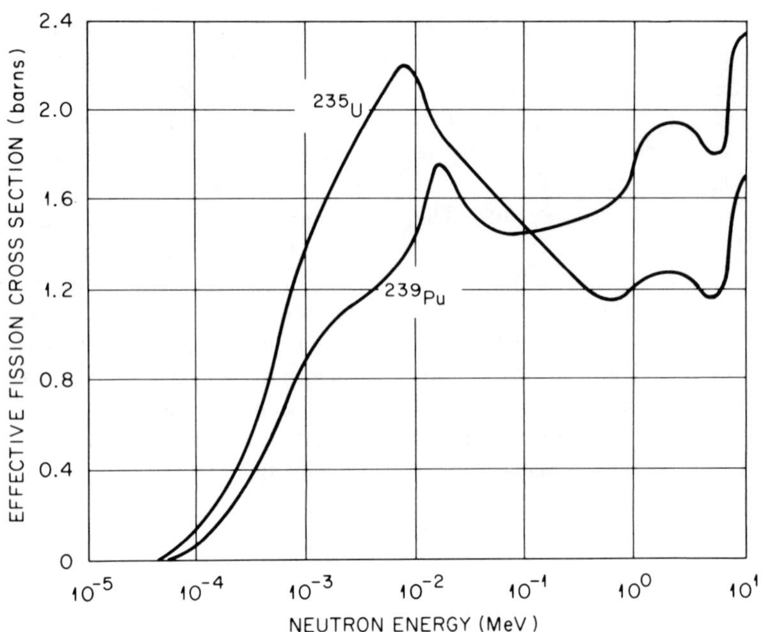

FIGURE 5-26.   Effective fission cross section of ^{235}U and ^{239}Pu, encapsulated in 1.65 g/cm^2 of ^{10}B, as a function of neutron energy. (After Rago et al., 1970.)

two detectors, there is less directional dependence for ^{237}Np than for ^{232}Th. Varying the distance between the detector and the surface of the phantom has no significant effect on the sensitivity (Sohrabi and Becker, 1972).

Several attempts have been made to modify the energy dependence of such detectors, in particular to adjust the detector response to the values recommended by ICRP over an energy range which should be as wide as possible. For example, an alloy consisting of 0.5% (by weight) of natural uranium and 99.5% ^{232}Th was suggested (Prêtre et al., 1968). This system responds to neutrons approximately rem-equivalent from thermal to a few eV, and again above ∿2 MeV, but is almost insensitive for the large neutron energy range between those energies. Other possible mixtures of different fissile materials do not promise much better results.

A slightly different approach takes into account the backscattering of moderated neutrons from the human body (Tatsuta, 1970). Using two detectors, one with natural and the other with slightly enriched UO$_2$, and a thermal shield (Cd) only on the exterior side, a somewhat better combination is obtained. Its undersensitivity is limited to the 10 keV to 1 MeV energy range. In

still another modification (Rago et al., 1970 and Keirim-Markus et al., 1973), the thermal neutron oversensitivity of detectors that contain ^{235}U and ^{239}Np is reduced by total encapsulation of the fissionable material in a 1/E thermal neutron absorber. A shield of 1.65 g/cm^2 ^{10}B results in an energy dependence, as given in Figure 5-26.

These examples may not represent the actual limits to the problem of adjusting the energy response, but it is unlikely that greatly improved characteristics may be obtained in a simple, small detector. Of course, a highly sensitive thermal neutron detector (^{235}U) can be placed in the center of a bulky moderator assembly to obtain a smooth energy dependence between thermal and 10 MeV. However, the use of such a combination is restricted to locality monitoring.

Due to the above-mentioned complications, so far there have been only few larger-scale applications of detectors containing fissionable materials in routine personnel neutron dosimetry. At the Swiss Reactor Institute, a dosimeter containing a ^{237}Th and a ^{235}U aluminum alloy has been employed for several years (Figure 5-27). Its evaluation procedures, including spark-counting, have been rationalized so that a technician can

FIGURE 5-27. Closed and open view of a track etching personnel dosimeter containing a ^{232}Th and a ^{235}U foils, which has been in routine use for several years at the Swiss Reactor Center in Würenlingen. (After Prêtre and Heusi, 1972.)

read about 40 dosimeters per day (Prêtre and Heusi, 1972 and Prêtre, 1973).

At the Jülich Nuclear Research Center in Germany, an electroplated ^{232}Th dosimeter is used (Heinzelmann and Schüren, 1972). A routine dosimeter containing thin layers of ^{232}Th and ^{235}U is also considered in East Germany (Dörschel, 1971) and Czechoslovakia (Trousil et al., 1973); for the personnel at the AGS in Brookhaven, ^{232}Th is used (Distenfeld et al., 1972). A ^{237}Np dosimeter for a small high-risk group of employees is being introduced at ORNL and various systems are being considered in the Soviet Union (Bochvar et al., 1969 and Obaturov et al., 1972).

There have also been some special applications of fission-fragment registration in personnel dosimetry, such as in the hand monitoring of persons working with plutonium in glove boxes. A finger-ring dosimeter that is used at ORNL (Becker, 1971) contains a thin ^{232}Th foil attached to the top and a 10 μm Kimfol foil to be evaluated by spark-counting, which is attached to a retainer ring and pressed gently against the thorium foil by

a lead disk. A similar device is being used at Karlsruhe, Germany. Its design and response characteristics have recently been described in great detail (Burgkhardt et al., 1972 and Buijs et al., 1973), particularly considering its potential for work with transuranium elements.

Stationary detectors that have a reasonably flat response over a wide neutron energy range from thermal to about 10 MeV, can be prepared by simply placing a sensitive thermal neutron detector in the center of a moderator sphere. This has been done with polyethylene and paraffine spheres using uranium or ^{239}Pu in contact with a fission-fragment detector as thermal neutron detector in the center (Mijnheer et al., 1973), or with water-filled bottles and cellulose nitrate covered with lithium or boron compounds (Kumamoto, 1973).

Table 5-4 lists the neutron registration sensitivities of some nuclear reactions. The major applications of threshold techniques are neutron flux and energy measurements in or around reactors, and charged particle studies around accelerators. In the latter case, a number of additional high energy threshold reactions can be

Sensitivities of Some Fission Reactions Used in Fast Neutron Dosimetry

Neutron energy (MeV)	Fissionable material	Approximate sensitivity (tracks/n)*
Thermal	Natural U	3.5×10^{-5}
Fission	Natural U	4.2×10^{-6}
Thermal	^{235}U	3.5×10^{-3}
Fission	^{235}U	1.6×10^{-4}
Fission	Th	4.0×10^{-7}
Fission	^{237}Np	1.8×10^{-6}
~4 (Pu/Be)	Natural U	4.5×10^{-6}
~4 (Pu/Be)	Th	1.2×10^{-6}
14	Natural U	1.4×10^{-5}
14	^{237}Np	1.8×10^{-6}
14	Th	4.5×10^{-6}
230	Ta	5.0×10^{-8}
230	Be	2.0×10^{-6}
230	Au	4.5×10^{-7}
5,500	Ta	6.1×10^{-7}
5,500	Bi	1.1×10^{-6}
5,500	Au	9.3×10^{-7}
5,500	U	3.6×10^{-6}

*For radiator thickness exceeding the fission fragment range, precise number depends somewhat on detector sensitivity, etching, and counting conditions.

(After Becker, 1965 and Wollenberg and Smith, 1969a, b.)

used. High energy nucleon fission cross sections of U, Bi, Au, and Ag have been determined with fission track detectors (Hudis and Katcoff,1969).

Figure 5-28 presents some data for the GeV energy range which have been used for nucleon energy determination at the CERN proton synchrotron (Wollenberg and Smith, 1969). The 50-MeV threshold reaction in bismuth has been applied to the selective measurement of high energy neutrons with a detector area of ∿100 cm². A sensitivity of ∿0.4 tracks/mrem, corresponding to a detection limit of <0.1 mrem of 90-MeV neutrons, was reported (Heinzelmann and Schüren, 1970).

In similar experiments at SLAC (Svensson, 1970), ^{238}U and ^{232}Th have been employed. ^{232}Th is recommended not only because of its lower spontaneous fission rate, but also because its nucleon fission cross section stays at a constant level of ∿0.6 barn up to energies of at least 350 MeV. If the measurements are carried out in the presence of an intense high energy (>0.2 GeV) photon field, corrections for photo-fission have to

be made. These corrections may become substantial at energies >10 GeV (the photo-fission cross section at 13 GeV is ∿0.25 to 0.4 barn). Spallation reactions in heavy nuclei like Pb and Au have also been used for high energy (18 GeV proton) measurements (Debeauvais et al. 1967). Unlike fission, spallation is, of course, dependent on the direction of nucleon incidence as well as the atomic number of the elements involved (Anon., 1973).

Detailed information on the response of different fissionable materials to various neutron sources, in particular to the common isotopic sources and more or less modified fission spectra from reactors, can be found in a number of the references at the end of this chapter (Becker, 1965; Bhatt, 1966; Prêtre et al., 1968; Debeauvais et al., 1969; Rago et al., 1970; Hashimoto et al., 1970; Remy et al., 1970; Köhler, 1970; Jozefowitz, 1971b; Guenther, 1971; and Anon., 1971). Obviously, it is quite easy to obtain a "histogram" of the neutron energy distribution by employing either fissile materials with different thresholds, the same material encapsulated in different neutron shields, or materials with different isotopic composition. The results agree well with those obtained using other techniques. The fissionable material may, incidentally, also be used in solution (Danis et al., 1972).

The detectors have been used also for fission density distribution measurements inside fuel elements or near interfaces (Tuyn, 1967 and 1969). Important advantages of track etching detectors for those applications are the high spatial resolution (<10 μm) and the insensitivity to substantial beta and gamma radiation levels. A limitation for in-core measurements can be the high neutron flux inside a reactor, which may result in track densities which are impossible to resolve in an optical microscope. In such cases, track counting of briefly etched, carbon-covered foils in an electron microscope can extend the dynamic range to >10⁶ tracks/mm² (Besant, 1970).

There are numerous special applications of fission-fragment track etching in neutron dosimetry, of which only a few examples will be mentioned. The flux around the target of a 14-MeV source has been mapped (Nakanishi and Sakanoue, 1969). In an interesting neutron flux recording device that runs unattended for at least one year, a polycarbonate band is slowly drawn

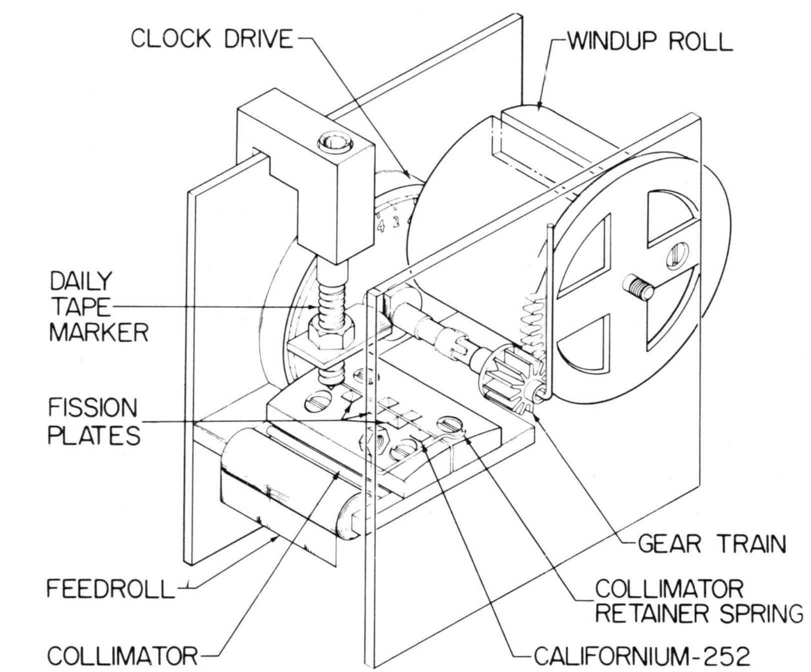

FIGURE 5-28B. Automatic neutron flux and time-of-exposure recorder for reactor surveillance. (After Weidenbaum et al., 1970.)

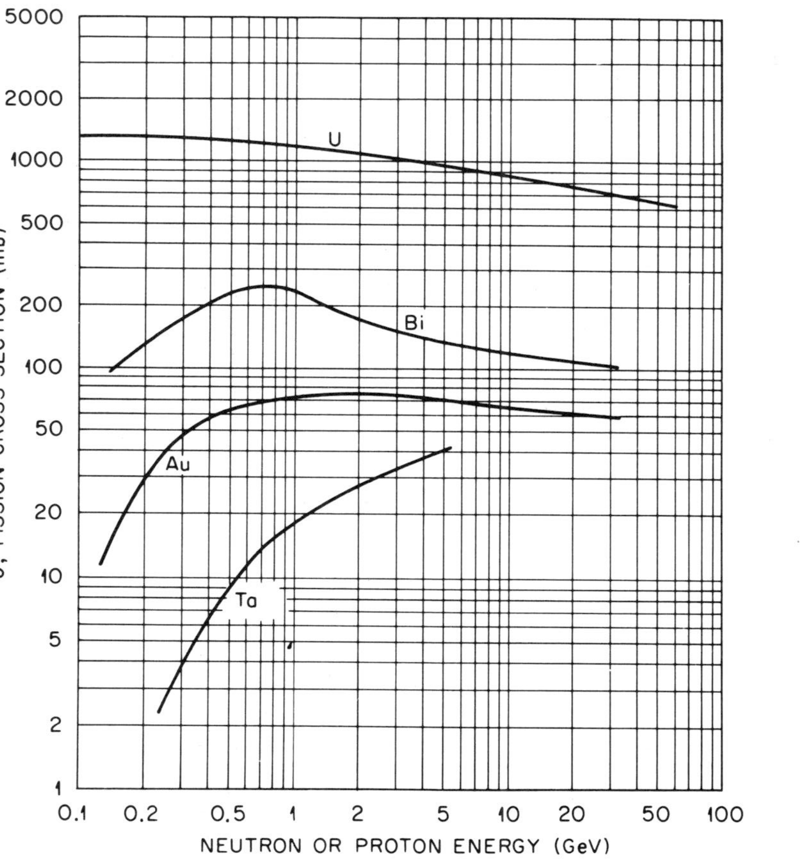

FIGURE 5-28A. Fission cross section in different nuclides as a function of high neutron or proton energies. (After Wollenberg and Smith, 1969.)

over small pellets of ^{235}U, ^{238}U, and ^{237}Np (Figure 5-28 B). Etching of the band reveals a complete record of the time variations in neutron flux plus some information on its energy distribution. The device has been developed for reactor surveillance by the U. S. Arms Control and Disarmament Agency (Weidenbaum et al., 1970). Another application is the measurement of neutron depth-dose distributions, in tissue-equivalent materials. Thorium (Becker, 1969a) as well as natural U and ^{235}U (Dörschel and Stolz, 1970a) has been used. Track etching detectors were also used as a standard flux measurement technique for in vivo neutron activation studies (Ågard et al., 1971).

The sensitivity of fission-fragment neutron dosimeters, although basically limited by the constant 1.16×10^{-5} tracks/n/barn, can be adjusted to needs within rather wide limits. Optimum sensitivity can be obtained under the following conditions:

1. The area to be counted is as large as possible and free of disturbing background (using spark counters, even polycarbonate foil areas exceeding 100 cm^2 are essentially free of background counts);

2. The fissile element has a high cross section and is used undiluted in a thickness which corresponds to the maximum range of the fission-fragments; and

3. The thin detector foil is sandwiched between two layers of the fissile element, thus doubling its response.

Under optimum conditions, fast-neutron doses of <0.1 mrad may be measured with a reasonable precision in large detectors, and doses around 1 mrad in small (2 to 3 cm^2) detectors which may be included into personnel dosimeters. Thermal neutron fluxes around 10^3 n/cm^2 can be measured using ^{235}U foils of 100 cm^2 area (Congel et al., 1972). If, on the other hand, the sensitivity has to be reduced, reversal of the above conditions and additional measures such as absorbing layers of Mylar foil between the radiator and the detector are advisable.

Use can also be made of the low inherent U and Th concentration in inorganic or organic track detectors (the natural U and Th content in glasses

and minerals exceeds that of polymers). However, if polymers are exposed to a high fast-neutron dose (more than several krad), their bulk etching rate rapidly increases (Boyett and Becker, 1970). This effect can lead to a rapid dissolution of the foil which complicates the automatic counting of fission-fragment tracks.

If the neutron flux is known, track etching may be also used as a sensitive tool for the determination of low concentrations of fissionable materials. The concentration (c) of the fissionable material in a thin layer can be determined from the number of tracks T, if the neutron flux Φ is known, by

$$c = \frac{TM}{GN_A \sigma \Phi t_E}$$

where N_A is Avogadro's number, G is a geometry factor that accounts for the number of charged particles produced per neutron capture and the fraction of these that register in the track-detecting material, and M and σ are the atomic weight and reaction cross section of the nuclide to be measured. For natural uranium M equals 238.03, σ equals 4.2×10^{-24} cm^2 and G is 1 since two fission-fragments are produced per reaction and half of these reach the detector foil if the deposit is thin. The factor t_E corrects for the less than 100% detection efficiency (invisible or unsparkable shallow tracks) and depends on the detector material and evaluation technique, but is usually about 0.90 to 0.95 in organic detectors.

The method was first suggested in 1963 for the determination of low natural U concentrations in minerals (Price and Walker, 1963). It has since been studied in more detail by several investigators, who extended its use to the analysis of minerals; water* (Fleischer and Lovett, 1968; Dörschel and Stolz, 1970b; Hashimoto, 1971; Carpenter, 1972; Piesch and Weng, 1972); aerosols in air (McEachern et al., 1971); and biological samples such as urine, blood (Carpenter and Cheek, 1970); bones (Hamilton, 1971); and seaweed (Su, 1972). A particular advantage of this method is that it also provides detailed information on the spatial distribution of the fissile material in the sample.

If a spark counter is employed for evaluation (Becker and Johnson, 1970), the sensitivity is further increased. For example, the spot in Figure 5-29 resulting from a drop of evaporated tap water

---

*Normal uranium concentrations found in drinking water, rain, and sea water are in the 10^{-6} to 10^{-8} g/l range.

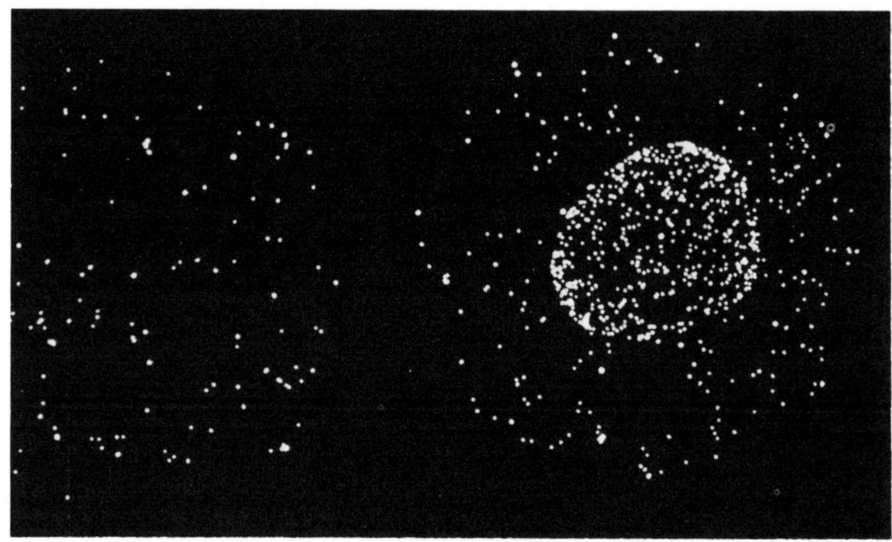

FIGURE 5-29.   Sparked replicas: (left) distilled water (1 cm^3) and (right) tap water (1 cm^3) evaporated on polycarbonate films, followed by exposure to a high thermal neutron flux in a reactor, etching, and spark counting. (After Becker and Johnson, 1970.)

corresponds to the natural U concentration. To avoid absorption of fission-fragments in thick layers of residues, solvent extraction of the fissile elements is advisable in case of such liquids as urine, blood, or sea water. The technique of neutron-induced autoradiography (NIAR) has great potential in other areas. For example, it is now routinely used for the determination of plutonium in bones (Jee, personal communication). Fission-fragment registration (in combinations of ^{235}U foils and organic foils such as red-dyed cellulose nitrate) has also been used extensively in neutron radiography (Berger and Kraska, 1967; Berger and Lapinski, 1972; and Morley, 1972).

There are numerous other applications of fission-fragment track etching, for example, in the precise determination of fission cross sections (Rago and Goldstein, 1967), absolute fission rates and fission ratios (Gold et al., 1968; Besant and Ipson, 1970; and Jowitt, 1971), fission product yields (Armani et al., 1970), for the determination of effective neutron temperatures in a thermal reactor (Su and Liu, 1970), spectrometry of fission-fragments (Fiedler and Schreck-Köllner, 1972 and Khan et al., 1972), or for the analysis of U/Th mixtures (Nishiwaki et al., 1971a).

5-5. Applications of Alpha Particle and Recoil Nuclei Registration

Nuclides having a high (n,α) cross section, e.g., ^{6}Li and ^{10}B, have often been employed as radiators for thermal and epithermal neutron dosimetry. With various radiators such as a dosimeter glass containing boron and lithium or a plastic containing B_4C, sensitivities between 0.5 and 6 x 10^{-5} tracks/thermal neutron have been obtained with Makrofol E (lower sensitivities), Triafol B and T (medium), and cellulose nitrate which exhibits the highest sensitivity (Becker, 1969b).

Others used LiF TLD's as radiator in an interesting combination of a ^{6}LiF/^{7}LiF and a track etching albedo dosimeter (Tymons et al., 1973). By using the sparking technique for the alpha track counting in large detector areas extremely high sensitivities can be obtained. Of course, thermal neutrons are rarely of direct practical interest in medicine, biology, or radiation protection. However, some special applications of thermal neutron dosimetry via (n,α) reactions deserve mentioning.

Using a large-area combination of ^{10}B or ^{6}Li radiators with cellulose acetobutyrate detection foils, sensitivities of up to $\sim$1.3 x 10^{-2} tracks per thermal neutron have been reported in the measurement of naturally occurring thermal neutrons over land and water (Roberts et al., 1970). A step-wedge consisting of an alpha-attenuating material simplifies the evaluation of such dosimeters (Kastner and Oltmann, 1971). Intermediate energy neutrons are measured in a

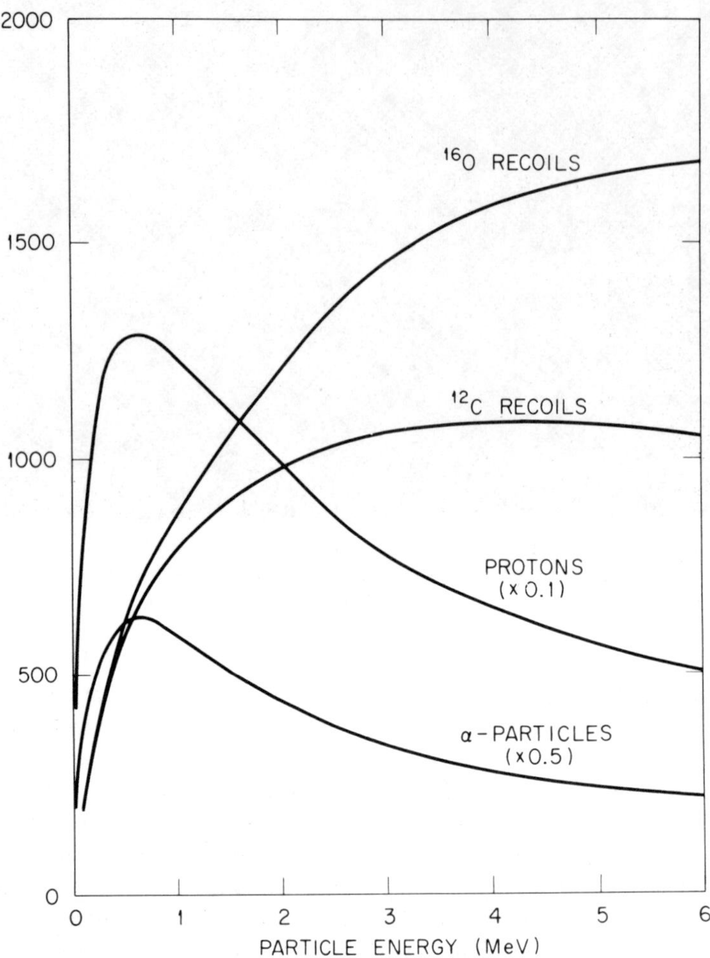

FIGURE 5-30A. LET of charged particles formed by direct fast-neutron interaction with cellulose acetobutyrate as a function of particle energy. (After Spurny, 1972.)

Cd-B plastic combination (Dayashankar and Venkataram, 1969); (n,α) reactions, for instance, in $Li_2B_4O_7$-coated red-dyed cellulose nitrate foils (Barbier, 1971) or in LiF-covered foils of this type (Tymons et al., 1973) have been suggested as a sensitive detector of thermal neutrons.

Also, a polymer that has boron uniformly dispersed throughout its volume has been suggested (Schultz, 1970).

If a sensitive polymer is exposed to fast neutrons, numerous tracks of recoil nuclei (depending on sensitivity and composition of the detector and the neutron energy, usually carbon, oxygen, or nitrogen plus some alpha tracks from (n,α) reactions at higher neutron energies) can be revealed by extended etching. In Figure 5-30A, the LET of the various types of particles that are produced in a polymer by direct neutron interaction is given as a function of particle energy. The effect has attracted much interest for dosimetric purposes during the last five years (Medveczky and Somogyi, 1966; Becker, 1967; and Dragu and Nicolae, 1971) because it avoids many of the disadvantages of fissionable materials in fast-neutron dosimetry mentioned above. On the other hand, recoil tracks are small and more difficult to count visually than fission-fragment tracks.

Recoil particle tracks are distributed homogeneously throughout the thickness of the polymers. External "radiators" having a composition different from that of the polymers, including materials undergoing (n,α) threshold

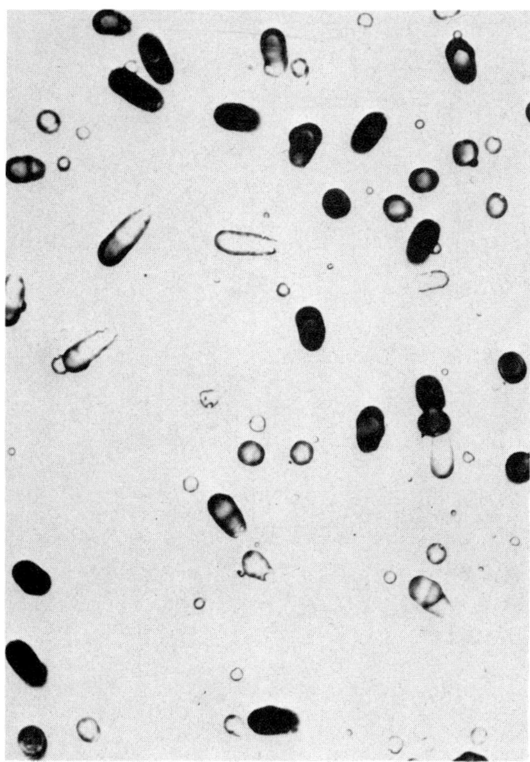

FIGURE 5-30B. Micrograph of fast-neutron-induced recoil particle tracks in a polycarbonate foil.

Detector foil	HPRR (fission)	Neutron energy 2.7 MeV	14 MeV
Makrofol E	0.5	1.9	1.8
Triafol T	0.7	1.3	1.9
Triafol B	0.6	0.6	2.0
Cellulose nitrate		1.4	2.8

*In tracks/neutron x 10^{-5}
(After Becker, 1969a.)

reactions at higher energies, only slightly modify the "inherent" sensitivity (Becker, 1969a; Frank and Benton, 1970; and Fängewisch and Scharmann, 1972). If exposed in a helium atmosphere, the sensitivity of a cellulose nitrate foil to fission neutrons is increased by a factor of 12 (Becker, 1969c). The sensitivity of such detectors, if exposed in a normal atmosphere, amounts to about 0.5 to 3 x 10^{-5} tracks/n, depending on the neutron energy and the detector foil (Table 5-5).

The neutron energy dependence of at least some materials such as cellulose acetate seems to follow the first collision dose quite well down to at least 0.5 MeV (Figure 5-31). The lower neutron-energy threshold of recoil track production has not yet been established exactly, but appears to be around 0.8 MeV for the less sensitive polycarbonates (Jozefowicz, 1971a and 1972) and probably around 0.3 to 0.7 MeV for more sensitive detector materials.

In this process the size distribution of etch pits depends on the energy of perpendicularly incident neutrons (Figure 5-30B); it has been suggested to use this dependence (which can also be measured by indirect methods such as light scattering) to obtain additional information on the neutron energy (Tuyn and Broerse, 1970). The etch pit diameter distribution also depends, however, on the direction of neutron incidence, which limits the practical value of this method.

Only after extended etching does the recoil particle etch pit count reach a maximum value, representing equilibrium between the revealing of tracks from a greater depth, and apparent disappearance of shallow pits (Becker, 1967 and Nishiwaki et al., 1971b). As can be seen in Figure 5-32, the response decreases by about 33% if the angle of neutron incidence approaches 90° (Becker, 1967). For lower neutron energies, this directional dependence seems to increase, and there is also some influence of the etching time on the directional response (Geisler and Heinzelmann, 1970 and Piesch, 1970b).

Principal applications of this "direct interaction" type of dosimeter have been in personnel accident neutron dosimetry (in at least two major nuclear installations, Karlsruhe, Germany, and Oak Ridge, polymer foils have been incorporated in dosimetry belts to be worn by high-risk personnel for accident exposure measurements) and neutron depth-dose studies in phantoms (Becker, 1969a and Tuyn and Broerse, 1970b). A typical result for the latter is given in Figure 5-33. Tissue equivalent composition is one important feature of the detector for such applications. Another is high spatial resolution, for example, for studies of interfaces such as bone/tissue. No fading has been observed at room temperature, and less than 10% within one week at 60°C in various common

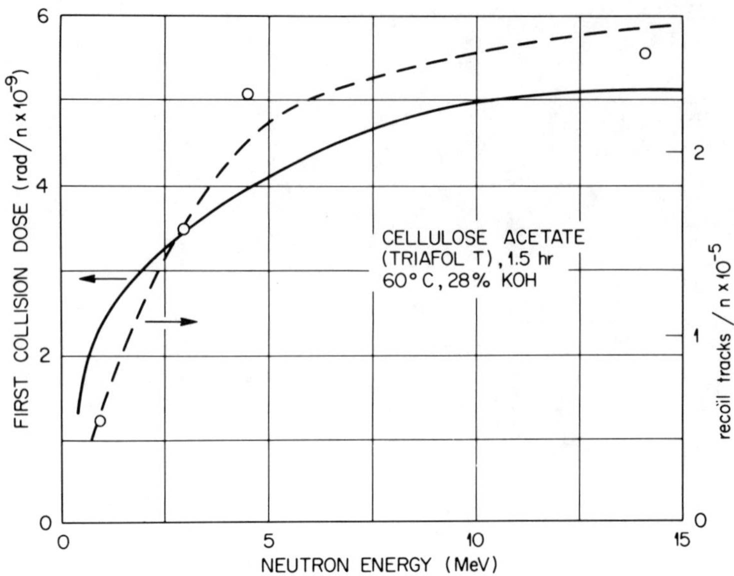

FIGURE 5-31.   First collision dose and recoil track density in cellulose triacetate as a function of neutron energy. (After Becker, 1969.)

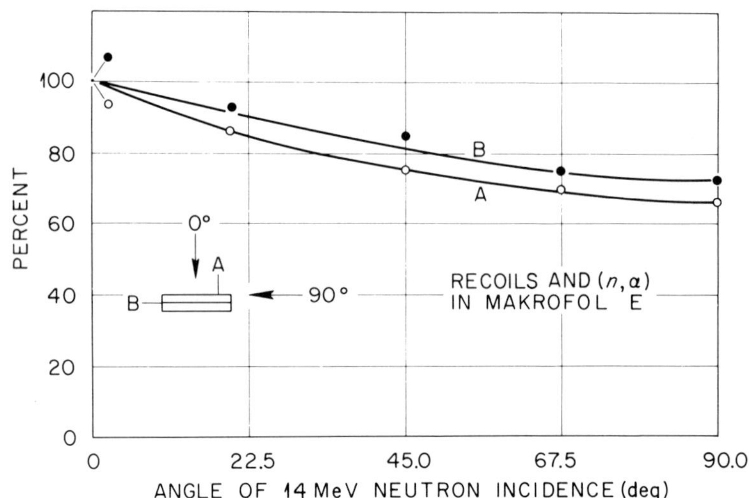

FIGURE 5-32.   Number of visible recoil particle tracks in polycarbonate as a function of 14 MeV fast-neutron incidence angle. (After Becker, 1969a.)

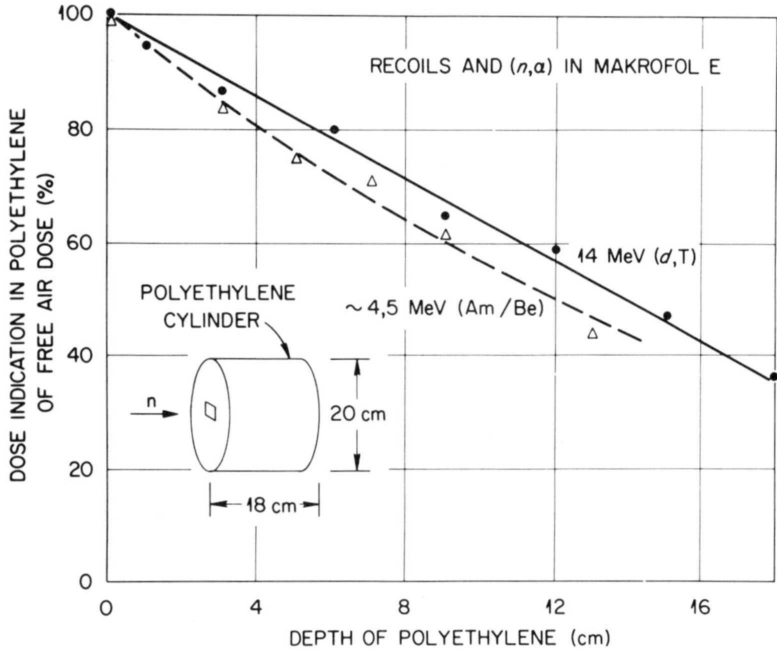

FIGURE 5-33.   Measurement of depth dose in a polyethylene cylinder, using direct interaction of fast neutrons with polycarbonate foil. (After Becker, 1969a.)

polymers (Becker, 1969a). Unfortunately, however, the exposed and etched foils have to be evaluated by visual counting because the efficiency of spark counters for recoil particle tracks which are produced by neutrons in the more common energy range of < 15 MeV is low (the short-range recoils only rarely produce preforations in the foils). Other automatic evaluation techniques such as laser scattering may hold some promise.

For the selective measurement of high energy (>10 MeV) neutrons by direct interaction, the observation of characteristic three-pronged stars which are produced by the reaction ^{12}C (n,n', 3 ^{4}He) may be used (Frank and Benton, 1970, and Price et al., 1971). The sensitivity of this method is $\sim$4 x 10^{-8} tracks/n for 14.5 MeV neutrons. In another method for the selective measurement of high energy neutrons (Frank and Benton, 1972), a thin "degrader" film (usually gold) is placed between the recoil nuclei radiator foil and a cellulose nitrate detector which is not sensitive enough for registration of the alpha particles from (n,α) reactions. Direct interaction with nuclei in polymers has also been used for studies of negative pion beams (Benton et al., 1970). Tracks of

heavier ions such as ^{40}Ar can, of course, be monitored directly with foils.

Another application is in the integrating dosimetry of alpha particles. Only alpha particles of an energy not exceeding about 5 MeV can be registered in the most sensitive cellulose nitrate foils. Alpha particles having higher energies have to be "slowed down" in the polymer before they reach the threshold LET required for preferential etchability. For higher alpha particle energies, a thin absorber layer between source and detector may increase the initial sensitivity of the foil due to an energy degradation (LET increase) of the alpha particles (see, for example, Somogyi and Srivastava, 1971; Haider and Jacobi, 1972; and Spurny, 1972b). This effect is demonstrated in Figure 5-34, which shows the relative alpha particle track density in Makrofol as a function of the degrading absorber layer thickness between foil and source.

One particularly interesting application of alpha particle dosimetry is the measurement of radon and radon daughter products which are attached to aerosol particles, either in the normal atmosphere, for instance, at higher altitudes, or, more importantly, in uranium mines. An alpha-sensitive

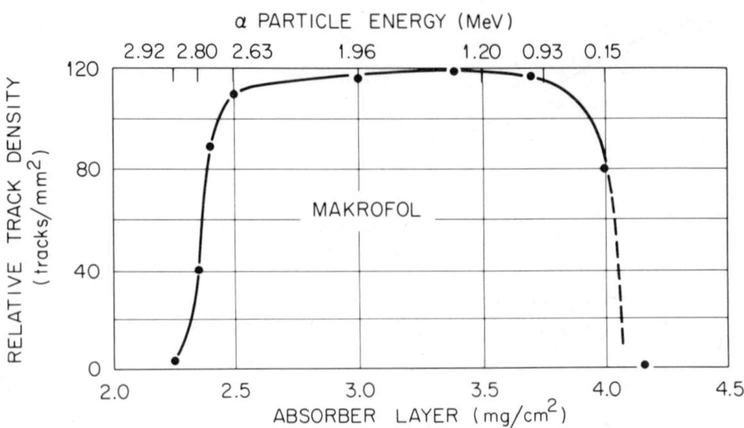

FIGURE 5-34.   Relative alpha particle track density in Makrofol as a function of the total absorber thickness between a monoenergetic alpha radiation source and the foil surface. (After Dutrannois, 1971.)

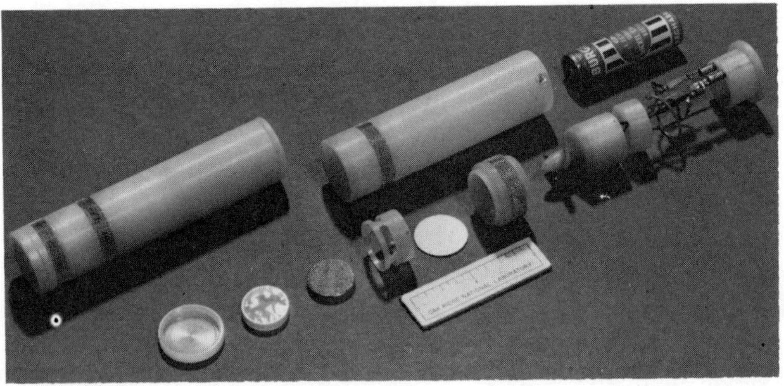

FIGURE 5-35.   Closed and exploded view of radon-progeny personnel dosimeter for uranium miners. (After Auxier et al., 1971.)

foil such as cellulose triacetate or nitrate may be exposed directly to an atmosphere containing alpha-emitting gaseous or solid compounds (Becker, 1969a; Lovett, 1969; Anno and Commanay, 1971; Alter and Price, 1972; Kurosawa, 1972; and Domanski et al., 1972). Such techniques have been used successfully for measuring the radon concentration in ground gas over large areas, thus facilitating the detection of uranium ore deposits (Gingrich and Lovett, 1972). The variable ratio between radon and its daughter products in uranium mines makes a selective measurement of the daughters which are attached to aerosol particles and which are responsible for the induction of lung cancer desirable.

This has been accomplished (Becker, 1968c; Auxier et al., 1971; Chapuis et al., 1972; and Haider and Jacobi, 1972) in an "active" monitoring system to be worn by each miner (Figure 5-35). Air is forced through a filter by a small propeller, which is operated by a constant speed motor with a rechargeable battery. Aerosols and attached daughter products are continuously collected at the filter surface during the work period of a miner. Alpha particles emitted are registered by a frame-mounted cellulose nitrate foil opposite the filter. Varying the distance between filter and detector foil can help to "collimate" the alpha particle flux and adjust the energy distribution to obtain optimum detection efficiency (Figure 5-36).

Both filter and detector foils are replaced after

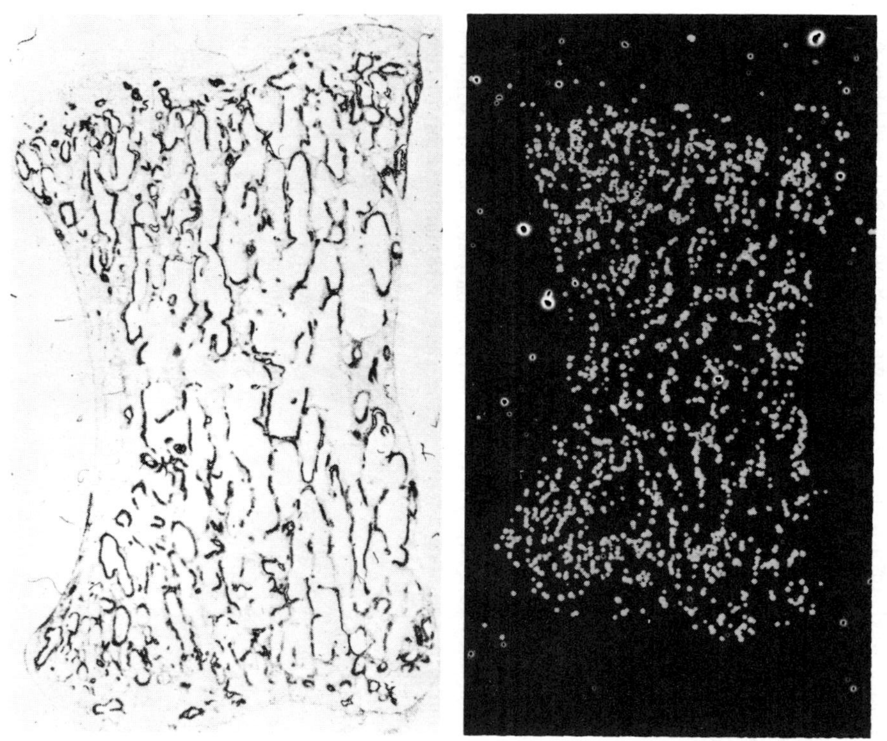

FIGURE 5-36. Variation of the original alpha particle spectrum from Rn daughter products by increasing the distance between source and detector in air. (After Chapuis et al., 1972.)

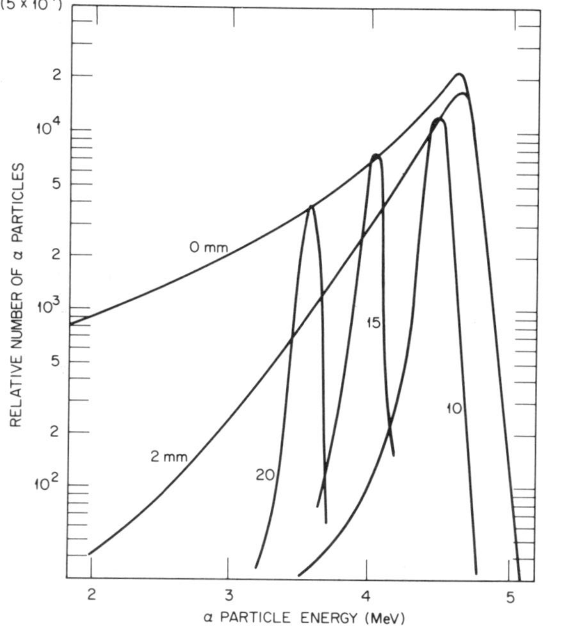

FIGURE 5-37. Alpha-autoradiograph of a beagle bone sample containing ^{226}Ra with a 15 μm cellulose nitrate foil; (left) after an exposure time equivalent to that required for a photographic alpha-autoradiogram; and (right) spark image of the sample (slightly reduced scale) after reduction of the exposure time by a factor of 30. (After Becker and Johnson, 1970.)

an adequate period, such as a week or a month, and the exposed foil is etched and spark counted. The high inherent sensitivity can be decreased by reduction of the foil sensitivity, the air flow rate, or the filter area. Humidity and dust have no adverse effects on the dosimeters. The weight of the unit is less than 50 g and it can easily be attached to the miner's safety helmet. The battery is recharged between shifts in a charging rack. These detectors have been tested successfully under laboratory and mine conditions (White, 1970) and are now also used for studies in salt mines, homes, etc.

Finally, quantitative alpha autoradiography has to be mentioned as an important application of track etching. Cellulose nitrate foils may be used for the determination of the concentration and distribution of alpha emitters such as ^{239}Pu or ^{226}Ra in bone in a way comparable to the use of photographic nuclear track emulsions (Bleaney, 1969; Becker and Johnson, 1970a; Cole et al., 1970; and Simmons and Fitzgerald, 1970) or, if higher sensitivity is desired, in combination with spark counting (Figure 5-37). The exposure time required for obtaining high resolution autoradiography with good contrast is about the same as with a photographic emulsion, but track etching poses less fading and background problems. On the other hand, the etching conditions may have to be adjusted to the alpha particle energy for optimum results, and presensitizing is advisable in the less sensitive cellulosics (Somogyi and Gulyas, 1972).

By the use of foils having different sensitivities, some alpha particle energy discrimination is possible (Somogyi and Srivastava, 1970 and Novotny, 1970). Unfortunately, this method cannot be used for distinguishing between various alpha emitters which may be quite different in the actual biological hazard which they represent because the resolution is insufficient. In analogy to neutron-induced autoradiography of fissionable elements, neutron-induced alpha autoradiography may be used for determining the spatial distribution of such elements as boron and lithium in other materials which do not undergo (n,a) or (n,f) reactions (Cervasek, 1969) or for the evaluation of paper chromatograms (Iwata et al., 1971). There also have been numerous applications of alpha particle track etching in nuclear physics, for example, in the investigation of (p,a) reactions (Selim and Abo Elkheir, 1973).

REFERENCES

Abmayr, W., Burger, G., Gais, P., and Paretzke, H. G., Automatic Measurements of Tracks in Solid-State Track Detectors with a Spark Counter, an On-Line Flying Spot and TV-Device, and with Coherent Optics, Proc. 8th Int. Symp. Nucl. Photography and Solid-State Track Detectors, Bucharest, 1972.

Agard, E. T., Jervis, R. E., and McNeill, K. G., Neutron dosimetry with nuclear track detectors applied to in-vivo neutron activation analysis, *Health Phys.,* 21, 625, 1971.

Allkofer, O. C., *Teilchendetektoren,* Verlag K. Thiemig, München, 1971.

Allkofer, O. C., Enge, W., Heinrich, W., and Röhrs, H., Preliminary Results of Measurements of Heavy Primaries in the Region of Supersonic Transport Using Plastic Stacks, *Proc. Int. Congr. Protect. Accelerator and Space Radiation,* CERN 71-16, Vol. 1, CERN, Geneva, Switzerland, 1971, 512.

Alter, H. W. and Price, P. B., Radon detection, U. S. Patent 3,665,194, 1972.

Anno, J., Use of Cellulose Nitrate for Alpha-Particle Detection, NP-18411 and Ph.D. Thesis, University of Toulouse No. 943.

Anno, J., Commanay, L., and Teyssier, J. L., Registration and counting of alpha particle tracks in cellulose nitrate, *Radioprotection,* 5, 147, 1970.

Anno, J. and Commanay, L., Dielectric track detectors — alpha particle detection using cellulose nitrate, *Ann. Phys. Biol. Med.,* 5, 65, 1971.

Anon. Tentative method for fast-neutron flux measurements by track-etch technique, Annual Book of ASTM Standards, Part 30, 1971, 1191.

Anon. An Evaluation of the Potential of Thick Nuclear Emulsion for Use as a High-Energy Neutron Personal Dosimeter, Health Physics Group, Lawrence Berkeley Lab., Berkeley, Calif., 1973.

Armani, R. J., Gold, R., Larsen, R. P., and Roberts, J. H., Applications of Solid-State Track Recorders in Absolute Fission Product Yield Measurements, Proc. 7th Int. Colloq. Corpuscular Photography and Solid Detectors, Barcelona, 1970.

Audran, R. G. L. and Renard, G. J. A., Product for Neutronography, French Patent 2,092,830, 1972.

Auxier, J. A., Becker, K., Robinson, E. M., Johnson, D. R., Boyett, R. H., and Abner, C. H., A new radon progeny personnel dosimeter, *Health Phys.,* 21, 126, 1971.

Avan, L. et M., Blanc, D., and Teyssier, J. L., Ionographie – Emulsions, detecteurs solides de traces. Doin ed., Paris/France, 1972.

Barbier, J., Some recent progress in neutron radiography and ionizing particle dosimetry using cellulose nitrate as detecting material, *J. Photogr. Sci.,* 19, 108, 1971.

Barbier, J. and Renard, G., Improvement in the performance of cellulose nitrate films for the dosimetry of ionizing particles, in *New Trends in Radiation Protection,* French Radiat. Protect. Soc., Montrouge, 1972, 475.

Baroni, G., DiLiberto, S., Petrera, S., Romano, G., and Sgarbi, C., Technical Study of the Efficiency of Nitrocellulose Detectors, Proc. 8th Int. Conf. Nucl. Photography and Solid-State Track Detectors, Bucharest, 1972a.

Baroni, G., DiLiberto, S., Romano, G., Sgarbi, C., and Tabasso, M. C., Effect of etching interruptions of track formation in plastics, *Nucl. Instr. Meth,* 98, 221, 1972b.

Baumbach, H., Application of solid-state track detectors for autoradiography, using induced radioactivity, *At. Kozlem.,* 13, 93, 1971.

Becker, K., Nuclear Track Registration in Dosimeter Glasses for Neutron Dosimetry in Mixed Radiation Fields, USNRDL-TR-904, 1965; also see *Health Phys.,* 12, 769, 1966.

Becker, K., Photographic, glass or thermoluminescence dosimetry, *Health Phys.,* 12, 955, 1966.

Becker, K., Neutron Personnel Dosimetry by Nonphotographic Nuclear Track Registration, Proc. ENEA Symp. Rad. Dose Measurements, Stockholm, 1967.

Becker, K., Nuclear track registration in solids by etching, *Biophysik,* 5, 207, 1968a.

Becker, K., The effect of oxygen and humidity on charged particle registration in organic foils, *Radiat. Res.,* 36, 107, 1968b.

Becker, K., Personnel Radon Dosimeter, U. S. Patent 3,505,523, 1968c.

Becker, K., Alpha particle registration in plastics and its applications for radon, thermal, and fast neutron dosimetry, *Health Phys.,* 16, 113, 1969a.

Becker, K., Applications of the Track Etching Process in Radiation Protection, Proc. Int. Conf. Track Registrat. in Solids, Clermont-Ferrand, 1969b, V-2.

Becker, K., Direct Fast Neutron Interactions with Polymers, in ORNL-4446, Oak Ridge National Laboratory, Oak Ridge, Tenn., 1969c, 266.

Becker, K. and Johnson, D. R., Nonphotographic alpha autoradiography and neutron-induced autoradiography, *Science,* 167, 1370, 1970.

Becker, K., Personnel Neutron Dosimetry Research at ORNL, Proc. Sec. AEC Workshop on Personnel Neutron Dosimetry, New York, BNWL-1616(UC-48), Battelle Northwest Lab., Richland, Wash., 1971.

Becker K., Dosimetric applications of track etching, in *Topics in Radiation Dosimetry,* Vol. 1, Attix, F. H., Ed., Academic Press, London, 1972a.

Becker, K., Applied Dosimetry Research, Annual Report, in ORNL-4811, 1972b, 69.

Becker, K., Progress in Solid-State Personnel Fast Neutron Dosimetry, Proc. IAEA Symp. Neutron Monitoring for Radiol. Protection, Vienna, 1972c.

Becker, K., Nagpal, J. S., Cheka, J. S., and Sohrabi, M., Fading Characteristics of Various Film and Solid-State Dosimeters in a Warm and Humid Climate, Paper 58, 17th Ann. Meet., Health Phys. Soc., Las Vegas, 1972.

Benton, E. V., A study of Charged Particle Tracks in Cellulose Nitrate, USNRDL-TR-68-14, 1968.

Benton, E. V. and Henke, R. P., Heavy particle range energy relations for dielectric nuclear track detectors, *Nucl. Instr. Meth.,* 67, 87, 1969.

Benton, E. V., Curtis, S. B., Raju, M. R., and Tobias, C. A., Studies of the Negative Pion Beams by Means of Plastic Nuclear Track Detectors, *Proc. 7th Int. Colloq. Corpuscular Photography and Solid Detectors,* Vol. 2, Barcelona, 1970.

Benton, E. V., Method for development of volume tracks in dielectric nuclear track recorders, *Nucl. Instr. Meth.,* 92, 37, 1971.

Benton, E. V. and Henke, R. P., Heavy Cosmic-Ray Exposure of Apollo 14, 15 and 16 Astronauts, Tech. Report nos. 14, 16, and 20, Department of Physics, University of San Francisco, Calif., 1971-72.

Benton, E. V., Curtis, S. B., Henke, R. P., and Tobias, C. A., Comparison of measured and calculated high-LET nuclear recoil particle exposure of Biosatellite III, *Health Phys.,* 23, 149, 1972.

Benton, E. V. and Henke, R. P., On Charged Particle Tracks in Cellulose Nitrate and Lexan, Int. Tech. Rep. No. 19, Department of Physics, University of San Francisco, Calif., 1972.

Berger, H. and Kraska, I. R., A track-etch plastic-film technique for neutron imaging, *Trans. Am. Nucl. Soc.,* 10, 72, 1967.

Berger, H. and Lapinski, N. P., Improved Sensitivity and Contrast, Track Etch Thermal Neutron Radiography, Trans. Am. Nucl. Soc. Ann. Meet., Las Vegas, 1972, 123.

Besant, C. B. and Ipson, S. S., Measurement of fission ratios in zero power reactors using solid-state track recorders, *J. Nucl. Energ.,* 24, 59, 1970.

Bhatt, R. C., Measurement of Neutron Fluence by Fission Fragment Damage Track Detection, AECL-2657, 1966.

Bitter, F., Fiedler, G., and Wallnik, H., Die Automatische Auswertung von Kernspur-und Festkörperdetekoren, *Nucl. Instr. Meth.,* 51, 241, 1967.

Blanc, D., Les Détecteurs Solides des Traces, *Radioprotection,* 5, 37, 1970.

Blanford, G. E., Jr., Walker, R. M., and Wefel, J. P., Track Etching Parameters for Plastics, *Radiat. Eff.,* 3, 267, 1970.

Bleaney, B., The radiation dose-rates near bone surfaces in rabbits after intravenous or intramuscular injection of ^{239}Pu, *Br. J. Radiol.,* 42, 51, 1969.

Blok, J., Humphrey, J. S., and Nichols, J. E., Hole detection in polymer films and in plastic track detectors, *Rev. Sci. Instr.,* 40, 509, 1969.

Bochvar, I. A., Vasileva, A. A., Gimadova, T. I., Keirim-Markus, I. B., Kraitor, S. N., Kushnerev, A. Y., Sergeeva, N. A., Uspenskii, L. N., Chernova, O. N., and Yakubik, V. V., Handling of Radiation Accidents, IAEA, Vienna, 1969, 235.

Boyett, R. H. and Becker, K., LET effects on the chemical resistance of irradiated polymers, *J. Appl. Polym. Sci.,* 14, 1654, 1970.

Boyett, R. H., Johnson, D. R., and Becker, K., Some studies on the chemical damage mechanism along charged particle tracks in polymers, *Radiat. Res.,* 42, 1, 1970.

Buijs, K., Vaane, J. P., Burgkhardt, B., and Piesch, E., Operational Experience with a Finger Dosimeter for Fast Neutrons, Paper IAEA/SM-167/2, Proc. Symp. Neutron Monitor. Radiat. Protect., IAEA, Vienna, 1973.

Burger, G., Grünauer, F., and Paretzke, H., The Applicability of Track Detectors in Neutron Dosimetry, Proc. Symp. Adv. Radiat. Detectors, IAEA, Vienna, 1970, 349.

Burgkhardt, B., Piesch, E., Buijs, K., and Vaane, J., Ein Fingerdosimeter für die Messung von Neutronenstrahlung, Proc. Symp. Strahlenschutz am Arbeitsplatz, Karlsruhe, 1972, 183.

Carpenter, B. S. and Cheek, C. H.,Trace determination of uranium in biological material by fission track counting, *Anal. Chem.,* 42, 121, 1970.

Carpenter, B. S. and LaFleur, P. D., Nitrogen Determination in Biological Samples Using the Nuclear Track Technique, Trans. Am. Nucl. Soc. Ann. Meet., Las Vegas, 1972, 118.

Carpenter, B. S., Quantitative applications of the nuclear track technique, *Microscope,* 20, 175, 1972.

Cervasek, J., Alpha particle recording with the use of cellulose acetate, *Jad. Energ.,* 15, 413, 1969.

Chapuis, A. M., Dajlevic, D., Duport, P., and Soudain, G., Dosimetrie du Radon, Proc. 8th Int. Conf. Nuclear Photography and Solid-State Track Detectors, Bucharest, 1972.

Chave, A. and Monnin, M., Etude Comparee des Effets des Gas sur Characteristiques de Detection, *Proc. 7th Int. Colloq. Nucl. Photogr. and Solid-State Track Detectors,* Vol. 1, Barcelona, 1970, 205.

Childs, C. B., Heavy-Ion Passive Dosimetry with Silver Halide Single Crystals, NASA-TM-X-2440, University of North Carolina, Chapel Hill, 1972, 138.

Cohn, C. E. and Gold, R., Computer-controlled microscope for automatic scanning of solid-state nuclear track recorders, *Rev. Sci. Instr.,* 43, 12, 1972.

Cole, A., Simmons, D. J., Cummins, H., Congel, F. J., and Kastner, J., Application of cellulose nitrate films for alpha autoradiography of bone, *Health Phys.,* 19, 55, 1970.

Comstock, G. M., Fleischer, R. L., Giard, W. R., Hart, H. R., Jr., Nichols, G. E., and Price, P. B., Cosmic ray tracks in plastics: the Apollo helmet dosimetry experiment, *Science,* 172, 153, 1971.

Congel, F. J., Roberts, J. H., Dreis, D., Kastner, J., Oltman, B. G., Gold, R., and Armani, R. J., Automatic system for counting etched holes in thin dielectric plastic, *Nucl. Instr. Meth.,* 100, 247, 1972.

Crawford, W. T., Desorbo, W., and Humphrey, J. S., Enhancement of track etching rates in charged-particle irradiated plastics by a photo-oxidation effect, *Nature,* 220, 1313, 1968; also see British Patent 1,240,767, 1968.

Cross, W. G., Developments in Threshold Detectors for Personnel Dosimeters, in: Nuclear Accident Dosimetry Systems, IAEA Panel Proc. Series PL-329, 1970, 117.

Cross, W. G. and Tommasino, L., Detection of Low Doses of Fast Neutrons with Fission Track Detectors, Paper 12th Ann. Meet. Health Phys. Soc., Washington, D. C., 1967.

Cross, W. G. and Tommasino, L., Electrical detection of fission fragment tracks for fast neutron dosimetry, *Health Phys.,* 15, 196, 1968.

Cross, W. G. and Tommasino, L., Rapid reading technique for nuclear particle damage tracks in thin foils, *Radiat. Eff.,* 5, 85, 1970.

Cross, W. G. and Tommasino, L., Improvements in the Spark Counting Technique for Damage Track Neutron Dosimeters, Proc. Symp. Neutron Dosimetry in Biology and Medicine, Neuherberg, Germany, 1972.

Danis, A., Tommasino, L., and Oncescu, M., Neutron Fluence Measurements by Fission Track Methods Using the Fissionable Element in Solutions and Glasses, Proc. 8th Int. Conf. Nucl. Photography and Solid-State Track Detectors, Bucharest, 1972.

Dayashankar and Venkataram, G., Monitoring of Intermediate-Energy Neutrons, Proc. IAEA Symp. Radiat. Protect. Monitoring, Bombay, 1969, 47.

Debeauvais, M., Maurette, M., Maury, J., and Walker, R. M., Registration of fission-fragment tracks in several substances and their use in neutron detection, *Int. J. Appl. Rad. Isotopes,* 15, 289, 1964.

Debeauvais, M., Stein, R., Ralarosy, J., and Cüer, P., Spallation and fission fragments of heavy nuclei induced by 18 GeV protons registered by means of solid plastic detectors, *Nucl. Phys.,* A90, 186, 1967.

Debeauvais, M., Stein, R., Remy, G., Ralarosy, J., and Tripier, J., Neutron spectrometry with the aid of visual ionographic detectors, *Rev. Phys. Appl.,* 4, 259, 1969.

Distenfeld, C., Klemish, J. R., Jr., and Mayhew, M., Developmental Study of Personnel Neutron Dosimetry at the AGS, Paper 23, 17th Ann. Meet. Health Phys. Soc., Las Vegas, 1972, and BNL-Rep. 17452, Brookhaven National Laboratory Upton, N. Y.

Doke, T., Progress in solid-state track detectors, *Radioisotopes,* 21, 121, 1972.

Domanski, T., Chruscielewski, W., and Liniecki, J., (1972) Studies on determination of radon and its daughter products in the air using triacetate cellulose foils, *Health Phys.,* in press.

Dörschel, B., Auswertung geätzter Festköperspurdetektoren durch α-Absorptionsmessungen, *Kernenergie,* 12, 303, 1969.

Dörschel, B. and Stolz, W., Neutronen-Gewebedosimetrie in gewebeäquivalenten Phantomen, *Kernenergie,* 13, 290, 1970a.

Dörschel, B. and Stolz, W., Urangehalts-Bestimmung von Wasser mit Hilfe der Festkörperspurmethode, *Radiochem. Radioanal. Lett.,* 4, 277, 1970b.

Dörschel, B., Neutronen-Routinedosimetrie mit Festkörperspurdetektoren, *Kernenergie,* 14, 360, 1971.

Dragu, A. and Nicolae, M., Application of Cellulose Nitrate in Fast-Neutron Dosimetry, Report IFA-PN-24, Inst. Fiz. Atom., Bucharest, 1971.

Dutrannois, J., Utilisation de Detecteurs Solides Visuels en Dosimetrie de Rayonnements de Hante Energie, *Proc. Int. Congr. Protection Against Accelerator and Space Radiation, CERN-71-16,* Vol. 1, CERN, Geneva, 1971, 271.

Enge, W., Beaujean, R., and Nicken, H. P., The Influence of Humidity on Track Registration in Cellulose Nitrate Plastic Detectors, *Proc. 7th Int. Colloq. Nucl. Photogr. and Solid-State Track Detectors,* Vol. 1, Barcelona, 1970, 175.

Enge, W., Grabisch, K., Bartholomä, K. P., and Beaujean, R., The Mechanism of Etching Plastic Detectors, Proc. 8th Int. Symp. Nucl. Photography and Solid-State Track Detectors, Bucharest, 1972a.

Enge, W., Bartholomä, K. P., Beaujean, R., Heinrich, W., Fukui, K., Horneck, G., and Bücker, H., Plastic Track Detectors in the Biostack Experiment M-211 on Board of Apollo 16, Proc. 8th Int. Conf. Nucl. Photography and Solid-State Track Detectors, Bucharest, 1972b and *Atomkernenergie,* 20, 231, 1972b.

Fängewisch, G. L. and Scharmann, A., Neuere Ergenbnisse bei der Verwendung von Kernspurdetektoren in der Neutronendosimetrie, Proc. Int. Symp. Neutron Dosimetry, Neuherberg, Germany, 1972.

Fain, J., Monnin, M., and Montret, M. Heavy Ion Effects in Polymers, Third Symp. on Radiat. Chem., Tihany, Hungary, 1971.

Fain, J., Monnin, M., and Montret, M., Spatial Energy-Density Distribution Around Ion Path in Polymers, Proc. 8th Int. Symp. Nucl. Photography and Solid-State Track Detectors, Bucharest, 1972.

Fiedler, G. and Schreck-Köllner, H., Special Glasses as Energy Detectors for Fission Fragments, Trans. Am. Nucl. Soc. Ann. Meet., Las Vegas, 1972, 128.

Fleischer, R. L. and Price, P. B., Tracks of charged particles in high polymers, *Science,* 140, 1221, 1963.

Fleischer, R. L., Price, P. B., and Walker, R. M., Tracks of charged particles in solids, *Science,* 149, 383, 1965a.

Fleischer, R. L., Price, P. B., and Walker, R. M., Ion explosion spike mechanism for formation of charged-particle tracks in solids, *J. Appl. Phys.,* 36, 3645, 1965b.

Fleischer, R. L. and Lovett, D. B., Uranium and boron content of water by particle track etching, *Geochim. Cosmochim. Acta,* 32, 1126, 1968.

Fleischer, R. L., Price, P. B., and Walker, R. M., Nuclear tracks in solids, *Sci. Am.,* 220, 30, 1969a.

Fleischer, R. L., Price, P. B., and Woods, R. T., Nuclear particle identification in inorganic solids, *Phys. Rev.,* 188, 563, 1969b.

Fleischer, R. L. and Hart, H. R., Jr., Fission Track Dating, Techniques and Problems, Report no. 70-C-328, General Electric Co., Schenectady, N. Y., 1970.

Fleischer, R. L., Hart, H. R., Jr., and Giard, W. R., Particle track identification: application of a new technique to Apollo helmets, *Science,* 170, 1189, 1970.

Fleischer, R. L., Viertl, J. R. M., Price, P. B., and Aumento, F., A chronological test of ocean-bottom spreading in the North Atlantic, *Radiat. Eff.,* 11, 193, 1971.

Fleischer, R. L. and Hart, H. R., Fission track dating: techniques and problems, in *Calibration of Hominoid Evolution,* Bishop, W. W. and Miller, J. A., Eds., Scottish Academic Press, 1972.

Fleischer, R. L., Alter, H. W., Furman, S. C., Price, P. B., and Walker, R. M., Technological Applications of Science: The Case of Particle Track Etching, Report 72CRD107, General Electric Co., Schenectady, N.Y., 1972a.

Fleischer, R. L., Viertle, J. R. M., and Price, P. B., Stretch Biological Filters, Rep. No. 72CRD148, General Electric Co., Schenectady, N.·Y., 1972b.

Fotland, R. A., Technical Rep. AFAL-TR-67-362, Horizons Inc., Cleveland, Ohio, 1968.

Frank, A. L. and Benton, E. V., High energy neutron flux detection with dielectric plastics, *Radiat. Eff.,* 3, 33, 1970a.

Frank, A. L. and Benton, E. V., Neutron and Gamma-Ray Measurements with Nuclear Track Detectors, Final Report, DASA No. 2573, University of San Francisco, Calif., 1970b.

Frank, A. L. and Benton, E. V., Development of a High-Energy Neutron Detector, Report DNA No. 2918F, Department of Physics, University of San Francisco, Calif., 1972.

Geisler, F. and Philips, P. R., An improved method for locating charged particle tracks in thin plastic sheets, *Rev. Sci. Instr.*, 43, 283, 1972.

Gingrich, J. E. and Lovett, D. B., A Track Etching Technique for Uranium Exploration, Trans. Am. Nucl. Soc. Ann. Meet., Las Vegas, 1972, 118.

Girshin, A. B., Kartuzhanskii, A. L., Martysh, G. G., Privaova, V. E., and Shur, L. L., On Some Registering Characteristics of Plastic Particle Detectors in Vacuum and in Gaseous Media, *Proc. 7th Int. Colloq. Nucl. Photogr. and Solid-State Track Detectors,* Vol. 2, Barcelona, 1970, 825.

Goland, A. N. and Mateosian, E. D., Effect of energetic electron irradiation upon charged-particle track registration in polycarbonate films, *Nucl. Instr. Meth.,* 106, 295, 1973.

Gold, R., Armani, R. J., and Roberts, J. H., Absolute fission rate measurements with solid-state track detectors, *Nucl. Sci. Eng.,* 34, 13, 1968.

Gold, R. and Cohn, C. E., Analysis of automatic fission track scanning in nuclear track recorders, *Rev. Sci. Instr.,* 43, 18, 1972.

Guenther, G., Work on Solid-State Track Detectors, KFK-1461, Kernforschungszentrum, Karlsruhe, Germany, 1971.

Haider, B. and Jacobi, W., Entwicklung von Verfahren und Geräten zur langzeitigen Radon-Überwachung im Bergbau, BMBW-FB K 72-14, Hahn-Meitner Institute, Berlin, 1972.

Hamilton, E. I., The Concentration and Distribution of Uranium in Human Skeletal Tissues, to be published.

Haseganu, D., Heavy Ion Registration in Solid Track Detectors, IFA-PN-23, Inst. Fiz. Atomica, Bucharest, 1971.

Haseganu, D. and Nicolae, M., Dosimetry of Heavy Ions by Means of Nuclear Emulsions and Solid-State Track Detectors, CERN-Report 71-16, Vol. 1, CERN, Geneva, 1971, 260.

Haseganu, D., Etch kinetics of heavy ion tracks in solid-state track detectors, *Rev. Roum. Phys.,* 17, 1023, 1972.

Hasegawa, H., Matsuo, M., Yamakoshi, K., and Yamawaki, K., Alpha particle track detection with celluloid films, *Radioisotopes,* 17, 419, 1968.

Hashimoto, T., Iwata, S., Nishimura, S., Nakanishi, T., and Sakanoue, M., Measurements of Reactor Neutron Flux by Fission Track Method, Proc. 9th Jap. Conf. Radioisotopes, 1970, 231.

Hashimoto, T., Determination of the uranium content in sea water by a fission track method with condensed aqueous solution, *Anal. Chim. Acta,* 56, 347, 1971.

Heinzelmann, M., Richtungsabhängigkeit eines Dosimeters mit nichtfotografischer Spaltfragment-Registrierung, Direct-Informat. 1/68, in *Atompraxis,* 14, 1968.

Heinzelmann, M. and Schüren, H., Spaltfragmentdosimeter für hohe Neutronenenergien, Jül-670-ST, KFA Jülich, Germany, 1970, 147.

Heinzelmann, M. and Haschke, W., Feststellung günstiger Ätzbedingungen zur nichtfotografischen Kernspurregistrierung, Report Jül-787-ST, Kernforschungsanlage Jülich, Germany, 1971a.

Heinzelmann, M. and Haschke, W., Geeignete Verfahren zur Empfindlichkeitserhöhung bei der nichtfotografischen Kernspurregistrierung, Report Jül-881-ST, KFA Jülich, Germany, 1971b, 97.

Heinzelmann, M. C. and Schüren, H., Thorium-Spaltfragmentdosimeter in der Neutronendosimetrie, Proc. Int. Symp. Neutron Dosimetry, Neuherberg, Germany, 1972.

Henig, G., Schopper, E., Granzer, F., Dardat, K., et al., Radiobiological and Dosimetric Applications of AgCl-Detectors, Proc. 8th Int. Conf. Nucl. Photography and Solid-State Track Detectors, Bucharest, 1972.

Henke, R. P., Benton, E. V., and Heckman, H., Sensitivity enhancement of plastic nuclear track detectors through photo-oxidation, *Radiat. Eff.,* 3, 43, 1970.

Henke, R. P. and Benton, E. V., Geometry of tracks in nuclear track detectors, *Nucl. Instr. Meth.,* 97, 483, 1971.

Hudis, J. and Katcoff, S., High Energy Fission Cross Sections of U, Bi, U, and Ag Measured with Mica Track Detectors, BNL-13124, 1969.

Iwata, S., Hashimoto, T., and Nishiwaki, Y., Rapid Separation of Natural Radionuclides by Electromigration and Detection of the Alpha Emitters based on Alpha Particle Tracks, in Proc. Int. Symp. on Rapid Methods for Measurement of Radioactivity in the Environment, Neuherberg, Germany, 1971.

Johnson, D. R. and Becker, K., Mechanism and Applications of Nuclear Track Etching in Polymers, ORNL-TM-2826, Oak Ridge National Laboratory, Oak Ridge, Tenn., 1970.

Johnson, D. R., Boyett, R. H., and Becker, K., Sensitive automatic counting of alpha particle tracks in polymers and its applications in dosimetry, *Health Phys.,* 18, 424, 1970.

Jowitt, D., Measurement of ^{238}U/^{235}U fission ratios in ZEBRA using solid state track recorders, *Nucl. Instr. Meth.,* 92, 37, 1971.

Jozefowicz, K., Energy threshold for neutron detection in a Makrofol dielectric track detector, *Nucl. Instr. Meth.,* 93, 369, 1971a.

Jozefowicz, K., Fission Track Detectors in Neutron Fluence Measurements, INR-1332, Inst. Nucl. Res., Warsaw, 1971b.

Jozefowicz, K., Determination of the Energy Threshold for Neutron Detection In Polymer Foil Track Detectors, Proc. 8th Int. Symp. Nucl. Photography and Solid-State Track Detectors, Bucharest, 1972.

Kashukeev, N., Ribarov, S., Neova, I., Taneva, T., and Todorov, N., Recording and monitoring of neutron fluxes by means of fission-fragment glass detectors, *Izv. Fiz. Inst. Aneb. Bulg. Akad. Nauk.,* 20, 247, 1971.

Kastner, J. and Oltmann, B. G., Neutron Dosimeter Including a Step Wedge Formed of an Alpha-Attenuating Material, U.S. Patent 3,604,921, 1971.

Katz, R. and Kobetich, E. J., Formation of etchable tracks in dielectrics, *Phys. Rev.,* 170, 401, 1968.

Katz, R., Sharma, S. C., and Homayoonfar, M., Detection of energetic heavy ions, *Nucl. Instr. Meth.,* 100, 13, 1972; The structure of particle tracks, in Topics in Radiation Dosimetry, Attix, F. H., Ed., Academic Press, London, 1972, Chap. 6.

Keirim-Markus, I. B., Koroleva, T. V., Kraitor, S. N., and Uspenskii, L. N., Characteristics of a personal neutron track dosimeter, *At. Energ.,* 34, 11, 1973.

Kerr, G. D. and Strickler, T. D., The application of solid-state nuclear track detector to the Hurst threshold detector system, *Health Phys.,* 12, 1141, 1966.

Khan, H. A., Semi-automatic scanning tracks in plastics, *Radiat. Eff.,* 8, 135, 1971.

Khan, H. A. and Durrani, S. A., Prolonged etching factor in solid-state track detection and its applications, *Radiat. Eff.,* 13, 257, 1972.

Khan, H. A. and Durrani, S. A., Efficiency calibration of solid-state nuclear track recorders, *Nucl. Instr. Meth.,* 98, 229, 1972b.

Khan, H. A. and Durrani, S. A., Electronic counting and projection of etched tracks in solid-state nuclear track detectors, *Nucl. Instr. Meth.,* 101, 583, 1972c.

Khan, H. A., Durrani, S. A., and Fremlin, J. H., Fission Fragment Energy Spectrum Determination by Solid-State Nuclear Track Detectors, Proc. 8th Int. Symp. Nucl. Photography and Solid-State Track Detectors, Bucharest, 1972.

Khovanivich, A. I., Pikalov, G. L., and Kryvokrysenko, I. F., Using a photoelectric colorimeter for counting charged particle tracks on the surface of glass detectors, *At. Energ.,* 29, 367, 1970.

Köhler, W., Application of solid-state track recorders for neutron spectra measurements in a reactor irradiation facility, *Radiat. Eff.,* 3, 231, 1970.

Koroleva, T. V. and Kraitor, S. N., Efficiency of track detectors for fission fragments, *At. Energ.,* 31, 52, 1971.

Kosanke, H. D., Track Etching Registrant for High-Temperature Applications, Trans. Am. Nucl. Soc., Ann. Meet., Las Vegas, 1972, 124.

Koshijima, T., Practical method for neutron detection by fission track registration, *Nippon Igaku Hoshasen Gakkai Zasshi,* 31, 928, 1971.

Kumamoto, Y., Alpha tracks in cellulose nitrate as detector elements for spherical moderator type neutron monitors, *Health Phys.,* 24, 558, 1973.

Kurosawa, R., On the application of solid-state track detector to alpha track monitor in uranium mines, *Hoken Butsuri,* 7, 105, 1972.

Lark, N. L., Spark scanning for fission fragment tracks in plastic foil, *Nucl. Instr. Meth.,* 67, 137, 1969.

Lecerf, M. and Peter, J., Glass detectors, *Nucl. Instr. Meth.,* 104, 189, 1972.

Lovett, D. B., Track etch detectors for alpha exposure estimation, *Health Phys.,* 16, 623, 1969.

Marchetti, M., Tommasino, L., and Casnati, E., Damage Track Registration Related to the Thermal Stability of Polymers, Proc. 8th Int. Conf. Nucl. Photography and Solid-State Detectors, Bucharest, 1972.

Maurette, M., Track formation mechanism in minerals, *Radiat. Eff.,* 3, 149, 1970.

McEachern, P., Myers, W. G., and White, F. A., Uranium concentrations in surface air at rural and urban localities within New York State, *Environ. Sci. Technol.,* 5, 700, 1971.

Medveczky, L. and Somogyi, G., Paper 6th Int. Conf. Corpuscular Photography, Florence, 1966; *At. Kozlem.,* 8, 226, 1966.

Mijnheer, B. J., Pauw, H., van Herk, G., and Aten, A. H. W., Fluence and dose equivalent determination in a neutron field by means of moderating spheres containing fission track recorders, *Health Phys.,* 24, 423, 1973.

Monnin, M., Etude de l'Interaction des Particules Lourdes Ionisantes avec les Macromolécules à l'Etat Solide, PNCF 68-RI-9, University of Clermont-Ferrand, France, 1968.

Monnin, M., Mechanism of track formation in polymers, *Radiat. Eff.,* 5, 69, 1970.

Monnin, M. and Isabelle, D. B., Solid track detectors and their applications in biology, *Ann. Phys. Biol. Med.,* 4, 95, 1970.

Morley, J. A., Two techniques to increase contrast of track etching neutron radiographs, Trans. Am. Nucl. Soc., Ann. Meet., Las Vegas, 1972, 120.

Nakanishi, T. and Sakanoue, G., Measurement of 14 MeV neutron flux density by fission track method, *Radiochem. Radioanal. Lett.,* 2, 313, 1969.

Nicolae, M., On Some Factors Affecting the Cellulose Nitrate and Acetate Sensitivity to Particle Track Registration, Proc. 8th Int. Symp. Nucl. Photography and Solid-State Track Detectors, Bucharest, 1972.

Nishiwaki, Y., Kawai, H., Morishima, H., Iwata, S., and Tsuruta, T., Rapid Method of Confirmation of Fissionable Materials with Etch-Pit Method, in Proc. Int. Symp. on Rapid Methods for Measurement of Radioactivity in the Environment, Neuherberg, Germany, IAEA, Vienna, 1971a.

Nishiwaki, Y., Tsuruta, T., and Yamazaki, K., Detection of fast neutrons by etch-pit method of nuclear track registration in plastics, *J. Nucl. Sci. Technol.,* 8, 162, 1971b.

Nishiwaki, Y., Kawai, H., Morishima, H., Koga, T., and Okada, Y., Studies on the Spark Counting Techniques of Etched Nuclear Tracks on Plastic Foils for Neutron Monitoring, Paper IAEA/SM-167/22, Proc. Symp. Neutron Monitor. Radiat. Protect., IAEA, Vienna, 1973.

Nomura, R., Yokota, R., Imagawa, H., Muto, Y., and Tomishima, H., Measurement of the Fast Neutron Fluxes in an Experimental Critical Facility, Part 1, JAERI Memo 4337, 1971.

Novotny, J., Alpha dosimetry and spectrometry by solid-state nuclear track detector, *Radioisotopy,* 11, 777, 1970.

Obaturov, G., Rebanov, V., Chumbarov, Y., and Shalin, V., Emergency Neutron Track Dosimeter, Paper IAEA/SM-160/74, Proc. Symp. Dosimetry Techniques Applied to Agriculture, Industry, Biology, and Medicine, IAEA, Vienna, 1972.

Oltman, B. G., Congel, F., Capinski, N. P., and Berger, H., The Use of Sparking Techniques to Enhance Track-Etch Radiographs, Trans. Am. Nucl. Soc. Winter Meeting, 1970, 531.

Oosterkamp, W. J., Schaar, J., and Van Velze, P. L., Automatic Counting of Solid-State Fission Track Recorders, Trans. Am. Nucl. Soc. Winter Meeting, 1970, 526.

Paretzke, H. G., Mechanisms of Damage and Etching of Particle Tracks in Plastics, Proc. 8th Int. Symp. Nucl. Photography and Solid-State Track Detectors, Bucharest, 1972a.

Paretzke, H. G., A Simple Method for the Determination of Etch Product Concentration, Proc. 8th Int. Symp. Nucl. Photography and Solid-State Track Detectors, Bucharest, 1972b.

Peterson, D. D., Improvement in Particle Track Etching in Lexan Polycarbonate Film, *Proc. 7th Int. Colloq. Nucl. Photography and Solid-State Track Detectors,* Vol. 1, Barcelona, 1970, 181.

Piesch, E., Anwendung der Interferenz-Kontrast-Mikroskopie zur Kernspurregistrierung in Festkörpern, *Zeiss Informat.,* 18, 58, 1970a.

Piesch, E., Development of New Neutron Detectors for Accidental Dosimetry, Paper SM-143/31, Proc. Symp. Adv. Radiat. Detectors, IAEA, Vienna, 1970b.

Piesch, E. and Weng, P. S., Application of Fission Track Etching Process for Determination of Low-Level Uranium Concentration in Liquids and Aerosols, KFK-1552, Kernforschungszentrum, Karlsruhe, Germany, 1972.

Platzer, H., Abmayr, W., and Paretzke, H. G., Evaluation of dielectric track detectors with diffracted laser light, *Atomkernenergie,* 20, 162, 1972.

Prêtre, S., Tochilin, E., and Goldstein, N., A Standardized Method for Making Neutron Fluence Measurements by Fission Fragment Tracks in Plastics. A Suggestion for an Emergency Dosimeter with rad Response, *Proc. First Int. Cong. IRPA, Rome, 1966,* Part I, Pergamon Press, Oxford, 1968, 491.

Prêtre, S., Measurement of Personnel Neutron Doses by Fission Fragment Track Etching in Certain Plastics, Proc. Int. Conf. Nucl. Track Registrat, in Insulating Solids, Clermont-Ferrand, 1969; *Radiat. Eff.,* 5, 103, 1970.

Prêtre, S. and Heusi, K., Routine- und Unfall-Personendosimetrie mit Festkörper-Spaltspur- Detektoren für thermische und schenelle Neutronen mit automatischer Auswertung, Report TM-SU-149, EIR Würenlingen, Switzerland, 1972.

Prêtre, S., Design of a Personal Neutron Dosimeter for Routine and Emergency Use, Based on Automatic Fission Track and Spark Counting, Paper IAEA/SM-167/14, Proc. Symp. Neutron Monitor. Radiat. Protect., IAEA, Vienna, 1973.

Prevo, P. R., Dahl, R. E., and Yoshiwaki, H. H., Thermal and fast neutron detection by fission-track detection in mica, *J. Appl. Phys.,* 35, 2636, 1964.

Price, P. B. and Walker, R. M., Electron microscope observation of etched tracks from spallation recoils in mica, *Phys. Rev. Lett.,* 8, 217, 1962.

Price, P. B. and Walker, R. M., A simple method of measuring low uranium concentrations in natural minerals, *Appl. Phys. Lett.,* 2, 23, 1963.

Price, P. B., Fleischer, R. L., and Miller, A. A., Study of High Resolution Plastic Track Detectors of Heavy Cosmic Rays, Report N-69-35895, 1969.

Price, P. B., Walker, R. M., and Fleischer, R. L., Method of Measuring Fast Neutrons by Observing Multipronged Tracks of Charged Particles Formed in Cellulose Nitrate, U. S. Patent 3,564,250, 1971.

Rago, P. F. and Goldstein, N., Thorium fission cross section for neutrons between 12.5 and 18 MeV using fission fragment damage tracks in lexan, *Health Phys.,* 13, 654, 1967.

Rago, P. F., Goldstein, N., and Tochilin, E., Reactor neutron measurements with fission foil-lexan detectors, *Nucl. Appl.,* 8, 302, 1970.

Reimer, G. M., Wagner, G. A., and Carpenter, B. S., The thermal stability of fission tracks in the standard reference material glass standard (NBS), *Radiat. Eff.,* 15, 273, 1972.

Remy, G., Ralarosy, J., Tripier, J., Debeauvais, M., and Stein, R., Dosimetry and spectrometry of fission and thermonuclear neutrons by means of visual plastic detectors, *Radiat. Eff.,* 5, 221, 1970.

Roberts, J. H., Parker, R. A., Congel, F. J., Kastner, J., and Oltman, B. G., Environmental neutron measurements with solid-state track recorders, *Radiat. Eff.,* 3, 283, 1970.

Sakanoue, M., Nuclear track methods in solids and its development. Part I. Fundamentals and the tracks in minerals and glasses, *Kagaku No Ryoiki,* 24, 124, 1970a.

Sakanoue, M., Nuclear track methods in solids and its development. Part II. Discrimination of incident particles and of their energy and application of plastics, *Kagaku No Ryoiki,* 24, 207, 1970b.

Schaefer, H. J., Benton, E. V., Henke, R. P., and Sullivan, J. J., Nuclear track recordings of the astrounauts' radiation exposure on the first lunar landing mission Apollo XI, *Radiat. Res.,* 49, 245, 1972.

Schopper, E. and Haase, G., Die Rolle der Silberhalogenide als Teilchenspur-Detektoren, *Phot. Korresp.,* 107, 155, 1971.

Schott, J. U., Schopper, E., Granzer, F., et al., Detecting Properties of AgCl(Cd) Detectors for Ionizing Particles, *Proc. 8th Int. Symp. Nucl. Photography and Solid-State Track Detectors,* Bucharest, 1972.

Schreurs, J. W., Friedman, A. M., Rokop, D. J., Hair, M. W., and Walker, R. M., Calibration of U-Th glasses for neutron dosimetry and determination of uranium and thorium concentration by the fission tracks, *Radiat. Eff.,* 7, 231, 1971.

Schultz, W. W., Track density measurement in dielectric track detectors with scattered light, *Rev. Sci. Instr.,* 39, 1893, 1968.

Schultz, W. W., 1/v Dielectric Track Detectors for Slow Neutrons, KAPL-M-7121, 1970; *J. Appl. Phys.*, 41, 5260, 1970.

Seitz, M. G., Walker, R. M., and Carpenter, B. S., Improved methods for measurement of thermal neutron dose by the fission track technique, *J. Appl. Phys.*, 44, 510, 1973.

Selim, Y. S. and Abo Elkheir, A. A., Plastics as solid-state track detectors for (p,a) reactions, Paper, Sec. Cairo Solid-State Conference, 1973.

Sherwood, H. F., Enhancing Radiation Damage for Nuclear Particle Detection, U.S. Patent 3,501,636, 1970.

Simmons, D. J. and Fitzgerald, K. T., Application of Cellulose Nitrate Films for Alpha Autoradiography of Bone. II. Resolution and Sensitivity of Cellulose Nitrate Films, ANL-7760, Part 2, Argonne Nat. Lab., Argonne, Ill., 1970, 208.

Sohrabi, M. and Becker, K., Track Etching in Personnel Fast Neutron Dosimetry, ORNL-TM-3605, Oak Ridge Nat. Lab., Oak Ridge, Tenn., 1971; see also *Nucl. Instr. Meth.*, 104, 409, 1972.

Somogyi, G., A theory for the development of track holes in solid-state nuclear track detectors, *Atomki Közlem.*, 9, 77, 1967.

Somogyi, G. and Strivastava, D. S., Investigations on alpha radiography with plastic track detectors, *Atomki Kozlem.*, 12, 101, 1970a and *Proc. 7th Int. Colloq. Nucl. Photography and Solid-State Detectors*, Vol. 2, Barcelona, 1970a, 711.

Somogyi, G. and Srivastava, D. S., Effect of Irradiation in Vacuum on the Optical Density Curves for *a* Particles in Plastics, *Proc. 7th Int. Colloq. Nucl. Photography and Solid-State Track Detectors*, Vol. 2, Barcelona, 1970b, 719.

Somogyi, G., Varnagy, M., and Medveczky, L., Influence of etching parameters on sensitivity of plastics, *Radiat. Eff.*, 5, 111, 1970.

Somogyi, G. and Srivastava, D. S., Alpha radiography with plastic track detectors, *Int. J. Appl. Rad. Isotopes*, 22, 289, 1971.

Somogyi, G., Influence of Thermal Effects on the Track Registration Characteristics of Plastics, Proc. 8th Int. Symp. Nucl. Photography and Solid-State Track Detectors, Bucharest, 1972a.

Somogyi, G., Effect of Ozone Atmosphere on the Detecting Properties of Plastic Track Recorders, Proc. 8th Int. Conf. Nucl. Photography and Solid-State Track Detectors, Bucharest, 1972b; see also *Radiat. Eff.*, 16, 233, 1972.

Somogyi, G. and Gulyas, J., Methods for improving radiograms in plastic detectors, *Radioisotopy*, 13, 549, 1972.

Somogyi, G. and Szalay, A. S., Track-diameter kinetics in solid-state nuclear track detectors, *Atomki. Kozlem.*, 14, 113, 1972 (English version *Nucl. Instr. Meth.*, in press).

Song, K. S. and Lee, S. S., Detection of ^{235}U fission fragments by the solid state track recorder using cellulose nitrate, *J. Korean Nucl. Soc.*, 2, 223, 1970.

Sowinski, M., Stephan, C., Czyzewski, T., and Tys, J., Crystalline quartz as a detector for heavy charged particles, *Nucl. Instr. Meth.*, 105, 317, 1972.

Spurny, F., Tracks of Charged Particles Formed in Polymer Foils by Fast Neutrons; Implications to Fast Neutron Dosimetry, Proc. 8th Int. Conf. Nucl. Photography and Solid-State Track Detectors, Bucharest, 1972a.

Spurny, F., Detection of alpha particles by thin Triafol TN foil. *Jad. Energ.*, 18, 315, 1972b.

Stern, R. A. and Price, P. B., Charge and energy information from heavy ion tracks in Lexan, *Nature Phys. Sci.*, 240, 82, 1972.

Stolz, W. and Dörschel, B., Neutronen-Personendosimetrie mittels Spaltfragment-Registrierung in Glimmer, *Kernenergie*, 12, 244, 1969.

Stone, D. R., Identification of fission fragment tracks in Lexan after pretreatment to high doses of alpha particles, *Health Phys.*, 16, 772, 1969.

Su, C. S. and Liu, T. C., Effective neutron temperatures in reactor core measured by fission track detectors, *Nucl. Instr. Meth.*, 89, 233, 1970.

Su, C. S., The determination of uranium concentrations in seaweeds by nuclear track detectors, *Radiat. Eff.*, 14, 109, 1972.

Svensson, G. K., A Feasibility Study of the Use of Track Registration From Fission Fragments for Neutron Personnel Dosimetry at the 20 GeV SLAC, Paper 2nd Int. Congr. IRPA, Brighton, 1970.

Tatsuta, H., Evaluation of Dose Equivalent by Fission Fragment Detectors, Paper No. 122, 2nd Int. Congr. IRPA, Brighton, 1970.

Thiel, K. and Herr, W., Determination of uranium distributions in terrestrial and lunar rocks by means of fission track and microprobe techniques, *Kerntechnik*, 14, 513, 1972.

Tommasino, L., Electrochemical Etching of Damaged Track Detectors by H.V. Pulse and Sinusoidal Waveforms, Proc. 7th Int. Colloq. Corpuscular Photography and Solid Detectors, Barcelona, 1970; CNEN Report RT/PROT (71) 1, 1970.

Trousil, J., Singer, J., and Marsal, J., Track Detectors in Czechoslovak National Personnel Dosimetry Service, Paper IAEA/SM 167/5, Proc. Symp. Neutron Monitor. Radiat. Protect., IAEA, Vienna, 1973.

Tuyn, J. W. N., Solid-state nuclear track detectors in reactor physics experiments, *Nucl. Appl.*, 3, 372, 1967.

Tuyn, J. W. N., A Simple Model for the Development of Tracks in Solid-State Nuclear Track Detectors, Report PE-158, Kjeller, Norway, 1968.

Tuyn, J. W. N., Measurement of fission density distribution within fuel elements in the vicinity of grey absorber control elements, *Nukleonik*, 12, 183, 1969.

Tuyn, J. W. N. and Broerse, J. J., Analysis of Etch-Pit Size Distribution in Makrofol Polycarbonate Foil After Fast-Neutron Exposure, Proc. 7th Int. Colloq. Corpuscular Photography and Solid Detectors, Barcelona, 1970.

Tymons, B. J., Tuyn, J. W. N., and Baarli, J., A System for Neutron Dosimetry in Mixed Radiation Fields, Paper IAEA/SM-167/20, Proc. Symp. Neutron Monitor. Radiat. Protect., IAEA, Vienna, 1973.

Uzgiris, E. E. and Fleischer, R. L., Amber-Particle Track Registration, Track Stability, and Uranium Content, Report 71-C-193, General Electric Res. Lab., Schenectady, N. Y., 1971.

Uzgiris, E. E. and Fleischer, R. L., Etched laser filament tracks in glasses and polymers, *Phys. Rev. A,* 7, 734, 1973.

Varnagy, M., Szegedi, S., and Nagy, S., Observations of proton tracks by a plastic detector, *Nucl. Instr. Meth.,* 89, 27, 1970.

Varnagy, M., Szabo, J., Juhasz, S., and Csikai, J., Determination of track parameters by diffraction method using laser light, *Nucl. Instr. Meth.,* 106, 301, 1973.

Veprik, Y. M., Perelygin, V. P., Romanenko, V. P., Tretyakova, S. P., and Vinogradov, Y. A., Effect of chemical factors on the sensitivity of polymeric detectors, *Instr. Exp. Technol.,* 4, 1015, 1970a.

Veprik, Y. M., Kartuzhanskii, A. L., Marennyi, A. M., and Popov, V. I., On the Mechanism of Cellulose Films Sensitivity to Ionizing Particle Track Registration, *Proc. 7th Int. Colloq. Nucl. Photography and Solid-State Detectors,* Vol. 1, Barcelona, 1970b, 287.

Veprik, Y. M., Kartuzhanskii, A. L., Marennyi, A. M., and Popov, V. I., Possible mechanism of track recording from ionizing particles by cellulose films, *Zh. Nauch. Prikl. Fotogr. Kinematogr.,* 17, 349, 1972.

Wagner, G. A., The Geological Interpretation of Fission Track Ages, Trans. Am. Nucl. Soc. Ann. Meet., Las Vegas, 1972, 117.

Walker, R. M., Price, P. B., and Fleischer, R. L., A versatile, disposable dosimeter for slow and fast neutrons, *Appl. Phys. Lett.,* 3(2), 28, 1963.

Weidenbaum, B., Lovett, D. B., and Kosanke, H. D., Flux Monitor Utilizing Track-Etch Film for Unattended Safeguard Application, Trans. Am. Nucl. Soc. Winter Meet., 1970, 524.

White, O., Jr., Evaluation of MIT and ORNL Radon Daughter Dosimeters, HASL-70-3, 1970.

Wollenberg, H. A. and Smith, A. R., Energy and Flux Determinations of High-Energy Nucleons, UCRL-19364, 1969; Proc. 2nd Int. Conf. Accelerator Dosimetry, Stanford, 1969.

Yokota, R., Nakajima, S., and Muto, Y., Registration of fission-fragment tracks in the Th- and U-doped phosphate glasses and its possible application to neutron dosimetry, *Nucl. Instr. Meth.,* 61, 119, 1968.

Young, D. A., Etching of radiation damage in lithium fluoride, *Nature,* 182, 375, 1958.

6. VARIOUS OTHER SYSTEMS

6.1. Silicon Diodes for Fast-neutron Dosimetry

Neutron-induced nuclear reactions, elastic scattering of heavy charged particles and, to a much smaller degree, high energy electrons can produce measurable changes in the electrical or optical properties of various semiconductors. The most common type of radiation damage is the displacement of lattice atoms, resulting in the production of pairs of vacancies and interstitials (Frenkel defects). In case of silicon crystals, the minimum required displacement energy E_m is between 12 and 30 eV. The energy for the production of Frenkel defects E_{fd} for electrons can be described by the equation:

$$E_{fd} = [E_m \frac{M_2 m_0 c^2}{2 m_0} + (m_0 c^2)^2]^{1/2} - m_0 c^2 \qquad (6.1)$$

with

M_1 = mass of the incident particle (electron),
M_2 = mass of the silicone atom,
m_0 = electron rest mass, and
c = velocity of light.

For heavy particles such as neutrons, this equation is simplified to

$$E_{fd} = \frac{E_m (M_1 + M_2)^2}{4 M_1 M_2} \qquad (6.2)$$

It can be shown that electrons below ~ 0.16 MeV in energy cannot produce displacements and that a high energy neutron can produce cascades of thousands of secondary, tertiary, etc. displaced silicon atoms which act as recombination centers and traps in the forbidden band and change the charge carrier lifetime. This results in an increase in the forward resistance as well as changes in the photoelectric and spectroscopic properties of the crystal.

The small dimensions, almost total sensitivity to X-, gamma, and beta radiation and thermal neutrons, and the extremely easy evaluation by a simple electrical resistance measurement make such detectors potentially very interesting fast-neutron dosimeters. Initial work at the Battelle Memorial Institute under contract with the U.S. Army Signal Corps led to the development of the first dosimeters of this type in the late 1950's and early 1960's (Battelle, 1958–61; Mengali et al., 1959; Gorton, 1962; and Kramer, 1965). Further work on improving the sensitivity and reproducibility of the system by the Phylatron Corp., Columbus, Ohio, now Precision Systems Co. (Swartz, 1963; Thurston et al., 1966; and Swartz and Thurston, 1966), resulted in a commercial system consisting of three diodes of different sensitivities (pH 15, 30, and 50[®]) and a reader that permits forward resistance measurements at a constant current (Figure 6-1).

The success of these developments triggered research in several other countries, including England, France, and Germany, where AEG/Telefunken produces such diodes under the code name HL 668[®] (see also Arnold, 1968) and Sweden (Wik, 1965 and 1967), where cooperation between the government Defense Research Institute FOA and the HAFO Corp. resulted in another commercially available system (Fast Neutron Dosimeter 5422 and Reader 3806, sold by Studsvik AB Atomenergi, Nyköping, Sweden). These systems as well as more recent British diodes have been subject to further improvements and careful tests and comparisons (Becker, 1967c; Svansson et al., 1968; DeCosnac et al., 1968; Batchelor, 1971; Widell, 1972; Wall, 1972; Oliver and Hamm, 1973; and Krüger et al., 1973).

The silicon diode fast-neutron dosimeter is basically a n^+pp^+ junction, which is produced by diffusing phosphorus and boron into the opposite ends of slightly p-type float zone silicon. The n^+ and p^+ regions are heavily doped ($>10^{18}$ impurity atoms per cm^3), while the base region contains $\sim 10^{14}$ acceptor atoms per cm^3. The doped regions are about 25 μm thick, while the base region is up to 1.2 mm thick in the more sensitive diodes. A typical relationship between forward current and applied voltage for such a sensitive diode is given in Figure 6-2.

Obviously, the sensitivity in terms of voltage change per rad depends on neutron dose as well as on several other parameters, such as forward current and diode characteristics.

The change in forward resistance $dU/d\Phi$ of a diode for a given neutron fluence depends on the damage constant, K, the carrier lifetime, (Φ), and

"

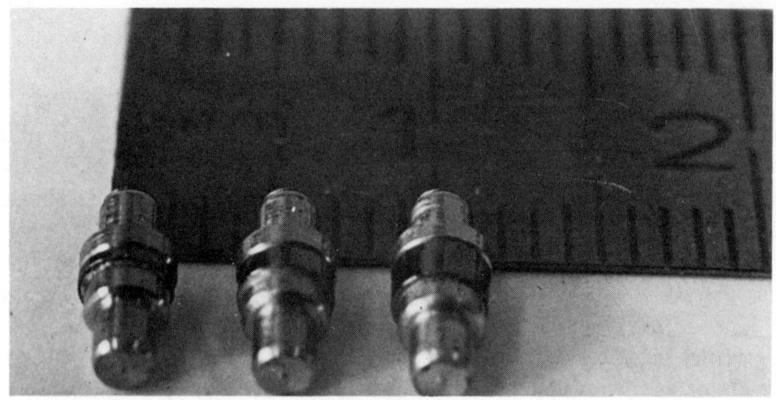

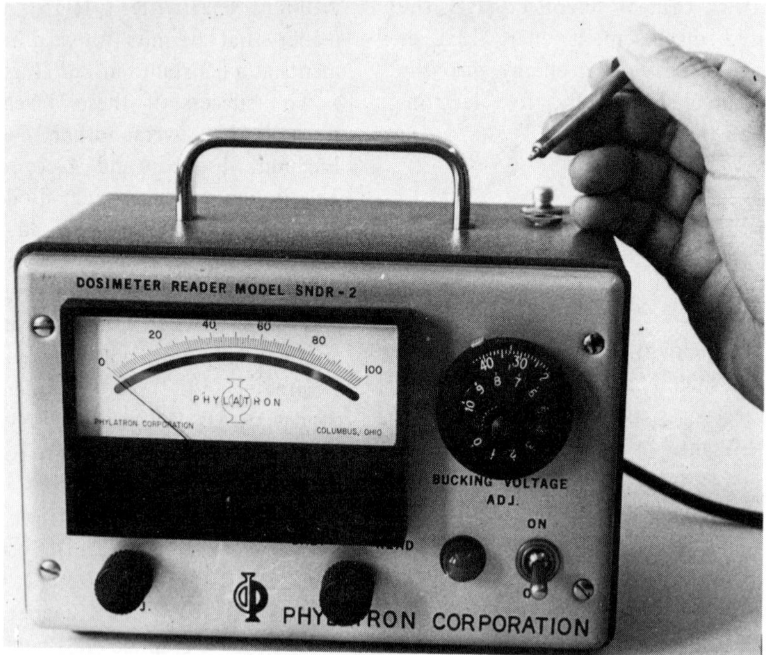

FIGURE 6-1. Phylatron silicon diode fast-neutron dosimeters (top) with increasing sensitivities from left to right and simple commercial readout instrument.

the ratio between the diode base width d and the diffusion length L, (Φ) (Swartz and Thurston, 1966):

$$\frac{dU}{d\Phi} = K \cdot (\Phi) \cdot f \frac{d}{L(\Phi)}, \qquad (6.3)$$

and the carrier lifetime is related to the initial carrier lifetime Φ_0 by

$$(\Phi) = \frac{\Phi_0}{1 + K\Phi_0} \qquad (6.4)$$

Obviously, the response will depend on the base width; diodes of different sensitivities are produced by modification of d.

The recommended currents for dose measurements are in the 0.02 to 0.1 A range. Higher currents increase the apparent sensitivity in $\Delta V/$rad; as they proportionally increase the predose voltage, there is a real sensitivity gain. Furthermore, high currents heat the diode, causing secondary changes in the resistance. This effect is illustrated in Figure 6-3 for different dose-levels at 100 mA. The Swedish reader restricts the duration of the current flow of 0.4 sec. Because of the strong temperature dependence of the resistance

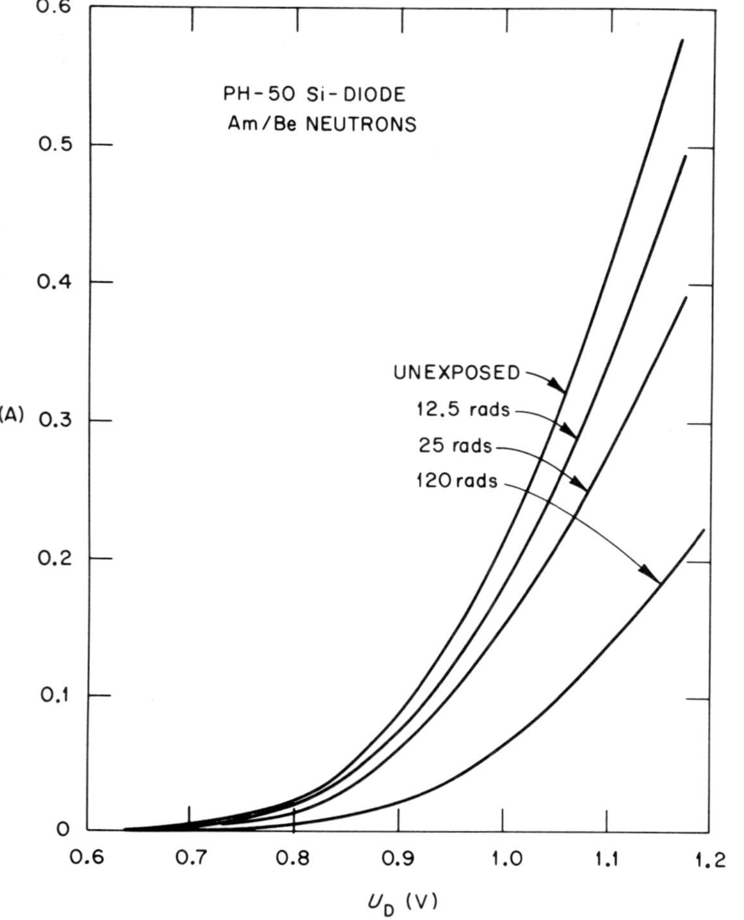

FIGURE 6-2. Forward current as a function of forward voltage for unirradiated and neutron irradiated silicon diodes. (After Becker, 1967.)

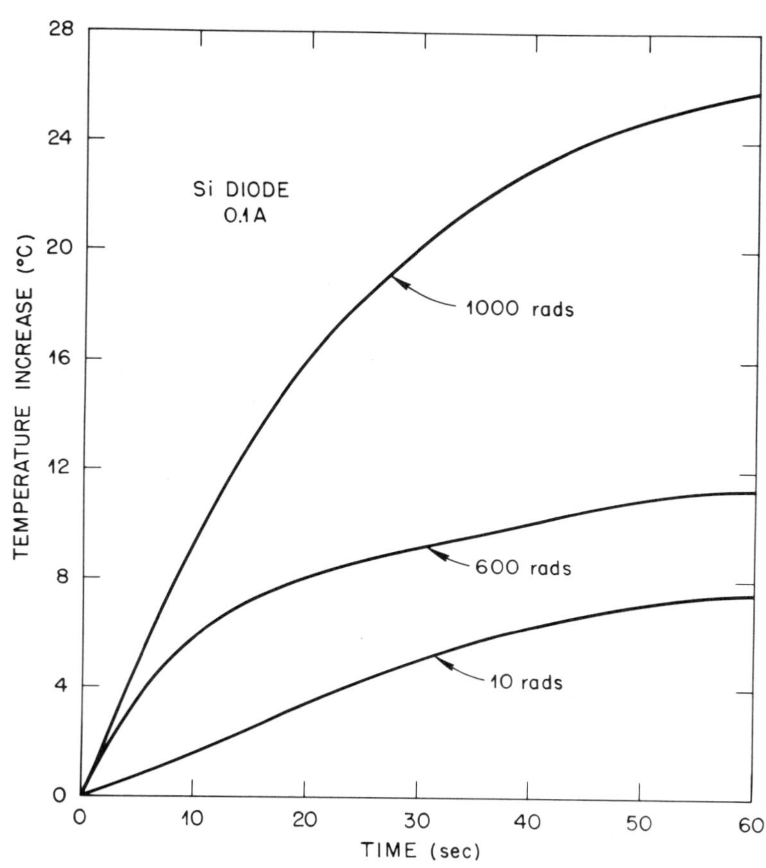

FIGURE 6-3. Temperature increase in typical silicon diodes due to 100 mA current as a function of time. (After Wall, 1972.)

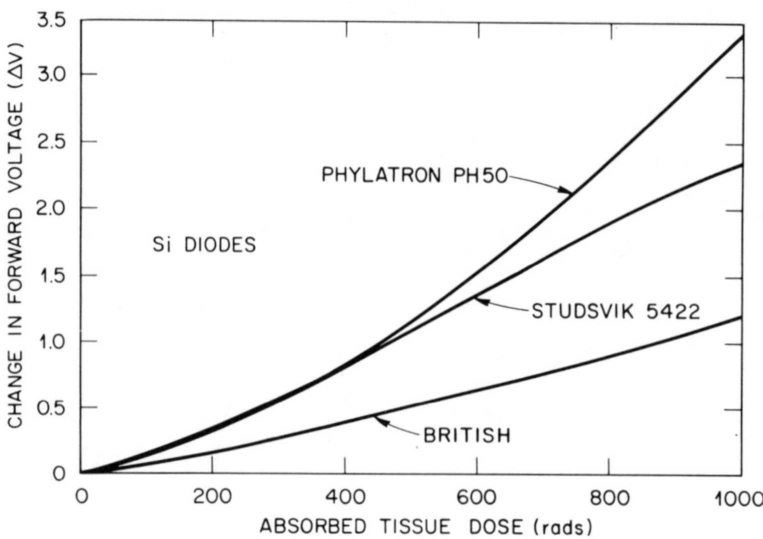

FIGURE 6-4.   Forward voltage change as a function of neutron dose in three types of silicon diodes. (After Wall, 1972.)

reading (depending on the dose level, ⁻8 to +0.4%/°C in British diodes — Wall, 1972 — and ∿0.5%/°C in HAFO and German diodes at 1,000 rad — Krüger et al., 1973), it has been suggested to provide the reader with a thermostat or add a second, insensitive diode as a control to each dosimeter.

The relationship between neutron dose and response is not linear (Figure 6-4), thus requiring the use of calibration curves. At 100 mA the predose voltage of a typical sensitive diode is ∿1 V, and the neutron sensitivity ∿1 mV/rad. If the voltmeter has a precision of ∿0.1%, 10 rad can be measured with an accuracy of ∿10%. So far it has been impossible to measure neutron doses of less than 1 to 5 rad with resonable precision. Considering the efforts that have already been made to increase sensitivity and/or precision, it is unlikely that this lower limit will be greatly improved. The upper dose limit for the most sensitive diodes is about 700 rad; for less sensitive diodes such as PH-15, it is 15,000 rad.

A complicating factor is the relatively large fluctuation in predoses within the same batch of diodes. Various authors report standard deviations between 7 and 10%, but individual diodes may read as much as 25% above or below the average (Widell, 1972). Individual predose adjustments are, therefore, highly advisable.

The neutron-induced lattice damage is partially reversible. The thermal stability of the radiation damage apparently varies somewhat with details of the manufacturing process. For the U.S. diodes, 8% fading was found between 1 and 30 min after exposure (Becker, 1967c), while Wall (1972) found only 3.5% during 1 day and 26% after 100 days of storage at room temperature. Under similar conditions, 15% fading was reported after 100 days in U.K. diodes, and 18% in Swedish diodes. After 10 days, some authors found 12% fading in the Swedish diodes (Svansson et al., 1968, and Widell, 1972), while others (Krüger et al., 1973) reported 20 to 23% in Swedish and German diodes. Perhaps the disagreements are largely due to inconsistencies in the conditions of irradiation, storage temperature, and the time of taking the first post-irradiation measurement (the initial fading during the first ∿30 min is quite rapid — Becker, 1967c). The fading rate tends to increase slightly with the dose-level.

The fading proceeds more rapidly at increased temperature; a short temperature treatment of the irradiated diodes after exposure (Krüger et al., 1973, suggest 2 min at 100°C) helps to complete the initial fading, thus "stabilizing" the radiation effect in the diodes. Several authors have also attempted to anneal the detectors for reuse. As can be seen in Figure 6-5, 1 hr at 300°C anneals almost 90% of the radiation damage in Phylatron® (Becker, 1967) as well as British diodes (Wall, 1972). Repeated annealing has little, if any, effect on the inherent sensitivity, but increasing predose

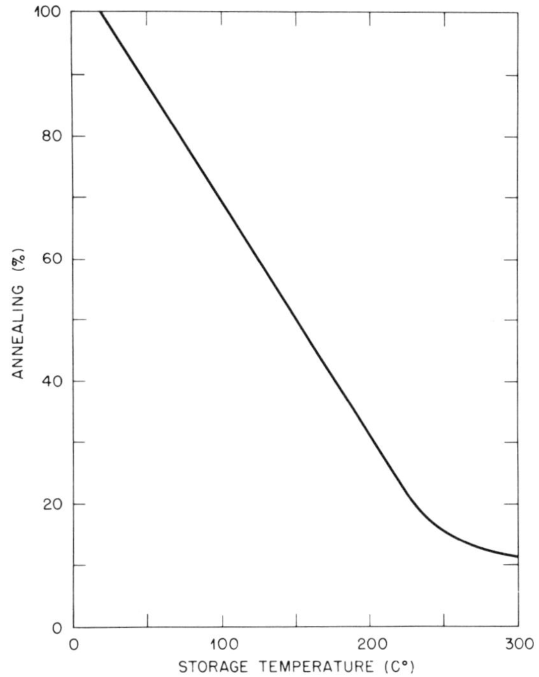

FIGURE 6-5.   Annealing of neutron-induced radiation damage in silicon diodes as a function of the temperature at which they have been stored for 1 hr. (After Becker, 1967.)

may reduce the practical sensitivity. The protective plastic coating of the Swedish diodes disintegrates at temperatures lower than 300°C.

The maximum annealing temperature for these diodes is ∿120°C, resulting in ∿33% recovery (Oliver and Hamm, 1973). They can, therefore, not be reused as easily as other, more heat-resistant types.

The neutron energy response of silicon diodes has been the subject of numerous studies (for recent compilations, see Widell, 1972; Wall, 1972; and Krüger et al., 1973). Widely quoted early measurements (Gamertsfelder et al., 1963) that reported an increase between 0.2 and ∿3 MeV, followed by a decrease, were obviously in error. Figure 6-6 shows the average of the calculated and/or measured values of seven authors. It can be seen that the response is equivalent to the neutron dose in rad (or kerma) within about ±20% between ∿0.3 and 15 MeV.

Below 350 keV, no measurements have been reported, but theoretical calculations (DeCosnac et al., 1968) indicate large deviations from tissue equivalence. Consequently, correction factors have to be applied in the intermediate energy neutron range. According to Batchelor (1971), the difference between silicon diode and ionization chamber reading may amount to almost a factor of two in greater depths of a fission neutron irradiated phantom, where one would expect a low average neutron energy. According to one report (Krüger et al., 1973), the German AEG/

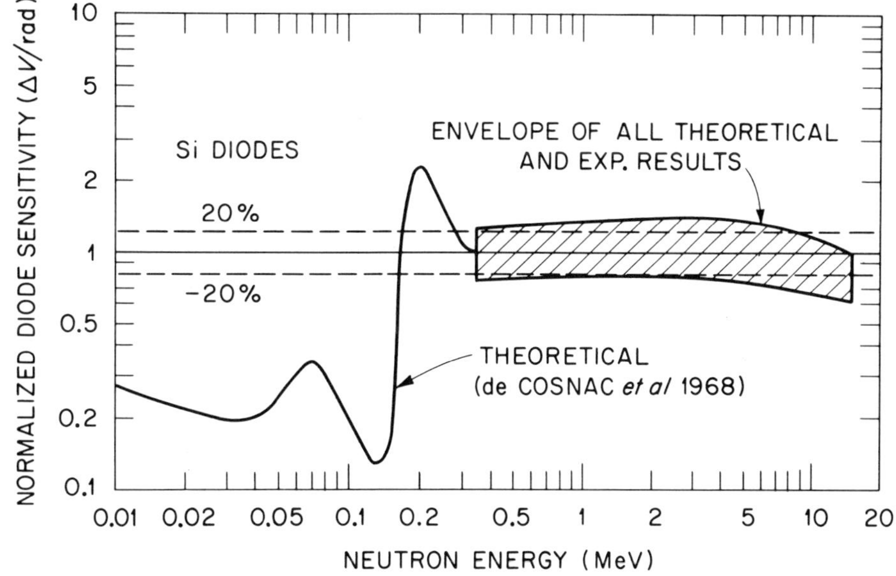

FIGURE 6-6.   Measured and calculated diode sensitivities (expressed as change in forward resistance per rad, normalized for response at 0.35 MeV) as a function of neutron energy. (Compiled by Wall, 1972.)

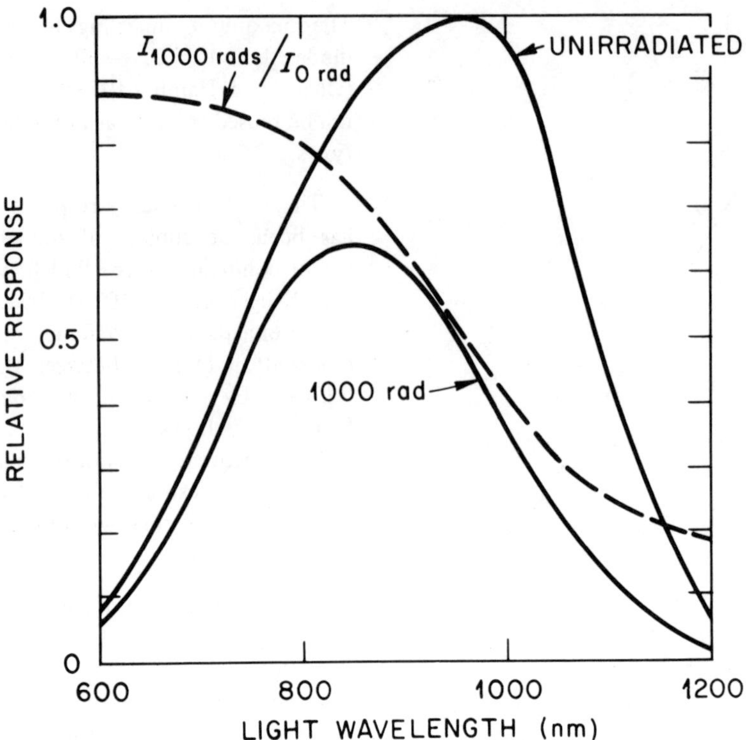

FIGURE 6-7. Spectral response characteristic of unirradiated and neutron-exposed silicon photodiodes and (dotted line) ratio of the two curves. (After Rösch et al., 1972.)

Telefunken exhibits an unexplained, much larger energy response than Swedish and U.S. diodes.

The overall accuracy of the system is less than desirable. Among the contributing factors are

a. fluctuations in predose and sensitivity within a batch (usually about ±7 to 10%), which may be even larger from batch to batch;

b. errors in the voltage measurement (approximately ±5 to 10% at 10 rad, 0.5 to 1% at 100 rad);

c. errors due to the temperature dependence (assuming fluctuations in the readout temperature of ±5°C, this would amount to 40% for 10 rad, and 1 to 2% between 100 and 1,000 rad);

d. fading ($\sim$1 to 12% during 10 days, 15 to 18% during 100 days at room temperature); and

e. neutron energy dependence, and, perhaps, some directional dependence.

Depending on the type of diode, dose-level, readout conditions, individual calibration of the diodes, etc., the total error will vary between about 15 and 50% at 10 rad and 8 to 15% at 100 rad (Wall, 1972).

Thus, limited precision, restricted reusability, a dynamic range of only two orders of magnitude, poor sensitivity, and high costs (about $8 per unit) are distinct disadvantages of silicon diodes. On the positive side, there are a flat energy response in the 0.3 to 14 MeV range, negligible response to other types of radiation, small size, and easy, fast evaluation. These properties make silicon diodes of interest in military and accident personnel dosimetry, and for field mapping in neutron radiobiology or therapy.

Somewhat different approaches may, however, change this situation. In one interesting recent study (Rösch et al., 1972), it has been suggested to make use of the neutron-induced change in the optical response characteristics of n^+p (or p^+n) junction silicon photodiodes for dosimetry. As is demonstrated in Figure 6-7, the light response of such diodes is most strongly affected at long wavelengths. Under optimized readout conditions, the change amounts to 2.5%/rad for neutrons (with a reported detection limit of less than 1 rad), and less than 2 x $10^{-3}\%$/rad for gamma radiation. Fading is less than 10% between 2 and 50 hr after exposure, and the variations in predose amount to <10%.

6.2. Photographic Emulsions

Unlike in other fields of solid-state dosimetry, no dramatic new developments have taken place in photographic film dosimetry during the last few years. As there are several fairly recent compilations on the basic aspects of the photographic process and the preparation of photographic emulsions (Mees and James, 1966; Katz and Fogel, 1972; and Zelikman and Levi) as well as the photographic effects of ionizing radiation and film dosimetry (Ehrlich, 1961; Becker, 1966; Dudley, 1966; Herz, 1969; IAEA, 1970; and McLaughlin, 1970), the treatment of this complex subject can be restricted to a short description of the principles as far as they relate to dosimetric applications, and to a summary of some of the more important developments within the last few years.

6.2.1. Photographic Process

Few physicochemical processes have been as thoroughly investigated as the effect of photons on silver bromide crystals in gelatin. Large industrial, governmental, and university laboratories in many countries have worked for decades on the understanding of the photographic process and the improvement of the properties of photographic emulsions. More specifically, more than a thousand publications have been devoted to the effects of ionizing radiation on photographic emulsions (for a bibliography, see Becker, 1964a and b, 1967a and b, and 1971). Somewhat surprisingly, important characteristics which seriously limit the usefulness of film as a dosimeter are, however, still not widely known.

6.2.1.1. Preparation of Emulsions

The photographic film consists of silver halide microcrystals (grains) that are embedded in a special type of gelatin made from animal bones. The gelatin has been partially replaced by synthetic polymers in some recent experimental emulsions (Nikodemovna and Bradna, 1970). The grain diameter is not uniform, varying from less than 0.01 μm to more than 5 μm. It is possible to make the size distribution curve fairly narrow, with all grains having approximately the same size and shape (Figure 6-8A). This is attempted in the fine-grained, silver-rich emulsions for nuclear track recordings. More commonly, the sizes and shapes are rather nonuniform (Figure 6-8B), with the average grain diameter ranging from $\sim$0.02 μm in very slow, high resolution emulsions, to about 2 μm in high speed optical or x-ray emulsions.

Shape and photographic properties of the microcrystals are strongly affected by their chemical composition. The grains of sensitive emulsions as used in film dosimetry usually consist of silver bromide with a small addition of silver iodide, as well as trace amounts of other elements such as sulfur and gold, which are distributed either throughout the crystal or on its surface and play an important role in the formation of the latent image. Because of the importance of trace elements, composition and preparation of the gelatin, in particular its content of some organic sulfur compounds, also greatly influence the properties of the emulsion.

Gelatin is hygroscopic and contains varying amounts of water, depending mostly on relative humidity and the quality of the film packaging (as a high humidity content is undesirable because it accelerates fading, fogging, and bacterial growth, most manufacturers try to exclude it by carefully sealing fresh, dry films into plastic coated metal foil). The gelatin acts as a protective colloid for the silver halide; as a binding agent that can absorb large amounts of water, thus giving the processing chemicals access to the grains; as an acceptor for the halogen that is liberated in the photographic process; and as a sensitizer for the grains.

A photographic emulsion is a product of very complex physical and chemical operations which are still guarded by commercial secrecy to a large degree. It is generally not advisable to attempt the preparation of emulsions in a laboratory which is not specifically equipped for this purpose. The operation consists of several steps, beginning with the precipitation of the silver halide. This is done either by dissolving a stoichiometric excess of the alkali halide in a gelatin solution and adding silver salt (usually silver nitrate) under accurately controlled conditions, or by adding complex-forming NH_4OH to the silver nitrate solution, resulting in an ammoniacal emulsion. The precipitation rates and techniques vary, depending on the desired results. The precipitation, which takes place at a temperature of about 50 to 70°C, is followed by a period of storage for 30 to 60 min at this

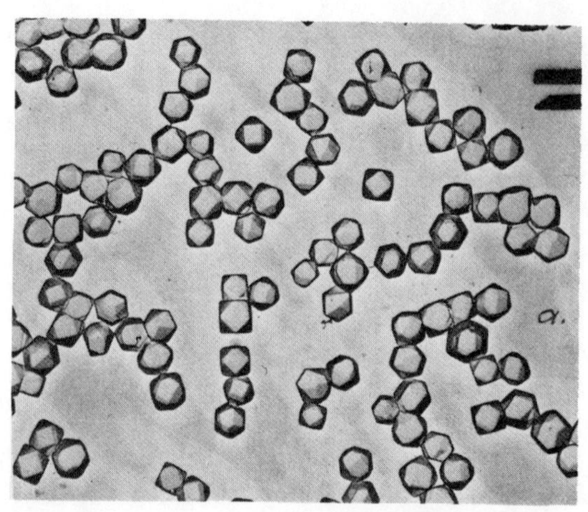

FIGURE 6-8.   Electron micrographs of preshadowed carbon replicas of (a) grains in a nuclear track emulsion and of (b) grains in a sensitive x-ray emulsion. (After Mees and James, 1966.)

temperature to permit the initially very small crystals to grow to the desired grain size (Oswald ripening).

During the next step, the jelly-like gelatin is diced into small cubes or shredded into "noodles" for extensive washing with water to remove excess soluble salts. Then the emulsion is melted, thus allowing the addition of more gelatin and various sensitizers such as gold thiocyanate, stannous chloride, lead or cadmium ions, and/or organic dyes. The emulsion is kept at about 50°C for one or two more hours during this "after ripening," which establishes the desired sensitivity with a minimum of fog (grains which are developable without being exposed to radiation).

During this period, other chemicals are also added to increase the flexibility (glycerol) or the mechanical strength, to prevent bacterial attack, reduce fogging and electrostatic charge build-up, and aid the coating. Finally, the emulsion is coated on a carrier (film base) which has been treated with surface-active reagents to establish adhesion. Some characteristics of several common dosimetry emulsions are summarized in Table 6.1.

The film base usually consists of a 100 to 200 μm layer of cellulose acetate, but at least one

TABLE 6-1

Characteristics of Some Dosimeter Films

	Kodak Personal Monitoring Film Type 3[®]		DuPont Type 519[®]*	DuPont Type 555[®]*	Ilford PM-1[®]
	Sensitive emulsion	Insensitive emulsion			
Response range (rad γ radiation)	0.02–5	10–1000	0.05–10	0.05–10	0.03–8
Grain diameter (μm)	1–3	0.1–0.5	1–2	0.5–1.5	1–2
Thickness of emulsions (μm)	30	–	20	10	–
Silver content (mg/cm^2)	9.0	0.6	1.6	2.2	2
Halogen content (mg/cm^2) Br	6.7	0.4	1.1	1.5	1.2
I	0.17	0.03	0.03	0.06	0.04
Cl	0.02	<0.01	0.02	<0.01	0.02

*Once very popular, these films are not in production any more. Several other commercial dosimeter (or dental) films made by Agfa-Gevaert, Fuji, Kodak-Pathe, etc. are not listed in this table. (After McLaughlin, 1970.)

commercially available dosimeter film (the French Kodak-Pathe Type 1[®]) is coated on cardboard. The usual x-ray emulsions used in film dosimetry consist of $\sim$50% by weight AgBr (containing a few percent AgJ) and $\sim$50% gelatin, which is coated on both sides of the carrier to result in dry layers of 5 to 20 μm thickness each (when saturated with water, it can be 5 times as thick), corresponding to $\sim$1 to 5 mg/cm^2 of AgBr, or 10^9 to 10^{12} grains/cm^2. In nuclear track emulsions, the AgBr/gelatin ratio may be as high as 5, and the thickness much larger. Finally, a supercoating of a protective thin (0.5 to 1 μm) gelatin layer is added. Uniform, absolutely dust-free coating is very important in the delicate production of good films.

In some films such as the Kodak Personal Monitoring Film Type 2[®], two emulsions of different sensitivity are coated on the two opposite sides of the carrier. The sensitive layer can be scraped off for evaluation of the less sensitive emulsion, when its optical density (O.D.) is too high for accurate densitometry. Thus, the rather limited dose-range of the film (two to three orders of magnitude) is extended. Other dosimeter film packs contain for the same purpose two or more different films (only in the Kodak-Pathe Type 1 film there are four different emulsions coated on the same carrier).

The processes that are involved in the production of a visible and measurable optical density as a result of exposure to visible or ionizing radiation can be described in two steps:

1. The physicochemical process leading to the production of a developable latent photographic effect and

2. The chemical reduction procedure that converts the radiation-exposed grain into metallic silver (development), removes the unexposed silver halides (fixing), thus stabilizing the processed film, and the consecutive quantification of the relationship between measured optical density and radiation exposure.

6.2.1.2. The Elementary Photographic Process

The minute changes that occur in a silver halide crystal during or immediately after exposure to low energy electrons, which are the basis of the effect of visible light, X-, beta, and gamma radiation as well as ionizing particles, have been the subject of intense studies for over 100 years. According to the generally accepted theories by Mott and Gurney that have been further developed by Mitchell (1957 and 1962), the developable latent "image" consists of extremely small silver particles which are distributed inside and/or over the surface of the silver bromide crystal.

They are produced by an ionic and an electronic process which occurs almost simultaneously when sufficient energy is transferred to the system. The most interesting types of energy transfer are, of course, light and ionizing radiation, but pressure, heat, certain chemicals, etc. also can render grains developable. The concurrent occurrence of the two processes has been experimen-

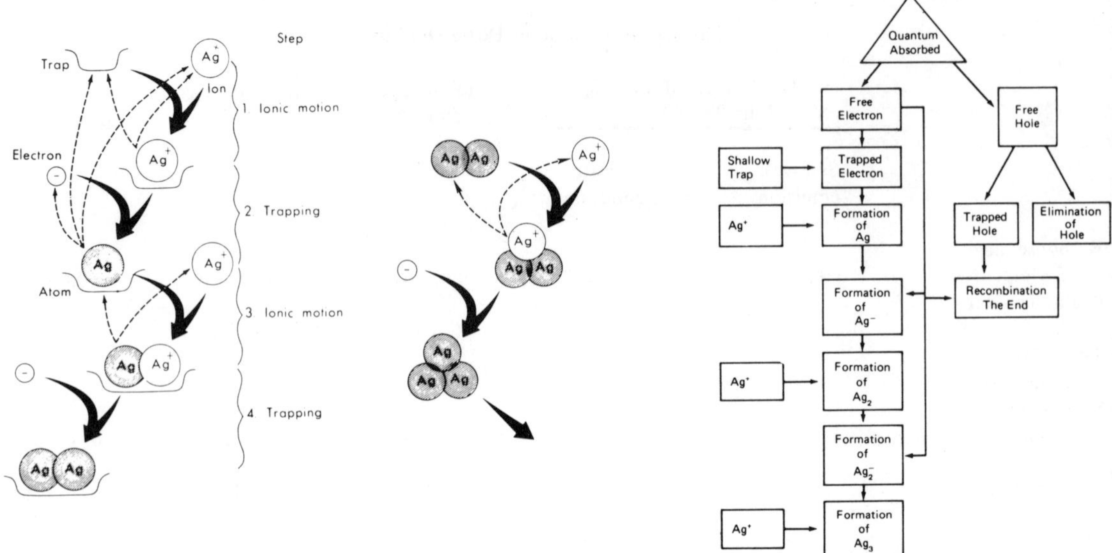

FIGURE 6-9.   Schematical diagram of nucleation stage of latent image formation in a silver halide grain (left), of the further growth of the silver aggregate leading to the stabilization of a developable silver center (middle), and of the same processes represented by a flow diagram (right), according to Mitchell's modifications of the Gurney-Mott theory. (After Mees and James, 1966 and Katz and Fogel, 1972.)

tally demonstrated by such techniques as applying strong electric fields to silver halide crystals before, during, or after exposure; by exposure at very low and elevated temperatures and fractionated heating; by measurements of the photoconductivity of large AgBr crystals as a function of temperature; and by electron microscopy of developed grains.

In the optically unsensitized AgBr lattice, quanta or electrons with an energy exceeding 2.5 eV (corresponding to light of a wavelength <480 nm) are capable of raising electrons from the valence into the conduction band. These electrons, released from bromine ions, move freely through the lattice and can either recombine with a hole (defect electron) or become captured in an electron trap, which may consist of a silver atom on a surface imperfection with an excess positive charge. Such traps in pure silver bromide crystals are, however, shallow, and the electrons become easily detrapped and recombine with holes. Much more important are the deep electron traps due to impurities such as gold or other heavy metal ions, structural disorders in the crystal, and Ag_2S complexes at the surface which are either created permanently, leading to sensitization, or temporarily by so-called hypersensitization. Storage of a

film in mercury vapor can, for example, temporarily double its sensitivity.

To prevent the recombination of the electron with the bromine ion, the hole also has to be trapped. This occurs through the action of halide acceptors both inside and outside the grain (in the gelatin). Coupled electron-hole pairs (excitons) probably also play some role in these processes. In any case, an ionic process that immediately follows the electronic process is required for the stabilization of the sublatent development centers, as indicated in Figure 6-9.

The common structural disorder in AgBr at around room temperature is of the Frenkel type, i.e., there are silver ions in interstitial positions leaving gaps in the cation sites without a corresponding disorder in the anion sites. As the interstitial silver ions are highly mobile, the primary process $Ag^+ + e \rightleftharpoons Ag$ continues to $Ag^+ + Ag \rightleftharpoons Ag_2^+$. This aggregate is stabilized by its positive charge, and capable of capturing another electron:

$$Ag_2^+ + e^- \rightleftharpoons Ag_2,$$
$$Ag_2^+ + Ag^+ \rightleftharpoons Ag_3^+, \text{ etc.} \tag{6.5}$$

The latent image nuclei that are growing in this manner remain thermodynamically unstable until they have reached a size for which the surface energy is small enough to stabilize them against thermal dissociation. Further growth can then occur at the expense of smaller, substable units. This critical size is probably around four silver atoms. Although one such silver speck is sufficient to catalyze the development process, there are usually several per developable grain. Highly sensitive grains may require at least four photons in the visible range, or about 10 eV, to become developable. On the average, however, this number is in the neighborhood of 20 photons, or 50 eV.

For high energy electrons, the minimum energy that is required is much higher, namely, about 1 keV. Obviously, much energy is "wasted" due to such processes as the highly disperse distribution of individual ionizations throughout the grain, ionization occurring in deeper shells of the silver and the bromide, frequent "overexposure" of grains with high LET electrons, and loss of absorbed energy from the grain and the emulsion through secondary and tertiary electrons and fluorescence x-rays. In contrast, with visible light, the photon absorption is more likely to occur close to the sensitivity centers at the grain surface.

Several effects can easily be understood on the basis of this model. As most emulsions including relatively insensitive fine-grained ones are prepared with maximum possible grain sensitivity, the minimum number of visible-light photons (or electrons of a given energy) required to render a grain developable is usually quite constant. Depending on the absorption mechanism for the type of radiation under consideration (mostly a surface mechanism for visible light, and a volume mechanism for high energy electrons released by gamma radiation) the probability for collecting the minimum required energy for developability and consequently the macroscopic sensitivity will increase approximately with the projected area (second power of diameter) or the volume (third power) of the grains.

It is also easy to explain the fact that photographic emulsions do not rigidly integrate exposures, in particular visible light (Schwarzschild effect, or reciprocity failure). At extremely high light intensities, the photoelectron supply in the crystal can be so large that the associated ionic processes, i.e., the migration of silver ions to the centers of sensitivity, cannot take place fast enough for the accommodation of all electrons. Electrons will be captured in shallow traps or recombine immediately; numerous unstable silver centers are formed deeply in the grain which are not accessible by the common fast-acting surface developers. These processes result in a sensitivity loss for short time (or high-intensity) exposures.

Fortunately, many modern high speed emulsions show little, if any, high-intensity reciprocity failure. Also, all effects of ionizing radiation may be considered ultrashort-time exposures because an electron traverses a grain in about 10^{-14} sec. At higher optical densities, however, there may be a substantial dose-rate dependence for x- and gamma rays, particularly with incomplete processing. This is reflected not only in different maximum densities, but also in the different dose-levels at which solarization (Figure 6-10) occurs, and in strange "memory" effects, with the measured O.D. depending on the sequence of X- and gamma radiation exposures (Ehrlich and McLaughlin, 1961).

Of greater practical importance, at least as far as exposures to visible light are concerned, is a rather pronounced low-intensity reciprocity failure in all emulsions: The unstable subcenters, which normally would become stable development centers, thermally dissociate when the average time interval between individual adsorption acts becomes comparable to, or exceeds the halflife of the subcenters at a given temperature. Furthermore, chemical oxidation of the subcenters occurs under the influence of moisture and oxygen; this effect (fading) will be discussed later. At very low temperatures, the low-intensity reciprocity failure disappears. Also, many types of relatively densely ionizing radiation interacting with sensitive emulsions will not show much of a low-intensity reciprocity failure. They "overexpose" the grain since each incident particle (or photon) leads to a sufficient concentration of stable centers.

However, when visible light is involved, for example, if fluorescent intensifying or converter screens are employed for the more sensitive detection of ionizing radiation, reciprocity failure may be large. As can be seen in Figure 6-11, the magnitude and direction of reciprocity failure are different for the surface and the interior of the grains. Processing in a normal commercial developer usually results in maximum sensitivity at an exposure time to light of about 1 sec. There are, in principle, two methods of reducing this effect, which can easily lead to errors amounting to several orders of magnitude:

1. Since the deeper, more effective electron

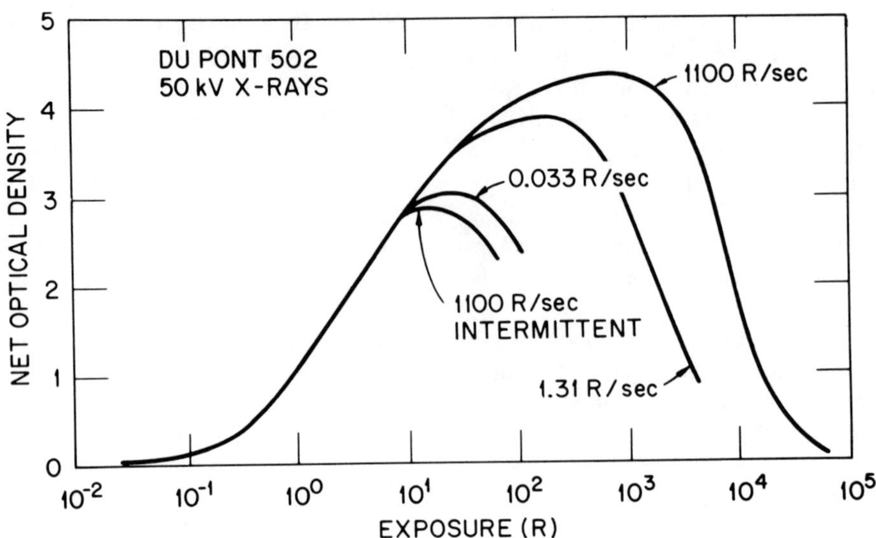

FIGURE 6-10. Characteristic curves of a DuPont 502 film for 50 kV X-radiation at different dose rates. (After Ehrlich, 1956.)

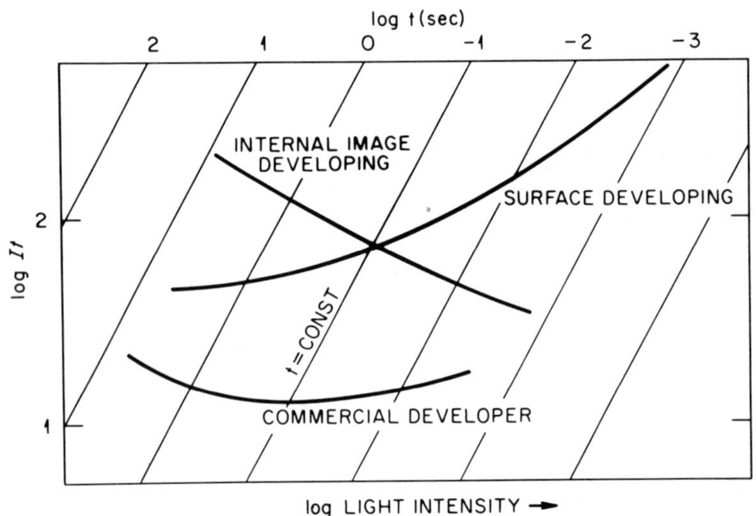

FIGURE 6-11. Reciprocity curves of a photographic emulsion for exposure to light and different types of development; t = exposure time and I = light intensity. (After Berg, 1954.)

traps are located at the grain surface, preferential surface development and suppression of the development of smaller centers by developer additives such as 2-mercapto-oxalozines reduce the effect, but unfortunately also the sensitivity.

2. Very fast, coarse-grained emulsions with maximum sensitization can be produced with such a high concentration of deep surface traps that surface development results in little long-time reciprocity failure (for details, see Becker, 1966).

It can also be easily explained on the basis of the latent image forming mechanism described above that photographic emulsions are less sensitive at low temperatures since ion mobility decreases with temperature. For example, just above $0°K$, only 20% of their sensitivity at room temperature has been observed. This effect should not be neglected even in the limited temperature range common in personnel dosimetry: An x-ray exposure that produces an O.D. of 1.5 in an x-ray film at 0°C was shown to produce an O.D. of 1.7 at 40°C (Eggert and Luft, 1931); the consequence may be a substantial error in dose estimates.

6.2.1.3. Processing

The first and most important step in the processing of photographic film is the conversion of the latent image into visible and easily measurable quantities of silver. This developing process is basically an electrochemical electrode process, the anode being the part of the developable silver speck that adjoins the solvent, and the cathode the part adjoining the AgBr lattice. One can also look at it as a catalytic process. The point of equilibrium in the equation

$$Ag^+ + e^- \text{ (developer)} \rightleftharpoons Ag + \text{oxidized developer} \qquad (6.6)$$

always is on the right-hand side, i.e., with sufficiently long development time (or in the absence of gelatin), even grains without developable centers will be reduced.

The silver specks, however, strongly accelerate the equilibrium process through the following mechanism: interstitial silver ions and silver ions from the solution, as well as developer ions or molecules, are absorbed at the developable center (chemisorption), with the developer acting as electron donor. The electrons discharge the adsorbed silver ions; the center will increase in size until no other silver ions are available or the developer is exhausted. This process amplifies the original amount of silver by a very large factor (10^8 to 10^{10} times).

It is immaterial for the result whether the silver ions originate from the solution or from the crystal. In the first case, complete removal of the silver halide from the exposed emulsion (fixing) takes place during processing. The same silver is later agglomerated on the centers from the solution. This is called *physical development*. When the silver supply originates from within the same crystal, the process is termed *chemical development*. This is the more common type in rapid acting x-ray developers.

However, it has been shown that chemical development is always accompanied by some physical development: the silver bromide is partially dissolved as a complex, and reprecipitated on silver specks from the solution, which may reach a high silver concentration in the immediate vicinity of the grain. The contribution of physical develop-ment to the total process is particularly pronounced in so-called fine-grain developers which contain large quantities of complex-formers such as sulfites, rhodanides, or *p*-phenylendiamine.

The physical shape of the deposited silver strongly depends on the type of development. In chemical development, the silver aggregates usually look like a network of threads, which may be so entangled that the original grain shape is almost retained. Microscopic inspection frequently shows a larger diameter for the developed than for the undeveloped grains. With increasing influence of physical development, the silver threads become shorter and thicker, attaining an almost spherical shape in the case of purely physical development.

The redox potential of the developer should be at least 100 mV below the silver potential. In the case of organic developers, the pH of the solution strongly influences their oxidation capacity. The activity of most organic developers increases with increasing pH, concentration, temperature, and agitation. A usual commercial x-ray developer of the type that is widely used for dosimeter films consists of developing agents, accelerator and buffer, silver solvent and preservatives, and other additives.

The "classic" ortho- or para-substituted benzene derivatives, particularly substituted hydroxy-benzene and amino-benzene, are still widely used as developing agents. The best known are Metol® (methyl *p*-amino phenole sulfate), which produces a high emulsion speed but low contrast, and hydrochinone (*p*-dihydroxyben-zene), which is known for low speed, but high contrast. If used in proper combination, these two compounds become more active than each consti-tuent. This phenomenon of superadditivity is also observed with other pairs of compounds. For example, Metol has been replaced by the more potent Phenidone® (1-phenyl 3-pyrazolidone) in many of the modern concentrated liquid devel-opers. The oxidation of these substances leads through semiquinoid stages to rather complicated compounds, which react with other constituents of the developer. As the oxidation products usually have a dark coloration, deterioration of a developer is indicated by a shift in color from yellowish to dark brown.

Accelerator and buffer, consisting of alkali such as potassium and/or sodium hydroxide or carbon-ate to obtain the optimum pH, and borates and phosphates for stabilizing this pH are also impor-

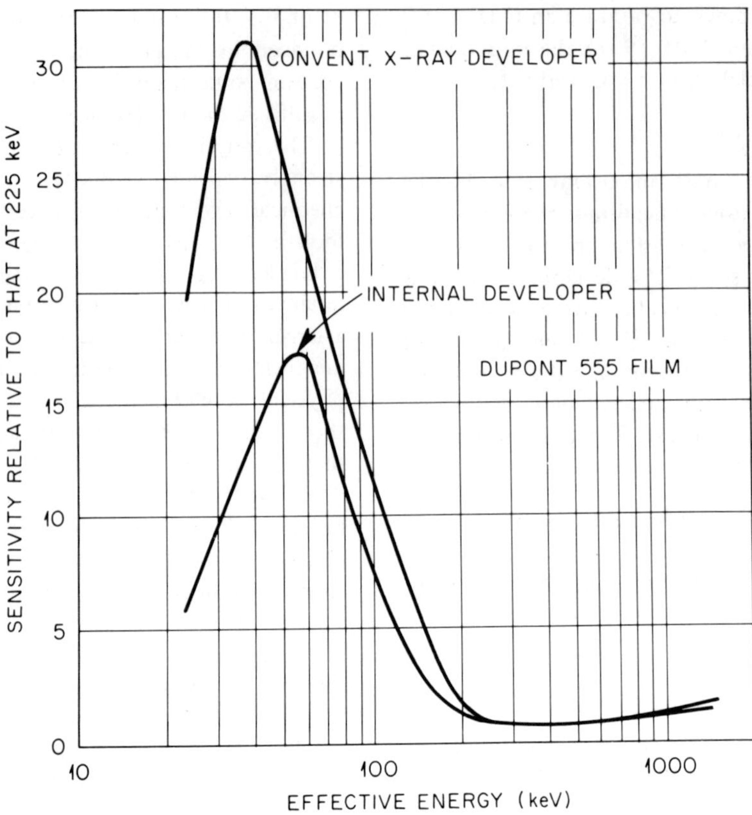

FIGURE 6-12.   Energy dependence of the response of DuPont 555® Dosimeter film if developed in a conventional x-ray developer and developed in an internal latent image developer. (After McLaughlin, 1970.)

tant constituents of a developer. Chemicals such as potassium or sodium sulfite are used in rather large quantities in x-ray developers. They act as a solvent for the silver bromide, laying bare the silver centers in the interior of the grain for partially physical development. Sulfite also protects the organic developers from rapid air oxidation by forming sulfonic acids. It is sometimes assisted in this task by the addition of other antioxidants such as ascorbic acid.

Other additives that can frequently be found in commercial developers include restrainers or anti-foggants to reduce development of unexposed grains (potassium or sodium bromide, and benzotriazole are such compounds) or complex-forming agents such as ethylenediamine tetra-acetic acid disodium salt, to prevent precipitation of salts in hard tap water used in diluting the developer.

Numerous unconventional developers and developing techniques have been suggested for special purposes. It is, for example, possible to modify the composition of the developer, and developing conditions, in such a way that either primarily the surface centers, or only the internal centers of the grain are developed. One can also "peel" the grain by gradually dissolving surface layers. Many dosimetric characteristics such as sensitivity, maximum optical density, reciprocity failure, and even the energy dependence of response (Figure 6-12) are strongly affected by the type of developing.

Very thick nuclear track emulsions may be developed by soaking them in the developer at very low temperatures, warming to the proper temperature for the time of actual development, and washing the chemicals out again at low temperature. Another special technique that has found some interest is the use of monobath developers, which provide developing and fixing with the same solution. They are usually based on Phenidone® and thiosulfate (Becker, 1962, and Ehrlich, 1965). This rapid and simple processing

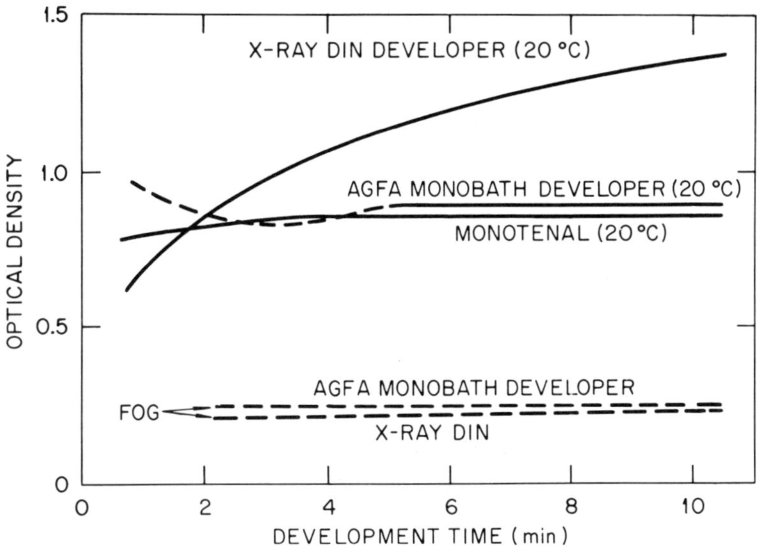

FIGURE 6-13. Dependence of fog and radiation-induced optical density in a sensitive x-ray emulsion as a function of development time, for a conventional x-ray developer and for two types of monobath developers. (After Becker, 1962.)

technique reduces the sensitivity somewhat (Figure 6-13), which is probably the reason why it never found widespread practical application.

In a conventional x-ray emulsion and developer at 20 to 25°C, development is completed after about 3 to 10 min. Completion is indicated by a plateau with a small slope in the O.D. vs. development time curve of an exposed film.

In practical film dosimetry, care should be taken that

1. the developer is not excessively exposed to air;

2. temperature and time of the development are kept fairly constant; and

3. the developer and/or films are sufficiently agitated to prevent local exhaustion of the developer, temperature gradients, and other irregularities. This can be done by manually moving the film racks horizontally and vertically once or twice every minute, or (better) with a commercial system which provides bursts of nitrogen gas bubbles rising through the tank about every 10 sec.

The developer can be replenished by adding appropriate amounts of special replenisher solution, of a composition similar, but not identical to, the original developer. It usually contains a larger concentration of developing agents and no re-

strainer. Processing is followed by rinsing carefully for several minutes in flowing water (or, better, in diluted acetic acid) to neutralize and remove the residual developer. The fixing solution, which removes undeveloped AgBr from the emulsion, consists mostly of sodium thiosulfate and some sodium metabisulfite to maintain the required low pH.

The unreduced silver halides become soluble by the formation of complex salts such as $Na[Ag(S_2O_3)_2]$, $Na_3[Ag(S_2O_3)_2]$, $Na[Ag_3(S_2O_3)_2]$, and $Na_5[Ag_3(S_2O_3)_4]$. These salts and the residual fixer have to be removed completely during the final rinsing to obtain a film whose optical density is stable for at least several years (the residual sodium thiosulfate content should be less than ~ 0.003 mg/cm^2). Finally, the film may be rinsed briefly in distilled water or a surface-active compound to prevent the formation of "water spots" on the dried film. It should be dried in dust-free air below $\sim 60°C$ to prevent melting of the wet emulsion.

6.2.1.4. The Optical Density

It is, of course, possible to determine the silver content of the developed emulsion directly, for example, by chemical analysis, neutron activation, or x-ray fluorescence. It is also possible to count the individual silver grains in the microscope. In practice, however, such techniques are only

239

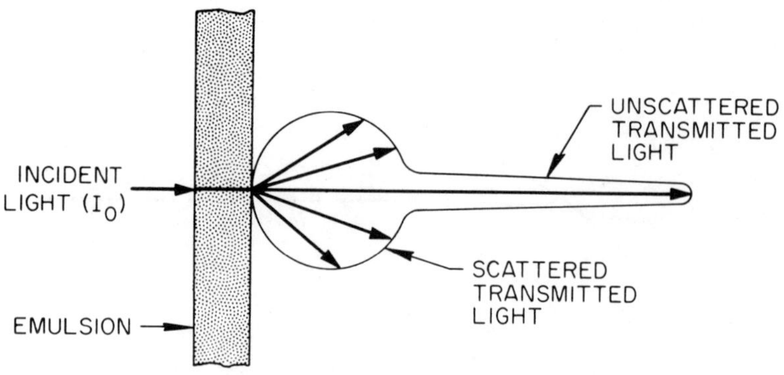

FIGURE 6-14.   Separation of the parallel beam of incident light into a diffuse scattered part, and an unscattered transmitted light fraction. (After Becker, 1966.)

sporadically used, mostly for optical densities that cannot be measured easily by photometric means. For optical densitometry, one could determine either the *opacity* I_0/I, or its reciprocal value, the *transparency* I/I_0, where I_0 is the incident and I the transmitted light intensity. More commonly, the optical density is used, which is defined as the logarithm of the opacity, O.D. = log I_0/I. This value is almost proportional to the number of developed grains per unit area of film; it permits direct addition of the densities if two or more films are superimposed.

Optical density usually is measured with a photoelectric densitometer which is adapted to the requirements of such measurements. This is important because Beer's law is not valid for photographic emulsions: When a parallel beam of light falls on a darkened film (Figure 6-14), part of the transmitted light will be diffusely scattered by the emulsion, while another part passes undisturbed. The value of the fraction of diffusely scattered to unscattered light depends on shape, size, and distribution of the grains, optical density, emulsion thickness, and other factors such as water content. The optical density of a wet emulsion can be quite different from that of the same emulsion when dry.

The quotient of the optical densities obtained in measuring geometries excluding diffuse scatter and those including such scatter is called the Callier coefficient. It usually is small for very fine-grained films, but can attain 1.7 in fast, coarse-grained emulsions. As the medium diameter of the developed grains depends on the photon energy, the Callier coefficient in the same film may vary substantially with radiation energy (Herz, 1949). The eye observes only the unscattered part. Thus, it is quite possible that the optical densities of two films which exhibit the same visual density turn out to be quite different when measured with a densitometer, which also "sees" the scattered light and vice versa.

There are three basic methods for optical density measurements: One can measure the "specular," nondiffuse density; the diffuse density (parallel incident beam, measurement of all including the scattered transmitted light); and the double-diffuse density (incident light also diffuse). The two latter techniques are the more common ones in densitometry, but as the optical conditions in two different brands of densitometers are rarely identical (for example, the scattering indicatrix of light-diffusing opal glass may be more or less circular) one should be aware of the limitations in comparing "absolute" optical densities.

The upper range is usually not limited by the light detector in the densitometer (some instruments with photomultipliers could easily read an O.D. of 6), but by the reliability of such measurements. Even a microscopic light leak, for example, a bubble in the emulsion, easily falsifies such measurements. The meaningful limits of transmission densitometry are usually around an O.D. of 3 to 4. One system that is used by the French C.E.A. employs reflection densitometry with an even more limited range.

The area of measurement in densitometry should normally be smaller than the metal filter area in the film badge, but not smaller than a few millimeters in diameter because this would introduce larger statistical fluctuations. Usually two to four density measurements are taken at different spots of the film behind each filter; the average of those readings is used for the subsequent dose estimation. Nowadays, most larger film badge services use semiautomatic or automatic densitometers (for typical examples from different countries, see Madden, 1969; Cottignies, 1968; Rae, 1968; Belletti and Zaglio, 1968; and Buschlen, 1972).

The knowledge of the relationship between silver content, grain size, grain density, and optical density can be of some interest. Various formulas have been suggested, representing more or less fitting approximations for chemical and/or physical development. A particularly simple equation, for the common case that the thickness of the developed grain equals about one third of its largest diameter after chemical development, is

$$O.D. = 0.4\ A_e/d_e, \tag{6.7}$$

where A_e is the silver content in g/cm^2, and d_e the mean diameter of the developed grains in cm.

Another interesting relation is that between optical density and absorbed energy in the emulsion. The absorbed energy may be a large fraction of the incident energy, as is the case of visible light, low energy beta radiation or short-range ionizing particles, or a very small fraction, as is the case of penetrating neutron or high energy photon radiation. The probability of the absorption of a photon decreases with increasing photon energy, yet, most absorbed x- or gamma ray photons make more than one grain developable. It has been shown (Bromley and Herz, 1950) that about 15 grains become developable in a fast x-ray emulsion for each 250 keV photon which is absorbed, but almost 100 for each ^{60}Co photon. This is largely due to the fact that absorption of a high energy photon produces numerous high energy electrons, which in turn produce secondary and tertiary electrons over a considerable distance in the emulsion.

Equations have been suggested to describe quantitatively the relationship between film density and quantum or electron fluence. Unfortunately, many of the simplifying assumptions that have to be made to arrive at a simple equation (single-hit processes to render a grain developable, uniform projection area of all grains without change during development, no absorption or scattering in the emulsion, and perpendicular radiation incidence) need modifica-tion under the practical conditions. For example, multihit processes and a spread in grain-size and sensitivity both reduce the contrast. For more realistic treatments of these complex relationships, see Tellez-Plasencia (1958) and Frieser and Klein (1960).

To arrive at a meaningful measure of the sensitivity of an emulsion to a particular type of radiation, several parameters have to be standardized. Obviously, the developing and densitometry should be kept constant and the background (fog) density has to be subtracted from the measured O.D. Furthermore, the fact has to be taken into account that the maximum O.D. that results from the total conversion of the emulsions AgBr into silver is different in different emulsions. The maximum optical density of x-ray emulsions after complete development may be as low as 3 or 4, or as high as 10 or 12, depending on the silver content. For a typical response of a widely used dosimeter film, see Figure 6-15. Usually the O.D. is a linear function of dose up to about 20% of the maximum O.D.

A "standard optical density" as a measure of an emulsion's sensitivity could be defined as the net measured density at a suitable fixed level less than 20% of the net maximum optical density, divided by this value (Becker, 1961). Because of the difficulties in measuring the maximum O.D. in many films, however, sensitivities are commonly expressed by comparing the exposure of gamma and/or X-radiation which is required to yield a certain O.D. above fog. It has to be stressed that the values thus obtained may vary substantially from batch to batch for the same film type, and depend on the exact conditions of measurement (for a more detailed description of suggested calibration conditions, see Barber, 1966).

6.2.2. Properties of Photographic Films

Obviously, some properties of the photographic film that are of great importance for its use in "normal" optical photography (for instance, its spectral response) are rather unimportant for film dosimetry. On the other hand, factors such as storage stability, which are not of great concern to the optical photographer, demand careful attention if the film is to be used as a long-term integrating dosimeter.

6.2.2.1. Storage Stability

One of the crucial disadvantages of the photo-

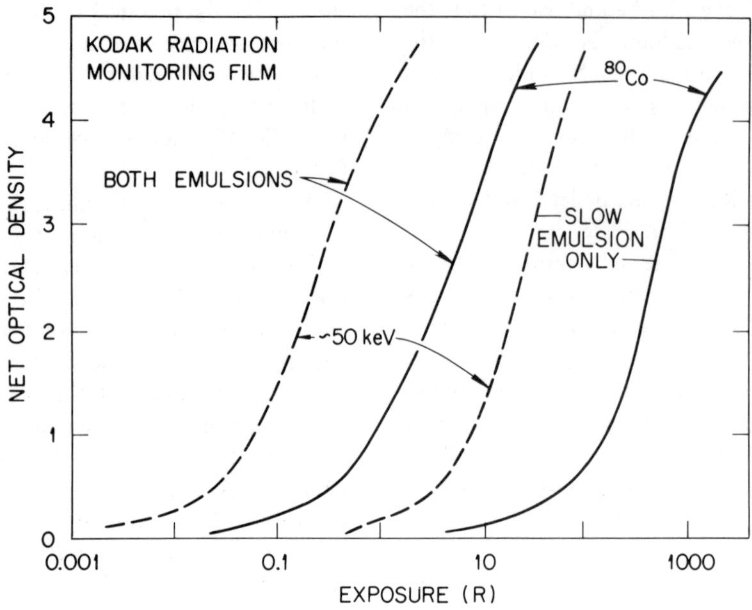

FIGURE 6-15. Response of a British dosimeter film with two emulsions of different sensitivity to X- and gamma radiation. (After Herz, 1969.)

graphic film, severely limiting its usefulness in many areas of application but especially in personnel dosimetry, is its limited storage stability. Extended storage of the fresh films, in particular in the humid and warm environment that is fairly common in large parts of the world, leads to

a. fogging (increase of the background optical density, O.D.) partly due to the natural radiation background, · but mostly caused by thermally induced formation of developable silver centers, or thermally accelerated chemical fogging;

b. irreversible changes in sensitivity and other response characteristics (Ehrlich, 1961);

c. "sticking" of the emulsion layer to the film wrapping, with partial or total destruction of the emulsion occurring when the pack is opened in the darkroom (Figure 6-16); and

d. microbiological growth in the gelatin, an excellent culture medium for various common bacteria and fungi, which may also result in total destruction of the sensitive layer.

It is a fairly common observation that dosimeter films that had been left near hot radiators in the glove compartments of cars parked in the sun in summer, etc., exhibit a high O.D. which can be mistaken for a gamma radiation exposure. The mechanism of thermal fogging and the role played by humidity are not yet completely understood, but it appears that little thermal fogging occurs at temperatures below 45 to 50°C for shorter storage periods, while above this level it rapidly accelerates (Figure 6-17A) and maximum thermal fogging may be expected at relative humidities around 50%, probably because latent-image regression (fading) compensates for the spurious latent-image formation at higher humidities (Kathren, 1966).

Much more serious than these effects, which are usually fairly easy to detect, is the fading that occurs after exposure, partly because of thermal dissociation of the latent development centers which consist of aggregates of $\sim$4 to 10 silver atoms in the silver halide crystal, but mostly (accounting for about 90% of the total fading) because of the combined chemical action of oxygen and humidity on these latent-image specks. One can visualize, in a somewhat oversimplified way, the process as a silver-catalyzed reaction $2 H_2O + O_2 \rightarrow 2 H_2O_2$, in which the resulting H_2O_2 destroys the catalyst, namely, the silver development centers, which are usually located at the microcrystal (grain) surface even if the original ionization took place in the grain interior. Fine-grained emulsions with their larger surface area frequently exhibit more fading than coarse-grained ones.

Being a chemical process, fading is strongly

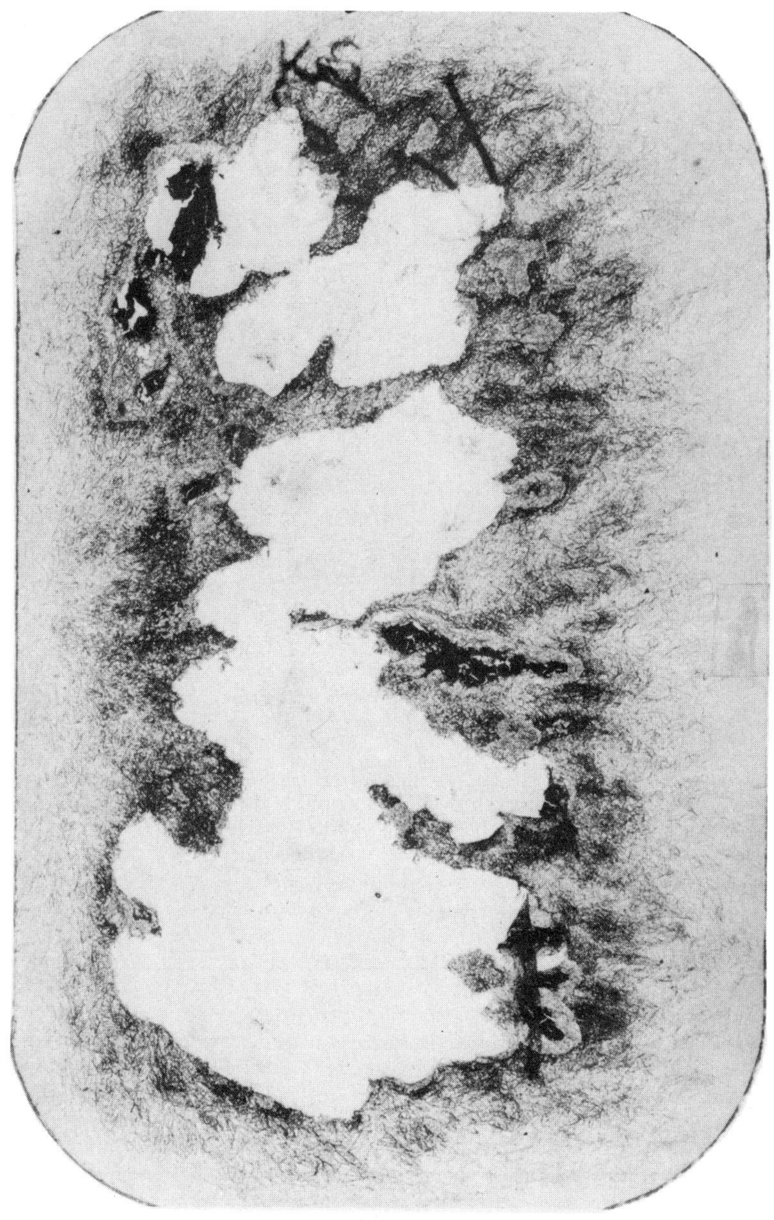

FIGURE 6-16. Developed Kodak Personal Monitoring Film Type A, heavily damaged by bacteria and fungi during several weeks of storage at 30°C and 95% relative humidity.

accelerated by increases in the storage temperature. The kinetics is not simple, but the fading of the radiation-induced O.D. can usually be approximated as a linear function of the logarithm of time. The superimposed effects of fading and fogging may lead to an "equilibrium" O.D. in dosimeter films which is independent of their radiation exposure. This effect, as observed in the insensitive emulsion of a Kodak Type 2[®] film pack during storage at 30°C and 95% relative humidity (the less stable sensitive emulsion was totally destroyed after this period), is illustrated in Figure 6-17B.

Fading has been known since 1910, and many publications have been devoted to this subject, of which only a few can be quoted here. Several reports indicate that only relatively little fading takes place under "normal" conditions in coarse-

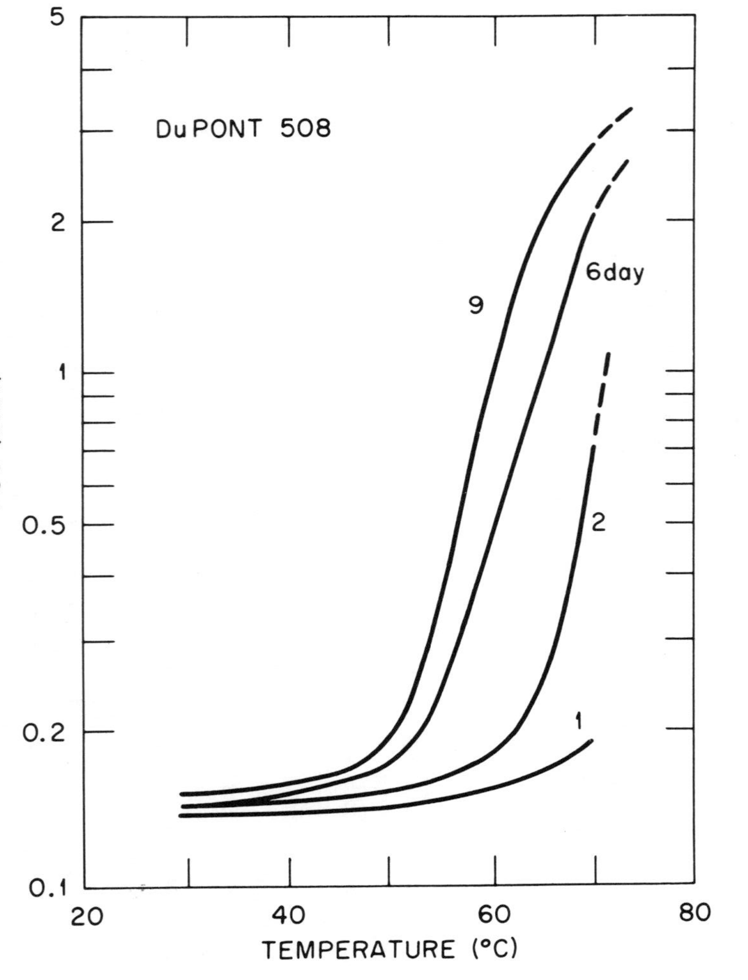

FIGURE 6-17A. Thermal fogging of an unirradiated dosimeter film during storage for various times at different elevated temperatures. (After Kathren, 1966.)

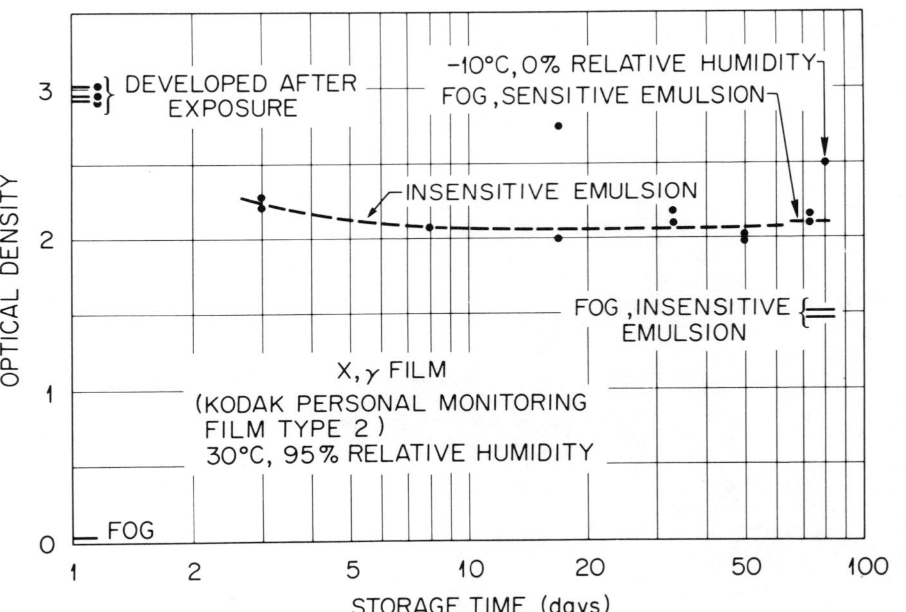

FIGURE 6-17B. Fading and fogging in the insensitive emulsion of a Kodak PM Type 2® film during three months of storage in a simulated tropical climate (30°C, 95% relative humidity). (After K. Becker, unpublished.)

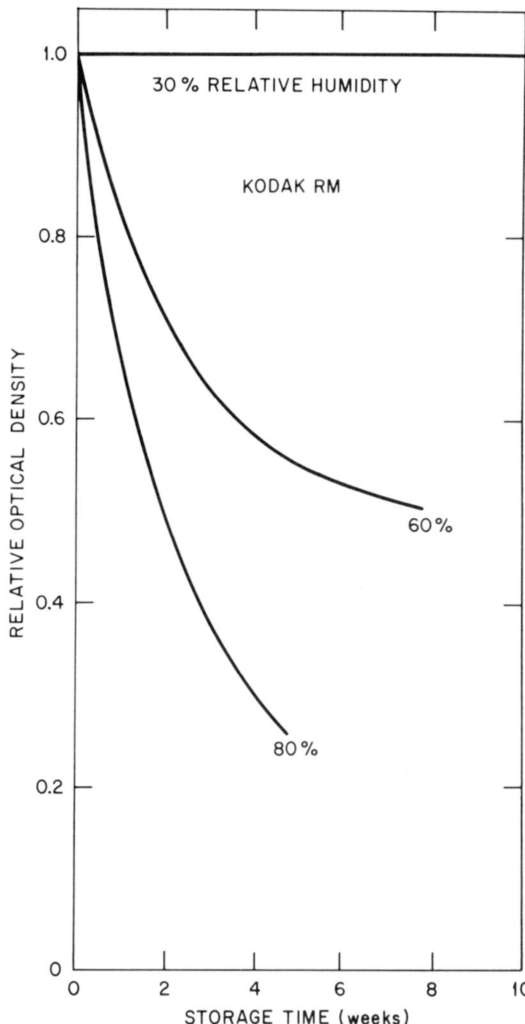

FIGURE 6-18A. Latent image fading in a Kodak RM® dosimeter film at 30°C and different relative humidities. (After Heard et al., 1964.)

grained, highly sensitive x-ray films which are commonly used as sensitive x and γ-ray detectors in personnel dosimetry. In most of these reports, the storage climate is not well specified, but it may be assumed that most of the tests have been carried out in a fairly dry, comfortable laboratory climate. At high relative humidities, however, substantial fading has also been observed in such films in laboratory as well as in field tests, for example, in the Kodak RM® film (Figure 6-18A) and in the Kodak Personnel Monitoring Type 2® film (Figure 6-18B). In another study, during 10 days at 25°C or during 6 days at 35°C (Figure 6-18C) 100% fading in moist air has been found. In the latter study, the apparent gamma dose

reading actually became negative after about one week because the background fog was also reduced by fading.

Besides its dependence on humidity and temperature, the fading rate also depends on the dose-level and on photon energy. In one experiment involving a slow x-ray film, for example, no fading was observed during one month after a low energy x-ray exposure, but 30% fading occurred after ^{60}Co gamma radiation exposure (Herz, 1969). Most published data on dosimeter film stability, in particular older ones, are not very reliable, partly because of differences in the detailed experimental conditions, but also because emulsions and film packaging are frequently modified by the manufacturers (with these changes not necessarily resulting in improved performance).

Fading is much more pronounced in the fine-grained nuclear track emulsions which are still widely used in fast-neutron personnel dosimetry. The effect is well documented, but as the experimental conditions have been often ill-defined, or are for other reasons (such as the use of different types of emulsions) not comparable, there has been some disagreement in the reported fading rates and the effect of protective measures (see, for example, Watson, 1951; Cheka, 1954; Hart and Hale, 1956; Cook, 1958; Amadesi et al., 1960; Lehman, 1961; Becker, 1963; Portal, 1963; Cusimano, 1963; Turner, 1966; Cavallini and Busuoli, 1967; Meyer, 1968; Schimmerling and Sass, 1968; Zelac, 1968; Kahle et al., 1969; Wachsmann, 1969; Arnett, 1969; Nishi et al., 1970; Distenfeld and Klemish, 1972; Becker, 1973; Jasiak and Musialowicz, 1973; and Krishnamoorthy et al., 1973).

One will also obtain quite different results for the fading of the O.D. on the one hand, and the disappearance of visible tracks on the other hand, because a long, dense track may still be visible after some of its grains have disappeared. The visibility of tracks also strongly depends on their length (e.g., on the effective neutron energy), and the fog density (neutron to gamma dose ratio).

Furthermore, the films may not have been equilibrated with their environment at the beginning of a fading test: As can be seen in Figure 6-19, the visible track density in a film in equilibrium with an average laboratory climate (20°C, 75% relative humidity) decreases in 10 days by a factor of 2. Equilibrium is the more realistic

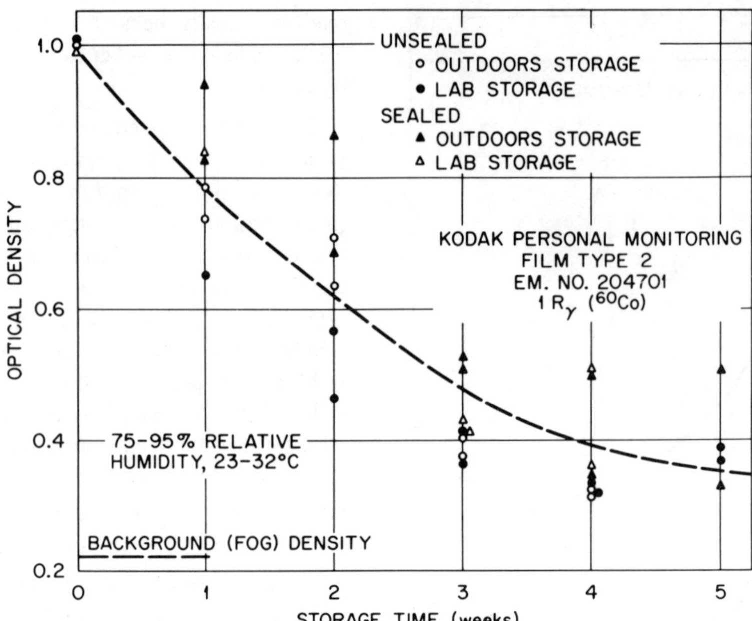

FIGURE 6-18B.   Optical density of a gamma-irradiated Kodak Personal Monitoring Film Type 2[®] (sensitive emulsion) as a function of storage time between exposure and (simultaneous) processing of the films, storage in standard laboratory atmosphere, and in a protected space outdoors with or without additional sealing in a thin polyethylene bag in a subtropical climate. (After Becker, 1972.)

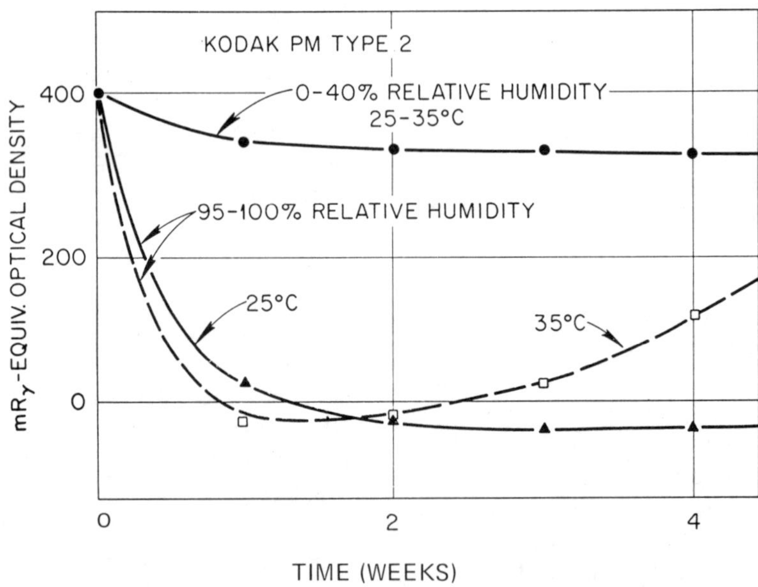

FIGURE 6-18C.   Fading and fogging of the optical density (expressed in mR gamma radiation equivalent) in the Kodak Type 2[®] film during storage at 25 and 35°C and at low and high relative humidity. (After Christensen et al., 1973.)

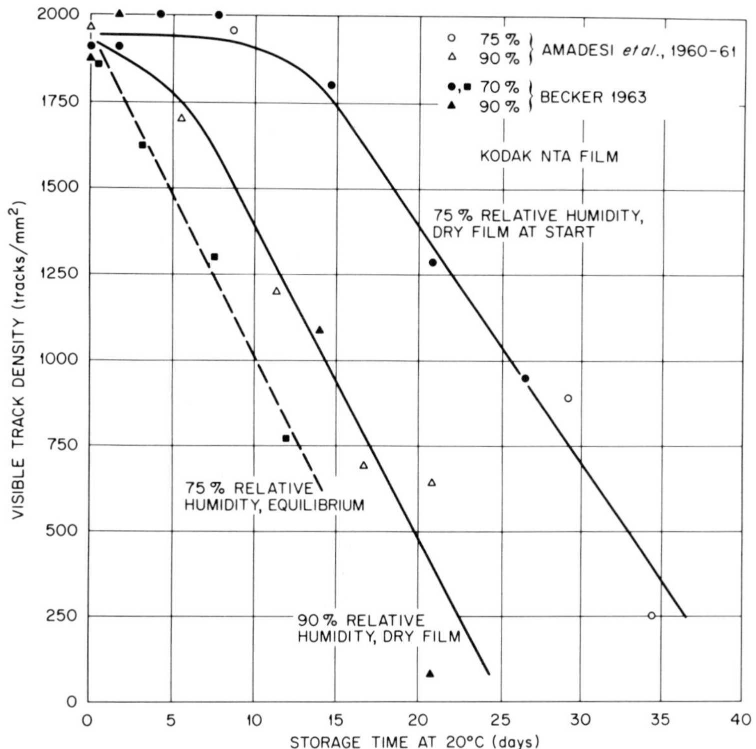

FIGURE 6-19. Visible recoil proton track density in Kodak NTA® film as a function of storage time at 20°C and 75% and 90% relative humidity, if the film (without additional wrapping) is irradiated after careful drying, and at 75% relative humidity, if permitted to equilibrate prior to neutron irradiation. (After Becker, 1973.)

condition because a personnel dosimeter has usually been worn for some time prior to the neutron exposure. But if one starts with a dry film, it may require a few days before the humidity fully penetrates the wrapping. In Figure 6-20, the track fading of equilibrated films is given for a constant temperature (22°C) for different relative humidities. These data are in good agreement with other recent measurements (for example, Nishi et al., 1970).

The accelerating effect of increased temperature on fading is illustrated in Figure 6-21. As can be seen, at 35°C (which is not an uncommon temperature in subtropical and tropical countries), 50% track fading occurs in one week even at the very low humidity of 38%, instead of the ∿82% relative humidity necessary to induce this fading rate at 22°C. Field tests confirm these laboratory results: In a subtropical climate, 50% fading was observed in equilibrated track films without additional wrapping within ∿1.5 days, and ∿90% of

the radiation effect had disappeared after 1 week (Figure 6-22).

It is, in principle, possible to reduce fading somewhat by modifications of the emulsion or suppression of surface development centers (the internal image is more stable). More effective has been the exclusion of oxygen, or, even better, of humidity. Unfortunately, gelatin is a rather hygroscopic substance which absorbs large quantities of water (at 75% humidity, 20% by weight — Figure 6-23A).

Many investigators and several commercial dosimeter film suppliers (Figure 6-23B) have tried to exclude humidity by sealing dry films into polymer foils. The permeability of polymers to water vapor varies widely, and strongly depends on various factors such as their density, crystallinity, orientation and crosslinking of the molecules, and additives. As can be seen in Table 6-2, ethyl cellulose is more permeable than certain types of Teflon by a factor of more than 10^4. Many

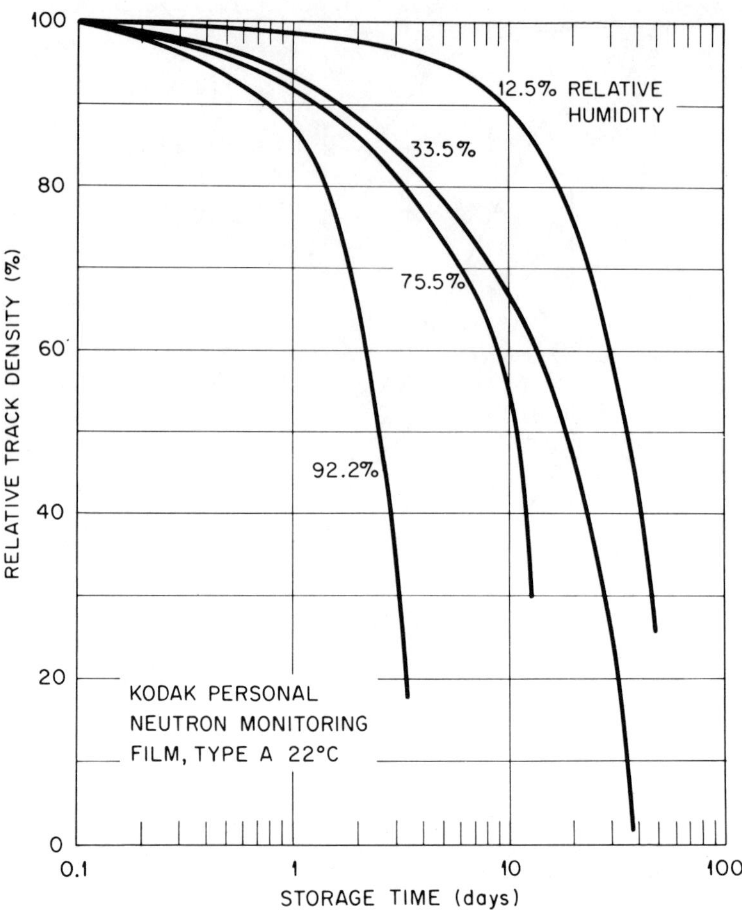

FIGURE 6-20. Relative visible recoil proton track density in Kodak NTA film exposed to Po/Be neutrons, as a function of storage time at different humidities. (After Becker, 1963.)

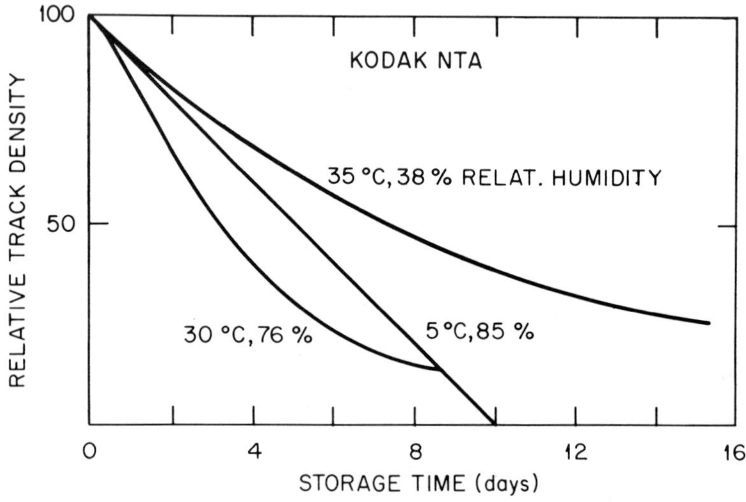

FIGURE 6-21. Relative track density in Kodak NTA film for storage at different temperatures and humidities. Krishnamoorthy et al., 1973.)

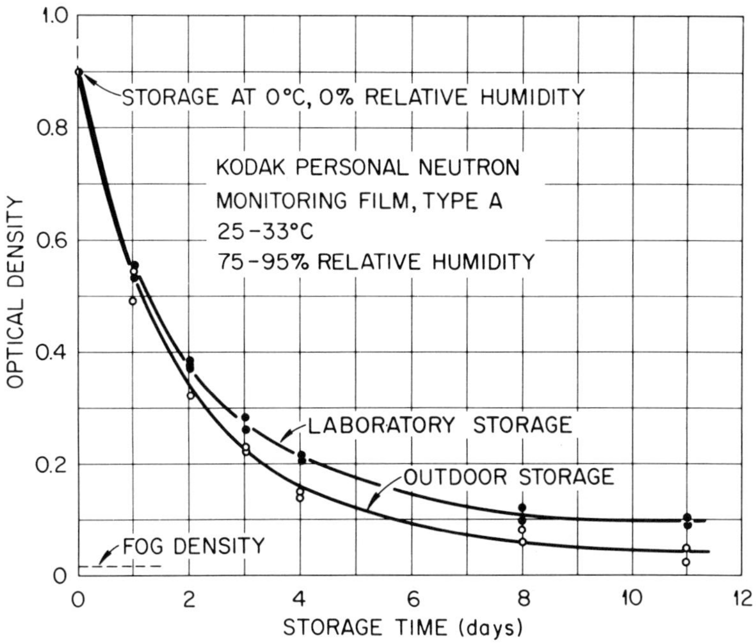

FIGURE 6-22.   Fading of latent image in a Kodak NTA film in a semitropical climate. (After Becker, 1973.)

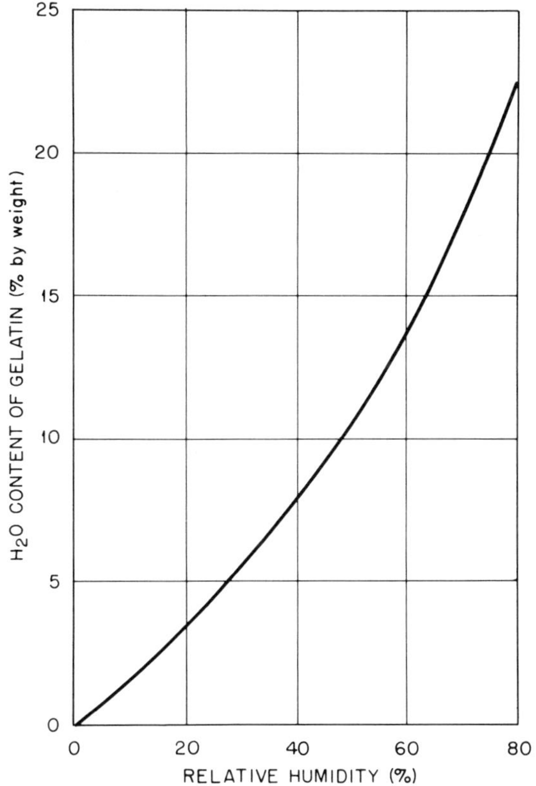

FIGURE 6-23A.   Water content of the gelatin in a nuclear track emulsion at 20°C as a function of relative air humidity. (After Powell et al., 1959.)

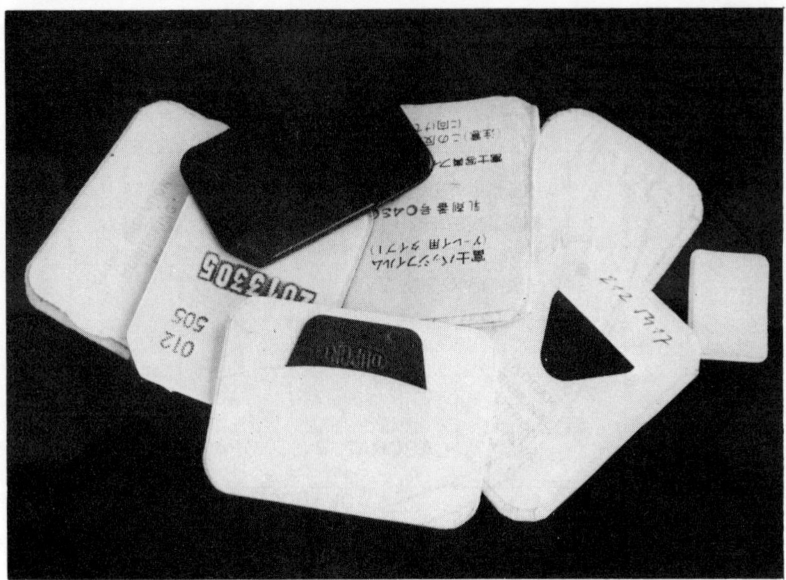

FIGURE 6-23B. Various U.S., German, and Japanese dosimeter films with different types of wrappings, including (left) a small plastic-sealed film for finger-ring dosimeters; the production of some of these films has been discontinued.

TABLE 6-2

Permeability of Some Polymers to Water Vapor at 25°C

Polymer	$P \times 10^{10}$*
Ethyl cellulose	12,000
Cellulose nitrate	6,300
Cellulose acetate	5,500
Polycarbonate	1,400
Polystyrene	1,200
Polyvinyl chloride	156
Polyethylene terephthalate	130
Polyethylene, low density	90
Polypropylene	51
Polyethylene, high density	12
Polyvinylidene chloride	0.5
Polytrifluorochloro-ethylene, amorphous	0.29

*Permeability coefficient, defined as the product of the amount of permeant and the film thickness, divided by the product of area, time, and pressure-drop across the film.
(After Yasuda, 1966.)

polymers are obviously rather "transparent" to water vapor, but even the most "impermeable" foil has a measurable permeability; the rate of water vapor transmission depends on temperature and on the vapor pressure gradient across the foil. The driving force is largest with a completely dry film inside and a high humidity outside at an increased ambient temperature (which are the conditions to be found in a tropical climate). Indeed, field experiments (Figure 6-18) show little difference in the fading of Kodak Type 2[®] films with and without sealing in thin "protective" polyethylene bags. Sufficiently thick plastic may, of course, retard the humidity more efficiently. In Figure 6-24A, the fading rate in an equilibrated Kodak NTA[®] film in the factory-provided wrapping is compared to that of such a film in a special thick polyethylene holder, with and without previous drying.

The relatively most efficient vapor barrier is a plastic-aluminum-paper compound ("pipe tobacco pouch paper"), in which the metal foil acts as the actual barrier, and the polymer is used only for heat-sealing with a wide ($\sim$5 mm) rim. The Kodak NTA[®] film should only be used for very short ($\sim$1 week) monitoring periods in a very dry and cool climate unless protected by such a sealing, which is commonly used by the photographic industry to protect fresh film packs. Unfortunately, only one photographic dosimeter is available with this type of packaging for the individual film. This is the Kodak-Pathé Type 1[®], which is not a nuclear track film, but a paper base on which different emulsions for X- and gamma radiation dosimetry (to be evaluated by reflection densitometry) are coated.

It is widely assumed that in the fresh dosimeter packs, as received from Kodak, the films have been desiccated prior to this packaging and sealing. In

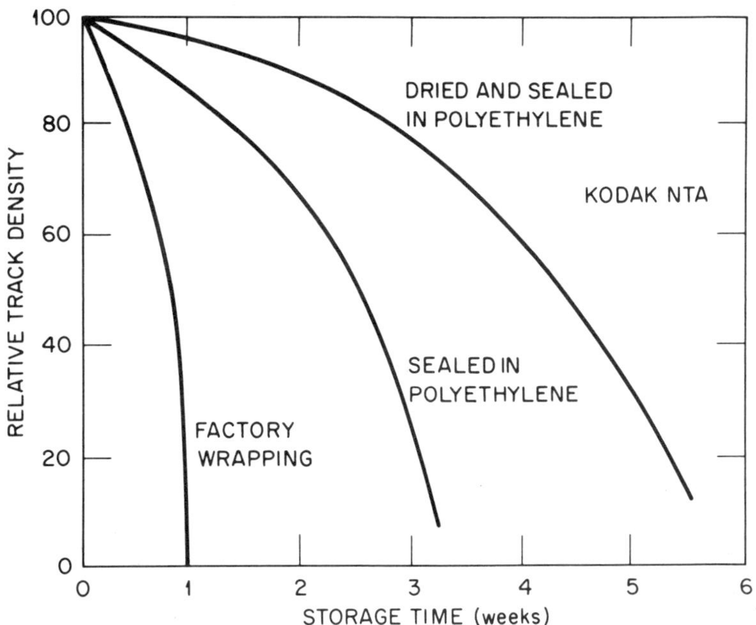

FIGURE 6-24A. Track fading in Kodak NTA film at ambient temperature and 100% relative humidity in normal wrapping, placed in a protective thick polyethylene holder, and with both film and holder carefully dried before sealing. (After Jasiak and Musialowicz, 1973.)

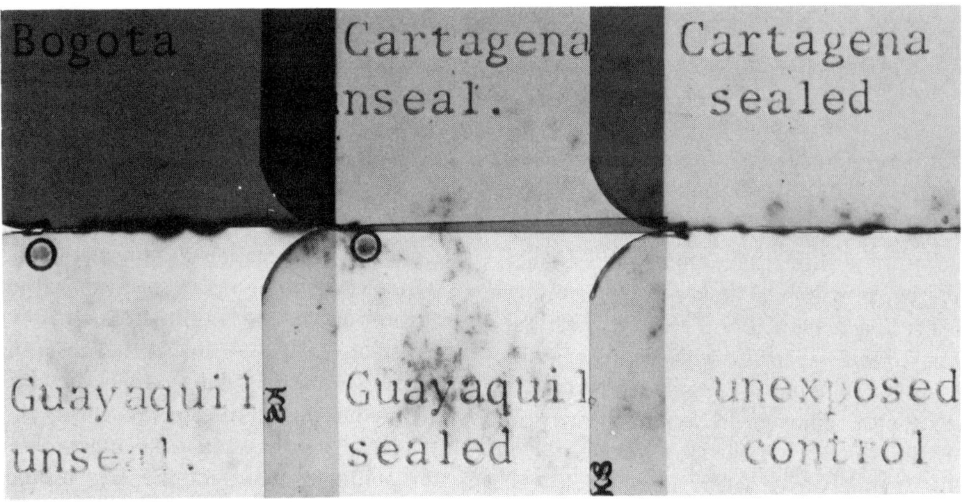

FIGURE 6-24B. Kodak PM Type 2[®] films that have been simultaneously exposed to 400 mR of gamma radiation and simultaneously developed after about three months of storage in a location with a fairly moderate climate (Bogota), and in warm and humid cities such as Cartagena, Colombia and Guayaquil, Ecuador, either in standard factory wrapping or with additional heat-sealing in "pouch paper." (After Becker, 1973.)

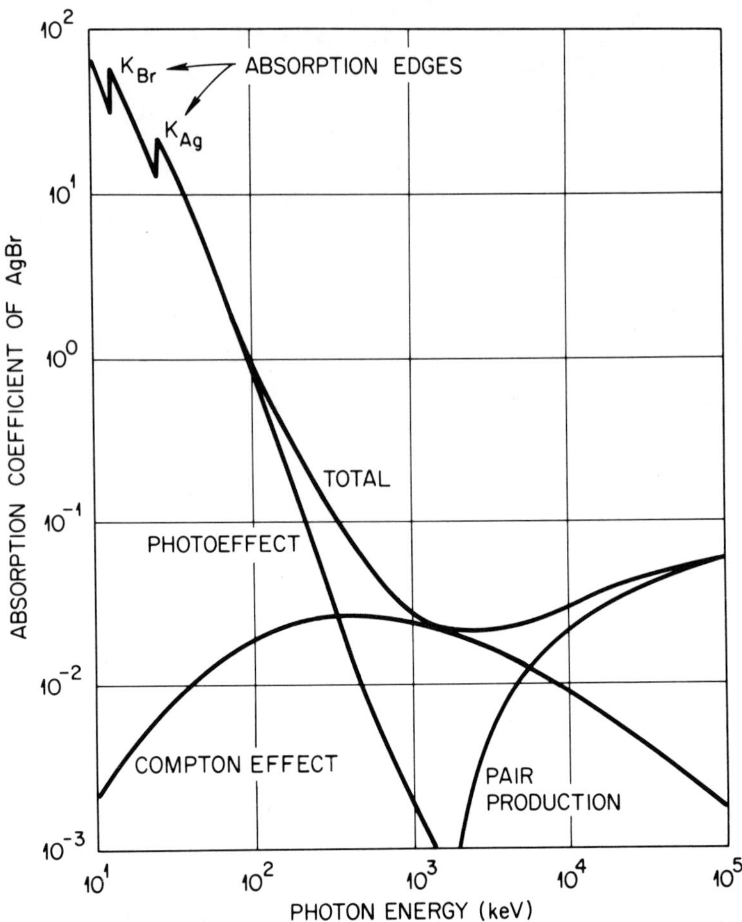

FIGURE 6-25. Calculated absorption coefficients for the different photon absorption mechanisms, and total energy absorption in silver bromide. (After Mauderli, 1957.)

fact, the films are sealed after being equilibrated at 21°C and 40 to 50% relative humidity and should, therefore, be stored in a desiccator before any additional sealing is carried out. Even very careful sealing in "pouch paper" will, however, not prevent substantial fading over extended periods in a severe tropical climate. In a recent study in several Latin American countries (Becker, 1973), 54 to 90% fading was observed in sealed Kodak PM Type 2® films during three months of storage in hot and humid locations such as Cartagena, Colombia or Guayaquil, Ecuador (Figure 6-24B). It may help somewhat to seal a desiccant into each individual film package for use in tropical countries.

6.2.2.2. Photon Response

Only certain highly sensitized emulsions have some sensitivity in the near or middle infrared region. With increasing photon energy, the sensitivity increases in the visible spectrum with increasing absorption in the AgBr, followed by loss of sensitivity in the UV due to the absorption in the gelatin. Emulsions that contain very little gelatin are used in this photon energy range.

As far back as 1897, Röentgen observed that the photographic effect of x-rays also depends on their energy. This energy dependence follows from the high concentration of high Z atoms (AgBr) in the sensitive layer. In Figure 6-25, the calculated contribution of the various absorption mechanisms to the total absorption coefficient in AgBr is plotted as a function of energy. In Figure 6-26 this curve has been divided by that of air to obtain the AgBr response per unit exposure as a function of photon energy, with discontinuities at the K_α

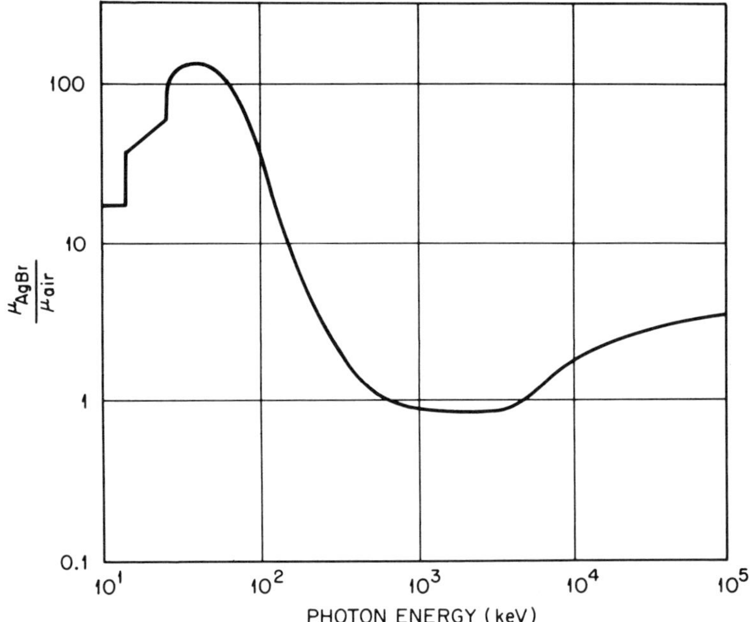

FIGURE 6-26. Calculated energy dependence of photon interaction with pure AgBr. (After Becker, 1966.)

absorption edges of bromine at 13.5 keV and of silver at 25.5 keV, and a flat minimum between $\sim$0.8 and $\sim$4 MeV.

The calculated ratio between maximum response per unit exposure at 40 keV and minimum response is $\sim$150, but in fact only factors of 12 to 50 are observed in emulsions (Figure 6-11). This considerable difference is due to the fact that the emulsion also contains gelatin and is surrounded by low Z materials (film base, wrapping). In the energy range of primary interest between about 10 and 2,000 keV, the effective ranges of the free electrons released by the photon absorption process are of the same order of magnitude as the grain diameter. Consequently, electrons that originate outside the grain contribute to a varying degree to the photographic process, while electrons released close to the surface of the AgBr grain may be absorbed in the gelatin and thus be lost to the photographic process.

Furthermore, the fact has to be considered that (in contrast to other solid-state detectors) the photographic grain is a "threshold detector," with a certain amount of energy required for developability, energy above and below this amount being "wasted." There have been various attempts to quantify the relationship between absorbed x-ray energy and developed optical density, also considering such factors as grain size and sensitivity distribution, electron scattering, etc. (Greening, 1951; Aglintsev, 1953; Tellez-Plasencia, 1958; Becker, 1961a and 1966; and others), but the problem is complex and has not yet been completely solved.

There have been substantial efforts to reduce the inherent energy dependence of photographic films relative to that of an ionization chamber. Indeed, the energy response decreases with

    a. increasing grain diameter;

    b. increased grain sensitivity;

    c. certain additives and activators including 0.5 to 5 mol % of Cd, Au, and certain sulfur compounds;

    d. "dilution" of the AgBr in the emulsion layer, resulting in an increase of the gelatin's contribution to the photographically effective electrons;

    e. suppression of surface development, and complete development of the internal image (perhaps also with physical development).

By combining all these factors, experimental films with an energy dependence of only a factor of 7 between maximum and minimum response have been prepared (Hoerlin et al., 1953). Unfortunately, they are not necessarily very stable and/or sensitive.

As in other solid-state detectors, it is not easy to measure the true energy dependence, and many of the data that can be found scattered in the literature have poor reproducibility. One should make sure that electron equilibrium is established at the higher photon energies (2.5 mm of plastic in front of a normally wrapped film may double its response to ^{60}Co radiation), and scattered radiation should be minimized. As the sensitivity and energy dependence may depend on the exposure level, developer, and dose-rate, these parameters should be kept constant. The effect of fading between exposure and processing should be minimized. Most importantly, the X-radiation should be filtered heavily, in case it is not mono-energetic K_a radiation.

In addition, the film sensitivity depends on the angle of photon incidence. Several authors (Greening, 1951; Becker, 1961b; Heard et al., 1964; McCarthy and Mejdahl, 1964; and others) report only little ($<$10%) changes in response between normal and 45° incidence, followed by a drop to about 80% at 80°, and as little as 50% of the frontal response (depending on photon energy and electron equilibrium) if the angle of photon incidence approaches 90° oblique.

Considering the strong inherent energy dependence of the film response, there are, in principle, two possible approaches to the film dosimetry of photon radiation. One may compensate the energy response with filters, or take other measures to obtain a more or less energy-independent indication. On the other hand, one may measure the density behind different absorbers and use this filter analysis as a "poor man's spectrometer" for a determination of effective energy to be used for correcting the reference O.D. measurement.* Obviously, the first method is simpler, while the second provides more information on the composition of the radiation field. Many efforts have been devoted to both methods, but most of the results are only of historic interest. Their description, therefore, will be brief.

The easiest method of flattening the energy response is the use of metal compensating filters.

The effect of such filters is not only due to the absorption of low energy photons, but is determined also by other factors such as back-scattering and whether or not electron equilibrium has been established in the filter. The electrons emitted from the metal filter will reach the emulsion if their energy is high enough. In fact, if an unprotected film is exposed in contact with a high Z metal foil to gamma radiation, only a small fraction of the observed photographic effect is due to electrons produced in the emulsion (Harrington et al., 1948).

At a given photon energy, the preferred direction of electron emission from filters on both sides of the emulsion strongly depends on the angle of photon incidence and the atomic number of the filter. With increasing photon energy, electrons are increasingly emitted in the forward direction. If the filters have a higher atomic number, however, the electrons are scattered in the material. Scattering becomes almost isotropic in lead (Figure 6-27). This makes lead a suitable "intensification foil." Density increases up to a factor of 3 have been observed if the Pb foil is in direct contact with the film (Schweitzer and Lloyd, 1963). If low Z electron absorbers of different thickness are placed between the emulsion and the high Z secondary electron emitter, the density ratios measured behind these absorbers can be used as a photon energy indicator (Böhler, 1958 and Spiegler and Davis, 1959).

A wide variety of filters and filter combinations have been suggested for flattening the energy response of various film types. A tolerably flat response above $\sim$150 keV can be obtained with an average sensitive x-ray emulsion by covering it with $\sim$1 mm of cadmium or tin. A flat response down to lower photon energies, which is highly desirable, is possible by other means such as

1. Optimized multifilter combinations. The most sophisticated described so far is a six-element filter for a DuPont 502® film. It results in a flat response down to $\sim$40 keV (Figure 6-28) for frontal exposure. Unfortunately, the energy compensation is much less impressive for non-normal radiation incidence, which results in a larger effective filter thickness for the photons to

---

*In his first communication on the discovery of x-rays in 1895, Röentgen described the basic idea of the filter-analytical method of film dosimetry: "In an attempt to establish a relationship between penetration and thickness of materials, I have made photographic exposures with the plate partly covered with thin layers of tin foil, using a gradually increasing number of layers."

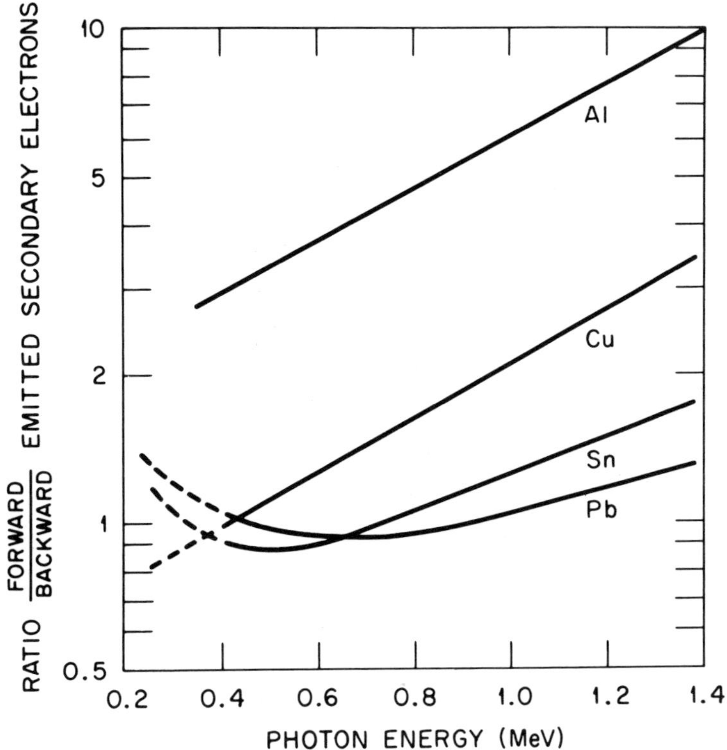

FIGURE 6-27. Ratio of forward to backward emitted secondary electrons for different filter materials as a function of photon energy. (After Hine, 1954.)

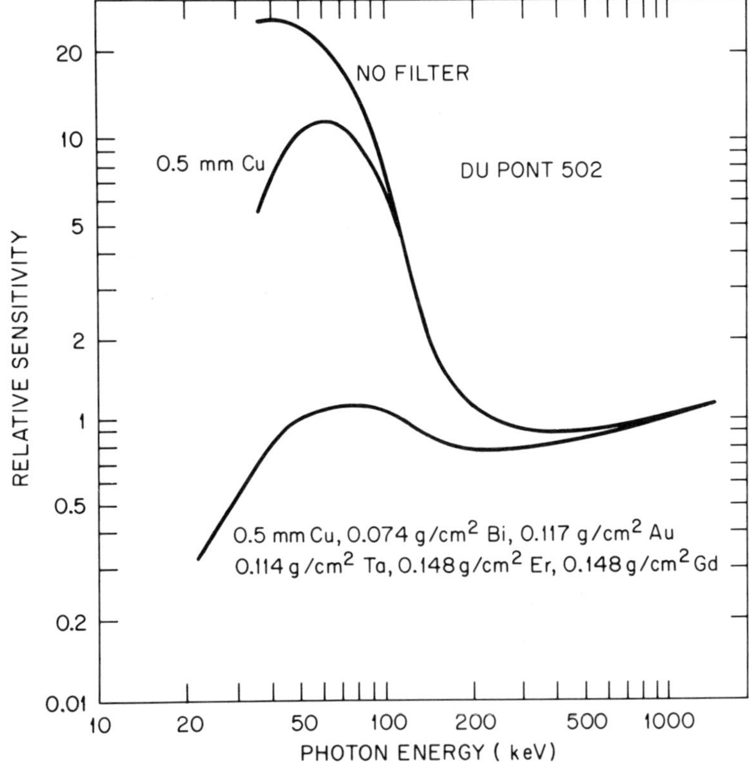

FIGURE 6-28. Photon energy dependence of DuPont 502 film without filter, covered with 0.5 mm copper, and with an optimized six-element filter above the film. (After Storm and Shlaer, 1965.)

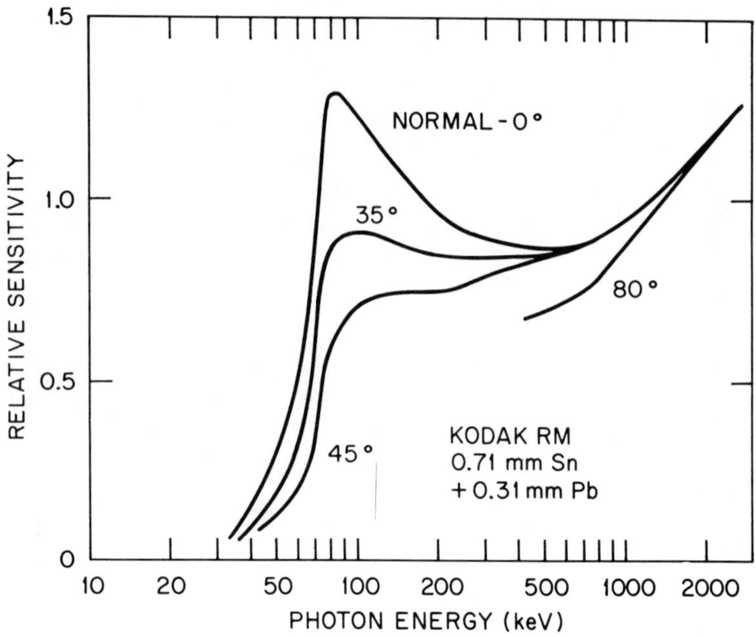

FIGURE 6-29. Energy response of a Kodak RM film covered with a combination filter (0.71 mm Sn + 0.31 mm Pb), for different angles of photon incidence. (After Heard and Jones, 1963.)

penetrate (Figure 6-29, also see Paic and Paic, 1968). Also, the filter combination requires readjustment for other films having a different inherent energy response. For methods of calculating the energy dependence of film-filter combinations, see Ehrlich and Fitch, 1951 and Storm and Shlaer, 1965.

2. Partial filtration. One can, for example, cover a film with a thin Cu foil plus a thick Pb filter with a conical hole in it and measure the average density behind the heavily and the weakly filtered area. The response can thus be smoothened down to $\sim$30 keV (Wachsmann and Stadelmann, 1961), but it is strongly dependent on direction and exposure level because maximum density will be reached behind the weak absorber much earlier than behind the strong absorber.

3. Exposing more than one film behind different filters, but reading their added densities. This method works well above $\sim$30 keV. The first device of this type consisted of a conventionally wrapped film; a sequence of 0.2 mm Pb, 2.07 mm Cu, and 0.1 mm Ta; an unwrapped film three times as sensitive as the first film; and a shield to protect the device from exposure from the back. Its response is given in Figure 6-30. A similar arrangement (Ehrlich, 1957) was symmetrical with

a total of three films. Its response characteristics, as well as of another simpler device employing two pieces of the same film (van Stekelenburg, 1958), are also satisfactory above $\sim$30 keV, but all these systems have in common a very pronounced angular dependence of the response. In the relatively sophisticated Ehrlich device, for example, the response drops to 50% at 45° and to 15% at 90° radiation incidence for 110 keV x-rays (at lower energies, the situation is even worse).

4. Calculation of an "energy-independent" exposure from the apparent exposures behind different filters. Various formulas have been devised, one of them resulting in an essentially flat response ($\pm$20%) between 20 keV and $\sim$3 MeV (Figure 6-31) if the densities behind three filters (0.1 mm dural + 0.3 mm Pb, 1 mm dural, and 300 mg/cm^3 plastic) are used for the calculation. Again, a strong angular dependence is a serious disadvantage of such techniques.

There also is another, completely different approach to energy independence, namely, fluorescence compensation. As can be seen in Figure 6-32, the energy responses of film and that of the light emission from some low Z organic scintillators (such as polyphenyls) can be made to

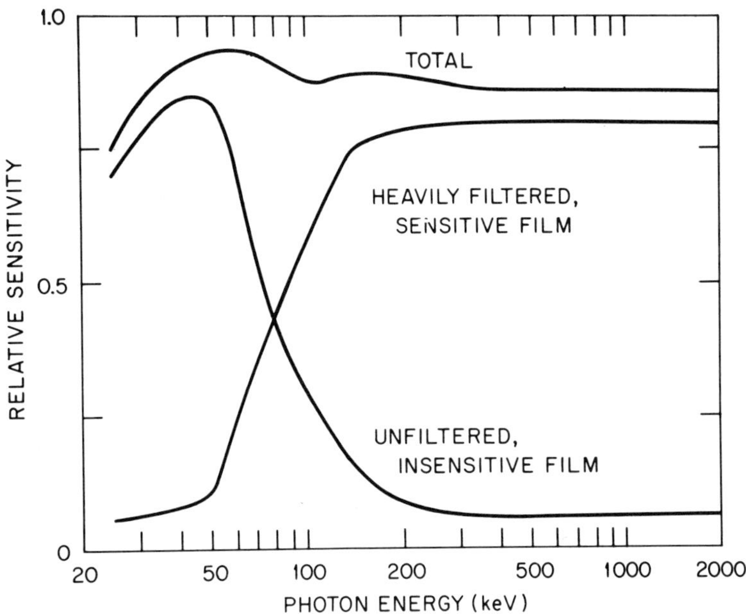

FIGURE 6-30. Energy dependence of response of an insensitive, unfiltered film, a sensitive, heavily filtered film, and the added densities of both films. (After Allisy, 1955.)

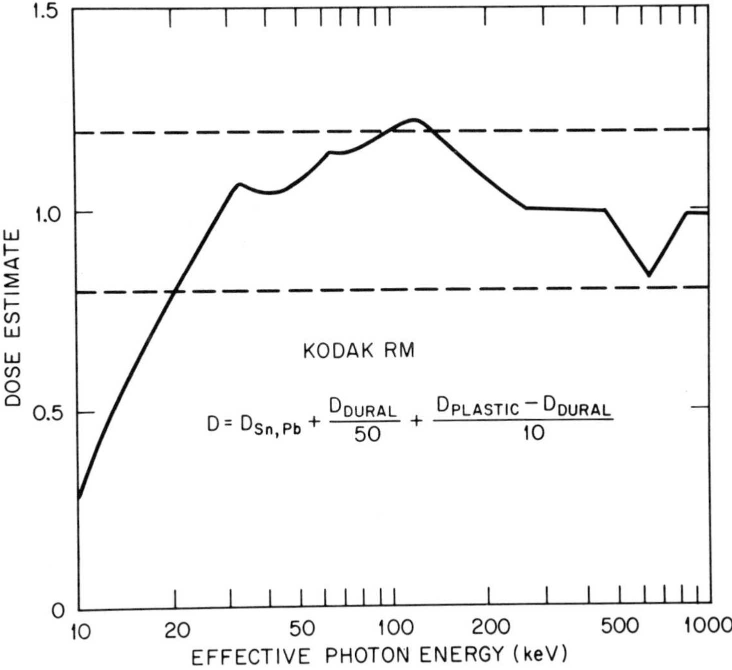

$$D = D_{Sn,Pb} + \frac{D_{DURAL}}{50} + \frac{D_{PLASTIC} - D_{DURAL}}{10}$$

FIGURE 6-31. Dose estimate as a function of effective photon energy, as calculated from the "apparent doses" behind various filters according to given formula. (After Jones and Marshall, 1964.)

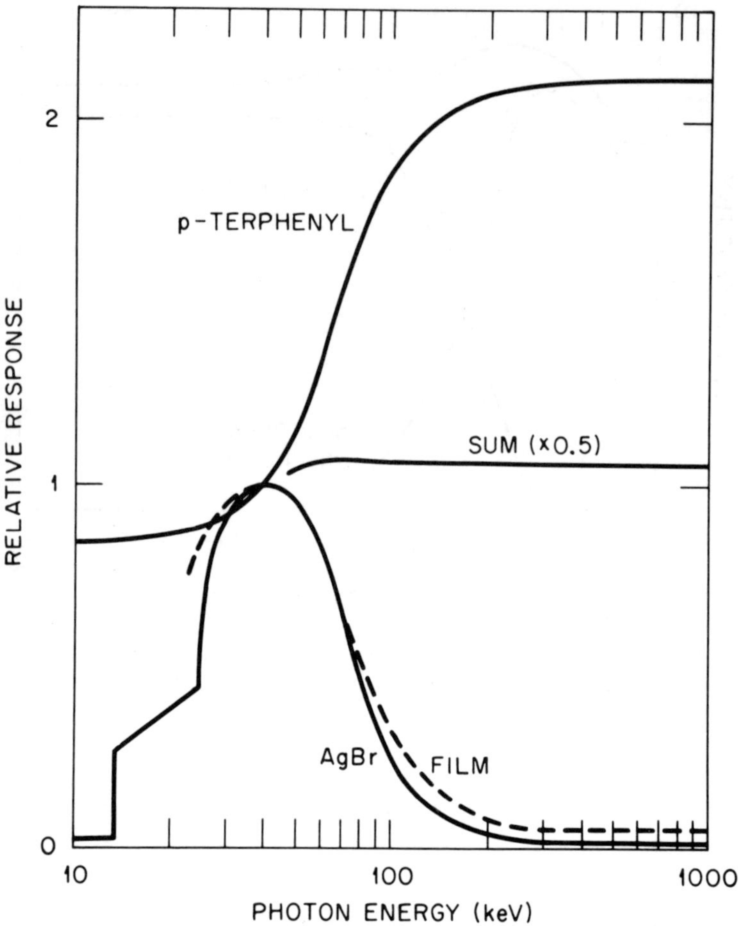

FIGURE 6-32.   Energy dependence of visible light emission in *p*-terphenyl, of the photographic effect in the emulsion, and the sum of both. (After Becker, 1961.)

compensate each other above ∿40 keV. Indeed, it has been shown by sandwiching a film between two organic scintillators that a high energy and directional independent response can be obtained (Hoerlin et al., 1953). The sensitivity is high and gamma doses as low as 0.5 mrad have been measured with such devices (Ehrlich and McLaughlin, 1959).

Among the disadvantages were that they were rather bulky and could only be opened in the darkroom. It was later shown that the scintillator could also be incorporated directly into the emulsion and remain there for processing and densitometry (Becker et al., 1960 and Becker, 1961a and b) with essentially the same results. Unfortunately, all such devices have in common some dose-rate dependence due to the contribu-

tion of visible light to the total photographic effect.

The most widely used method in film dosimetry nowadays is one in which the effective photon energy and homogeneity of the radiation is determined from the film contrast behind filters of different Z and/or thickness (filter analysis). The principles of this method have been known for over 40 years (Eggert and Luft, 1929), but the method became widely used only in the 1940's (Dorneich and Schäfer, 1942 and Langendorff et al., 1952). There have been, and still are, many dozens of different filter combinations in use. Usually a badge has three to four filters in addition to an unfiltered (open window) area and/or one or two areas with different thicknesses of plastic to aid in the evaluation of beta or soft x-ray exposures. Frequently, the filters in front of the film are slightly different from those in the back, permitting one to distinguish between the different directions of radiation incidence.

Number, thickness, composition, and arrangement of the filters can be varied within wide limits without much effect on the results as long as certain basic requirements are fulfilled: The filters should give sufficient contrast in the energy range of interest, they should be sufficiently large to enable one to avoid edge effects, and not be too expensive. Commonly used metal filters are Al, Cu, Sn/Cd, and Pb, in thicknesses between 0.05 and 1.5 mm. For the description of some relatively advanced film badges, see Thornton et al., 1961; Heard et al., 1964; McCarthy and Mejdahl, 1964; Storm and Shlaer, 1965; Cipperley et al., 1965; Anon., 1968; and Brady and Iverson, 1969. Two typical badge designs are given in Figure 6-33A and B.

The principle of evaluation can be described briefly, using as an example an older German badge with an open window and three copper filters of 0.05, 0.5, and 1.2 mm thickness (Dresel, 1956): First, the optical density behind the filters is determined as a function of exposure, covering the whole dose range of interest ($\sim$10 mR to $\sim$1,000 R), for both photon energies at maximum sensitivity ($\sim$40 keV) and minimum sensitivity (^{137}Cs, ^{226}Ra, or ^{60}Co). Badges are then exposed to such a dose that a reasonably precise density measurement can be obtained behind each filter, with radiations of different photon energies, for example, 10, 20, 30, 40, 50, 65, 80, 108, 270, 660, and 1,250 keV (see Table 1-5). From the densitometry of films, the following calibration curves are plotted:*

a. The net O.D. as a function of log dose for both maximum and minimum response.

b. The net densities behind the filters at the various energies are converted into "apparent dose values" with either one of the curves obtained in (a), which are plotted as a function of filter thickness, as indicated in Figure 6-34. For homogenous (monoenergetic) radiation, the plot should,

in theory, result in a straight line, the slope of which is a measure of the photon energy.

c. The dose value that is obtained by graphical extrapolation of the "apparent doses" behind 0.5 and 1.2 mm Cu thickness to 0 mm Cu (or of the 0.5 and 0.05 mm Cu "doses" to 0 mm) will differ by a certain amount from the actual dose. By dividing the actual dose by this dose, a correction factor is obtained. The log of the correction factors is then plotted against the apparent dose ratios behind the filters.

For the evaluation of an unknown exposure, the logarithms of the apparent doses are plotted, as in Figure 6-34. The result of the 0.5/1.2 mm Cu extrapolation is multiplied with the appropriate correction factor, resulting in an estimate for the high energy component of the exposure. The procedure is then repeated with the 0.5/0.05 mm and the 0.05/0 mm Cu extrapolations, representing the lower photon energies, and the three values are added. Of course, this evaluation can be simplified by the use of computers. It also helps to calibrate the densitometer directly in apparent dose units.

Unfortunately, this procedure is not only complicated, but also subject to serious errors under almost any but the most carefully controlled, highly idealized laboratory conditions. Only with perpendicular free-air exposures of badges with one photon energy, or with very simple mixtures of X- and gamma radiation in the $\sim$100 mrad to 10 rad range, and careful evaluation by experienced technicians, may one achieve relatively accurate dose readings within about ±20%.**

Under less favorable, practical conditions, errors exceeding a factor of three in both directions are not uncommon. A frequent source of errors is light leaks in the film wrapping, making the evaluation difficult or impossible; another source is irregularities during the development (aged chemicals or "lost" films in the developer). Radionuclides that permeate the film wrapping, such as tritiated water vapor, can irradiate the film

---

*To reduce the effect of fading and fogging, many film badge services expose their calibration films in the middle of the monitoring period, and store them under conditions similar to the average climate under which they are used. Of course, calibration and worn films should always be developed together.

**In research type applications of films in the laboratory, for example, in depth-dose studies in a well-known radiation field, very careful calibration can reduce the error to as little as ±5% (see, for example, Brodsky et al., 1965; Rassow et al., 1969 – 71; and ANSI, 1970).

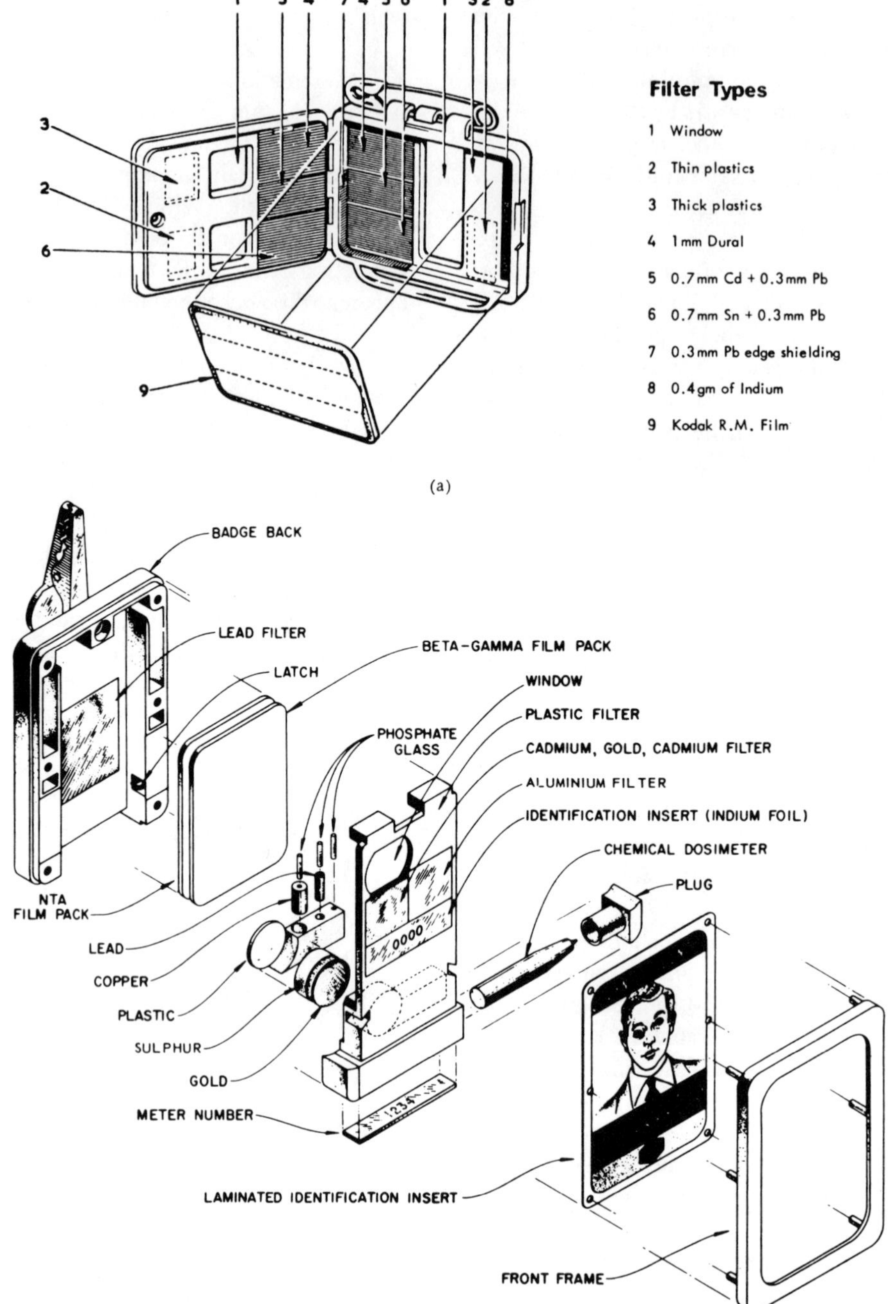

FIGURE 6-33A. Exploded view of two relatively advanced film badges as used (a) at the Atomic Energy Research Establishment in Harwell, England and (b) at Oak Ridge National Laboratory. (After IAEA, 1970.)

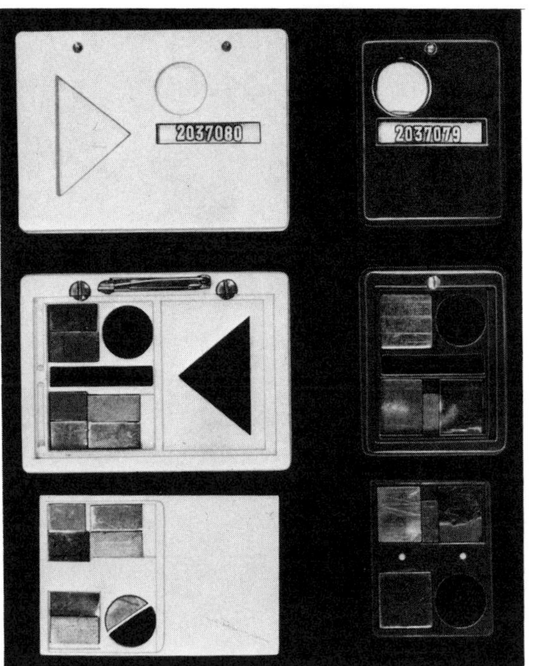

FIGURE 6-33B. Closed and open views of two badges which have been designed by Dresel and Wachsmann in West Germany in the 1950's and early 1960's and have since been copied in several countries, the large badge (left) containing a track film and Sn/Cd filter pairs for neutron dosimetry and the smaller badge (right) with Cu filters of different thickness and Pb for X- and gamma radiation.

internally;* various laboratory gases and fumes can give rise to accelerated fading or fogging. The evaluation procedure also becomes particularly inaccurate

1. In mixed beta/x-ray fields;

2. For small gamma doses (<10 to 100 mrad, depending on the film sensitivity) and at high doses close to the maximum O.D.; or in the sometimes ambiguous solarization region (>10 to 100 rad, depending on film type, developing techniques, and photon energy);

3. Whenever the photon energy is very low (<10 keV) or high (>2 MeV) (for the interpretation of film badges exposed to high energy gamma radiation from ^{16}N, see Nelson et al., 1966);

4. In mixed X- and gamma radiation fields of certain composition, such as a gamma to X-radiation ratio of $\sim$1/7 or less; and

5. When the direction of radiation incidence differs from normal because the effective thickness of the filters becomes larger. There is no general agreement on the most probable angle of incidence under practical conditions, but as it is unlikely to be 0°, some installations such as the AERE Harwell calibrate their films at 35°.

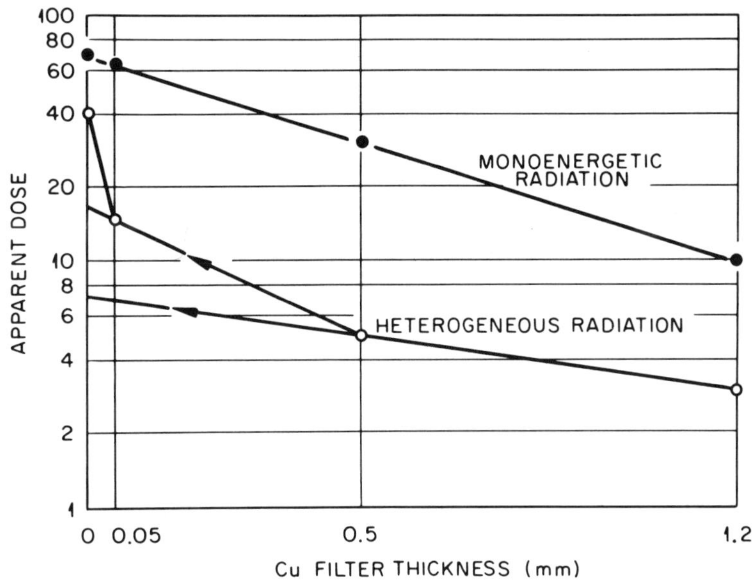

FIGURE 6-34. Apparent doses behind copper filters of different thickness for homogeneous and heterogeneous photons. (After Wachsmann, 1961.)

*For the effect of tritium on photographic films, see Becker, 1961c; David and Wasilewski, 1969; and Denham and Kathren, 1969.

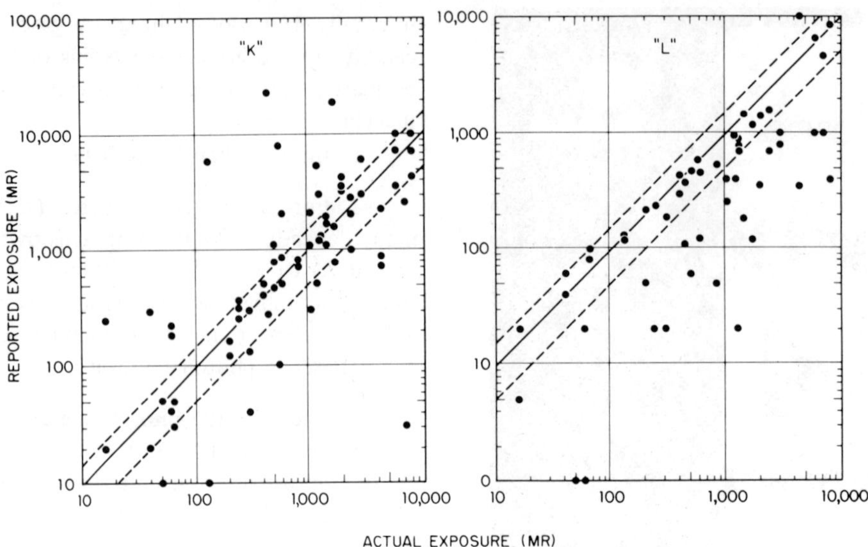

FIGURE 6-35A.   Relationship between actual exposure and reported film badge readings from two different film badge services. (After Gorson et al., 1965.)

Consequently, all realistic performance tests of film badges have shown them to be much less accurate than solid-state dosimeters such as advanced RPL or TLD systems, which are in the process of replacing the film badges in most advanced laboratories (Attix, 1972). The film badge has been tested by many investigators under more or less "realistic" conditions. In case of commercial film badge services, deficiencies in the operation of the service may be superimposed on the inherent inaccuracies of the photographic method.

A typical result of a performance test of film badge services is given in Figure 6-35A, showing that the interpretation of unknown exposures by film may deviate from the actual dose by up to a factor of 200, with about half of all readings outside of the ±50% accuracy band. One film badge service reported a 20 R reading for a 0.5 R exposure and a 30 mR reading for a 7 R exposure, and another one reports constantly 20 mR for actual doses between 15 mR and 1.5 R. For similar studies, see Egelhaaf et al., 1960; Schimmerling and Sass, 1968; Barber et al., 1968; Crosby, 1970; and many others. The film dosimeter has also been tested in comparison with other detectors including pocket ionization chambers, glass dosimeters (Piesch, 1967; Helm and König, 1969; Yokota, 1970; Tatsuta et al., 1972; Keirim-Markus et al., 1972; and also Figure 4-34), and/or TLD (Johnson and Attix, 1968; Cameron

and Suntharalingam, 1968; Ruault and Gatineau, 1968; Eastes, 1968; Scott, 1970; Crosby, 1971; Geiger, 1971; Fitzsimmons et al., 1972; and Yasar, 1972).

Among the positive aspects of film dosimetry is the fact that films can easily be marked to identify the wearer, either by applying pressure (care should be taken that this procedure does not damage the light-proof wrapping) or by very soft x-rays, with the larger part of the film being protected during exposure by a lead shield. The latter method has the disadvantage that the code or number cannot be read at maximum O.D.

Films also provide some information on the direction of radiation incidence from the "picture" of the filter shadows. Very sharp filter edges usually indicate a single exposure. Radioactive contamination can easily be detected from the dark spots on the film. It is also easy to scan large numbers of developed films for the few exposed ones by visual inspection. Most experts, however, agree that these advantages do not outweigh the disadvantages of poor accuracy and stability, the necessity for a darkroom, and elaborate evaluation procedures.

6.2.2.3. *Heavy Charged Particles and Neutron Response*

One has to distinguish between the "macroscopic" effects of heavy, directly or indirectly ionizing particles (e.g., the formation of measurable optical density), and their microscopic, particle-track producing properties. With regard to the first, a strong LET dependence of the produced O.D. has been observed. As can be seen in Figure 4-17, this LET dependence is more pronounced in the film than in other solid-state detectors, mostly due to the "threshold detector" response of the photographic grain. For protons below 5 MeV, for example, less than 10% of the gamma response is observed on a rad per rad basis. This is, of course, also the case for alpha particles in alpha autoradiography and other high LET particles.

The macroscopic effect of thermal neutrons on film is small, and can be attributed largely to the activation of the silver (^{108}Ag with a halflife of 2.3 min; ^{110}Ag and ^{110m}Ag with halflives of 253 days and 24 sec, respectively, are formed — Mercer and Golden, 1963). Although thermal neutron exposures are of very limited interest in practical radiation protection, most film badges have a provision to "measure" them, the most popular being a pair of equally thick ($\sim$0.5 to 1 mm) cadmium and tin filters. Having a similar Z, their absorption characteristics for photons are almost identical. In the presence of thermal neutrons, however, the film behind the cadmium filter will be darker due to the (n,γ) capture radiation from the filter. It has been determined that 1 rem of thermal neutrons below 0.4 eV in energy (9.6 x 10^8 n/cm^2) produces the same density behind Cd as 1.95 to 2.5 (calculated value: 2.1) rad of gamma radiation (see, for example, Caruthers and Story, 1964). Some other metal pairs including rhodium have also been tested as thermal neutron converter (for a review, see Becker, 1966).

The macroscopic effects of fast neutrons on film are small and difficult to measure. There have been various attempts to calculate these effects, and the obtained sensitivities vary, depending on neutron energy, film type, etc., between about 1 and 7% of the gamma radiation response. According to one of the more recent of these investigations (Ehrlich, 1960), 2.4 to 28 x 10^9 fast neutrons (3 MeV) and 3.6 to 5 x 10^9 thermal neutrons per cm^2 produced the same density as 1 R gamma radiation on two dosimeter film types, the neutron-to-gamma ratio depending largely on the mean grain diameter. In Figure 6-35B, the direct fast-neutron response of a typical dosimeter film is compared to that of other solid-state dosimeters including RPL glass, LiF TLD with and without hydrogenous radiator, and two TSEE dosimeters (MgO and BeO) covered with polyethylene.

Thermal neutrons can also be measured via track registration due to the ^{14}N (n,p) ^{14}C reaction in the gelatin, giving rise to about 2 x 10^{-5} tracks/cm^2 for each thermal neutron per cm^2, or 1 x 10^{-6} rad/track/cm^2 The sensitivity can be substantially increased if nuclear track emulsions (which differ from conventional x-ray emulsions by a higher concentration of silver halide, finer grains, greater thickness, and a more uniform grain size) are "loaded" with boron and/or lithium compounds.

With mixtures of lithium acetate and lithium citrate, up to 40 mg/cm^2 of Li can be introduced into the emulsion, and even higher concentrations may be obtained with powdered glass of the composition 75% B_2O_3, 15% Li_2O, and 10% B_2O_3 — Möller, 1960. The ^{6}Li (n,a)^{3}H reaction produces one alpha particle track (6.3 μm) and a triton track (35.5 μm), while the total track length of the lithium nucleus and alpha particle produced in the ^{10}B (n,a) ^{7}Li reaction is only $\sim$7 μm. These methods are much more cumbersome than optical density measurements. The track registration techniques are, in practice, essentially restricted to the detection of fast neutrons. One exception is a system described by Shapiro et al., 1970, in which Ilford K-1$^®$ plates loaded with 23 mg/cm^2 boron have been placed in the center of a spherical moderator for neutron area monitoring.

In a somewhat simplified way, the formation of visible nuclear particle tracks can be described, as indicated in Figure 6-36A: Electron tracks, being "thin" (not every grain along the track becomes developable) and erratic due to intense scattering, can only be seen with some difficulty in the more sensitive track emulsions. Tracks of high LET particles such as fission-fragments or alpha particles in the common energy range ($<$10 MeV), on the other hand, are dense, straight, and easily recognizable because all grains along the track become developable. Lower LET particles such as high energy protons produce a thin track until slowed down to a higher LET; neutrons have to produce secondary particles — mostly recoil protons in the gelatin — to become detectable. If the density of such recoil proton tracks is neither too high nor too low, if they are not obscured by a high density of randomly distributed " fog" grains, and if they have not yet thinned substantially due

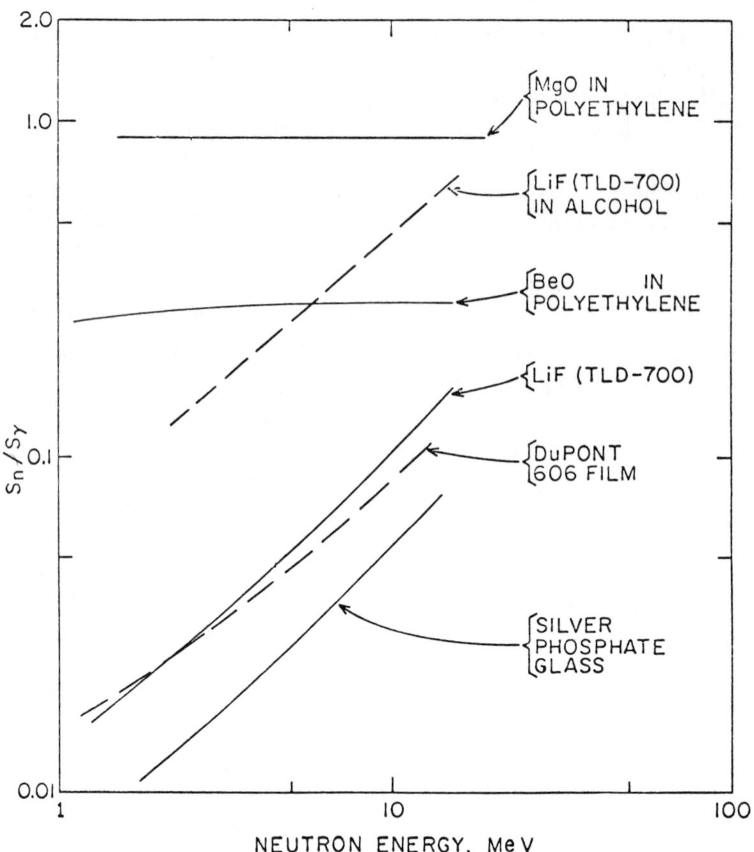

FIGURE 6-35B. Energy dependence of the fast-neuton response of a DuPont 606® dosimeter film, expressed as the ratio between neutron and gamma radiation sensitivities in rad, compared with other solid-state dosimeters such as RPL glasses, LiF TLD, and polyethylene-covered TSEE dosimeters BeO and MgO. (Compilation of data after Ritz and Attix, to be published.)

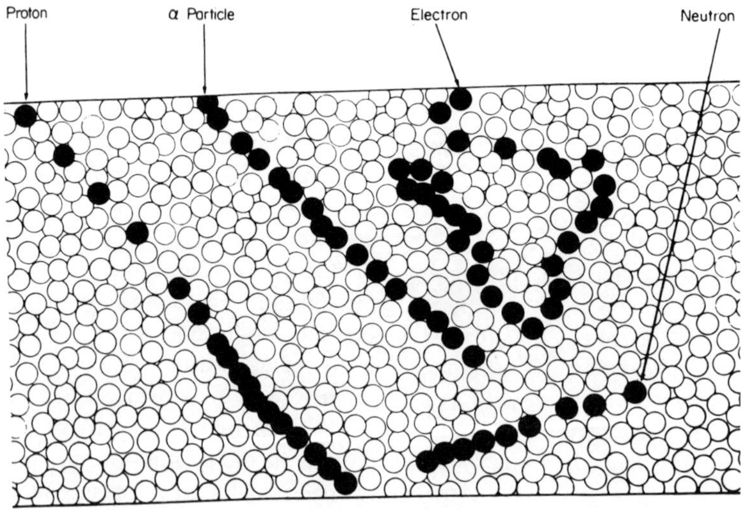

FIGURE 6-36A. Schematical diagram of the tracks produced by different types of particles in a nuclear emulsion. (After Herz, 1969.)

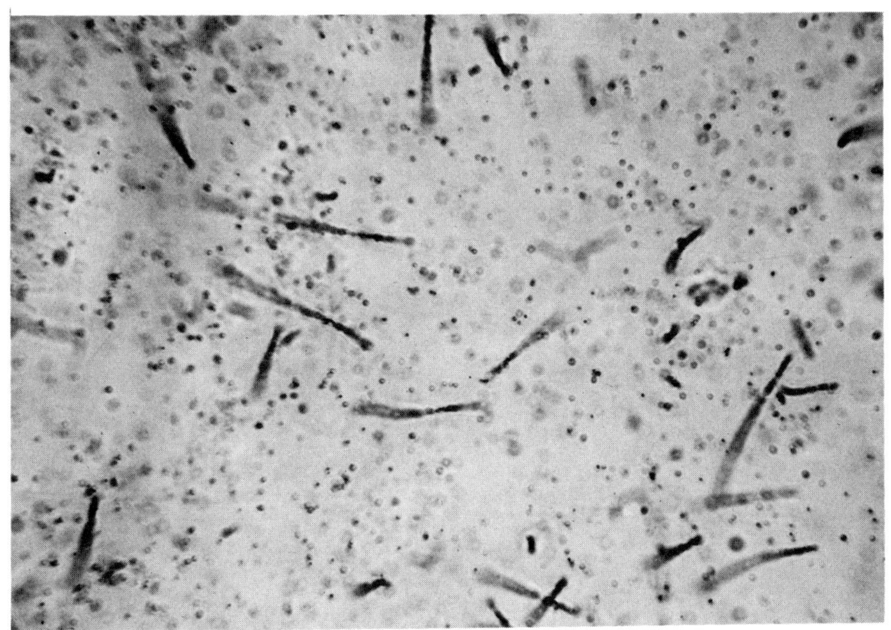

FIGURE 6-36B. Recoil proton tracks produced by 14 MeV neutrons in a Kodak NTA®
emulsion.

to fading, they may be counted in an optical microscope at a magnification of about 400 to 1,000 x (Figure 6-36B).

As nuclear track emulsions have played, and to a somewhat minor degree still play, an important role in nuclear and cosmic radiation physics, the registration characteristics of all known types of particles in a wide variety of emulsions have been the subject of exhaustive studies (for a review, see Barkas, 1963 and Avan et al., 1972) which cannot be discussed here. It should, however, be mentioned that the interpretation of track structure has changed somewhat in recent years, with more emphasis being placed on the effect of short-range secondary and tertiary electrons along the track (Katz et al., 1972). Based on this model, the detailed track structure can be simulated by computer. Figure 6-37 also illustrates that many grains are developed in a certain distance from the actual track.

Beginning in the 1940's and finding more widespread applications in the 1950's, the nuclear track emulsion (originally a "special fine grain alpha emulsion" made by Kodak, later a commercially available Kodak Personal Monitoring Film Type A®*) was introduced into fast-neutron personnel dosimetry by Cheka (1954a), who also was among the first to point out one of the most serious limitations of this technique, namely, fading of the latent tracks (Cheka, 1954b). The properties of this film have since been the subject of many studies, of which only a few can be mentioned here.

The most common track-inducing process is the production of recoil protons in the gelatin and, particularly at higher neutron energies, to an increasing degree in the paper or plastic wrapping of the film. The energy and direction of the recoil protons depend on energy and direction of the incident neutrons, the proton energy E_p being related to the neutron energy E_n and the collision angle θ by the simple equation

$$E_p = E_n \cos^2 \theta. \tag{6.8}$$

Thus, in principle, one can calculate the neutron energy distribution from the track length distribution, if one takes into consideration such complicating factors as the shrinkage of the emulsion during processing, the loss of tracks

---

*Eastman Kodak Co., Rochester, N. Y. Another variety of packaging with a slightly improved neutron energy dependence of response was available as Type B, but the production of this material has been discontinued.

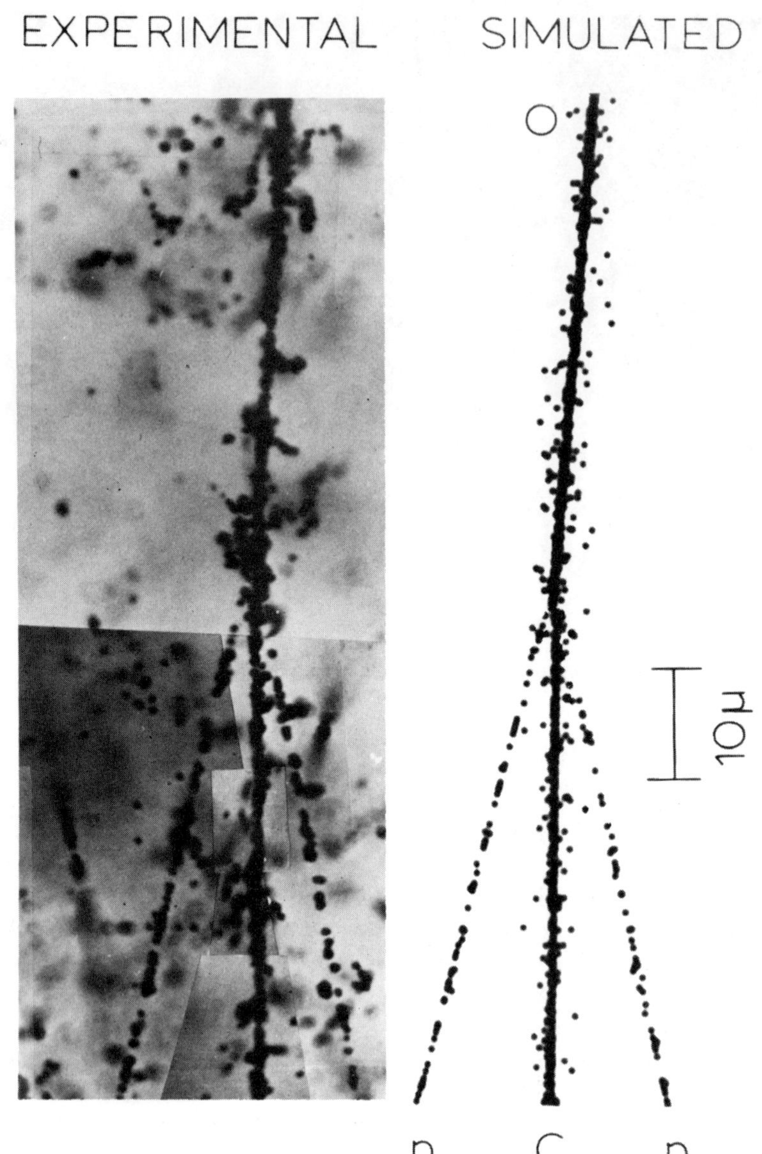

FIGURE 6-37. Recording of a nuclear event in a track emulsion as observed (left), and as simulated with a computer (right). (After Furtak and Katz, 1971.)

beginning or ending outside the emulsion, and tracks which are too short to be recognized. This method of analyzing the proton track length distribution, which has been applied in some special studies of neutron spectra (Amadesi et al., 1963 and Lehman and Fekula, 1965), is very time consuming and cannot be used in routine work.

The lower threshold of neutron detection depends on the grain size and track recognizability. In earlier calculations of the neutron energy response characteristics (see, for example, Cook, 1958; Lehman, 1961; and Piesch, 1963), it was assumed that "tracks" consisting of four or even of three grains can still be recognized. On this basis, one would expect that for neutrons of 2 MeV, only 5% of the produced proton tracks are overlooked, and that this percentage reaches 100% at 0.3 MeV, with an assumed practical threshold around 0.5 to 0.7 MeV.

It has, however, been shown recently that there is no relation whatsoever between the density of three- or four-grain tracks and neutron dose in

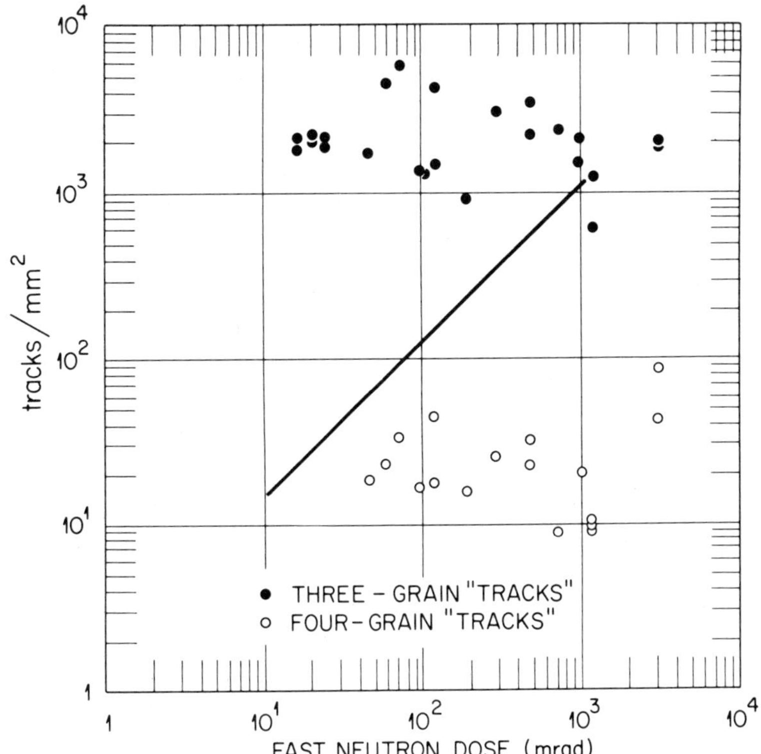

FIGURE 6-38.   Relationship between the number of "tracks" consisting of three and four grains and the actual fast-neutron dose in a Kodak NTA emulsion. (After Medveczky and Bornemisza-Pauspertl, 1970.)

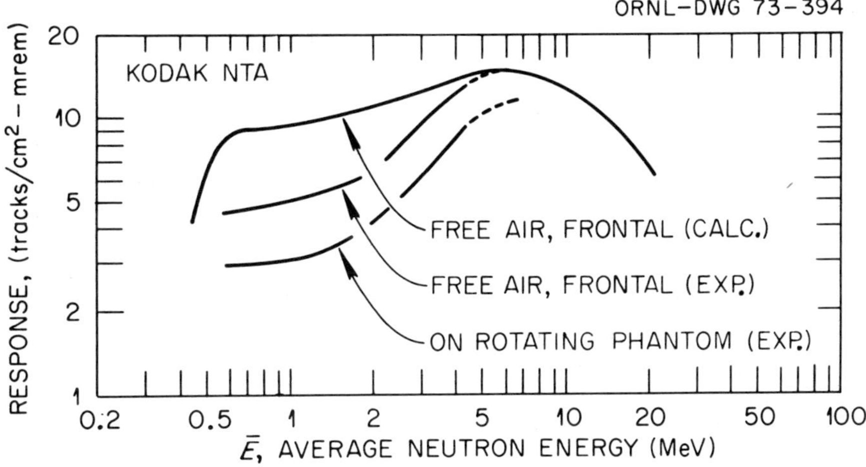

FIGURE 6-39.   Response of Kodak NTA film as a function of average fast-neutron energy with free-air frontal exposure of the films, on the surface of a rotating phantom, compared with Lehman's calculated response for the free-air, frontal irradiation with monoenergetic neutrons. (After Oshino, 1973.)

such emulsions (Figure 6-38). The practical threshold value should, therefore, probably be revised upward. As the NTA film is blind for all neutrons below at least 0.5 to 0.7 MeV, the experimental response as measured with nonmono-energetic neutrons from isotope sources, or fission neutrons, differs substantially from the calculated response for monoenergetic neutrons (Figure 6-39).

The sensitivity of track recording, if related to

the neutron dose equivalent in rem, is dropping rapidly at neutron energies exceeding $\sim$8 MeV. The response to energies up to 15 to 20 MeV can be somewhat improved with additional organic "proton radiators" in front of the film, whose thickness corresponds to the maximum range of the protons at these energies (for example, $\sim$5 mm of plastic for 20 MeV neutrons). There also have been suggestions to use a sequence of proton radiators (plastic or paper) and proton absorbers (aluminum) for improving the energy response. One such combination, as suggested by Cheka (1954a), consisting of 76 mg/cm^2 cellulose, 85 mg/cm^2 Al, 34.5 mg/cm^2 cellulose, 25 mg/cm^2 Al, and 28.5 mg/cm^2 acetyl cellulose, was found to produce an energy-independent response between 1 and 20 MeV within +100 and -50% (Littlejohn, 1960).

At very high nucleon energies, other nuclear reactions in the emulsion can be used for personnel dosimetry (Phillips and Champagne, 1965). In dosimeters worn around the CERN accelerators, for example, a dose of 100 mrad, or a dose equivalent of 180 mrem, is attributed to each "star" which is produced in the emulsion (Baarli and Dutrannois, 1967). In the personnel dosimetry of astronauts, the analysis of heavy particles in the cosmic radiation also becomes rather complex (see, for instance, Schäfer et al., 1972).

Visual track counting, which is usually carried out at magnifications from 400 (low track densities) to 1,000 x (oil immersion, high track densities), is slow, tedious, exacting, and subject to substantial systematic errors. Prolonged track counting can lead to eyestrain and headaches; short tracks tend to be overlooked by a tired person. Only about 50 films per day can be counted by a technician. Projection of the microscopic picture or contrast-amplifying TV techniques helps only little. Although there have been numerous attempts to automate track counting (for a compilation, see Becker, 1966), most of these cumbersome electro-optical systems were much too complicated, expensive, and inaccurate to be of any practical value. For a relatively advanced system, see Heard (1964) and Douglas (1970).

As the thickness of the sensitive emulsion layer (average 33 μm) varies from batch to batch, no "absolute" calibration of the track films is possible. Lehman (1961) suggested normalizing the response for the average thickness and using correction factors for individual emulsions. Some recent Kodak NTA film calibration values for normal, free-air irradiations in a low-scattering geometry are listed in Table 6-3. More data, many of them obsolete, can be found in earlier publications. The values are not very consistent because their inherent accuracy is limited by the accuracy with which the neutron flux is known, by counting statistics, subjective errors, and fluctuations in emulsion thickness and properties. For the widely available Po/Be source 1.4 x 10^4 tracks/cm^2 rem should be a reasonable approximation.

The useful dynamic range within the given neutron energy limitations, and excluding the effects of fading, is limited mostly by

a. the track densities that cannot be counted within a reasonable time and with tolerable precision; the lower limit, with only a few tracks per microscopic field of view at a low magnification, is around 300 tracks/cm^2 or $\sim$20 to 30 mrem; the upper limit, with the tracks increasingly overlapping and difficult to distinguish, is around 3 x 10^5 tracks/cm^2, corresponding to $\sim$25 rem; and

b. the background grain density (fog) due to photon radiation. Even 1 R of gamma radiation, or $\sim$50 mR of X-radiation, affect track count and counting accuracy; a gamma exposure of 2 to 5 R, or a correspondingly lower x-ray exposure, makes track recognition very difficult (Becker, 1963).

The reported values for the directional dependence of the NTA film vary widely, range from a loss in sensitivity of only 30% with oblique instead of normal neutron incidence observed by Cheka (1954a) to a loss of 66% reported by Jasiak and Musialowicz (1973). The numbers reported by other observers are between these extremes, with an average of about 50% (Becker, 1963). The measured angular response depends on neutron energy and exact irradiation geometry (Kathren et al., 1965). The asymmetrical packaging of the NTA film also results in a difference between frontal and back response, amounting to as much as 25% (Lehman, 1961). Some authors, therefore, have suggested rotation of the film during the calibrating exposure.

As one would expect from the well-known neutron-absorbing and backscattering properties of the human body, this directional response becomes much more pronounced if the track film is

Fast Neutron Response of the Kodak NTA-Film

Neutron source or energy (MeV)	Tract density*	Author
1.5 MeV	2×10^{-4} tracks/neutron	Seguin, 1973
4.5 MeV	5×10^{-4} tracks/neutron	Seguin, 1973
Am/Be	7×10^{-4} tracks/neutron	Seguin, 1973
13.6 MeV	9×10^{-4} tracks/neutron	Seguin, 1973
Fission	3.8×10^{-4} tracks/neutron	Piesch, 1964
Po/Be	6.2×10^{-4} tracks/neutron	Piesch, 1964
14 MeV	6.5×10^{-4} tracks/neutron	Piesch, 1964
Po/Be	1.56×10^{4} tracks/cm^2 rem	Jasiak and Musialowicz, 1973
14 MeV	0.75×10^{4} tracks/cm^2 rem	Jasiak and Musialowicz, 1973
Pu/Be	1.2×10^{4} tracks/cm^2 rem	Krishnamoorthy et al., 1973
Fission	1.4×10^{4} tracks/cm^2 rem	Piesch, 1964
Po/Be	1.55×10^{4} tracks/cm^2 rem	Piesch, 1964
14 MeV	1.05×10^{4} tracks/cm^2 rem	Piesch, 1964
2.5 MeV	1.2×10^{4} tracks/cm^2 rem	Portal, 1963
Po/Be	1.24×10^{4} tracks/cm^2 rem	Portal, 1963
^{238}Pu/F	0.46×10^{4} tracks/cm^2 rem	Oshino, 1973
^{252}Cf	0.55×10^{4} tracks/cm^2 rem	Oshino, 1973
^{238}Pu/Be	0.92×10^{4} tracks/cm^2 rem	Oshino, 1973
4.2	1.17×10^{4} tracks/cm^2 rem	Oshino, 1973

*Frontal, free-air exposure

TABLE 6-4

Effect of Thorax Phantom on Track Film Response

Neutron source or energy (MeV)	Relative track density*		Author
	Front	Rear	
Po/Be	110–130	27	Cheka, 1954
	132	19	Piesch, 1963
Ra/Be	108	57	Amadesi, 1961
Po/B	127	7.6	Cheka, 1954
Pu/Be	113	22	Oshina, 1963
^{252}Cf	120	15	Oshina, 1963
14	108	58	Piesch, 1963

*Normalized for perpendicular free-air exposure of the film; film-surface distance varies, but usually 1 to 2 cm.

attached to a phantom of the human trunk. The exact values depend on the type of phantom, phantom-film distance, and other factors. For Po/Be neutrons and on elliptical paraffin phantom, the front to back track density ratio is 5 for the films attached directly to the surface, 6.5 for 1 cm distance, and 8 for 2 cm (Becker, 1963). Of course, the main parameter determining this ratio is the neutron energy. Some data from the literature are compiled in Table 6.4 and Figure 6-40, indicating that the possible error due to the body influence, which may amount to only a factor of 2 for 14 MeV neutrons, is rather substantial at lower neutron energies.

Obviously, the nuclear track film is plagued by so many serious limitations that its use should be discontinued in all but a few exceptional situations in practical personnel monitoring, even if no better detectors are available. Its use can be actually dangerous because it seems to indicate the absence of neutron exposures even when they contribute substantially to the total dose. A misleading feeling of safety is clearly less desirable than to rely on indirect methods such as local neutron dose measurements with portable rem counters.

There have been several special applications of NTA films, notably in the detection of radon and radon daughter products (Jacobi and Koeppe, 1968). In a special film badge (Geiger, 1967 and Evans, 1972), the NTA film is separated from a central recess by a light-tight window sufficiently

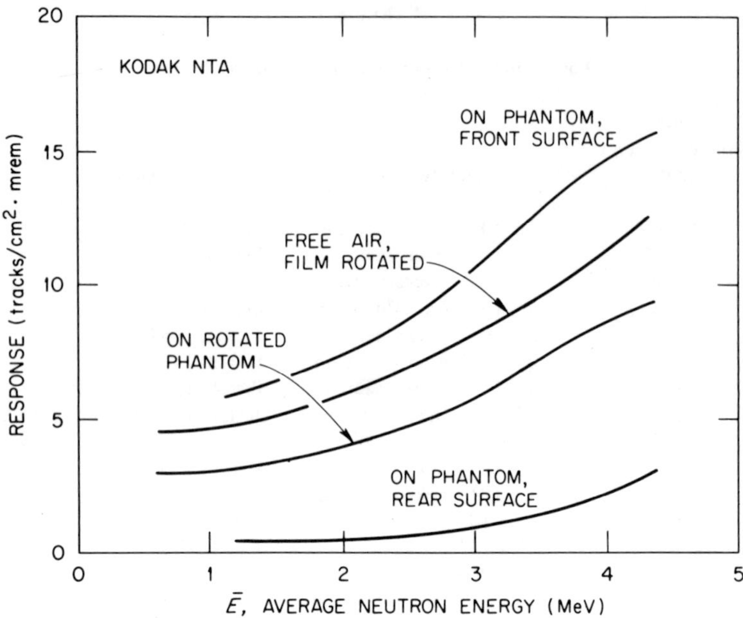

FIGURE 6-40. Energy dependence of Kodak NTA film after rotating free-air exposure, on the front and back of the phantom, and exposed on rotating phantom. (After Oshino, 1973.)

thin to permit the passage of alpha particles and radon-containing air to diffuse into the recess. A coefficient of variation of ±30% was claimed in early laboratory tests, but other laboratory and field tests did not produce satisfactory results (White, personal communication).

Other investigators (Costa-Ribeiro et al., 1969) used electrostatic collection of the radon daughters on a thin aluminum foil and a scintillator behind the foil to convert the alpha particle energy into visible light to be integrated photographically. Unfortunately, this results in a pronounced reciprocity failure. For the film dosimetry of radon daughters, see Markov et al. (1970).

6.2.2.4. Electron Dosimetry

A high energy electron traversing an emulsion loses part of its energy directly to the grains. Another, often photographically more active fraction of its energy is transferred through the action of low energy secondary and tertiary electrons. For example, a 1 MeV electron transfers $\sim$20% of its energy to silver bromide by means of 1 to 10 keV delta rays. Thus, grains in the path of an electron can become developable even when the calculated average energy loss per grain would be too small for producing developability.

For low electron energies and high grain sensi-

tivity, one can assume that each grain struck by an electron is rendered developable, and the net O.D. can be described by a simple Poisson distribution:

$$\frac{\text{O.D.}}{(\text{O.D.})_m} = 1 - e^{-KE} \tag{6.9}$$

where K is a sensitivity factor and E the number of incident electrons. For small densities not exceeding $\sim$20% of the $(\text{O.D.})_m$, the density will be a linear function of the electron flux (Tochilin and Golden, 1961). At high electron energies and/or low grain sensitivities and small grain size, multihit processes will become increasingly probable, and the equation becomes for r-hit processes (Frieser and Klein, 1960)

$$\frac{\text{O.D.}}{(\text{O.D.})_m} = 1 - e^{-KE} \left[1 + KE + \frac{(KE)^2}{2!} + \ldots \frac{(KE)^{r-1}}{(r-1)!} \right]$$
$$\tag{6.10}$$

The developed density is in most emulsions proportional to the absorbed energy in the electron energy range of 0.01 to 20 MeV (Glocker, 1960). If an emulsion is exposed to perpendicularly incident monoenergetic electrons of increasing energies, the response at first rises with

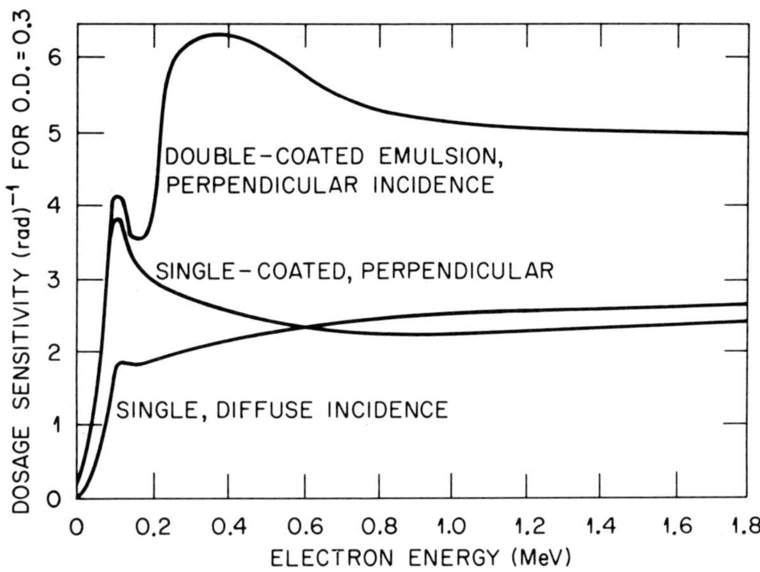

FIGURE 6-41.   Effect of monoenergetic, perpendicularly incident electrons on a single-coated and on a double-coated dosimeter film and the effect of a diffuse irradiation on single-coated film. (After Dudley, 1954.)

increasing penetration of the electrons into the emulsion (Figure 6-41). At approximately 100 keV, electrons have a range of 6 mg/cm², which corresponds to the thickness of an average emulsion if scattering is considered. From there on, the sensitivity decreases in a single-coated emulsion until the pathlength of the electrons in the emulsion and their LET become fairly independent of energy. In double-coated emulsions, electrons begin to penetrate the film base at ∿200 keV and to interact with the opposite emulsion layer, with the second maximum being less well defined because of scattering in the film base.

In practical dosimetry, electrons reaching the emulsion are neither monoenergetic nor monodirectional, but (at least in the case of radioactive beta emitters) have a continuous spectrum, which changes surprisingly little as the electrons pass through absorbers. A paper film wrapping of 30 mg/cm² corresponds to the approximate range of 0.15 MeV electrons. The beta radiation of several radionuclides such as ^{3}H, ^{14}C, and ^{35}S will, therefore, not produce any photographic effects unless the nuclide penetrates the film wrapping, for example, as tritiated water vapor. Therefore, film dosimetry of radiation workers dealing exclusively with solid compounds of these nuclides is meaningless, unless the bremsstrahlung from the

source becomes significant (i.e., for activities exceeding 1 to 10 Ci).

Increase in the absorber thickness between emulsion and beta source naturally strongly affects the response. In Figure 6-42, the relative sensitivity of a British Kodak RM film is given as a function of the maximum beta energy. Obviously, even in the unfiltered (open window) area of a badge, little response can be expected from the beta radiation of ^{45}Ca, ^{185}W, and other low energy beta emitters. Beta film calibrations are mostly carried out with uranium (thick foil, oxide layer, or compounds), which produces a rather soft and complex beta spectrum or with ^{90}Y/^{90}Sr.

In particular, in mixed radiation fields, photographic beta dosimetry tends to become highly inaccurate; the results should be considered semiquantitative at best. Numerous reports concerning the details of calibration and evaluation in "beta dosimetry" can be found in the literature (Tochilin and Golden, 1961; Kocher, 1962; Ehrlich, 1962; Miyanaga et al., 1963; McCarthy and Mejdahl, 1964; and Metalli et al., 1968).

The accuracy improves somewhat for the measurement of high energy electrons in the 3 to 100 MeV range. They produce a density corresponding to gamma radiation on a rad per rad basis, and the film badge can be calibrated fairly

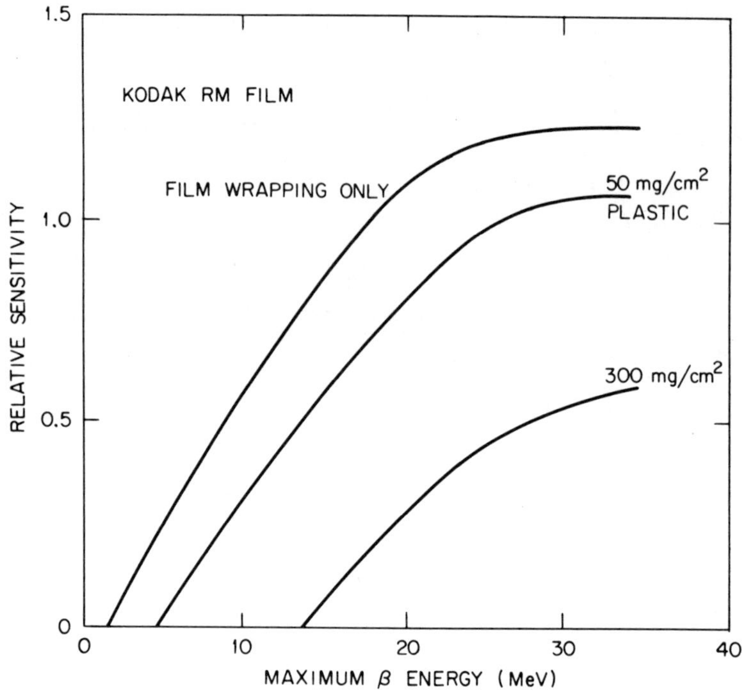

FIGURE 6-42. Sensitivity of the Kodak Radiation Monitoring Film®, without and with additional absorbers, to beta emitters with different maximum beta energies. (After Jones and Marshall, 1965.)

accurately for betatron electrons, for example, for depth-dose studies (Breitling and Seeger, 1963; Dutreix and Dutreix, 1969; Hayami et al., 1972; and others).

6.3. Optical Absorption Changes

Strange colorations of some minerals such as "blue" rocksalt and pleochroitic rings in minerals that can be attributed to radiation damage have been known for hundreds of years. It was also observed that glasses used for the storage of radium and old x-ray tubes also exhibited discoloration that could be annealed in many cases by a simple heat treatment. Many of our basic insights into the imperfections of the crystalline state have been gained by detailed studies of radiation-induced optical absorption bands in simple lattices (alkali halides) by the founding fathers of modern solid-state physics, such as Pohl and his co-workers in Germany and Seitz in the U.S., and some simple inorganic systems such as Al_2O_3 have been used in dosimetric studies (Jeltsch, 1969).

Later, it was discovered that irradiation at high dose-levels (10^3 to 10^7 rad) also leads to changes in simple organic crystals or, more importantly, in synthetic polymers. In some of them including polyethylene, polypropylene, polyvinyl chloride, polystyrol, and polyacrylamide, crosslinking of the macromolecules dominates, while in others (polyisobutylen, polymethacrylate, polytetrafluorethylene, polyvinylidene chloride) mostly a degradation of the macromolecules takes place. This leads to changes in their optical properties.

Obviously, it cannot be the purpose of this book to review the principles of radiation damage in crystals, inorganic glasses, and organic polymers, with all the lengthy discussions of solid-state physics and radiation chemistry that such a treatment would require. Instead, a brief review is given of some systems which have been specifically suggested for high-level radiation dosimetry during the past 10 or 20 years, without attempting to cover this large field comprehensively. For more detailed recent treatments of this subject, see Frank and Stolz, 1969; McLaughlin, 1970; Stolz, 1972; the Proceedings of a 1971 U.K. Panel on Gamma and Electron Irradiation-Radiation Dose and Dose Distribution in the Megarad Range; the Proceedings of a Symposium on Dosimetry Techniques Applied to Agriculture, Industry,

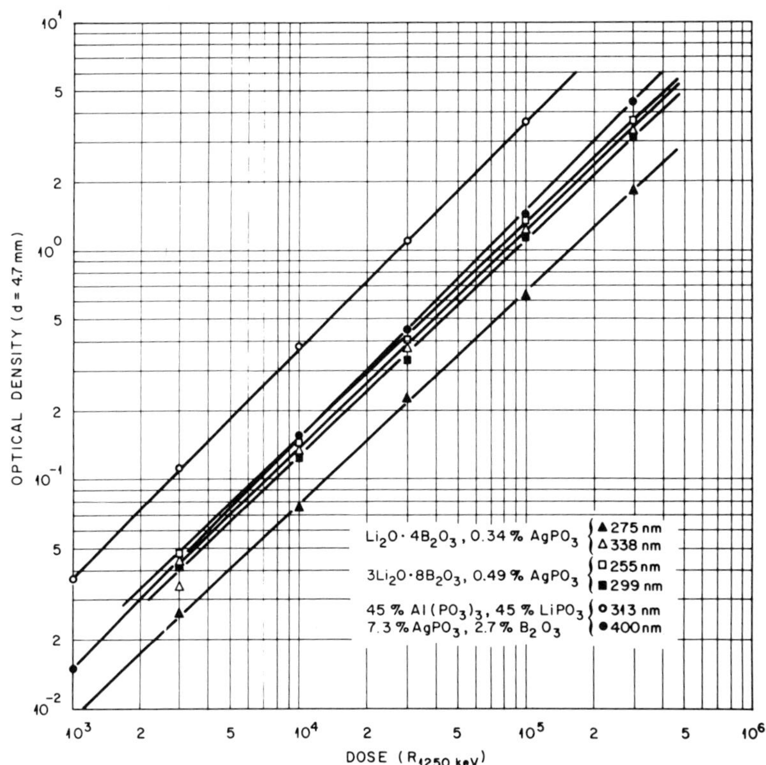

FIGURE 6-43. Optical density increase in gamma-irradiated silver-activated phosphate glass (Toshiba RPL glass) and low Z borate glass at different wavelengths. (After Cheka and Becker, 1968.)

Biology, and Medicine, IAEA, Vienna, 1972; bibliographies by Becker, 1964b, 1965b, 1967a and b, 1968a and b, and 1971; and various recent review papers such as Kantz et al., 1969; Hussman and McLaughlin, 1971; McLaughlin, 1971; Kets-Neels et al., 1972; McLaughlin et al., 1972; and Broszkiewicz and Bulhak, 1972.

6.3.1. Discoloration of Inorganic Glasses

The first reported use of a glass for high-level dosimetry dates back to 1951 (Day and Stein), but the antimonate glass used in these early studies in the 10^5 to 10^8 rad gamma radiation range was rather unstable. In 1955, Schulman et al. suggested the measurement of optical absorption changes which occur in the 10^4 to 10^6 rad range in silver-activated phosphate glasses for high-level gamma radiation measurements. They and others (Davidson et al., 1956), who studied the same glass which had been developed primarily as an RPL dosimeter for the U.S. Navy (see Chapter 4), compared the optical absorption spectrum of an irradiated with an unirradiated glass and found that the measurement of some wavelengths produced a more linear and thermally more stable effect than at others.

Most change occurred in the unactivated glass at $\sim$540 nm and in the Ag-activated glass at $\sim$315 nm. The effect of rapid fading of the radiation-induced UV absorption band may be reduced either by delaying the measurement or by a short stabilizing heat treatment which anneals most of the less stable color centers; 10 min at 130°C has been suggested for this purpose. Of course, these glasses exhibited a strong energy dependence, which can be partially compensated by proper metal filters.

Other, more advanced silver activated phosphate and borate glasses with a lower energy response (Figure 4-8) and improved stability were studied by Becker (1965a) and Cheka and Becker (1969). By measurements of different wavelengths between 255 and 550 nm, a sensitivity range from $\sim$$10^2$ to 10^7 rad gamma radiation can be covered (Figures 4-14, 6-43). Sensitivity, linearity, and stability depend on silver content, heat treatment,

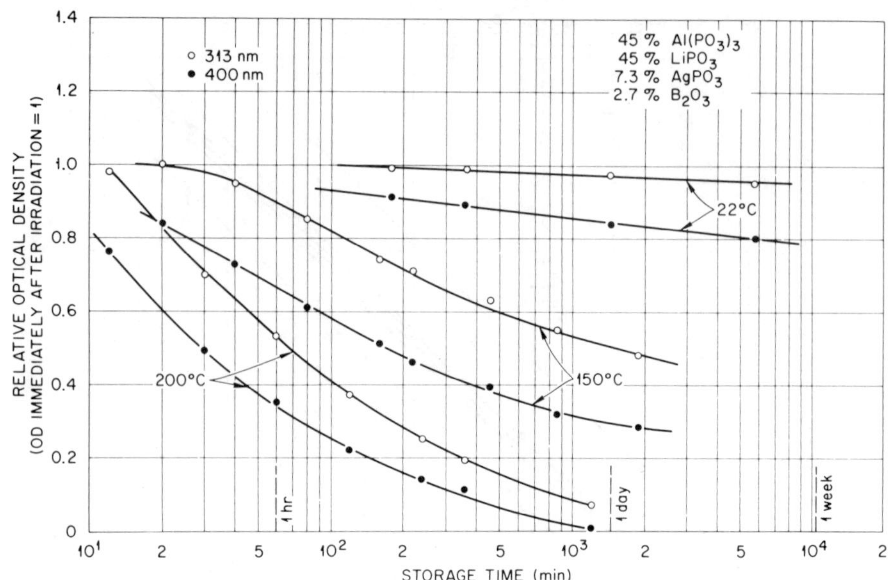

FIGURE 6-44. Relative net optical density (normalized for reading immediately after exposure) at two wavelengths and three different storage temperatures in conventional Toshiba RPL glass. (After Cheka and Becker, 1968.)

storage temperature, glass base composition, and on the wavelength at which the absorption change is measured (Figure 6-44).

In the standard Toshiba-type RPL glass, heating for $\sim$30 min at 400°C totally anneals the discoloration (Figure 4-15). The fading stability is highest at low silver concentrations ($\sim$0.1% Ag). Intense UV exposure also bleaches the absorption band(s). Only little qualitative changes take place in the absorption spectrum of the phosphate glass during bleaching (Figure 6-45), but rather substantial changes have been observed in the low Z borate glass. It may be used at temperatures up to 200°C if the proper silver concentration and wavelength are selected (Cheka and Becker, 1969).

Conventional glasses including microscope slides also have been used in the 10^5 to 10^6 rad range, but exhibit rapid fading (Taimuty et al., 1958). Numerous other different glasses have been studied in attempts to increase the dynamic range and stability or reduce the energy response. For example, a cobalt borosilicate glass (Bausch & Lomb type F-0621; composition: 62.5 mol % SiO_2, 10.6% Na_2O, 20.8% B_2O_3, 6.0% Al_2O_3, and 0.1% CO_3O_4) was reported to exhibit less fading (1 to 2% in one day) and have a lower atomic number than the earlier high Z silver phosphate glasses used by Schulman et al. (1955). This glass exhibits a linear response in the 10^3 to 10^6 rad range (Kreidl and Blair, 1956). The precision that can be obtained is around 2%. Its properties, including photon energy response (Myers and Kathren, 1966), temperature effects (Lee and Ziemer, 1970), and dose-rate dependence (Gibson et al., 1968 and Martin et al., 1971), have been carefully studied (see also Gibson and Stützer, 1968).

Bishay (1961) developed a bismuth lead borate glass containing As_2O_3 with a very pronounced photon energy dependence, but a wide dynamic range (10^4 to 10^9 rad) and very little fading. Another, manganese-activated arsenic borate glass that develops a stable absorption peak at 515 nm (Bishay and Arafa, 1967) is less energy dependent, but requires a post-irradiation heat treatment. Other manganese and antimonate glasses had been found in previous studies to be not sufficiently stable.

A bismuth silicate glass (Pye et al., 1964) covers the dose-range 10^6 to 10^{10} rad (linear up to 10^9 rad), but requires post-irradiation heat treatment. Another glass containing 1 to 5% by weight of antimony oxides, and 0.1 to 0.8% MnO_2 is unaffected by heat up to 300°C (Hood, 1965). Others (Sqentirmay et al., 1965) suggest a soda-lime-silicate glass.

Several other glasses have also been recommended. Among the more interesting ones is

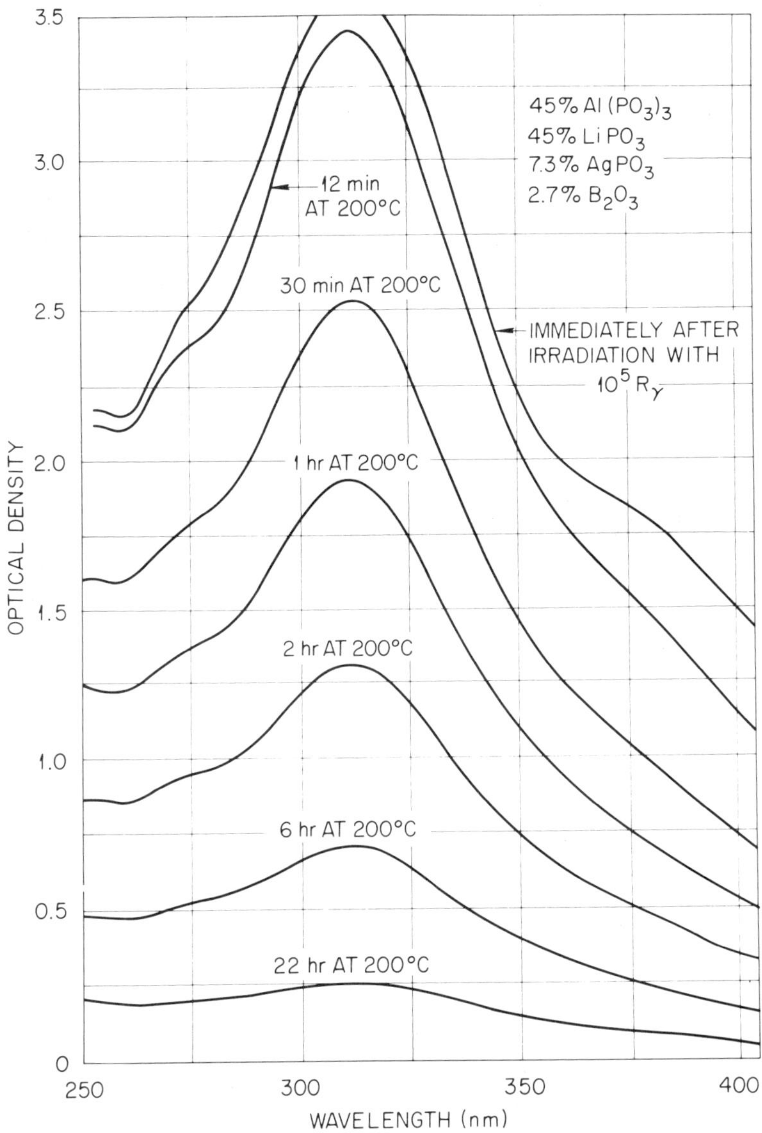

FIGURE 6-45.   Radiation-induced change in the UV absorption spectrum of a 4.7-mm-thick sample of Toshiba RPL dosimeter glass immediately after exposure to 10^5 R of ^{60}Co gamma radiation and after storage for different times at 200°C. (After Cheka and Becker, 1968.)

a glassy magnesium metaphosphate for the 5×10^4 to 10^7 rad range (Kügler, 1959), and a commercial glass containing SiO_2, B_2O_3, Na_2O, CaO, MgO, and small amounts of Cr_2O_3 which color the unirradiated glass green (made by VEB Jenaer Glaswerke Schott & Gen., Jena, East Germany). Irradiation produces a progressive bleaching of the 550 nm absorption band in this glass (Figure 6-46). Its fading amounts to 6% during one week and 10% during one month.

Of course, the "sensitivity," e.g., minimum change in the optical density at a given wavelength which can be detected with reasonable accuracy, can be increased or decreased within rather wide limits by changing the pathlength of the light in the glass, in the simplest case by varying its thickness. In Figure 6-47, the calibration curves of a commercial German dosimeter glass (PDG-11 made by Schott & Gen., Mainz — Jeltsch and Graf, 1968 and Jahn, 1969) are given for different

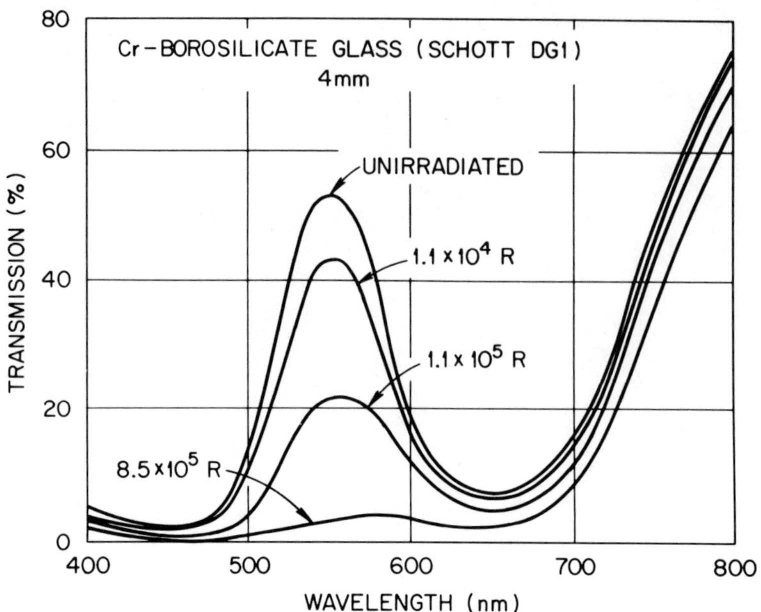

FIGURE 6-46.   Optical absorption spectrum in a green commercial dosimeter glass (Schott, Jena, East Germany) for different gamma irradiations. (After Frank and Stolz, 1969.)

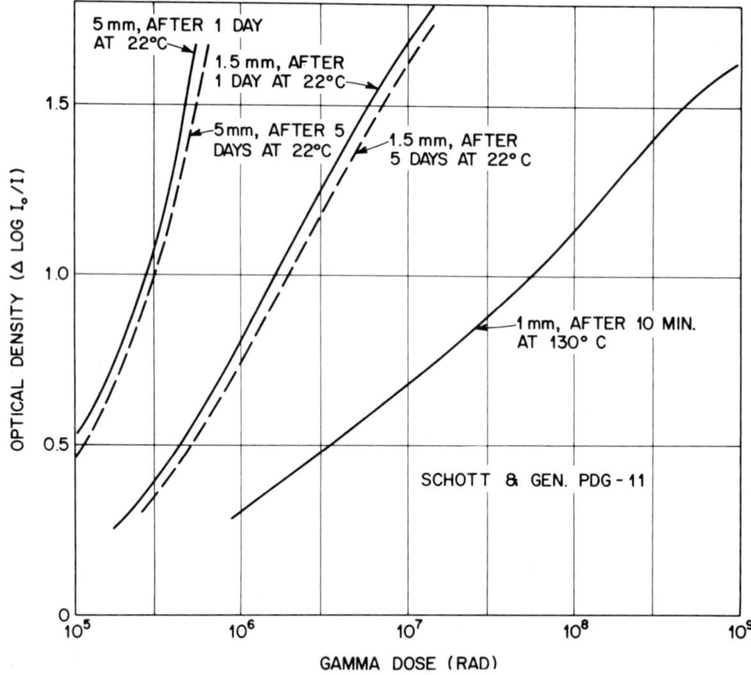

FIGURE 6-47.   Change in optical density of commercial dosimeter glass (Schott, Mainz, West Germany), as a function of gamma radiation dose for different glass thicknesses and storage times. (After Jeltsch and Graf, 1968.)

thicknesses. In more sophisticated devices, sensitivity may, for example, be increased by multiple reflection of the light beam, or by using long rods of the glass. Among the more exotic glass systems that have been suggested is a porous glass impregnated with dyes bleached by irradiation.

6.3.2. Organic Glasses, Films, and Gels

In Table 6-5, some radiation effects of dosimetric interest in irradiated clear polymers are compiled. In transparent, undyed polymers, changes occur mostly in the absorption in the UV region, extending into the visible region at higher dose-levels due to radiation-induced crosslinking and/or degradation, with radical as well as ionic processes involved. Formation of stable free radials and polyene double bonds are mostly responsible for the absorption changes.

As in glasses, one uses the difference in the optical density of an irradiated and an unirradiated, otherwise identical, sample as a measure of absorbed dose. As a rule, thick samples are used close to the lower end of the dynamic range, and thin foils close to its upper end in the 10^9 to 10^{10} rad range. The upper limit is determined by saturation of the discoloration, formation of bubbles of radiolytic gases (H_2, CO, CO_2, and CH_4, plus HCl and HOCl in chlorinated polymers) in the plastic, and increasing brittleness of the foils.

The visible discoloration of the polymer is usually yellowish-brown. In polyvinyl chloride, the color changes run from green to yellow to reddish-brown. For a detailed description of the use of one particular polyvinyl chloride, manufactured under the name OCA 5960[R] by the United Lamination Co., Mayfield, Pa., in high-level dosimetry, see Artandi (1970). Depending on their molecular stability, the plastics content of monomers, stabilizers and other additives, and on environmental factors (in particular the presence of oxygen is known to affect the mechanism of radiation damage because it reacts with free radicals, but intense light, humidity, and various vapors and gases are also known to affect sensitivity and/or stability), the dynamic range which can be covered varies with different plastics by orders of magnitude. Teflon[R] is one of the more stable and polyvinyl chloride is one of the radiation-sensitive polymers.

The radiation response of the same plastic may also vary substantially, both qualitatively and quantitatively, depending on type and manufacturer. For example, polymethyl methacrylate (Perspex[R], Lucite[R], Plexiglas[R]) exhibits absorption bands at 272, 280, and 290 nm after gamma irradiation, according to a study by Orton (1966), while others report for different types of the same polymer absorption bands at 250 nm (Perspex type E) or at 265 nm for Perspex HX[R]* (Rizzo et al., 1967; Whittaker and Lowe, 1967; Fleming, 1968; and Chadwick, 1972). Still others (Berry and Marshall, 1969) recommend 305 nm as the best wavelength for O.D. measurements in the same Perspex HX, to be used as a relatively stable (Figure 6-48) reference dosimeter for electron and gamma radiation (Figure 6-49). Sources of error in this material have been studied by Bolton (1969) and Chadwick et al. (1972).

Polyethylene terephyhalate (Melinex[R], Mylar[R]) is usually measured at 325 nm (Ritz, 1961 and Oshima and Tanaka, 1967). In cellulose acetate (diacetate, if 54% of the hydroxyl groups are substituted by acetate; triacetate with $\sim$60% substitution), optical density changes between 250 and 290 nm have been used (Figure 6-50), for example, for electron depth-dose measurements (Aiginger and Hubeny, 1965). The optical density at 270 nm was found to be most stable, with no change occurring during one month at room temperature. Low density polyethylene (Lupolen 184OD and 180OH[R] made by BASF, Ludwigshafen, Germany) has been used for food sterilization dosimetry in the 2.5 to 5.5 Mrad range (Mehringer, 1971).

The plastics may be used as foils of varying thickness, pressed pellets, strippable plates, or as a thin coating. Unsupported polymer foils of some materials such as Mylar can be made as thin as 1 μm. Multiple reflection effects may interfere with the evaluation of such thin foils (Bishop and Benson 1972). Sometimes extremely thin films are desirable. Their production is difficult because of an effect called nucleation (formation of crystalline spherulates), but special techniques, such as gas-phase photodeposition on metal surfaces, permit production of layers of only $\sim$0.1 μm thickness.

---

*Made by Imperial Chemical Industries, Welwyn Garden City, Herts., England. Sheet-to-sheet reproducibility is often exceedingly poor.

TABLE 6-5

Optical Absorption Changes in Some Clear Polymers and Polymer-dye Systems of Dosimetric Interest

Compound	Trade name	Dose-range (rad gamma)	Wavelength for evaluation (nm)	Effective atomic number	Stability	References
Polymethyl-methacrylate	Perspex[®] Lucite[®] Plexiglas[®] Transpex[®] PMMA[®]	$10^5 - 10^7$	250–305	6.5	5–10%/month	Boag, 1963; Orton, 1966; Rizzo et al., 1967; Kostalas, 1967; Whittacker and Lowe, 1967; Fleming, 1968; Berry and Marshall, 1969; Bolton, 1969; Tanaka et al., 1970; Ellis, 1971; Dealler, 1971; Chadwick, 1972; and Chadwick et al., 1971a
Melamine		$4 \times 10^4 - 10^7$	Near UV		Unstable	Birnbaum et al., 1955
Polyethylene	Lupolen[®]	$10^5 - 4 \times 10^7$	UV, infrared			Lentsch and Finston, 1968; Handley and Gressenbachen, 1970; and Mehringer, 1971
Cellulose acetate butyrate						Rizzo and Krishnamurthy, 1969
Polyethylene terephthalate	Mylar[®] Melinex[®] Hostaphan[®]	$10^6 - 10^9$	325	6.5	No fading	Ritz, 1961 and Oshima and Tanaka, 1967
Polytetrafluoro-ethylene	Teflon[®]					Scherer and Kline, 1967 and Tamura et al., 1970
Cellulose acetate	Triafol[®] Supraphan[®] Ultraphan[®]	$10^6 - 10^8$	270	6.9	No fading	Kugler and Scharmann, 1959 and Aiginger and Hubeny, 1965
Polyvinylidene chloride	Saran[®]	$5 \times 10^4 - 10^7$	260	15.3	Increase, dose dependent	Harris and Price, 1961

TABLE 6-5 (Continued)

Optical Absorption Changes in Some Clear Polymers and Polymer-dye Systems of Dosimetric Interest

Compound	Trade name	Dose-range (rad gamma)	Wavelength for evaluation (nm)	Effective atomic number	Stability	References
Polyvinyl chloride	PVC[®] Suprotherm[®]	$10^5 - 10^7$	396	13.9	Strong absorption increase after exposure	Maul et al., 1961; Behrens, 1966; Popovic, 1967; and Gupta and Bhat, 1970
Polystyrene		$10^6 - 10^8$			50% in 4 days	Fowler and Day, 1955
Polyvinyl fluoride		$10^6 - 3 \times 10^7$	3.5			Windley and Elleman, 1967 and 1968
Transstilbene in Polystyrene Polyvinyl alcohol						Harrah, 1969 Napali and Cortellessa, 1968
Red-dyed polymethyl methacrylate	Perspex[®] Red 4034[®]		615		Increase	Day and Stein, 1961; Brown, 1969; and Whittaker, 1970 and 1971
Dyed cellophane		5×10^5 to 1.5×10^7	Varies			Henley and Richman, 1956; Goldblith and Mateles, 1958; Menkes and Goldstein, 1966; Stolz, 1966; and Rauch and Andrew 1966
Polyvinyl alcohol with methyl orange		$2 \times 10^5 - 10^7$	Visual			Hübner et al., 1970
Leuco-compounds of triphenyl methane dyes in polystyrene		5×10^3 to 5×10^7	Varies (640 for malachite green)		8% fading/month	Stolz and Prokert, 1965 and Prokert and Stolz, 1970

TABLE 6-5 (Continued)

Optical Absorption Changes in Some Clear Polymers and Polymer-dye Systems of Dosimetric Interest

Compound	Trade name	Dose-range (rad gamma)	Wavelength for evalua-tion (nm)	Effective atomic number	Stability	References
Leuco-compounds of triphenyl methane dyes in gelatin coated on paper		$\sim 10^4 - 10^8$	Varies			Chalkley, 1964; McLaughlin, 1970; Eisen 1972; and Eisen et al., 1972
Azo dyes in polyvinyl chloride		$10^6 - 10^7$	Visual			Farrell, 1963
Azo dye, halogenated hydrocarbon in paraffin	HAP®	5×10^2 to 4×10^3	Visual		With phenole added no fading	Potsaid and Irie, 1961
Bromcresol purple, trichlorethylene in agar		$\sim 10^5$				Pestaner and Gevantman, 1958
Methylene blue or methylene yellow in polyphenyl alcohol		$10^4 - 10^6$	Bleaching			Lawrentowitsch et al., 1965
Methylene blue in agar or gelatin		$4 \times 10^2 - 10^5$				Day and Stein, 1950
Methylene blue in polyvinyl alcohol						Oster and Broyde, 1959
Leuco-dyes of di- or triarylmethane, hexachloroethane, etc.		$\sim 10^4$	Varies			Pförtner, 1962
Methyl red in agar		$10^3 - 10^4$	540			Andrews et al., 1957

TABLE 6-5 (Continued)

Optical Absorption Changes in Some Clear Polymers and Polymer-dye Systems of Dosimetric Interest

Compound	Trade name	Dose-range (rad gamma)	Wavelength for evalua-tion (nm)	Effective atomic number	Stability	References
pH paper		$10^5 - 10^6$	Visual			Huff and Mayo, 1968
Starch-iodide in agar		$10^4 - 10^5$				Devantman et al., 1957
Phenol-indo-2.6 dichlorophenol in gelatin or agar		$10^4 - 6 \times 10^4$	Bleaching			Day and Stein, 1957
Resazurine in gelatin or agar		$\sim 10^5$				Proctor and Goldblith, 1950
Methyl orange in polyvinyl alcohol						Hübner, 1971

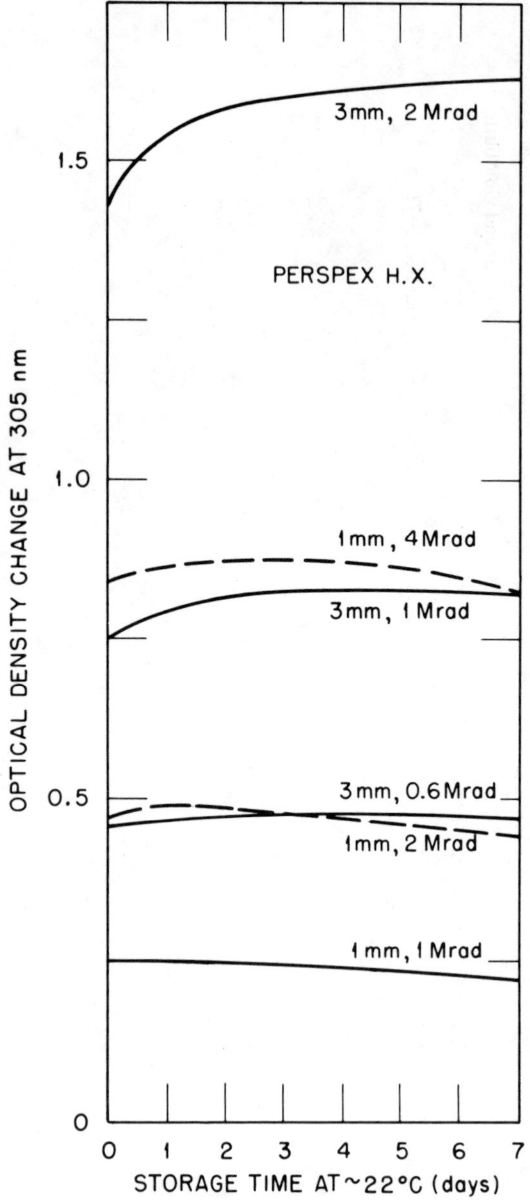

FIGURE 6-48.   Fading and build-up of optical density at 305 nm in clear Perspex HX® for different gamma radiation dose-levels and plastic thicknesses. (After Berry and Marshall, 1969.)

As many high dose-level irradiations for industrial purposes, including wood impregnation with polymers, paint curing, graft polymerization, and crosslinking, employ plastics which exhibit easily measurable changes in their physical properties, these processes can be considered their own dosimeters, with no need for an additional radiation measuring device.

Basic advantages of clear plastics, as well as the plastic-dye compounds discussed below, are that they are simple to use and to evaluate, inexpensive, and close to tissue equivalence in their energy response. The remaining negative or positive photon energy dependence can, in principle, be completely compensated for a wide energy range by mixing adequate amounts of chlorinated and nonchlorinated polymers. The facts that very thin foils can be used, and that high resolution scanning with a micro-densitometer is possible, make foils particularly attractive for depth-dose studies with nonpenetrating radiation (as an example, the measured depth-dose of monoenergetic electrons in polystyrene is given in Figure 6-51).

Main disadvantages are a very limited linear response range (because of the exhaustion of reaction components, radiation-induced chemical reactions are rarely linear) and more or less pronounced changes with time due to the limited halflife or delayed formation of many of the free radicals and other radiolytic compounds. As in many other systems exhibiting thermal fading or a slow increase after the end of the exposure, a stabilizing heat treatment partially solves the latter problem.

Another requirement, the need for a spectrophotometer for measurements in the UV region, is easily avoided by employing plastics which contain organic dyes or chromogens. Such systems also provide an immediate visual indication of radiation dose and they are usually more sensitive than the clear polymers. Three main reactions in such systems have been used for dosimetry:

1. radiation-induced destruction (bleaching) of dyes in plastics, such as red Perspex;

2. dyes which change color in chlorinated polymers as a result of HCl production, such as Kongo red in polyvinyl chloride; and

3. transformation of proto-dyes, such as the leuco-compounds of triphenylmethane dyes in polystyrol, into the actual dyes.

With methylene blue or other dyes in agar gels (with sodium benzoate added as a scavenger for hydroxyl radicals), 400 rad of gamma radiation were reported to produce a measurable decrease of 0.02 in optical density (Day and Stein, 1950). The response was linear up to $\sim 10^4$ rad. Films of methylene blue in polyvinyl alcohol have been bleached by exposure to visible light; gamma

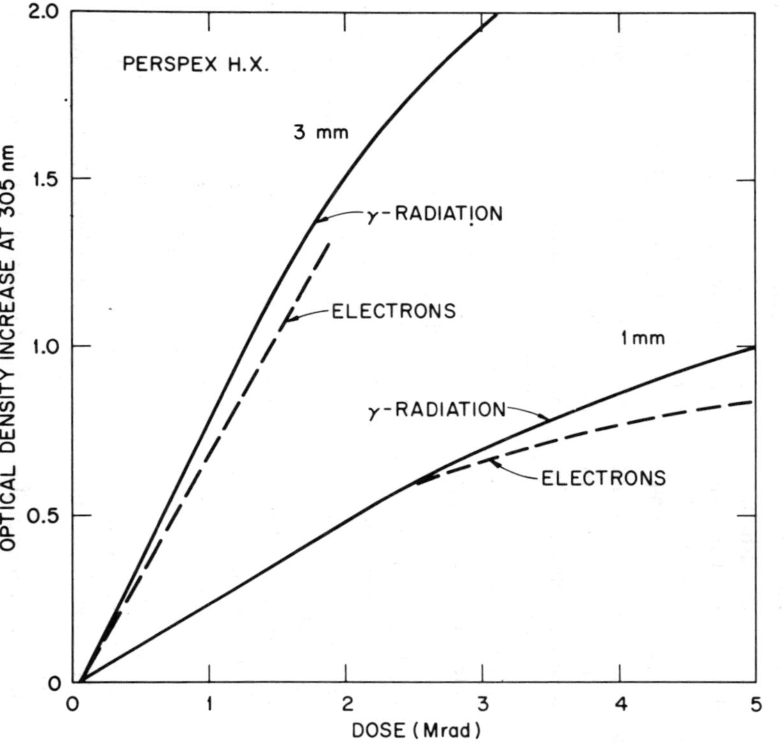

FIGURE 6-49.   Response of Perspex HX measured one day after exposure to gamma and electron irradiation. (After Berry and Marshall, 1969.)

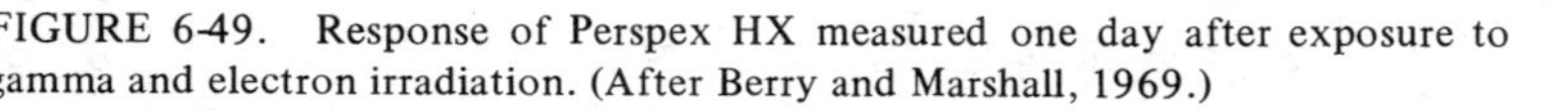

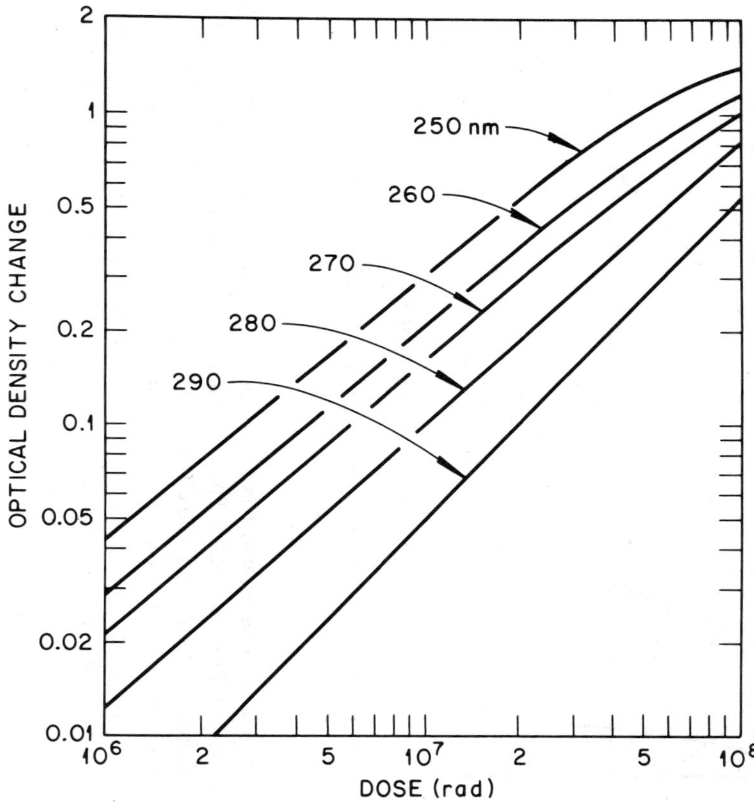

FIGURE 6-50.   Change in the optical absorption of a 30 μm cellulose acetate foil at different wavelengths as a function of X-radiation dose of ∿80 keV effective energy. (After Hofman, 1963.)

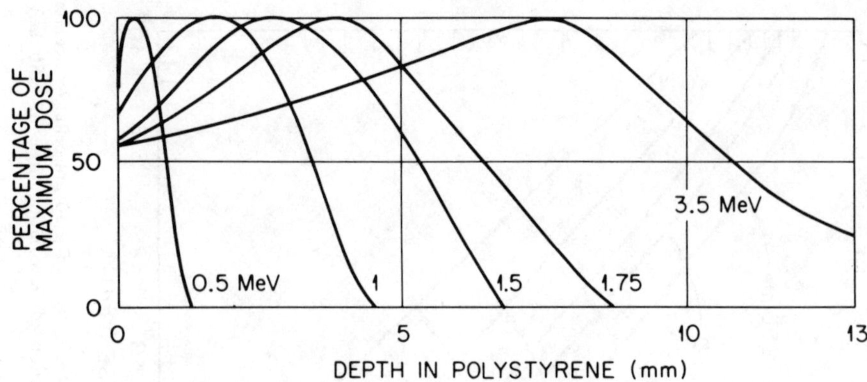

FIGURE 6-51. Depth-dose of monoenergetic, normally incident electrons in polystyrene. (After Prokert and Stolz, 1970.)

irradiation restored the blue color, as little as 1 rad being detectable (Oster and Broyde, 1959). The G-value (number of changed molecules per 100 eV of absorbed energy) in this system was 10^4, indicating the presence of chain reactions. A stability of such systems usually is poor. Later, somewhat similar foils containing di- or triaryl-methane leuco-dyes, hexachloroethane, plus other additives were made by Pförtner (1962). In these foils the liberation of chlorine ions regenerates the dye. They were insensitive to visible light.

Another system that attracted some attention consisted of slabs of halogenated hydrocarbon, an azo dye such as methyl yellow in chloroform, and paraffin (Potsaid and Irie, 1961). The light yellow color of this "HAP" system changes in the 500 to 4,000 rad range (with a thickness of 1 to 2 cm; doses as little as 5 rad have been detected) into red, but it was also light-sensitive and thermally unstable. Others used a system with similar sensitivity consisting of trichlorethylene and bromcresol purple in a gel (Pestaner and Gevantman, 1958).

The radiation-induced bleaching of methylene blue, yellow, or orange in thin polyvinyl alcohol foils has been used in the 10^4 to 10^6 rad range (Lawrentowitsch et al., 1965). Additions of boric acid made such materials sensitive to thermal neutrons in the 10^{12} to 10^{14} n/cm^2 range. There is little LET dependence of the response (G-value 1.6 for alpha particles and 1.8 for ^{60}Co gamma radiation), but fading amounts to $\sim$10% in two weeks. A similar system has been proposed as a threshold detector for medical irradiation sterilization (Hübner and Stolz, 1969).

One of the more widely used dosimeters of this type is a commercial red-dyed PMMA, available under the tradename Perspex Red 4034® (formerly Perspex Red 400®) from Imperial Chemical Industries, Welwyn Garden City, Herts., England. It is available in sheets of $\sim$3 mm thickness and begins to develop a new absorption band in the 600 to 700 nm region at doses exceeding $\sim$10^5 rad. The response in nonlinear doses and doses higher than 5 x 10^6 to 10^7 rad become more difficult to measure accurately (Day and Stein, 1961; Brown, 1969; and Whittaker, 1970). The broad, radiation-induced absorption band peaks at 615 nm. After exposure, an increase in optical density is observed during the first $\sim$15 to 25 days, depending on the dose-level and the wavelength at which the absorption is measured (Figure 6-52). Diffused oxygen is instrumental in this consecutive fading process, which begins to take place at the surface of the sample and proceeds into its interior.

Dyed cellophane foils have been another popular high-level dosimeter. A blue-dyed cellophane 25 μm thick was first used in 1956 by Kenley and Richman in the range between 5 x 10^5 and 1.5 x 10^7 rad, but the uniformity of the foil amounted to ±5 to 10% (Goldblith and Mateles, 1958). In more recent studies (Menkes and Goldstein, 1966 and Goldstein, 1970), Cinemoid® color films have been used which are available in over 60 different colors for theatrical lighting from Strand Electric and Engineering Co., London, England, in 50 x 60 cm foils 0.25 mm thick. The dose-range that is covered with the best-studied of the two foils, No. 25 and No. 48 (Figure 6-53), is about 5 x 10^5 to 5 x 10^7 rad.

No dose-rate effects have been observed up to

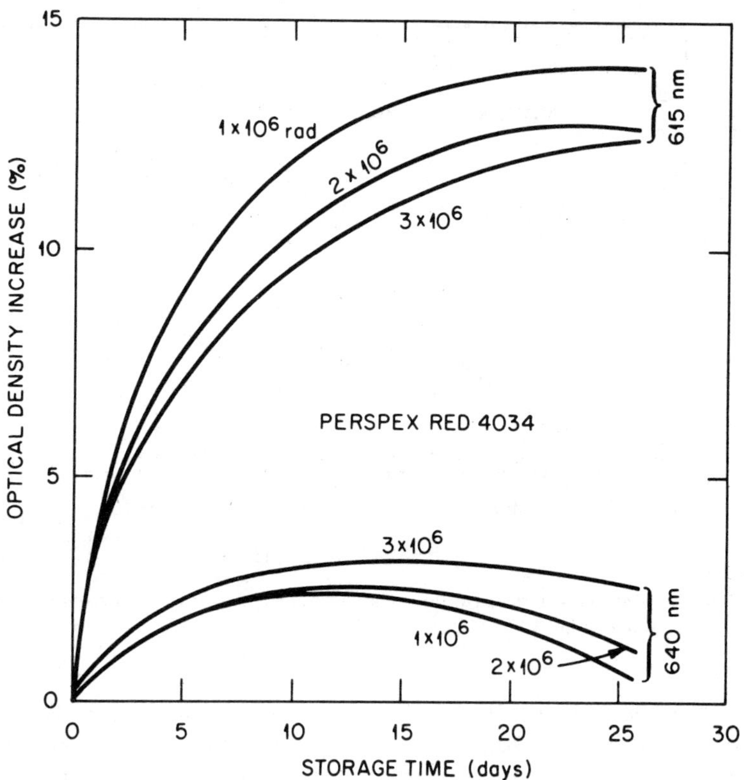

FIGURE 6-52.   Increase in the optical absorption of a commercial Perspex Red Type 4034 plastic after exposure for different dose-levels and wavelengths of measurements. (After Whittaker, 1970.)

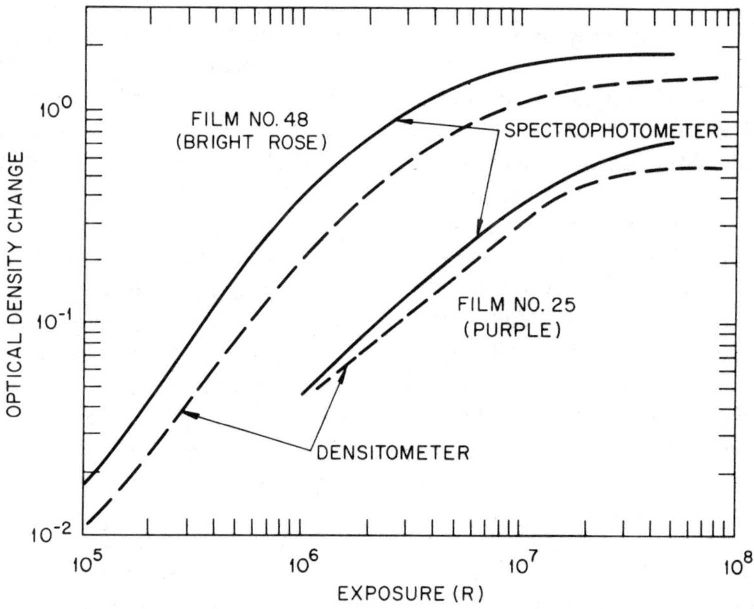

FIGURE 6-53.   Change in optical density as a function of gamma radiation exposure, with spectrophotometry of Film No. 25 at 437 nm (densitometry with a color filter with dominant transmission at 460 nm), and 530 nm (densitometry at ∿545 nm) for No. 48. (After Goldstein, 1970.)

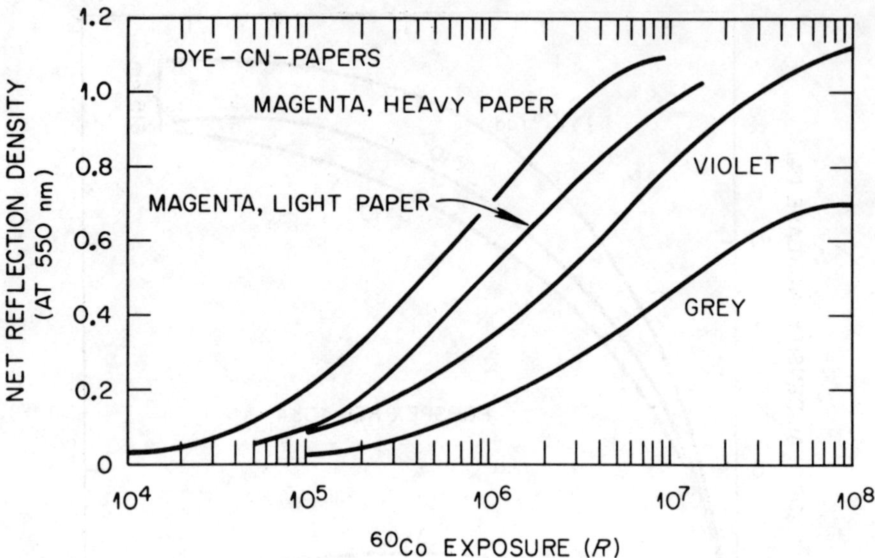

FIGURE 6-54.   Response curves of various papers impregnated with radiochromic dye-cyanide systems. (After McLaughlin, 1970.)

10^{14} R/sec at the 4 Mrad electron dose-level, but others found a dose-rate effect when 10 Mrad of electron energy were deposited in 30 to 40 nsec (Rauch and Andrew, 1966). Stability is claimed for the color changes up to 60°C (Stolz, 1966). Dyed cellophane foils are particularly suitable for depth-dose studies of short-range types of radiation.

Polyvinyl alcohol has frequently been used as a host for dyes. With tetrazolium blue and thicknesses between 50 μm and several millimeters, a dose-range from 5 x 10^3 to 5 x 10^6 rad may be covered, and there is little fading. If the foils are impregnated with ^{6}Li or ^{10}B, thermal neutron doses around 10 to 300 rad may be measured (Taplin and Malin, 1961). A foil has been made by evaporation of a solution of polyvinyl alcohol, chloral hydrate, sodium tetraborate, and methyl orange in water (Hübner and Stolz, 1969), which can serve as some sort of threshold detector for visualizing and adjusting the beam of accelerators (for the same purpose, polyvinyl chloride with azo dyes had been suggested earlier – Farrell, 1963).

Others prepared solid chemical systems from leuco-triphenyl methane dyes (malachit green, crystal violet), hexachlorcyclohexane, and tetrabromethane in polystyrene (Stolz and Prokert, 1965 and Prokert and Stolz, 1970); or leuco-cyanids of triphenylmethane dyes in gelatin (McLaughlin, 1970). The latter system has been studied particularly carefully. With coatings of 1 to 20 mg/cm^2, dose ranges of $\sim$$10^5$ to 10^8 rad are covered. With gelatin wafers 1 cm thick, the range is $\sim$$10^3$ to 10^6 rad.

Certain types of white paper have also been impregnated with such solutions since cellulose is a good activator for the sensitive molecule (Chalkley, 1964). Such papers have been produced with a highly reproducible response and sensitivities, depending on dye concentration and paper thickness, in the 10^4 to 10^8 rad range (Figure 6-54). There appears to be no dose-rate dependence of such systems up to at least 10^{15} rad/sec. This makes them ideal for pulsed-beam dosimetry (Eisen et al., 1972).

Numerous other systems have been described, but most did not find any widespread application. Among them is a porous glass (Corning 7930®) impregnated with dyes for the 10^6 to 10^7 rad range (Brocklehurst, 1959); even color changes in ordinary pH indicator paper have been suggested (Huff and Mayo, 1968). For a list of further proposed systems, see Table 6-5.

6.4. Miscellaneous Solid-state Detectors

There are numerous other radiation effects on solids which have, at one time or another, been suggested as potential dosimeters. Many of them turned out to be clearly inferior to already available techniques, while the capabilities of

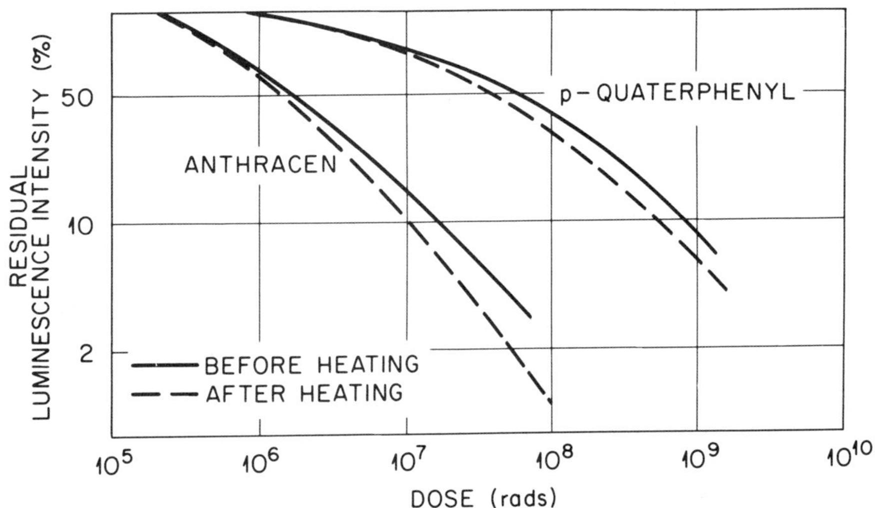

FIGURE 6-55. Luminescence degration (excitation wavelength 365 nm) in anthracene and *p*-quaterphenyl as a function of ^{60}Co gamma radiation dose before (dashed curve) and after 1 hr stabilization at 100°C. (After Attix, 1959.)

others have apparently never been fully explored. The following compilation, which is by no means complete, is intended to give an impression of the type of effects which have been studied.

Luminescence degration — It was first observed by Birks (1950) that intense bombardment of an anthracene crystal permanently and substantially decreases its scintillation efficiency due to the production of chemically altered species which disturb the intermolecular energy transfer (quenching). In mixed polymer scintillators, radiation damage in both the host and the scintillator quenches the luminescence output. In some materials such as naphthalene, a degration is observed during excitation at 254 nm, but a radiation-induced luminescence increase occurs for excitation at 365 nm due to the formation of a new luminescence band at 464 nm.

Schulman et al. (1957) and Attix (1959) first applied this principle to high dose-level measurements. They used pressed wafers of anthracene and *p*-quaterphenyl, and measured the fluorescence at 444 nm during excitation with the 365 nm mercury line before and after irradiation, thus covering a gamma radiation dose-range from $\sim$5 x 10^5 to $\sim$5 x 10^7 rad with anthracene, and 10^7 to 5 x 10^9 rad with *p*-quaterphenyl (Figure 6-55).

The luminescence degration follows a simple equation

$$\frac{I}{I_o} = \frac{1}{1 + AD^n},$$

(6.11)

with I_o being the luminescence intensity before, and I after irradiation, D the dose, A a characteristic constant for the material, and n an exponent smaller than or equal to 1. Not all luminescence peaks are equally affected by irradiation, but there are shifts in the peak ratios. The radiation-induced discoloration of plastics reduces the intensity of the exciting as well as the luminescence light, an effect which is superimposed to the actual luminescence degration.

The luminescence degration fades slightly (2 to 3% during 100 hr at room temperature). This process may be accelerated with a 1 hr heat treatment at 100°C, thus reducing the fading rate to less than 4%/month. No dose-rate effects have been observed in the 5 x 10^4 rad/hr to 1.6 x 10^8 rad/hr range. Among the advantages of luminescence degration dosimeters are low Z, good fast-neutron sensitivity because of their high hydrogen content (Richter and Schneider, 1968), low costs and easy evaluation, no annealing during repeated measurement, and no saturation.

Disadvantages are relatively poor accuracy (±10%) and sensitivity and nonlinear response. Self-sticking tapes containing organic scintillators, for example, made by embedding anthracene crystals in gelatine films (Ilic-Papovic and Hjortenberg, 1969), have been used in the automatic dose control during radiation processing of food. For a study on the structural dependence of luminescence degration, see Stolz and Schmitt

(1967); for other recent information on this subject, see Frank and Stolz (1969) and Doroshenko et al. (1971).

Electron spin resonance — Irradiation of many organic compounds creates rather stable free radicals detectable by electron spin resonance spectroscopy. In a-alanine, for example, doses in the 100 to 10^5 rad range can be measured with a precision of up to ±5% and only 5% fading occurs in three months (Bradshaw et al., 1962; Descours et al., 1970; and Berman et al., 1971). Amine salts of organic acids with a hydrogen content of $\sim$10%, making them quite sensitive to fast neutrons, have also been suggested for mixed field dosimetry (Peters, 1972).

Other compounds that have been studied (at very low temperatures) include ammonium dihydrogen phosphate and ammonium sulfate (Fuller et al., 1972). Attempts have also been made to use ESR signals in irradiated tissue, keratine (fingernails), and other substances for dosimetry, but have been met with limited success because of poor sensitivity and/or stability of the radiation-induced effects. Also the evaluation instrument is rather expensive (Manambelona, 1972).

Light scattering — Rayleigh scattering of red (He-Ne) laser light in transparent polymers has been suggested as a possible method for fast-neutron dosimetry with a potentially good discrimination against low LET (gamma) radiation. Preliminary tests with Lucite® (Wagner, 1972) in the $\sim$10 to 40 krad neutron dose-range show some promise, but it is not likely that this technique will become of interest to personnel dosimetry.

Thermally stimulated conductivity — Conductivity "glow curves" resulting from the mobility of de-trapped charge carriers during the heating of pre-irradiated solids have been reported for several inorganic materials such as zinc sulfide, tourmaline, and diamond (Henisch and Bose, 1968). Dose-dependent pyroelectric processes have also been observed in these materials. Thermally stimulated currents also occur in some polymers such as polyethylene (Ramsey, 1972). They are reasonably reproducible if voltage gradient and heating rate are kept constant. The lowest detectable dose in polyethylene was $\sim$500 rad, above which the response is linear up to at least 20 krad (Figure 6-56). As the "glow peak" is very low ($\sim$ 60 to 70°C), fading occurs very rapidly.

Thermally stimulated current — It has been

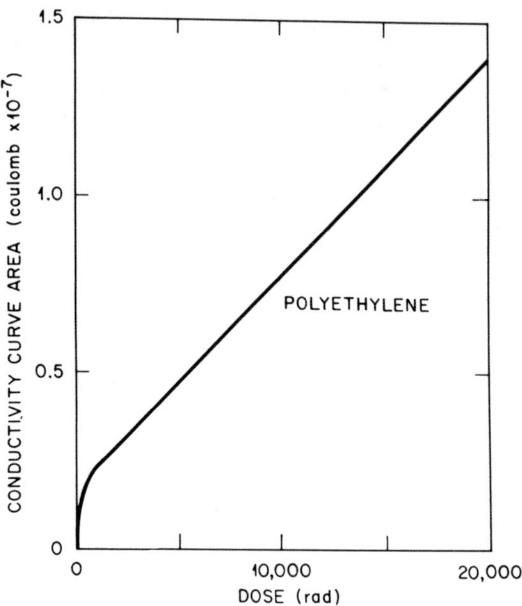

FIGURE 6-56. Integrated thermally stimulated conductivity (area under the "conductivity glow curve") as a function of dose in polyethylene. (After Ramsey, 1972.)

observed that heating discs of a pre-irradiated special borosilicate sealing glass, as used to make transistor heads, released trapped charge carriers (Mitchell and DeNure, 1972). Heating from one side only produces a temperature gradient and, consequently, a thermally stimulated current across the glass. Unfortunately, the studied system is not of dosimetric interest because of the low-temperature ($\sim$100°C) current peak producing very rapid fading (80% during one day at 23°C, or during 3 hr at 40°C) and because of poor sensitivity (lowest "measured" dose $\sim$10^3 rad). It may, however, be possible to find other, more suitable materials which exhibit the same effect.

Triplet exciton annihilation — The mobile triplet exciton bimolecular annihilation fluorescence (or delayed fluorescence) of anthracene (Aviakan and Merrifield, 1968) is known to be sensitive to radiation damage. Pearson et al. (1971) have studied this effect in fast-neutron exposed single anthracene crystals, applying the apparatus in Figure 6-57. With flash lamp excitation, the triplet decay constant is measured from the slope of the P.M. current as displayed on the scope. Due to its high hydrogen content, the system exhibits a relatively large fast-neutron to gamma response ratio. The observed fading rates are less than 5%/month. Unfortunately, chances for practical

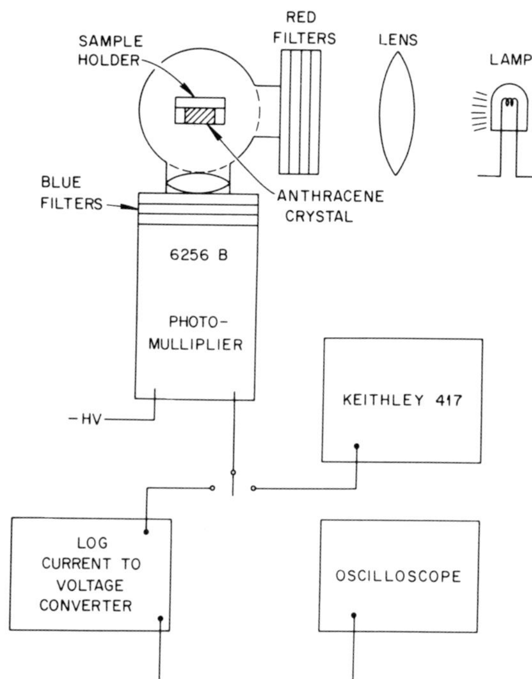

FIGURE 6-57. Schematical diagram of apparatus for measuring neutron-induced triplet exciton annihilation fluorescence changes in anthracene. (After Pearson et al., 1971.)

application are not good: Each of the required, relatively expensive crystals can be used only once, there are strong sample-to-sample response variations related to their impurity content, and the lowest detectable dose is in the neighborhood of 3 to 4 rad.

Electrets — The possible use of the radiation-induced depolarization of certain isolators (electrets) such as beeswax and some synthetic polymers has occasionally been considered as a means of radiation dosimetry for several decades. Much of the early work appears to be only of historic interest, mostly because of serious sensitivity and stability problems. More recently, electret effects have been described which appear

to be much more sensitive and stable. Fabel and Henisch (1971), for example, examined the surface potential of Nylon®, Mylar®, and Teflon®, and polystyrene with a contactless voltmeter probe and found the polarization in all except in Nylon stable under normal conditions. Doses of 1 mR were detectable.

Another study focused on inorganic systems (Podgorsak et al., 1972 and Podgorsak and Moran, 1973a and b). By pre-poling a high-temperature electret state, the crystals (pure LiF and CaF_2) become very sensitive to lower temperature, radiation-induced, and thermally activated depolarization phenomena. In the materials studied so far, the radiation-induced peaks are at very low temperatures (about -75 to $+60°C$).

There is a long list of other, more or less exotic radiation effects in solids which may or may not one day become of dosimetric interest. Some examples are chemical changes after exposure to 10^7 to 10^9 rad detected by Mössbauer spectroscopy (Takashima et al., 1972), changes in the bulk etching rate of irradiated polymers or glasses (Boyett and Becker, 1970; Krätschmer, 1972; and others), and radiation-induced changes in the mechanical strength and elasticity, electrical resistance, volume, tensile strength, weight, refraction index, dielectric constant (see, for example, Gardner and Toosi, 1967), wave propagation properties, etc. of irradiated glasses, crystals, or polymers. Others have used the weight loss in solid oxalic acid (Gal and Draganic, 1971) or dicarboxylic acids such as succinic acid (Gal and Premovic, 1972) in the 10^8 to 10^9 rad range as a dosimeter in reactors, or suggested diffuse reflectance spectrometry of amino acids, dipeptides, etc. for measurements in the 10 to 100 Mrad range (Kushelevsky, 1973). Most of these effects have, however, poor sensitivity and/or stability in common, and have been of limited interest in high dose-level measurements only in a few instances.

REFERENCES

Aglintsev, K. K., Photographic effect of x-rays, *Zh. Tekh. Fiz.,* 23, 1727, 1953.

Aiginger, H. and Hubeny, H., Comparison of ionization chambers and optical absorption measurements for the determination of electron depth doses, *Atomkernenergie,* 10, 479, 1965.

Allisy, A., La mesure des doses de rayons X ou gamma à l'aide d'émulsions photographiques, *J. Radiol. Electrol.,* 36, 249, 1955.

Amadesi, P., Rimondi, O., Sifaki, H., and Turtura, M., I servici di dosimetria personale del centro dosimetrico di Bologna, *Minerva Nucl.,* 4, 299, 1960.

Amadesi, P., Cavallini, A., and Rimondi, O., Calibration of fast neutron film badges, *Minerva Nucl.,* 5, 323, 1961.

Amadesi, P., Grimellini, N., Guenzi, N., and Rimondi, O., Absolute Spectrometry of Fission Neutrons Using Nuclear Track Emulsions, Proc. Symp. Personnel Dosimetry Techniques, Madrid, OECD/ENEA, Paris, 1963.

Andrews, H. L., Murphy, R. E., and LeBrun, E. T., Gel dosimeter for depth-dose measurements, *Rev. Sci. Instr.,* 28, 329, 1957.

Anon., The New LASL Film Badge and Personnel Neutron Dosimetry Packet, LA-3889, Los Alamos Scientific Lab., Los Alamos, N.M., 1968.

ANSI, Film Badge Performance Criteria, Am. Natl. Standards Institute, 1970.

Arnett, E. N., Latent Image Fading in Personnel Neutron Monitoring Film Exposed to a Moderated ^{238}PuO$_2$ Source, MLM-1570, Mound Lab., Monsanto Research Corp., Miamisburg, Ohio, 1969.

Arnold, K. M., Siliziumdioden als Neutronendosimeter, Proc. Int. Symp. Nuclear Electronics and Radioprotection, Toulouse, CONF-680302, 1968.

Artandi, C., Rigid vinyl film dosimeter, in *Manual on Radiation Dosimetry,* Holm, N. W. and Berry, R. J., Eds., Marcel Dekker, New York, 1970, 353.

Attix, F. H., High level dosimetry by luminescence degration, *Nucleonics,* 17(4), 142, 1959 and 17(10), 60, 1959.

Attix, F. H., Current look at thermoluminescent dosimetry in personnel monitoring, *Health Phys.,* 22, 287, 1972.

Avakian, P. and Merrifield, R. E., Triplet excitons in anthracene crystals — a review, *Mol. Cryst.,* 5, 37, 1968.

Avan, L., Blanc, D., and Teyssier, J. L., Ionographie — Emulsions, detecteurs solides de traces, Doin Ed., Paris, 1972.

Baarli, J. and Dutrannois, J., Purpose and Interpretation of Personnel Monitoring Data for High-energy Accelerators, Proc. Symp. Radiat. Dose Measurements, Stockholm, OECD/ENEA, Paris, 1967, 275.

Barber, D. E., Standards of Performance for Film Badge Services, U.S. Department of Health, Education and Welfare, Public Health Service Pub. No. 999-RH-20, 1966.

Barber, D. E., Brown, B. W., and James, K. E., Statistical analysis of data from film badge performance tests, *Am. Ind. Hyg. Assoc. J.,* 29, 482, 1968.

Barkas, W. H., *Nuclear Research Emulsions, Parts I and II,* Academic Press, New York, 1963.

Batchelor, A. L., Comparison of tissue-equivalent ionization chambers and silicon diodes for measuring fast neutron dose, *Phys. Med. Biol.,* 16, 451, 1971.

Battelle Memorial Institute Final Report on Personnel Neutron Dosimetry Systems, U.S. Army Signal Corps Contract No. DA 36-039 SC-73174, AD 216887, May 1958.

Battelle Memorial Institute Final Summary Report on Exp. and Res. Work in Neutron Dosimetry, U.S. Army Signal Corps Contract DA 36-039 SC-78924, AD 24523, June 1960.

Battelle Memorial Institute Final Summary Report on Exp. and Res. Work in Neutron Dosimetry, Phase I-III, U.S. Army Signal Corps Contr. No. DA 36-039 SC 78924, AD251971, AD 261974, and AD 265749, 1961.

Becker, K., Klein, E., and Zeitler, E., Ein wellenlängenunanhängig registrierender Dosismeßfilm, *Naturwissenschaften,* 47, 199, 1960.

Becker, K., Beitrag zur Filmdosimetrie energiereicher Quantenstrahlung, Teil I, *Fortschr. Röntgenstr.,* 95, 694, 1961a.

Becker, K., Beitrag zur Filmdosimetrie energiereicher Quantenstrahlung, Teil II, Experimentelle Ergebnisse, *Fortschr. Röntgenstr.,* 95, 839, 1961b.

Becker, K., Photographic determination of tritium in water, *Atompraxis,* 7, 358, 1961c.

Becker, K., Vereinfachung der Dosismessfilmauswertung durch Fixierentwickler, *Atomkernenergie,* 7, 234, 1962.

Becker, K., Fehlerquellen bei der Neutronen-Personendosismessung mittels Kernspuremulsionen, *Atomkernenergie,* 8, 74, 1963.

Becker, K., Photographic Film Dosimetry Bibliography, AEC-2-21-01, ZAED, Karlsruhe, Germany, 1964a.

Becker, K., Chemical and Solid-State Dosimetry Bibliography, AED-C-21-02, ZAED, Karlsruhe, Germany, 1964b.

Becker, K., High gamma dose response of recent silver-activated phosphate glasses, *Health Phys.,* 11, 523, 1965a.

Becker, K., Photographic, Chemical and Solid-State Dosimetry Bibliography, AED-C-21-03, ZAED, Karlsruhe, Germany, 1965b.

Becker, K., *Photographic Film Dosimetry,* Focal Press, London, 1966.

Becker, K., Photographic, Chemical, and Solid-State Dosimetry Bibliography, AED-C-21-04, ZAED, Karlsruhe, Germany, 1967a.

Becker, K., Photographic, Chemical, and Solid-State Dosimetry Bibliography, AEC-C-21-05, ZAED, Karlsruhe, Germany, 1967b.

Becker, K., Silicon Diodes for the Integrating Measurement of Fast Neutrons in Personnel Dosimetry, Proc. Symp. on Semiconductors in Radiat. Protection, Fachverband für Strahlenschutz, Neuherberg, Germany, 1967c.

Becker, K., Photographic, Chemical, and Solid-State Dosimetry Bibliography, AED-C-21-06, ZAED, Karlsruhe, Germany, 1968a.

Becker, K., Photographic, Chemical, and Solid-State Dosimetry Bibliography, AED-C-21-07, ZAED, Karlsruhe, Germany, 1968b.

Becker, K., Recent progress in radiophotoluminescence dosimetry, *Health Phys.*, 14, 17, 1968c.

Becker, K., Photographic, Chemical and Solid-State Dosimetry Bibliography, AED-C-08, ZAED, Karlsruhe, Germany, 1971.

Becker, K., The future of personnel dosimetry, *Health Phys.*, 23, 729, 1972.

Becker, K., Progress in Solid-State Fast Neutron Personnel Dosimetry, Proc. Symp. Neutron Monitoring Radiat. Protect., IAEA, Vienna, 1973a.

Becker, K., Long-term stability of film, TLD, and other integrating dosimeters in warm and humid climates, ORNL-TM, in press, 1973b.

Behrens, H., Determination of the Absorbed Energy in Vinyl Chloride Irradiated by X-rays, Proc. 2nd Tihany Symp. on Radiat. Chemistry, Dobo, J. and Hedvig, P., Eds., 1966.

Belletti, S. and Zaglio, E., Automatic densitometer for photographic dosimetry, *Radiol. Med.*, 54, 366, 1968.

Berg, W. F., in *The Theory of the Photographic Process*, Mees, C. E. K., Ed., New York, 1954.

Bermann, F., de Choudens, H., and Descours, S., Dosimetric applications of electronic paramagnetic resonance measurements of free radicals formed in amino acids — construction of a tissue-equivalent dosimeter based on alanine, in *Advances in Physical and Biological Radiation Detectors*, IAEA, Vienna, 1971, 311.

Berry, R. J. and Marshall, C. H., Clear Perspex H. X. as a reference dosimeter for electron and gamma radiation, *Phys. Med. Biol.*, 14, 585, 1969.

Birks, J. B., Scintillation efficiency of anthracene crystals, *Proc. Phys. Soc.* (London), A63, 1044, 1950.

Birnbaum, M., Schulman, J., and Seren, L., Use of melamine as a X-radiation detector, *Rev. Sci. Instr.*, 26, 457, 1955.

Bishay, A. M., A bismuth lead borate glass dosimeter for high-lead gamma measurements, *Phys. Chem. Glasses*, 2(2), 33, 1961; see also *Nucleonics*, 19(9), 88, 1961.

Bishay, A. M. and Arafa, S., A manganese arsenic borate glass dosimeter, *Bull. Am. Ceram. Soc.*, 46, 1102, 1967.

Bishop, W. P. and Benson, J. L., Operational Treatment of Multiple Reflection Interference in the Readout of Thin-Film Poly(halo)-styrene Dosimeters, SC-DR-72 0641, Sandia Lab., Albuquerque, N. M., 1972.

Boag, J. W., Solid-state methods of dosimetry, in *Actions Chimiques et Biologiques des Radiations*, 6th Series, Haissinsky, M., Ed., Massou, Paris, 1963, Chap. 1.

Böhler, G., Ein Filmdosimeter für die Strahlenschutzüberwachung beim Umgang mit radioaktiven Isotopen, *Kernenergie*, 1, 440, 1958.

Bolton, J. H., Irradiation temperature effects of clear Perspex dosimeters, *Int. J. Appl. Radiat. Isotopes*, 20, 685, 1969.

Boyett, R. H. and Becker, K., LET effects on the chemical resistance of irradiated polymers, *J. Appl. Polym. Sci.*, 14, 1654, 1971.

Bradshaw, W. W., Cadena, D. G., Crawford, G. W., and Spetzler, H. A. W., The use of alanine as a solid dosimeter, *Radiat. Res.*, 9, 589, 1962.

Brady, W. J. and Iverson, G. K., Combination personnel Dosimeter and Security Credential Holder, NYO-162-27, 1969.

Breitling, G. and Seeger, W., On the film dosimetry of fast electrons, *Strahlentherapie*, 22, 483, 1963.

Brocklehurst, R. E., High-Intensity Gamma Ray Dosimetry, Tech. Memo WCLT-59-115, Wright-Patterson Air Force Base, Ohio, 1959.

Brodsky, A., Spritzer, A. A., Feagin, F. E., Bradley, F. J., Karches, G. J., and Mandelberg, H. I., Accuracy and sensitivity of film measurements of gamma radiation — intrinsic and extrinsic errors, *Health Phys.*, 11, 1071, 1965.

Bromley, D. and Herz, R. H., Quantum efficiency in photographic x-ray exposures, *Proc. Phys. Soc.* (London), B63, 90, 1950.

Broszkiewicz, R. K. and Bulhak, Z., Plastic Film Dosimeters, Paper IAEA/SM-160/60, Proc. Symp. Dosimetry Techniques Applied to Agriculture, Industry, Biology, and Medicine, IAEA, Vienna, 1972.

Brown, W. R., Red acrylic dosimetry system, *Trans. Am. Nucl. Soc.*, 12, 61, 1969.

Buschlen, D., Automatic processing and handling of dosimeter film data by the EDF, in *New Trends in Radiation Protection*, French Radioprotect. Soc., Montrouge, France, 1972, 543.

Busuoli, G. and Cavallini, A., Comparison of lithium fluoride and film for personal dosimetry of x- and gamma-radiations, *Minerva Fisiconucl.*, 13, 265, 1969.

Cameron, J. R. and Suntharalingam, N., A Comparison of TLD and Film for Personnel Dosimetry, *Proc. First Int. Congr. on Radiat. Protection, Rome, 1966*, Vol. 1, Pergamon Press, Oxford, 1968, 451.

Caruthers, L. T. and Story, E. J., Capture gammas for neutron dosimetry with a film badge, *Health Phys.*, 10, 667, 1964.

Cavallini, A. and Busuoli, G., Role, Interpretation and Accuracy of Measurement of Fast Neutrons by Means of Photographic Emulsions, Proc. OECD/ENEA Symp. Radiat. Dose Measurements, Stockholm, 1967, 293.

Chadwick, K., The Choice of Measurement Wavelength for Clear HX Perspex Dosimetry, Paper IAEA/SM-160/3, Proc. Symp. Dosimetry Techniques Applied to Agriculture, Industry, Biology, and Medicine, IAEA, Vienna, 1972.

Chadwick, K., ten Bosch, J. J., and Verholst, W. F., The Mechanism Involved in the Temperature and Light Effect on the Clear Perspex Dosimeter, Paper IAEA/SM-160/4, Proc. Symp. Dosimetry Techniques Applied to Agriculture, Industry, Biology, and Medicine, IAEA, Vienna, 1972.

Chalkley, L., Photosensitive Combination of Cellulose with Hydrophobic Dye Cyanide and Process for Making It, U.S. Patent 3,122,438, 1964.

Cheka, J. S., Recent developments in film monitoring of fast neutrons, *Nucleonics,* 15(6), 40, 1954a.

Cheka, J. S., Nuclear emulsion instability, *Nucleonics,* 12(10), 58, 1954b.

Cheka, J. S. and Becker, K., High-level glass dosimeters with low dependence on energy, *Nucl. Appl.,* 6, 163, 1969.

Cipperley, F. V., Henry, R. W., et al., Personnel Monitoring Practices at the National Reactor Testing Station, Symp. Personnel Dosimetry Radiat. Accidents, IAEA, Vienna, 1965.

Cook, J. E., Fast Neutron Dosimetry Using Nuclear Emulsions, AERE/HP/R 2744, Atomic Energy Res. Establishment, Harwell, Didcot, England, 1958.

Costa-Ribeiro, C., Thomas, J., Drew, R. T., Wrenn, M. E., and Eisenbud, M., Radon detector suitable for personnel or area monitoring, *Health Phys.,* 17, 193, 1969.

Cottignies, S. M., Lecteur Automatique de Films Dosimètres, Proc. Int. Symp. Nuclear Electronics and Radiat. Protect., Toulouse, France, 1968.

Crosby, E. H., Apparent discrepancies in results from a major commercial supplier of film badges, *Health Phys.,* 18, 93, 1970.

Cusimano, J. P., Latent Image Fading in Personal Neutron Monitoring Films at the NRTS, Report IDO-12031, Nat. Reactor Test. Station, Idaho Falls, Idaho, 1963.

David, J. and Wasilewski, A., Tritium Personnel Dosimetry with Film Dosimeters, CONF-690540, 1969, 119.

Davidson, S., Goldblith, S. A., and Procter, B. E., Glass dosimetry, *Nucleonics,* 14(1), 34, 1956.

Day, M. J. and Stein G., Chemical effects of ionizing radiations in some gels, *Nature,* 166, 146, 1950.

Day, M. J. and Stein, G., Effects of x-rays upon plastics, *Nature,* 168, 644, 1951.

Day, M. J. and Stein, G., The action of ionizing radiations on aqueous solutions of methylene blue, *Radiat. Res.,* 6, 666, 1957.

Dealler, J. F. B., Electron dose and dose distribution measurements in the megarad region, in Radiation Dose and Dose Distribution Measurements in the Megarad Range, U.K. Panel on Gamma and Electron Irradiations, 1971, 30.

DeCosnac, B., Dulieu, P., and LeRalle, J. C., Mesure des Flux de Neutrons Rapides au Moyen de Diodes in Silicium Sensibles aux Dommages, CEA-CONF-1067, French CEA, 1968.

Denham, D. H. and Kathren, R. L., Effect of atmospheric tritium on photographic, thermoluminescent, and radiophotoluminescent dosimeters, *Health Phys.,* 16, 773, 1969.

Descours, S., Assayrenc, J., Bermann, F., Couderc, B., de Choudens, H., Delard, R., Rassat, A., and Servoz-Gavin, P., ESR Study of Free Radicals Created by Irradiation in Some Organic Substances, Application to Dosimetry of Free Radicals Produced in Alanine, CEA-R-3913, Centre d'Etudes Nucléaires, Grenoble, France, 1970.

Distenfeld, C. H. and Klemish, J. R., Developmental Study of Personnel Neutron Dosimetry at AGS, BNL-17452, Brookhaven Natl. Lab., Upton, N.Y., 1972.

Dorneich, M. and Schäfer, H., On radiation protection measurements by the photographic method, *Phys. Z.,* 13, 390, 1942.

Doroshenko, V. N., Chervetsova, I. N., and Kabaktschi, A. M., Measurement of absorbed radiation dose by degradation of luminescence in anthracene, *At. Energ.,* 31, 622, 1971.

Douglas, J. A., Assessment of Personal Fast Neutron Doses Using an Automatic Proton Track Counter, AERE-R-6220, Atomic Energy Res. Establishment, Harwell, Didcot, England, 1970.

Dresel, H., Filmdosimetrie bei Strahlenschutzmessungen, *Fortschr. Rontgenstr.,* 84, 214, 1956.

Dudley, R. A., Photographic detection and dosimetry of beta rays, *Nucleonics,* 12(5), 24, 1954.

Dudley, R. A., Dosimetry with photographic emulsions, in *Radiation Dosimetry,* Attix, F. H. and Roesch, W. C., Eds., Academic Press, New York, 1966, 326.

Dutreix, J. and Dutreix, A., Film dosimetry of high-energy electrons, *Ann. N.Y. Acad. Sci.,* 161, 33, 1969.

Eastes, J. D., Comparison of Thermoluminescent and Film Dosimetry in Routine Radiation Dosimetry, Proc. Sec. Int. Conf. Luminescence Dosimetry, CONF-680920, 1968, 814.

Egelhaaf, H., Heinemann, H., and Seidler, K. H., Zur Strahlenschutz-Überwachung nach der Filmschwärzungsmethode in Radium- und Isotopenlaboratorien, *Atompraxis,* 6, 407, 1960.

Eggert, J. and Luft, F., Ein neuartiges Röntgen-Filmdosimeter. Parts I and II, *Röntgenpraxis,* 1, 188, 1929.

Eggert, J. and Luft, F., Die Temperaturabhängigkeit des photographischen Prozesses, *Veröffentl. Wiss. Zentrallab. Agfa,* Vol. 2, Springer-Verlag, Berlin, 1931.

Ehrlich, M. and Fitch, S. H., Photographic x- and gamma-ray dosimetry, *Nucleonics,* 9(3), 5, 1951.

Ehrlich, M., Reciprocity law for x-rays; Parts I and II, *J. Opt. Soc. Am.,* 46, 797, 1956.

Ehrlich, M., A photographic personnel dosimeter for X-radiation in the range from 30 keV to beyond 1 MeV, *Radiology,* 68, 251, 1957.

Ehrlich, M. and McLaughlin, W. L., Photographic Dosimetry at Total Exposure Levels Below 20 mR, NBS Tech. Note 29, Nat. Bureau of Standards, Washington, D. C., 1959.

Ehrlich, M., The sensitivity of photographic film to 3 MeV neutrons and thermal neutrons, *Health Phys.,* 4, 113, 1960.

Ehrlich, M., Influence of temperature and relative humidity on the photographic response to ^{60}Co gamma radiation, *J. Res. Nat. Bur. Stand.*, 65C, 203, 1, 1961.

Ehrlich, M. and McLaughlin, W. L., Photographic response to successive exposures of different types of radiation, *J. Opt. Soc. Am.*, 51, 1172, 1961.

Ehrlich, M., The Use of Film Badges for Personnel Monitoring, Manual STI/PUB-43, IAEA, Vienna, 1962.

Ehrlich, M., Characteristics of dosimeter films processed in phenidone thiosulfate monobaths, *Photogr. Sci. Eng.*, 9, 1, 1965.

Eisen, H., 2 MeV electron depth dose measurements in aluminum, copper, and tin absorbers using radiochromic dye film, *Int. J. Appl. Radiat. Isotopes*, 23, 97, 1972.

Eisen, H., Rosenstein, M., and Silverman, J., Electron Dosimetry Using Chalkley-McLaughlin Dye-Cyanide Thin Films, Paper IAEA/SM-160/56, Proc. Symp. Dosimetry Techniques Applied to Agriculture, Industry, Biology, and Medicine, IAEA, Vienna, 1972.

Ellis, S. C., Problems in Spectrophotometry and their Influence in Radiation Measurements, in Radiation Dose and Dose Distribution Measurements in the Megarad Range, U.K. Panel on Gamma and Electron Irradiation, 1971, 18.

Evans, R. D., Film Badge Radiation Dosimeter for Detecting Radon, U.S. Patent 3,655,975, 1972.

Fabel, G. W. and Henisch, H. K., Polymer electret dosimetry, *Phys. Status Solidi*, 6, 535, 1971.

Farrell, J. J., Radiation-sensitive pink verifies product irradiation, *Nucleonics*, 21(11), 78, 1963.

Fitzsimmons, C. K., Horn, W., and Klein, W. L., Comparison of Film Badges and Thermoluminescent Dosimeters Used for Environmental Monitoring, SWRHL-93-R, West. Environ. Res. Lab., Las Vegas, 1972.

Fleming, S. J., Absorption spectrum of gamma-irradiated polymethyl methacrylate within the range 245 to 450 nm, *Polymer*, 9, 489, 1968.

Fowler, J. F. and Day, M. J., High dose measurement by optical absorption, *Nucleonics*, 13(12), 52, 1955.

Frank, M. and Stolz, W., *Festkörperdosimetrie ionisierender Strahlung*, BSB B.G. Teubner, Leipzig, 1969.

Frieser, H. and Klein, E., Contribution to the theory of the characteristic curve, *Photogr. Sci. Eng.*, 4, 264, 1960.

Fuller, G. E., Leach, R. K., Moran, P. R., and Fitzsimmons, W. A., Some Preliminary Paramagnetic Resonance Studies for Dosimetry Research, COO-1105-179, University of Wisconsin, Madison, 1972.

Furtak, T. E. and Katz, R., Simulation of particle tracks in emulsion, *Radiat. Eff.*, 11, 195, 1971.

Gal, O. S. and Draganic, I. G., Solid oxalic acid as a chemical dosimeter for mixed pile radiation in multimegarad range (10^8 to 10^9 rad), *Int. J. Appl. Radiat. Isotopes*, 22, 753, 1971.

Gal, O. and Premovic, P., Solid dicarboxylic acids as in-pile chemical dosimeters, *Int. J. Appl. Radiat. Isotopes*, 23, 541, 1972.

Gamertsfelder, C. C., Bramson, P. E., Endres, G. W. R., and Wilson, R. H., Some notes on Practical Neutron Dosimetry, Paper SM-36/69, Proc. Symp. on Neutron Detection, Dosimetry and Standardization, Harwell, 1963.

Gardner, D. G. and Toosi, M. T. A., Radiation-induced changes in the index of refraction, density, and dielectric constant of polymethyl methacrylate, *J. Appl. Polym. Sci.*, 11, 1065, 1967.

Geiger, E. L., Radon film badge, *Health Phys.*, 13, 407, 1967.

Geiger, E. L., TLD vs. Film. Paper Symp. Recent Developments Pract. Dosimetry and Standards, Nat. Bureau of Standards, Washington, D. C., 1971.

Gevantman, L. H., Chandler, R. C., and Pestaner, J. F., Tridimensional examination of chemical systems irradiated in gel media, *Radiat. Res.*, 7, 318, 1957.

Gibson, N. N., Brimmins, T. W., and Jacobson, J. R., Exposure-rate Response of Selected Gamma Dosimeters, Report AD-679028, Army Nuclear Defense Lab., Edgewood Arsenal, Md., 1968.

Gibson, N. N. and Stützer, M., Evaluation of Gamma Dosimetry in a Nuclear Reactor Environment, Report AD-672475, U.S. Army Nuclear Defense Lab., Edgewood Arsenal, Md., 1968.

Glocker, R., Das photographische Schwärzungsgesetz für Electronenstrahlen verschiedener Energie, *Z. Phys.*, 160, 568, 1960.

Goldblith, S. A. and Mateles, R. L., Cellophane dosimetry found imprecise, *Nucleonics*, 16(6), 102, 1958.

Goldstein, N., Cinemoid color films, in *Manual Radiation Dosimetry*, Holm, N. W. and Berry, R. J., Eds., Marcel Dekker, New York, 1970, 371.

Gorson, R. O., Suntharalingam, N., and Thomas, J. W., Results of a film badge reliability study, *Radiology*, 84, 333, 1965.

Gorton, H. C., Semiconductors as Radiation Detectors, Battelle Technical Review, July 1962.

Greening, J. R., The photographic action of x-rays, *Proc. Phys. Soc.*, 64B, 977, 1951.

Gupta, B. L. and Bhat, R. M., Use of Caliplast PVC Films for Gamma Ray Dosimetry, *Proc. Chemistry Symp.* Vol. 2, Department of Atomic Energy, Bombay, 1970, 57.

Handley, R. and Grossenbacher, K. A., ^{90}Sr/^{90}Y Beta Dosage-Depth Relationships in Polyethylene Absorbers, UCB-34-P-147-X-3, University of California, Berkeley, 1970, 5.

Harrah, L. A., Chemical dosimetry with transstilbene doped polystyrene films, *Radiat. Res.*, 39, 223, 1969.

Harrington, E. L., Johns, H. E., Wiles, A. P., and Garetti, C., The fundamental action of intensifying screens in gamma radiography, *Can. J. Res.*, 26, Sect. F, 540, 1948.

Harris, K. K. and Price, W. E., A thin plastic radiation dosimeter, *Int. J. Appl. Radiat. Isotopes*, 11, 114, 1961.

Hart, R. S. and Hale, J. P., Fast Neutron Monitoring with NTA Film Packets, NAA-SR-1536, 1956.

Hayami, A., Mori, Y., and Azumi, I., Film dosimetry of betatron electron beams for radiotherapy planning, *Strahlentherapie*, 144, 22, 1972.

Heard, M. J. and Jones, B. E., A new film holder for personnel dosimetry, in *Personnel Dosimetry Techniques for External Radiations*, OECD/ENEA, Paris, 1963, 89.

Heard, M. J., Report on the Progress of Automation of Photographic Dosimetry, AERE M-1371, 1964; *J. Photogr. Sci.*, 12, 312, 1964.

Heard, M. J., Cook, J. E., and Holt, P. D., Photographic Radiation Dosimetry and the Development of the AERE-RPS Film Dosimeter, AERE-M 1370, U.K. Atomic Energy Res. Establishment, Harwell, Didcot, England, 1964.

Helm, W. and König, L. A., Comparison of the results of personnel dose surveillance with pocket ionization chambers and film dosimeters, *Atompraxis*, 15, 263, 1969.

Henisch, H. K. and Bose, D. N., Recent Experiments on X-Ray and UV Dosimetry by Means of Non-Luminescent Crystals, Proc. Sec. Int. Conf. Luminescence Dosimetry, CONF-680920, 1968, 234.

Henley, E. J. and Richman, D., Cellophane dye dosimeter for 10^5 to 10^7 Röentgen range, *Anal. Chem.*, 28, 1580, 1956.

Herz, R. H., Granularity measurements of x-ray emulsions exposed to x-rays with increasing quantum energy, *Photogr. J.*, 89B, 89, 1949.

Herz, R. H., *The Photographic Action of Ionizing Radiation*, Interscience, New York, 1969.

Hine, G. J., *Am. J. Roentgenol.*, 72, 293, 1954.

Hoerlin, H., Clark, R. H., Jones, P. D., Kaszuba, F. J., and Larson, E. T., Development of a Wavelength-Independent Radiation Monitoring Film, Final Report ANL-5168, Argonne Nat. Lab., Argonne, Ill., 1953.

Hofman, E. G., Messung hoher Röntgenstrahlen-Dosen durch optische Absorption in Gläsern und Kunststoffen, *Kerntechnik*, 5, 439, 1963.

Hood, H. P., Method for Stabilizing Color Glass Exposed to High-Level X-Radiation, U.S. Patent 3,173,350, 1965.

Hübner, K. and Stolz, W., Ein einfaches, visuell und objektiv auswertbares Schwellwertdosimeter für Sterilisierungsbestrahlungen medizinischer Artikel, *Isotopenpraxis*, 5, 230, 1969.

Hübner, K., Stolz, W., and Haberer, G., Ein Farbumschlag-Indikator zur Sichtbarmachung von Strahlungsfeldern und zur Strahljustierung an Grossβbestrahlungsanlagen, *Isotopenpraxis*, 6, 332, 1970.

Hübner, K., Use of the system methyl orange/polyvinyl alcohol for the detection of ionizing radiation, *Isotopenpraxis*, 7, 439, 1971.

Huff, J. B. and Mayo, R., Paper, 3rd Workshop on High-Level Dosimetry, Sponsored by USAEC at U.S. Nat. Bureau of Standards, 1968.

Hussman, E. K. and MacLaughlin, W. L., Dye films and gels for megarad dosimetry, in *Radiation Dose and Dose Distribution Measurements in the Megarad Range*, U.K. Panel on Gamma and Electron Irradiations, 1971, 35.

IAEA, Personnel Dosimetry Systems for External Radiation Exposure, Tech. Report Series No. 109, IAEA, Vienna, 1970.

Ilic-Papovic, J. and Hjortenberg, P. E., Anthracene gelatine films for luminescence degration dosimetry, *Int. J. Appl. Radiat. Isotopes*, 20, 541, 1969.

Jacobi, W. and Koeppe, P., [222]Rn-Empfindlichkeit von Personendosimetern, *Atomkernenergie*, 13, 429, 1968.

Jahn, W., Dosimetrie mit Gläsern, *Glastechn. Berichte*, 42, 176, 1969.

Jasiak, J. and Musialowicz, T., Personnel Fast Neutron Monitoring in Poland, Paper IAEA/SM-167/34, Proc. Symp. Neutron Monitor. Radiat. Protect., IAEA, Vienna, 1973.

Jeltsch, E. and Graf, W., Zur Gammadosimetrie an Kernreaktoren mit Dosimetergläsern, *Atomkernenergie*, 13, 425, 1968.

Jeltsch, E., Untersuchungen über Festkörperdosimeter für Neutronen, *Atomkernenergie*, 13, 289, 1968.

Johnson, T. L. and Attix, F. H., Pilot Comparison of Two Thermoluminescent Dosimetry Systems with Film Badges in Routine Personnel Dosimetry, *Proc. First Congr. Radiat. Protection, Rome, 1966*, Vol. 1, Pergamon Press, Oxford, 1968, 457.

Jones, B. E. and Marshall, T. O., The dosimetry of mixed radiations involving more than one energy or type of radiation, with the R.P.S./A.E.R.E. film dosimeter, *J. Photogr. Sci.*, 12, 319, 1964.

Jones, B. E. and Marshall, T. O., A practical method of personnel monitoring for beta-radiations using photographic emulsions, *J. Photogr. Sci.*, 13, 12, 1965.

Kahle, J. B., Arneit, E. N., and Meyer, C. T., Latent image fading in personnel monitoring neutron film, *Health Phys.*, 17, 735, 1969.

Kantz, A. D., Humpherys, K. C., Lucas, A. C., and Schlueter, W. A., Megarad Dose Response of Passive Dosimetric Materials, Report EGG-1183-2166, E. G. & G., Santa Barbara, Calif., 1969.

Kathren, R. L., Prevo, C. T., and Block, S., Angular dependence of Eastman Type A (NTA) personnel monitoring film, *Health Phys.*, 11, 1067, 1965.

Kathren, R. L., Thermal fogging of personnel monitoring film, *Health Phys.*, 12, 61, 1966.

Katz, J. and Fogel, S. J., *Photographic Analysis*, Morgan & Morgan, Hastings-on-Hudson, N. Y., 1972.

Katz, R., Sharma, S. C., and Homayoonfar, M., The structure of particle tracks, in *Progress in Radiation Dosimetry*, Attix, F. H., Ed., Academic Press, New York, 1972, Chap. 6.

Keirim-Markus, I. B., Kraitor, S. N., and Kuzmin, V. V., Comparison of neutron and gamma-ray personnel dosimeters at critical assemblies, *At. Energ.*, 33, 563, 1972.

Kets-Neels, G., Van de Voorde, M., and Tobback, P., Dosimetric Methods in Food Irradiation, Paper IAEA/SM-160/59, Proc. Symp. Dosimetry Techniques Applied to Agriculture, Industry, Biology, and Medicine, IAEA, Vienna, 1972.

Kocher, L. F., A Personnel Dosimeter Filter System for Measuring Beta and Gamma Doses in Mixed Radiation Fields, HW-71764, Battelle Northwest, Richland, Wash., 1962.

Kostalas, H. A., Clear PMMA as a dosimeter in the mixed gamma neutron field of a reactor, *Int. J. Appl. Radiat. Isotopes,* 18, 783, 1967.

Kramer, G., The Semiconductor Fast Neutron Dosimeter — Its Characteristics and Applications, Paper 12th Nucl. Sci. Symp., IEEE, San Francisco, 1965.

Krätschmer, W., Effects of Heavy Ion Radiation on Quartz Glass, Paper, 8th Int. Conf. Nucl. Photogr. and Solid-State Track Detectors, Bucharest, 1972.

Kreidl, N. J. and Blair, G. E., Recent developments in glass dosimetry, *Nucleonics,* 14(1), 56, 1956; see also *Nucleonics,* 17(10), 58, 1959.

Krishnamoorthy, P. N., Venkatakaman, G., Singh, D., Dayashankar, Comparison of Various Neutron Personnel Dosimeters, Paper IAEA/SM-167/78, Proc. Symp. Neutron Monitor. Radiat. Protect., IAEA, Vienna, 1973.

Krüger, H., Tumbrägel, G., Metzner, R., and Koch, H., Fast Neutron Dosimetry with Silicon Diodes Comparative Investigations of Different Commercial Diodes, Paper IAEA/SM-167/53, Proc. Symp. Neutron Monitor Radiat. Protect., IAEA, Vienna, 1973.

Kügler, J., Zur Verfärbung von Gläsern durch ionisierende Strahlung, *Atomkernenergie,* 4, 67, 1959.

Kügler, J. and Scharmann, A., Dosimetrie ionisierender Strahlung mit Kunststoffen, *Atomkernenergie,* 4, 23, 1959.

Kushelevsky, A., Diffuse reflectance dosimetry, *Int. J. Appl. Radiat. Isotopes,* 24, 127, 1973.

Langendorff, H., Spiegler, G., and Wachsmann, F., Strahlenschutzüberwachung mit Filmen, *Fortschr. Röntgenstr.,* 77, 143, 1952.

Lawrentowitsch, J. I., Lewon, A. I., Melnikowa, G. N., and Kabaktschi, A. M., *At. Energ.,* 19, 273, 1965; *Ukr. Khim. Zh.,* 31, 440, 1965.

Lee, P. K. and Ziemer, P. L., Effects of temperature on cobalt glass dosimeters, *Health Phys.,* 18, 673, 1970.

Lehman, R. L., Energy Response and Physical Properties of NTA Personnel Dosimeter Nuclear Track Film, UCRL-9513, University of California Radiat. Lab., Berkeley, 1961.

Lehman, R. L. and Fekula, O. M., Neutron Spectra Measured Inside Human Phantoms, Proc. Symp. Personnel Dosimetry Accidental High-Level Exposures, IAEA, Vienna, 1965.

Lentsch, J. W. and Finston, R. A., Increased x-ray dose adjacent to plane bone interfaces as measured in polyethylene, *Phys. Med. Biol.,* 12, 543, 1968.

Littlejohn, G. J., Photodosimetry Procedures at Los Alamos, LA-2494, Los Alamos Scientific Lab., Los Alamos, N. M., 1960.

Madden, E., Film Badge Reader and Card Punch System, ORNL-TM-2154, Oak Ridge Nat. Lab., Oak Ridge, Tenn., 1969.

Manambelona, J. R., Physical Dosimetry Based on Magnetic Resonance Spectra, Paper IAEA/SM-160/24, Proc. Symp. Dosimetry Techniques Applied to Agriculture, Industry, Biology, and Medicine, IAEA, Vienna, 1972.

Markov, K. P., Ryabov, N. V., and Stas, K. N., Use of Photographic Films for Individual Radiometry of the Daughter Products of Radon, *Tr. Soyuznogo Nack. Issled. Inst. Priborostr.,* No. 12, 184, 1970.

Martin, J. A., Jones, J. L., and Crehotsky, R. S., Relative response of dosimeters to ^{60}Co and 25 nsec pulsed electron irradiations, *Health Phys.,* 20, 267, 1971.

Mauderli, W., Dosimetrie von Röntgen-und Gammastrahlung mittels photographischer Filme, Parts I and II, *Fortschr. Röntgenstr.,* 86, 634, 1957.

Maul, J. E., Holm, N. W., and Draganic, I. G., *The Use of Polyvinylchloride Film for ^{60}Co Radiation Dosimetry,* Risö-Rep 31, Danish AEC Res. Establishment, Risö, Roskilde, 1961.

McCarthy, R. and Mejdahl, V., *Personnel Dosimetry Practice at Risö,* Risö-Rep. 78, Danish AEC Res. Establishment, Risö, Roskilde, 1964.

McLaughlin, W. L., Radiochromic dye-cyanide dosimeters, in *Manual on Radiation Dosimetry,* Holms, N. W. and Berry, R. J., Eds., Marcel Dekker, New York, 1970.

McLaughlin, W. L., Films, dyes and photographic systems, in *Manual on Radiation Dosimetry,* Holms, N. W. and Berry, R. J., Eds., Marcel Dekker, New York, 1970, 129.

McLaughlin, W. L., Chemical dosimeters for monitoring gamma-radiation doses of 1 to 100 krad, *Int. J. Appl. Radiat. Isotopes,* 22, 135, 1971.

McLaughlin, W. L., Hjortenberg, P., and Radak, B., Problems in Absorbed Dose Measurements in Thin Films, Paper IAEA/SM-160/32, Proc. Symp. Dosimetry Techniques Applied to Agriculture, Industry, Biology, and Medicine, IAEA, Vienna, 1972.

Medveczky, L. and Bornemisza-Pauspertl, P., Notes on the photographic film dosimetry of fast neutrons, *Acta Phys. Acad. Sci. Hung.,* 28, 217, 1970.

Mees, C. E. and James, T. H., Eds., *The Theory of the Photographic Process,* 3rd ed., Macmillan, New York, 1966.

Mehringer, W., A simple method of measuring radiation dose in the megarad range using low-density polyethylene film, *Phys. Med. Biol.,* 16, 311, 1971.

Mengali, J., Pastrell, E., Beck, R. W., and Beet, C. S., The Use of Diffused Junction in Silicon as Fast-Neutron Dosimeters, Proc. Sec. Conf. Nucl. Radiat. Effects on Semiconductors, Materials, and Circuits, 1959.

Menkes, C. K. and Goldstein, N., Color Films for Megarad Dosimetry, USNRDL-TR-1097, U.S. Naval Radiol. Defense Lab., San Francisco, Calif., 1966.

Mercer, T. T. and Golden, R., Dose to photographic emulsion due to activation of their silver content by thermal and epithermal neutrons, *Health Phys.*, 9, 187, 1963.

Metalli, P., Lesac, R. M., and McCannon, D., Calibration of film dosimeters with beta particles of different energies, *G. Fis. Sanit. Prot. Radiaz.*, 12, 28, 1968.

Meyer, C. T., Latent Image Fading in Personnel Neutron Monitoring Film, MLM-1490, Mound Lab., Monsanto Research Corp., Miamisbury, Ohio, 1968.

Mitchell, J. W., Photographic Sensitivity, *Rep. Progr. Phys.*, 20, 433, 1957; Die photographische Empfindlichkeit, *Photogr. Korresp.*, 1st Special Issue, 1957.

Mitchell, J. W., Some aspects of the theory of photographic sensitivity, *J. Phys. Chem.*, 66, 2359, 1962.

Mitchell, J. P. and DeNure, D. G., Thermally Stimulated Current Radiation Dosimeter, Paper, Ann. Meet. Health Phys. Soc., Las Vegas, 1972.

Miyanaga, I., Bingo, K., and Yamamoto, H., Sensitivities of Fuji badge film to beta rays and evaluation of beta ray dosage, *Nippon Genshiryoku Gakkaishi*, 5, 497, 1963; *Health Phys.*, 9, 677, 1963.

Möller, G., Use of Nuclear Track Emulsions into which Lindemann Glass has been Embedded for the Dosimetry of Slow Neutrons, Thesis, Technical University, Berlin, 1960.

Mott, N. F. and Gurney, R. W., *Electronic Processes in Ionic Crystals*, Clarendon Press, Oxford.

Myers, D. S. and Kathren R. L., Energy Dependence of Cobalt Glass, UCRL-50007-66-2, University of California Livermore Lab., Livermore, Calif., 1966, 23.

Napoli, C. and Cortellessa, G., New neutron detector films from polyvinyl alcohol, *Nucl. Instr. Meth.*, 59, 117, 1968.

Nelson, I. C., Baumgartner, W. V., and Faust, L. G., Interpretation of dose from ^{16}N radiation with the Hanford film badge dosimeter, BNWL-CC-484, Battelle Northwest Lab., Richland, Wash., 1966.

Nikodemonva, D. and Bradna, F., Utilization of polymer-gelatin nuclear emulsions for fast neutron dosimetry, *Radioisotopy*, 11, 739, 1970.

Nishi, T., Izumi, Y., and Hirasawa, Y., Effects of relative humidity on track fading and prevention of fading of the personal neutron monitoring film, *Hoken Buturi*, 5, 179, 1970.

Oliver, G. D. and Hamm, M., Improved techniques for fast neutron dosimetry with silicon diodes, *Phys. Med. Biol.*, in press.

Orton, C. G., Red perspex dosimetry, *Phys. Med. Biol.*, 11, 337, 1966.

Oshima, Y. and Tanaka, R., Radiation dosimetry by polyethylene terephthalate, *Oyo Butsuri*, 36, 515, 1967.

Oshino, M., Response of NTA personnel neutron monitoring film worn on a human phantom, *Health Phys.*, 24, 71, 1973.

Oster, G. and Broyde, B., A sensitive chemical dosimeter for ionizing radiation, *Nature*, 184, 545, 1959.

Paic, V. and Paic, M., Interpretation of the Angular Dependence of the Photographic Response to Gamma Rays of Bare Monitoring Films in Combination with Metallic Filters, Paper, First European Congr. on Radiat. Protect., Menton, France, 1968.

Pearson, D., Moran, P. R., and Cameron, J. R., Triplet Exciton Annihilation Fluorescence Changes Induced by Fast Neutron Radiation Damage in Anthracene, *Proc. Third Int. Conf. Luminescence Dosimetry, Risö-Rep. 249*, Vol. 3, Danish AEC, Risö, Roskilde, 1971, 1063.

Pestaner, J. F. and Gevantman, L. H., Depth dosimetry by means of a gel-incorporated chemical system, *Radiat. Res.*, 9, 166, 1958.

Peters, I. G., Mixed Field Radiation Dosimeter Materials of Amine Salts of Organic Acids, U. S. Patent 3,673,107, 1972.

Pförtner, K., German Patent 1,135,775, 1962.

Phillips, L. F. and Champagne, R. I., Calibration and Response of Personnel Monitoring Film Dosimeters to Synchrotron Stray Radiations, Proc. USAEC Symp. Accelerator Radiat. Dosimetry, Brookhaven Nat. Lab., Upton, New York, 1965.

Piesch, E., Zur Dosimetrie schneller Neutronen mit Kernspuremulsionen, *Atompraxis*, 9, 179, 1963.

Piesch, E., Intercomparison of Results for Film, Glass, and Ionisation Chamber Dosimeters worn together in Routine Personnel Monitoring, Proc. Symp. Radiat. Dose Measurements, OAECD/ENEA, Paris, 1967, 151.

Podgorsak, E. B., Fuller, G. E., and Moran, P. R., Radiation Response of a New Electret Effect in Ionic Solids, Proc. 142nd Meet. Electrochem. Society, Miami Beach, Fla., 1972.

Podgorsak, E. B. and Moran, P. R., Radiation dosimetry by a new solid-state effect, *Science*, 179, 380, 1973a.

Podgorsak, E. B. and Moran, P. R., Dynamics of the ionic space charge electret state in CaF_2, to be published, 1973b.

Popovic, J., Risö-Rep. 141, Danish AEC, Risö, Roskilde, 1967.

Portal, G., Dosimétrie des Neutrons Rapides au Moyen d'Emulsions à Protons de Recul, Personnel Dosimetry Techniques for External Radiation, OECD/ENEA Symp., Madrid, 1963, 219.

Potsaid, M. S. and Irie, G., Paraffin base halogenated hydrocarbon chemical dosimeters, *Radiology*, 77, 61, 1961.

Powell, C. F., Fowler, P. H., and Perkins, D. H., *The Study of Elementary Particles by the Photographic Methods*, Pergamon Press, Oxford, 1959.

Prokert, K. and Stolz, W., Dosimetrie ionisierender Strahlung mittels fester Farbstoffsysteme, *Isotopenpraxis*, 6, 325, 1970.

Pye, L. D., Hensler, J. R., and Snyder, A. W., Paper at Special Tech. Conf. on Nucl. Radiat. Effects, Seattle, Wash., 1964.

Rae, J. B., Optical Density Measurement by a Null Method for an Automatic Densitometer, AERE-R-5929, Atomic Energy Res. Establishment, Harwell, Didcot, England, 1968.

Ramsey, N. W., Dose Measurement by Thermally Stimulated Conductivity — A New Method of Dosimetry, Paper 22.7, Third Int. Conf. Medical Physics, Gothenberg, Sweden, 1972.

Rassow, J., Erdmann, U., and Strüter, H. D., (1969-71), Beitrag zur Filmdosimetrie energiereicher Strahlen, Parts I-V, Strahlentherapie, 138, 149; 140, 655; 141, 47; 141, 176; and 141, 336.

Rauch, J. E. and Andrew, A., *I.E.E.E. Trans. Nucl. Sci.*, NS-13, No. 6, 1966.

Richter, B. and Schneider, H., Lumineszenzschädigung organischer Leuchtstoffe durch schnelle Neutronen, Proc. Int. Symp. Nuclear Electronics and Radiat. Protection, CONF-680302, Toulouse, France, 1968.

Ritz, V. H., A note on Mylar film dosimetry, *Radiat. Res.*, 15, 460, 1961.

Rizzo, F. X., Cunningham, G., and Galanter, L., Evaluation of Perspex H. X. as a gamma-ray dosimetry material, *Trans. Am. Nucl. Soc.*, 10(1), 53, 1967.

Rizzo, F. X. and Krishnamurthy, K., Cellulose acetate butyrate dosimeter, *Trans. Am. Nucl. Soc.*, 12, 61, 1969.

Rösch, E., Weickelt, G., and Broström, D., Si-photodiodes as dosimeters for fast neutrons, *Nucl. Instr. Meth.*, 101, 47, 1972.

Ruault, P. A. and Gatineau, J. P., Comparison of the Results Obtained on Dose Measurements by Thermoluminescence or Films of 20 to 200 keV, Paper, First European Congr. on Radiat. Protection, Menton, France, 1968.

Schäfer, H. J., Benton, E. V., Henke, R. P., and Sullivan, J. J., Nuclear track recordings of the astronauts radiation exposure on the first lunar landing mission Apollo 11, *Radiat. Res.*, 49, 245, 1972.

Scherer, A. E. and Kline, D. E., Density-dose behavior of irradiated polytetrafluoroethylene, *J. Appl. Polymer Sci.*, 11, 341, 1967.

Schimmerling, W. and Sass, R. E., Experience with a commercial film badge service, *Health Phys.*, 15, 73, 1968.

Schulman, J. H., Klick, C. C., and Rabin, H., Measuring high doses by absorption changes in glass, *Nucleonics*, 13(2), 30, 1955.

Schulman, J. H., Etzel, H. W., and Allard, J. G., Application of luminescence changes in organic solids in dosimetry, *J. Appl. Phys.*, 28, 792, 1957.

Schweitzer, G. K. and Lloyd, E. A., The use of secondary electron emission in photographic film dosimetry, *Int. J. Appl. Radiat. Isotopes,* 14, 87, 1963.

Scott, A. G., The Effects of Energy and Angular Dependence on the Response of Film, Quartz-Fibre Electroscope and Thermoluminescent Personnel Dosimeters, AECL-2714, Atomic Energy of Canada Ltd., Pinawa, Manitoba, 1970.

Seguin, H., Preliminary Results of a Personnel Neutron Dosimeter Intercomparison Program, Paper IAEA—AM—167/77, Proc. Symp. Neutron Monitor. Radiat. Protect., IAEA, Vienna, 1973.

Shapiro, J., Smith, D. G., and Iannini, J., Experience with boron-loaded nuclear track plates as detector elements for spherical moderator type neutron monitors, *Health Phys.*, 18, 418, 1970.

Spiegler, G. and Davis, R., An improved method and filmholder for personnel monitoring, *Br. J. Radiol.*, 132, 464, 1959.

Sqentirmay, Z., Deszi, Z., and Patko, J., Acta Univ. Debrecen. Ludovico Kossuth Nom, *Phys. Chem. Ser. B*, 33, 1965.

Stekelenburg, L. H. M. van, New film badge enables cheaper x-ray monitoring, *Nucleonics*, 16(6), 83, 1958.

Stolz, W. and Prokert, K., Chemische Dosimetrie ionisierender Strahlung mit Leuco-Verbindungen in Polystyrol, *Kernenergie*, 8, 425, 1965.

Stolz, W., Dosimetrie intensiver Gamma-und Elektronen-Strahlungsfelder mit gefärbtem Zelloid, *Isotopenpraxis*, 2, 366, 1966.

Stolz, W. and Schmitt, W., Zur Strukturabhängigkeit der Lumineszenzschädigung organischer Molekülkristalle durch Gamma-Strahlung, *Ann. Phys.*, 19, 364, 1967.

Stolz, W., *Strahlensterilisation — Grundlagen und Anwendung in Medizin und Pharmazie*, J. A. Barth, Leipzig, 1972.

Storm, E. and Shlaer, S., Development of energy-independent film badges with multi-element filters, *Health Phys.*, 11, 1127, 1965.

Stützer, M. and Gibson, N. N., Cobalt Borosilicate Glass Gamma Dosimetry and Its Use at the U. S. Army Nuclear Defense Laboratory, Report AD-677373, Edgewood Arsenal, Md., 1968.

Svansson, L., Swedberg, P., Widell, C. O., and Wik, M., Silicon Diode Dosimeter for Fast Neutrons, AE-339, A. B. Atomenergi, Stockholm, 1968.

Swartz, J. M., A Feasibility Study of a High Sensitivity Fast Neutron Dosimeter, TID-22335, 1963.

Swartz, J. M. and Thurston, M. O., *J. Appl. Phys.*, 37, 745, 1966.

Taimuty, S. I., Glass, R. A., and Deaver, B. S., Jr., *High Level Dosimetry of Gamma and Electron Beam Sources*, Proc. 2nd Int. Conf. Peaceful Uses Atomic Energy, Vol. 21, *United Nations, Geneva, 1958*, 204.

Takashima, Y., Nakayima, Y., and Chandler, L., High Gamma Ray Dose Determination using Mössbauer Spectroscopy, Paper IAEA/SM-160/68, Proc. Symp. Dosimetry Techniques Applied to Agriculture, Industry, Biology, and Medicine, IAEA, Vienna, 1972.

Tamura, N., Oshima, Y., Yotsumoto, K., and Sunaga, H., Radiation dosimeter using peroxy radicals in irradiated polytetrafluoroethylene, *Jap. J. Appl. Phys.*, 9, 1148, 1970.

Tanaka, R., Mitomo, S., and Oshima, Y., Development of Irradiation Techniques. II. Dosimetry of Large Radiation Sources by Polymethyl Methacrylate, JAERI–Memo 4121, JAERI Tokai-Mura, Japan, 1970.

Taplin, G. V. and Malin, K., Solid-state chemical dosimetry system, *Radiat. Res.*, 14, 510, 1961.

Tatsuta, H., Moriuchi, Y., Katon, A., Yamachi, I., and Matsumoto, K., Intercomparison of the accuracy of dosimetry systems used in the different facilities, *Hoken Butsuri*, 7, 37, 1972.

Tellez-Plasencia, H., Études sur le Noircissement Photographique Produit par les Rayons X, *Sci. Ind. Photogr.*, 29, 249, 1958.

Thorton, W. T., Davis, D. M., and Gupton, E. D., The ORNL Badge Dosimeter and its Personnel Monitoring Applications, ORNL-3126, Oak Ridge Nat. Lab., Oak Ridge, Tenn., 1961.

Thurston, M. O., Swartz, J. M., Speers, R. R., and Closser, W. H., A Silicon Fast Neutron Dosimeter with a Wide Sensitivity Range, Paper AM-76/18, Proc. Symp. Neutron Monitor. Radiolog. Protect., IAEA, Vienna, 1966.

Tochilin, E. and Golden, R., Investigation on the relative beta to gamma response of photographic emulsions, *Health Phys.*, 4, 244, 1961.

Turner, C. A., Improving neutron film monitoring techniques, *Nucleonics*, 24(1), 62, 1966.

Wachsmann, F., The Customary Method in Germany for Personnel Monitoring with Film Badges, Proc. Symp. Selected Topics Radiat. Dosimetry, IAEA, Vienna, 1961.

Wachsmann, F. and Stadelmann, H., Unterdrückung der Energieabhängigkeit von Dosisfilmen durch Kombinationsfilter, *Photogr. Korresp.*, 97, 83, 1961.

Wachsmann, F., Organization, Methods and Results of Personnel Monitoring in the Federal Republic of Germany, Radiat. Protect. Monitoring, Public. STI/PUB/199, IAEA, Vienna, 1969, 407.

Wagner, J., Use of Light-Scattering in Fast Neutron Dosimetry, Report COO-1105-176, University of Wisconsin, Madison, 1972.

Wall, B. F., Fast Neutron Dosimetry Using Wide-Based $n^+ pp^+$ Silicon Diodes, Proc. Int. Symp. Neutr. Dosimetry in Biology and Medicine, Neuherberg, Germany, 1972.

Watson, E. C., Fast Neutron Monitoring of Personnel, HW-21552, 1951.

Whittaker, B. and Lowe, C. A., Photosensitivity of clear Perspex dosimeters, Int., *J. Appl. Rad. Isotopes*, 18, 89, 1967.

Whittaker, B., Red Perspex dosimetry, in *Manual on Radiation Dosimetry*, Holm, N. W. and Berry, R. J., Eds., Marcel Dekker, New York, 1970, 353.

Whittaker, B., Recent developments in polymethyl methacrylate dye systems for megarad dosimetry, in *Radiation Dose and Dose Distribution Measurements in the Megarad Range*, U.K. Panel on Gamma and Electron Irradiation, 1971, 11.

Widell, C. O., A Silicon Diode Dosimeter System, Proc. Int. Symp. Neutr. Dosimetry in Biology and Medicine, Neuherberg, Germany, 1972.

Wik, M., Neutron Sensitivity of a Conductivity Modulated Silicon Diode as a Fast Neutron Dosimeter, FOA-Report A 4461-4261, Swedish Defense Res. Inst., Stockholm, 1965.

Wik, M., Solid-State Neutron Dosimeter, Proc. Symp. on Semiconductor Radiat. Detectors, Fachverband f. Strahlenschutz, Neuherberg, Germany, 1967.

Windley, W. C. and Elleman, T. S., Gamma-ray dosimetry with polyvinylfluoride, *J. Nucl. Energy*, 21, 803, 1968.

Yasar, S., A Comparison of Film-Badge and Thermoluminescence Dosimeter for Personnel Monitoring, Report No. 88, Turkish Atomic Energy Com., Istanbul, 1972.

Yasuda, H., Permeability constants, in *Polymer Handbook*, Brandrup, J. and Immergut, E. H., Eds., Interscience, New York, 1966, V-13.

Yokota, R., Muto, Y., and Miyake, T., Performance of glass and film dosimeters in personnel monitoring, *Health Phys.*, 19, 316, 1970.

Zelac, R. E., Track fading in neutron personnel monitoring film under typical use conditions, *Health Phys.*, 15, 545, 1968.

Zelikman, V. L. and Levi, S. M., *Making and Coating Photographic Emulsions*, Focal Press, London.

ADDENDUM

Between the completion of the manuscript of this book early in 1973 and its proofreading stage in mid-1973 some additional relevant contributions came to the author's attention. They are briefly mentioned here in the sequence of chapters into which they would belong.

I. Introduction

Another new basic introduction into the field of radiation protection (Shapiro, 1973) recently became available, and a monograph on the health physics aspects of accelerators (with some discussion of the special dosimetric problems at very high energies) will soon be published (Patterson et al., 1973). A good, recent introduction into the specific problems of radiation monitoring instrumentation, units, national and international radiation protection guides, general measurement considerations, etc. can be found in Volume 3 of LBL-1 (Anon., 1972). A rather comprehensive guide to the English language health physics literature has been provided by Willis (1973).

There are several new studies on the exposure of the U.S. population to natural background radiation and the various factors affecting the exposure levels (Oakley, 1972; Oakley and Moeller, 1973; and Auxier, 1973). The exposure of the population and radiation workers in various countries to natural as well as artifical radiation sources is also the subject of Volume I of a new UNSCEAR Report (United Nations, 1972). In its Annex C, interesting comparisons of the occupational exposure can be found.

Some of the data from this report are given in Table 1, in which the mean annual external radiation dose of various types of radiation workers in 11 countries has been compared (although based on not very accurate film dosimeter data, it should be noted that the highest exposure is only 0.7 rad per year). Table 2 illustrates that only very few of the radiation workers in the atomic energy field receive more than 1 rad/year in any country. The total number of radiation workers varies between <0.1 per thousand of the population in many developing countries, and 2 to 4 per thousand in the most industrialized countries. The genetically significant contribution of occupational exposures to the total population exposure is presently very small

TABLE 1

Mean Annual External Radiation Exposure (in rad) of Radiation Workers

Country	Year	Medical	Dental	Industrial	Research and education	Atomic energy
Belgium	1968	0.29	0.08	0.66	0.48	0.32
Denmark	1968	0.18	0.005	0.36	0.05	0.07
France	1969–70		0.075	0.11	0.03	
Germany, East	1966	0.17		0.15	0.26	
Germany, West	1969	0.31		0.37	0.34	
India	1969	0.21		0.62	0.06	0.49
Israel	1969	0.07	0.07	0.08	0.04	0.06
Japan	1968	0.38		0.29	0.09	0.13
Sweden	1968	0.15		0.10	0.02	0.10
United Kingdom	1964	∿0.4	0.27	0.30		0.69 (1969)
U.S.	1969–70	0.34	0.12		0.16	

After United Nations, 1972.

TABLE 2

Percentage of Workers in Atomic Energy in Different External Radiation Dose Ranges

Dose range (rad/year)	Argentina (1970–71)	Canada (1968)	France (1968)	India (1969)	United Kingdom (1968)	U.S. (1968)
0–0.5	98.4	88.6	97.6	75.1	75.4	93.8
0.5–1	1.1					3.6
0.5–1.5		7.0	2.2	14.0	22.2	
1–5	0.5					2.6
1.5–5		4.4	0.2	10.4	2.5	
>5	0	0	0.01	0.4	0	0

After United Nations, 1972.

(0.1 to 1 mrad/year even in the most advanced countries).

Another interesting contribution further emphasizes the sad state of medical dosimetry technology in developing regions of the world (LeVan, 1973). In one large area, for example, one quarter of all radiotherapeutic installations have no dosimetry at all, in some large countries the number of medical physicists is zero. One way to help remedy this situation is the extensive use of national or international intercomparisons by mail, which is the subject of a recent publication (IAEA, 1973). In particular, the results of a worldwide survey of over 200 installations are given, using LiF:Mg,Ti as a transfer dosimeter (Moos et al., 1973 and Classen et al., 1973). Over 40% of the participating institutions still deviate by more than ±5% from the reference values, but repeated intercomparisons had a beneficial effect on reducing large deviations.

In the same publication (IAEA, 1973), the question of film dosimeter performance tests has been discussed in some detail by two experts (Ehrlich, 1973 and Langmead, 1973); a report of a working group on this subject is given. A questionnaire for international use is suggested to provide dates for possible future international performance tests. An annual international intercomparison study, mostly of accident individual monitoring in mixed neutron and gamma radiation fields, has been carried out for many years by the Health Physics Division of Oak Ridge National Laboratory. In recent years this program led to a similar program which is supported by IAEA, with meetings so far in Oak Ridge, Vaduc in France, and Vinca in Yugoslavia. The personnel dosimetry of ultraviolet radiation is also receiving more attention (Anon., 1972 and Matelsky, 1973).

II. TLD

There are several new reviews on the application of TL methods for the dating of ancient pottery (Michael and Ralph, 1971; Aitken, 1973; and Michels, 1973), and studies of various new TL phosphors of potential dosimetric interest including ZnS:Cu,Ag and/or Co (Grasser et al., 1971); ZnO:TM with a main glow peak at $\sim$230°C (Figure 1); and a mixture of dysprosium-activated $BaTiO_3$ and $SrTiO_3$ with a peak at $\sim$100°C (Elle et al., 1973). The latter material has a high dielectric constant; the fading in pre-irradiated phosphors is accelerated by microwaves. It could, therefore, be used in principle as a "microwave dosimeter."

The properties and applications of more conventional phosphors have been the subject of continued investigations. For example, Pohlit (1973) reported new measurements for the depth and half life of the various traps in LiF:Mg,Ti and CaF_2 (probably Mn-activated), and further discussed the effect of "trap transformation" and fading on the maximum obtainable accuracy with TLD-100 under the point of view of intercomparison studies (under optimized laboratory conditions, reproducibilities of as little as ±0.2% appear possible). Several other papers also deal with the use of LiF:Mg,Ti for international postal intercomparisons of radiation sources (Broerse and Puite, 1973; Classen et al., 1973; and Moos et al., 1973). The total uncertainty of the TL readings under field conditions is estimated to be around ±3%.

The effect of trap transformation on the

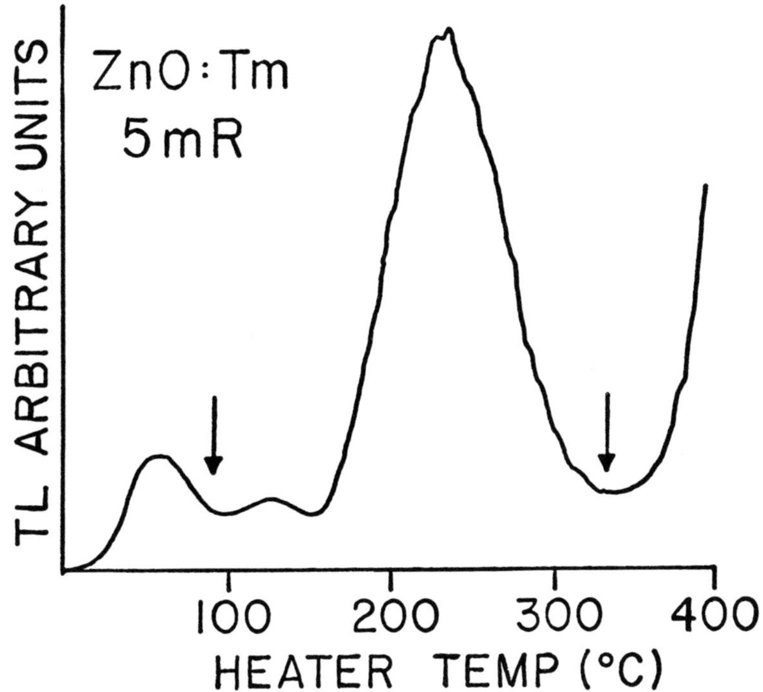

FIGURE 1.   Glow curve of ZnO:Tm. (After Pearson et al., 1973.)

apparent build-up and fading in LiF:Mg,Ti extruded ribbons, which are reused without the time-consuming "magic" annealing, can be minimized by a new readout technique, in which both the peak-height and the light sum under the peaks are measured and added (Rainbolt, 1973). A rather rapid fading of about 23% during 10 days at ambient temperature has been observed in LiF:Mg,Ti which is read in an automatic reader without the 24 hr, 80°C post-readout annealing (Cox and Lucas, 1973). Others report 20% fading during 1 month in the same system (Thompson et al., 1973). The 24 hr, 80°C treatment was found to reduce, as expected, this rather high fading rate, but also accelerated the aging of the Harshaw plastic card dosimeters into which the LiF chips are permanently sealed.

Tweezer-handling of the extruded LiF ribbons has been shown to produce scratches in the surface which change the optical properties, thus resulting

TABLE 3

TL Peak Parameters in LiF:Mg,Ti

Peak temperature ($^\circ$K at 3.2°/sec)	Trap depth (eV)	Halflife at 25°C	In untreated samples	After 400°C, 80°C "magic" annealing
			Fraction of total TL	
332	1.41 ± 0.1	2 min	0.09	–
363	1.77 ± 0.2	25 min	0.24	–
389	2.26 ± 0.2	10 hr	0.22	0.04
411	2.9	10^3 hr	0.18	0.32
427	3.62 ± 0.4	10^4 hr	0.25	0.61
–	–	–	0.02	0.03

After Pohlit, 1973.

in about 25% loss in the TL output after 60 reuse cycles. If handled with a nonscratching vacuum tool, $\sim$90% of the initial sensitivity still remains after >1,000 readout cycles, and the standard deviation of the readings remains within 2 to 4% (Cox et al., 1973).

The energy response and accuracy obtainable with extruded LiF have also been studied by others (Jacobson et al., 1973; Barber, 1973; Zendle et al., 1973; and Cox and Lucas, 1973). According to the latter, doses as small as 5^{mR} can still be measured with about ±5% accuracy. A much higher sensitivity is available with either CaF_2:Mn extruded ribbons in the new U.S. Navy system (0.1 mR are detectable, but "abnormal" fading amounts to $\sim$2.4% per decade of time, fairly independently of temperature between 20 and 100°C), or CaF_2:Dy, which exhibits 20% fading during 10 hr at 20°C (Lucas et al., 1973).

Extruded ^{6}LiF/^{7}LiF paris are finding increased use in "albedo" neutron personnel dosimetry. Systems of more or less complexity have been developed in various laboratories in Europe and the U.S. such as Savannah River, Rocky Flats, Los Alamos, Hanford, and Karlsruhe, usually containing four to five extruded ribbons (see, for example, Hankins, 1973). According to recent calculations (Alsmiller, personal communication), however, greater complexity does not substantially improve the poor energy response of such detectors with their rapidly decreasing sensitivity for neutron energies above $\sim$10 keV. A passive neutron spectrometer, consisting of a series of LiF distributed in equal distances along the central axis of a moderator cylinder, has been developed in England (Langworth, 1970).

Less conventional approaches to fast-neutron dosimetry include activation of the phosphor (Pearson et al., 1973); and heat-resistant compounds of finely powdered, highly sensitive TL phosphors with hydrogenous organics such as p-sexiphenyl (M.P. 450°C). Such compounds exhibit about 50% of their gamma radiation sensitivity on a rad/rad basis for fission as well as 14 MeV neutrons, but are somewhat light-sensitive (Becker et al., to be published). It appears that, besides the direct recoil proton effect, there is some additional optical or electronic energy transfer from the p-sexiphenyl (which is an excellent scintillator) to the TL phosphor.

A key problem in neutron dosimetry has always been the precise measurement of the energy response of the detectors, in particular in the intermediate energy range. The U.S. National Bureau of Standards is presently establishing facilities which will, when available, greatly simplify the calibration of all types of neutron dosimeters. The more important neutron sources to be used in addition to thermal, fission, and radioactive neutron sources are listed in Table 4. The dose-rate will, depending on the reaction and angle to the target, vary between about 0.1 and 30 rem/hr.

The light-induced fading of $Li_2B_4O_7$:Ag has been investigated by Odor and Ziemer (1973). In the instrumentation field, a new fully automatic readout system became commercially available (Figure 2). A very stable (0.5%), low-cost light-emitting gallium arsenide phosphide diode (Thorngate and Perdue, 1972) may have potentials as a light standard for TLD readers. Thermoluminescence as well as TSEE glow curves can easily be digitized for computer evaluation with a low-priced new device made by Elographics (Oak Ridge, Tenn.).

III. TSEE

An interesting Symposium on Exoelectron Phenomena took place June 28 and 29, 1973 in Detroit. There was, among many contributions related to the use of exoelectron phenomena in material damage studies (see, for example, Baxter, 1973), a report on the phenomenological theory of TSEE (Kelly, 1973). An improved proportional counter for TSEE studies was described by Ritz and Attix (1973). Evidently, the gas gain increases at elevated temperatures during readout, causing a major distortion of both pulse-height distribution

TABLE 4

Monoenergetic Neutron Sources to be Used in New NBS Calibration Facility

Source	Reaction	Neutron energy
Van de Graaff	Li (p,n)	0.01−0.65
	T (p,n)	0.1−2.25
	D (d,n)	1.75−6.5
	T (d,n)	12−20
Reactor	Sc + Mn antiresonance	0.002
	Fe antiresonance	0.025
	Si antiresonance	0.144

After Caswell, personal communication.

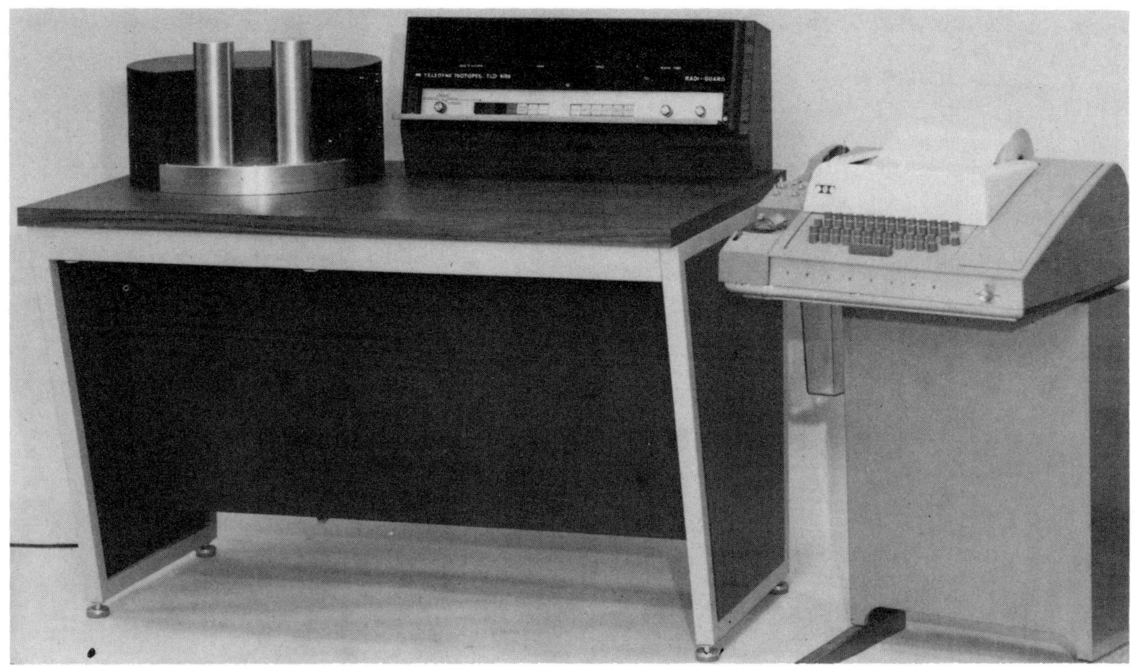

FIGURE 2. Fully automatic readout system for LiF-impregnated Teflon dosimeters, with printout of the dosimeter number, and consecutive evaluation of four different fields exposed behind different filters. (Courtesy of Teledyne/ Isotopes, Westwood, N.J.)

and count-rate plateau. These effects can be reduced by increasing the counting gas flow in the counter to break up the convection column of hot gas rising from the sample to the counting wire. Peterson and Regulla (1973) reported more details on lithium-activated BeO. For some detectors, sensitivities up to 8×10^8 counts/R $\cdot$ cm^2 were found, but the response was highly erratic. In one British BeO sample, exoelectron energies up to 100 eV have been observed.

There have been detailed investigations on the relationship between optical and thermal stimulation in CsCl, BaSO$_4$, SrSO$_4$, BeO, and CaSO$_4$ (Holzapfel and Nink, 1973). New results on TSEE in MgO (Kohnke and Mollenkopf, 1973) are explained on the basis of thermally released holes. A simple exoelectron detector for demonstrations of the effect to students has also been described (Neff et al., 1973).

One of the more interesting recent developments in TSEE dosimeter readout instrumentation was carried out by Williams (1972). Heat-sensitized ceramic BeO disks are used and read in a ionization chamber instrument in air. The mechanical design of the readout chamber is given in Figure 3, the block diagram of the whole instru-

ment in Figure 4. As can be seen in Figure 5, the response is linear up to gamma radiation doses of about 10^3 rad and saturates at $\sim 10^5$ rad. The photon energy dependence of the detectors was found to be flat within ±7% between 25 and 110 keV; there was no fading or sensitivity change within at least 1 month; the standard deviation for groups of about ten detectors was found to be within ±7%.

Progress has also been made at ORNL in increasing the reproducibility and stability of ceramic BeO detectors (Gammage and Cheka, 1973). Apparently, there are differences in sensitivity dependent on the lot number of the purchased product. The sensitivity shows a general increase over extended periods of heat treatment. The recommended recipe for detector preparation is to heat the detectors for 100 hr or more at 1,450°C and then to stabilize them in water (or water vapor at saturation pressure) for $\sim$100 hr before drying at 500°C. If impurities in the water are allowed to dry onto the BeO surface, spurious effects are introduced. Washing in alcohol is, therefore, recommended before the final drying.

Repetitive readings of a detector in one day can increase the response by up to 15%, probably due

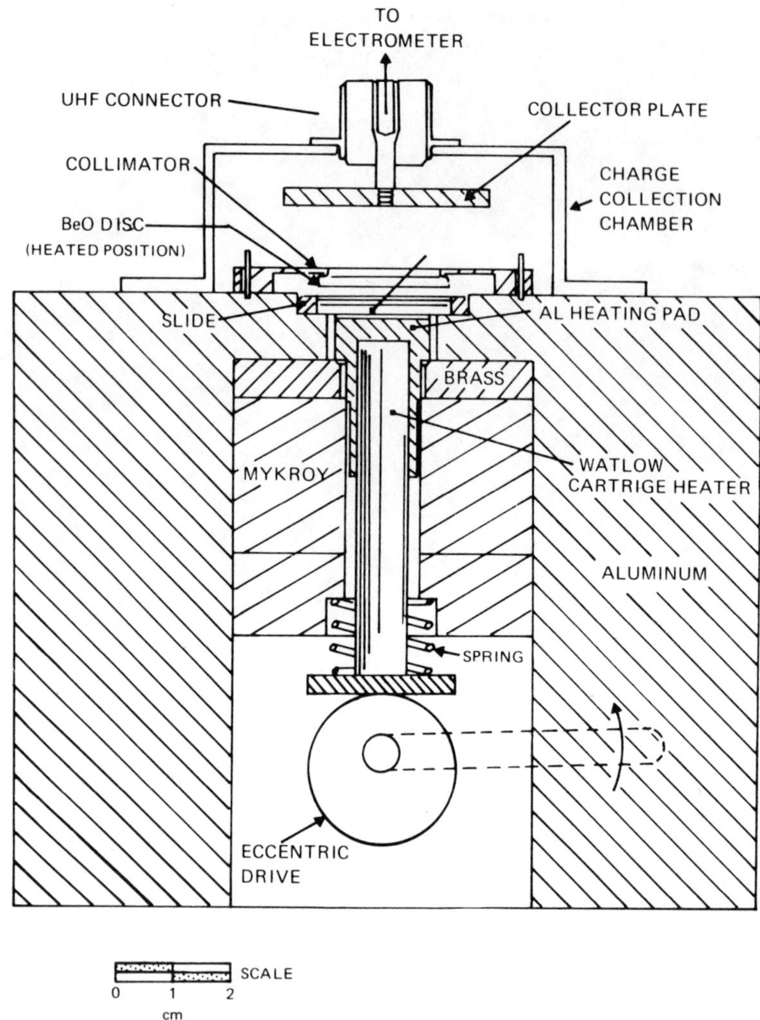

FIGURE 3. Vertical sectional view of TSEE reader for ceramic BeO discs. (After Williams, 1972.)

to outgassing of the detector surface. The effect is transient and after exposure to air for several hours the sensitivity returns to the original value. If a detector is read only once or twice per day, the response to a standard dose is invariant; for 5 detectors each read a total of 30 times over a period of 18 days, the average σ is 4%. Under these conditions, the intragroup σ has been in many instances within $\pm 10\%$. Sets of such well-calibrated, stabilized TSEE dosimeters have successfully been field-tested in personnel and environmental dosimeters at ORNL. Even with all precautions, however, there still are occasionally unexplainable, erratic results, and TSEE dosimetry cannot yet be considered a routinely easy matter.

Also studied in some detail was the angular fast-neutron response of BeO discs covered with hydrogenous radiators such as polyethylene (Becker and Abdel Razek, 1973). As can be seen in Figure 6, the sensitivity rapidly drops when approaching or exceeding an angle of neutron incidence of 90° (frontal exposure = 0°). A TSEE fast-neutron dosimeter can be transformed into a simple neutron "spectrometer" based on the strong neutron energy dependence of the recoil proton range, and consequently of the radiator thickness which is required to establish recoil proton equilibrium. With 14 MeV neutrons, for example, more than 2 mm of polyethylene are required to establish maximum recoil proton

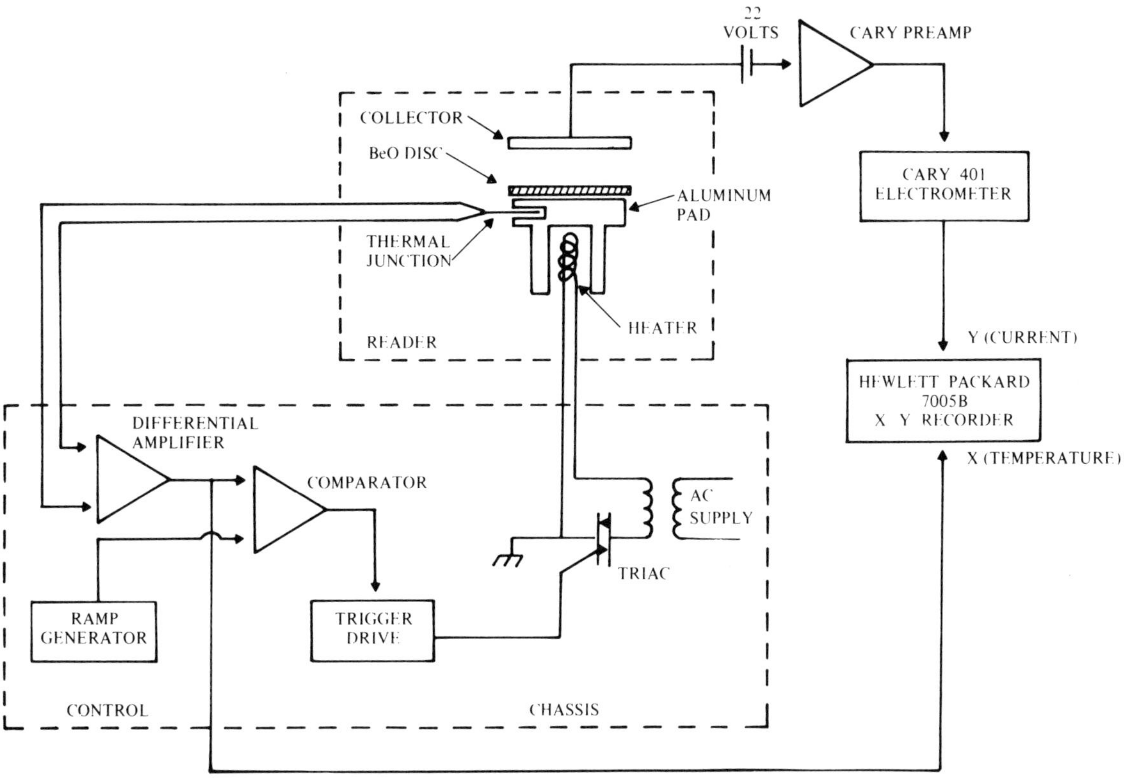

FIGURE 4.   Schematic diagram of TSEE reader system for high dose-level measurements. (After Williams, 1972.)

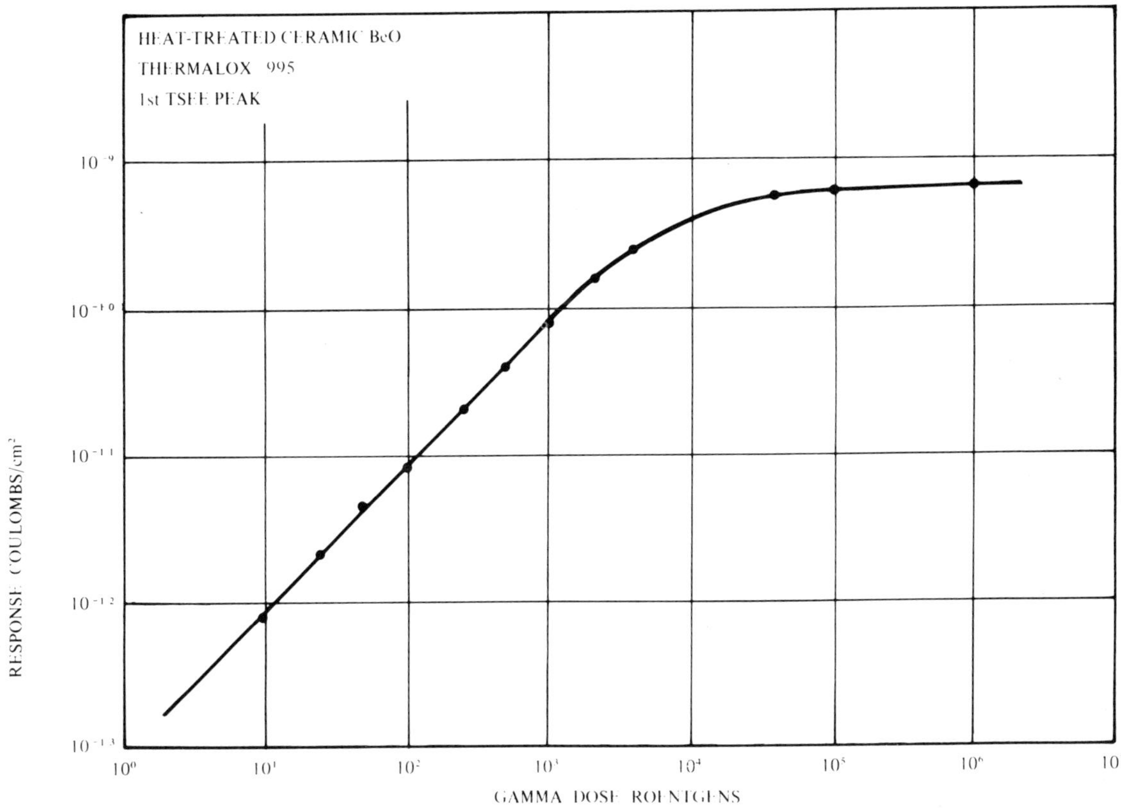

FIGURE 5.   Response of preheated ceramic BeO discs (Thermalox 995®) as a function of gamma radiation dose. (After Williams, 1972.)

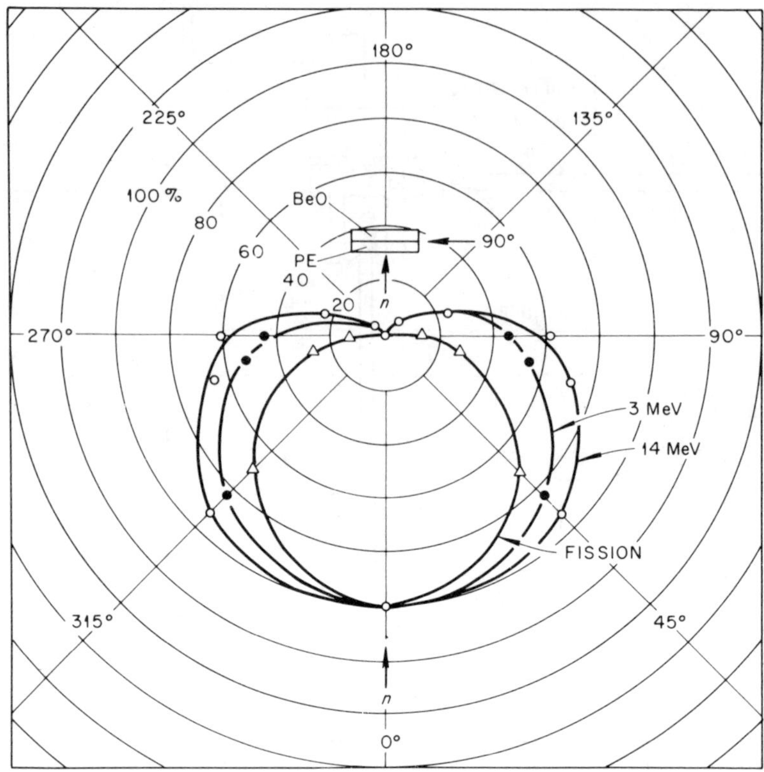

FIGURE 6.   Directional response of polyethylene-covered ceramic BeO disc for fast neutrons of different energies. (After Becker and Abdel Razek, 1973.)

response, while only ∿0.3 mm are necessary for the HPRR fission spectrum. If a sequence of radiators with different thicknesses is placed between the detector and a nonhydrogenous medium, the shape of the resulting curve also indicates whether the neutrons are monoenergetic or not (a wide energy distribution such as in the fission spectrum results in a less steep rise of the curve). A modification of this method consists of an arrangement in which recoil proton absorbers such as aluminum or Teflon® foils of varying thicknesses are placed between a thick recoil proton radiator and the detector (Figure 7).

IV. RPL

In a report from Poland (Wolska-Witer et al., 1972), preparation and properties of a silver-activated glass are described. With small blocks of this glass, gamma radiation doses between 0.2 and 1,000 R can be measured with a ±10% accuracy.

V. Track Etching

Several of the recent publications on track etching are related to dosimetric applications. In the heavy charged particle field, for example, stacks of cellulose nitrate foils have been used for the identification of 41.7 GeV argon ions produced at the Princeton particle accelerator (Fitz et al., 1972). Measurements of solar wind and higher energy solar particles during the Apollo 17 mission have been carried out with mica, glass, and Lexan® track detectors (Walker et al., 1973). Others (Lomanov et al., 1973) measured the neutron contamination of a clinical proton beam in phantoms with a 0.2 mm resolution employing ^{238}U foils and a glass fission-fragment detector.

According to a new study on the etching of polycarbonates in alkali hydroxides (Paretzke et al., 1973b), the main reaction in this process appears to be

$$\{OC_6H_4C(CH_3)_2C_6H_4O\overset{\overset{C}{\|}}{C}\} + 4OH^- \rightarrow$$
$$OC_6H_4C(CH_3)_2C_6H_4O^{2-} + CO_3^{2-} + 2H_2O.$$

After etching, the principal organic compound in the etchant is the anion of bisphenol A. At higher concentrations, a precipitate of the disodium salt

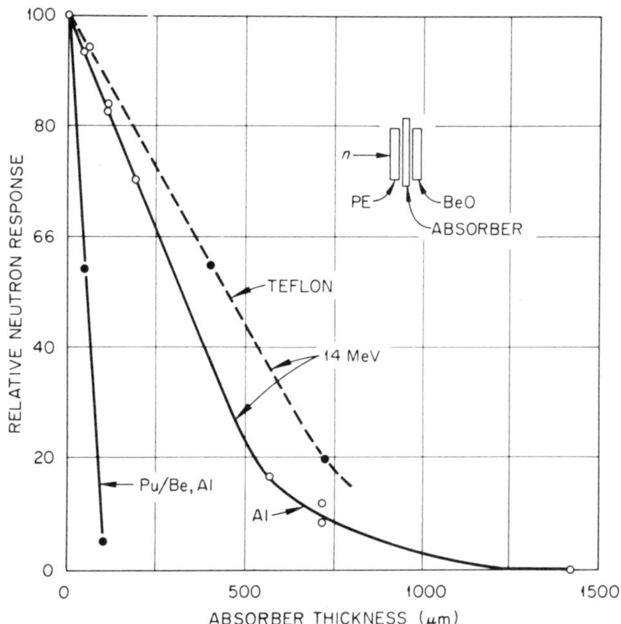

FIGURE 7. Fast-neutron response of radiator-covered ceramic BeO disc as a function of absorber thickness between radiator and detector. (After Becker and Abdel Razek, 1973.)

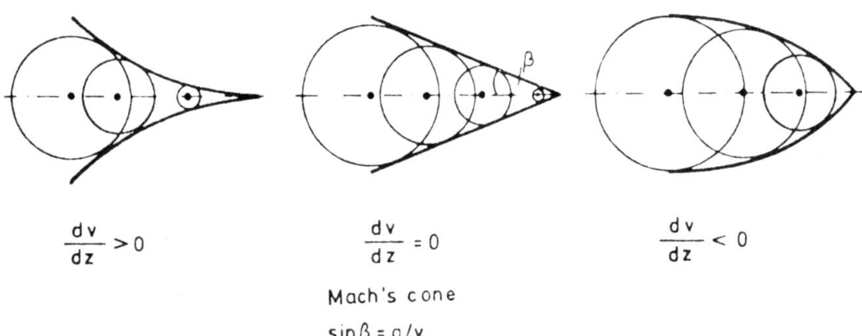

FIGURE 8. Schematic representation of wall shape for increasing, constant, and decreasing track etch rate r(z). (After Paretzke et al., 1973a.)

of bisphenol A is formed, and the anion of phenol may become a major component in the solution. With increasing age, air oxidation makes the composition more complicated. The etch product concentration in solution can be measured by monitoring its ultraviolet absorbance. The bulk etching rate of Makrofol KG® is independent of the etch product, while it increases with etch product concentration in Lexan. At 60°C, the maximum solubility of etch products in 6.25 N NaOH is 3 g/l.

It has been shown (Paretzke et al., 1973a) that a measurement of the etched track radius as a function of the amount of removed surface can be used for the determination of the differential track etching rate, and consequently for particle identification. This method offers advantages for particle tracks close to the detection limit such as proton, recoil or alpha tracks. The track evolution may be described by a simple model (Figure 8) as a function of only two terms, namely, the variable track etching rate v (z) as a function of position z along the axis of the damaged zone, and the bulk etching rate g.

A survey on track etching work going on in 1972 in 118 laboratories in 20 countries (71 of them in U.S. and West Germany) has recently been published (Griffith, 1973). Of these 118 laboratories,

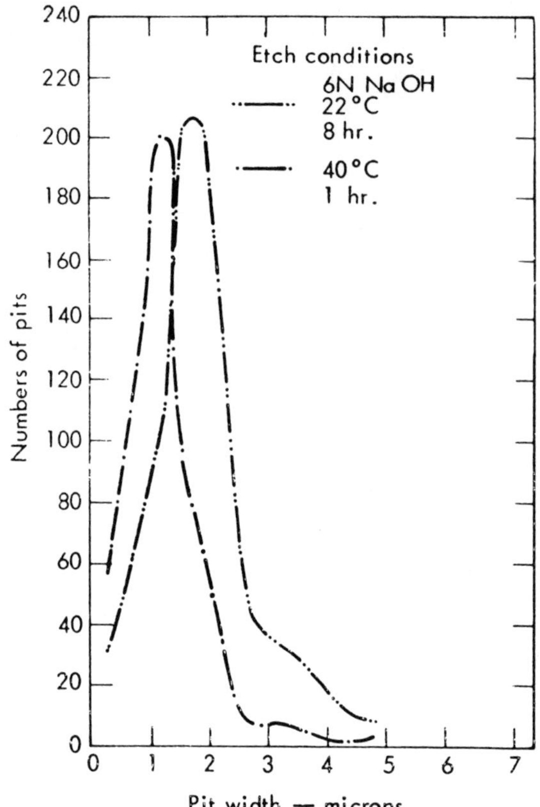

FIGURE 9. Recoil proton etch pit size distribution in cellulose nitrate (Kodak Pathé CA 8015®) exposed to ^{252}Cf fission neutrons. (After UCRL-50007-72-3, 1973.)

46 are interested in neutron dosimetry and 39 employ spark counting (24 of them for neutron dosimetry). Polycarbonate is the most popular detector material, followed by mica and cellulose nitrate.

There is a continuing discussion about the possible use of the fissionable materials in personnel fast-neutron dosimetry. Some laboratories such as E.I.R. in Würenlingen, Switzerland; Brookhaven National Laboratory; and Argonne National Laboratory in the U.S. actually issue dosimeters containing natural thorium in routine dosimetry. They feel that there are no serious administrative problems or radiation hazards associated with the distribution of small quantities of thorium (for example, 50 mg in the BNL badge producing about 5 mR/month in 0.5 cm distance — Distenfeld, 1973). In the Atomic Energy of Canada Laboratories in Chalk River, dosimeters are being considered which contain a few milligrams of ^{237}Np each, which is protected by a layer of 0.5 mg/cm^2 of gold (Cross, personal communication).

Other institutions have basic objections against radioactive materials in personnel dosimeters and focus on the improvement of recoil particle detection. In the more sensitive cellulose nitrate foils, recoil protons can be registered. Figure 9 shows the size distribution of recoil proton etch pits (the hump around 3.5 μm is apparently due to the heavier recoil nuclei such as carbon, oxygen, and nitrogen).

Unfortunately, recoil proton pits are small and difficult to identify or to count automatically. It has, however, been found (Becker and Abdel Razek, to be published) that the electrical breakdown in "overetched" thin cellulose nitrate foils strongly depends on their exposure to fast neutrons, apparently due to the spark-counting of recoil particle tracks at a point close to the normal breakdown of unexposed foils. Depending on the measuring conditions and neutron energy, several hundred counts per cm^2 and rad of fast neutrons have been obtained both in commercial 10 μm Kodak-Pathé LR 115® films, and in improved 15 μm foils made at ORNL from Hercules cellulose nitrate (Lupica, to be published).

VI. Other Detectors

A model explaining the strong temperature dependence of latent image formation in AgBr crystals based on Mitchell's theory has been developed (Sharma, 1972). The controversy about the merits and limitations of the photographic film dosimeter continues, and will probably continue for some time to come. Apparently, several film dosimetry services in Europe have carried out laboratory and/or field stability tests, which did in some cases lead them to conclude that "fading is no problem." As no details on these unpublished studies are known, they cannot be discussed here. On the other hand, there are more well-documented publications on the amount of fading, particularly in NTA films. Distenfeld and Klemish (1972), for example, report 50% track fading at 32°C and 71% relative humidity within 2 weeks, and 95% after 1 month. Similar data are given by Jasiak and Musialowics (1972).

The results of a storage test of exposed conventional Kodak Type 2® and two TLD phosphors in locations with different climates in Latin America are given in Table 5. Obviously, film dosimeters should not be used in four out of the seven

TABLE 5

Reading of Gamma-irradiated Film and TLD After Storage for 3 Months at Different Locations (Normalized for Controls Kept under Low-fading Conditions)

| | | | Kodak PM Type 2 | | TLD phosphors | |
Location	Average temp. ($^\circ$C)	Average relative humidity	Unsealed	Sealed	LiF:Mg,Ti (TLD−100)	$CaSO_4$:Dy
ORNL Control	−8	0	100	100	100	100
Bogota, Colombia	16	70	97.5	97.5	89	94.7
Cartagena, Colombia	28	90	21.3	46.3	77.2	97.6
Guayaquil, Ecuador	20.7	71	0	11.3	86.2	94.6
Santiago, Chile	18.6	63	103.5	97.6	87.5	106
Los Angeles, Chile	29	82	102.9	96.5	92.7	109.5
Buenos Aires, Argentina	29.0	88	29	92	90.5	100
Santa Fe, Argentina	31.0	78	67	97.5	82.6	97

After Becker, to be published.

locations without additional sealing, and even optimized sealing of the films in a plastic-aluminum compound does not prevent substantial fading at two of the locations (Becker, to be published). Procedures and criteria for film dosimeter performance tests have been discussed by Langmead (1973) and Ehrlich (1973).

A good new compilation of the more common plastic foils for high dose-level measurements is given in Table 6. An intercomparison of a clear Perspex-® HX dosimeter with different spectrophotometers for readout produced promising results (Chadwick et al., 1973). Other authors describe the application of organic polymer dosimeters such as tinted Lavsan® and Cellophane® (Kobatenko et al., 1971) and polyvinyl chloride (Kronenberg et al., 1972) for a variety of purposes. No significant dose-rate dependence between 10^3 rad/sec and 10^{15} rad/sec was found in dye-polychlorostyrene films (Chappell and Humphreys, 1972).

TABLE 6

Characteristics of Some Organic Polymers of Dosimetric Interest

Type	Formula	Approximate density (g/mc^3)	Typical response range (rad)	Typical wavelength for readout (nm)
Cellulose acetate butyrate	$(C_{15}H_{22}O_8)_n$	1.2	10^6 to 10^8	320
Cellulose acetates	$(C_6H_{10}O_5)_n$ $(C_{12}H_{16}O_8)_n$	1.2 to 1.4	3×10^5 to 10^9	280–380
Melamine	$(C_6H_{12}N_6O_3)_n$	1.5	10^5 to 4×10^7	380–420
Polyethylene	$(C_2H_4)_n$	0.95	10^6 to 10^8	236, 307, 1040, 1720, 3420, 5840
Polyamide	$(C_6H_{11}NO)_n$	1.1	10^8 to 10^{10}	315
Polyethylene terephthalate	$(C_{10}H_8O_4)_n$	1.4	10^6 to 10^9	325
Polymethyl methacrylate	$(C_5H_8O_2)_n$	1.2	10^5 to 10^7	305
Dyed polymethyl methacrylate	$(C_5H_8O_2)_n$	1.2	10^5 to 5×10^7	603, 620, 640, 651
Dyed polyvinyl alcohol	$(C_2H_4O)_n$	1.3	10^4 to 10^6	660
Polyvinyl chloride	$(C_2H_3Cl)_n$	1.4	10^5 to 10^7	395, 480
Dyed polyvinyl chloride	$(C_2H_3Cl)_n$	1.4	10^6 to 10^7	400, 615
Polyvinylvinylidene chloride	$(C_4H_5Cl)_n$	1.7	10^5 to 10^7	260
Polystyrene	$(C_8H_8)_n$	1.05	5×10^7 to 10^8	330, 420
Dyed polystyrene	$(C_8H_8)_n$ halogens	1.05	10^4 to 5×10^7	430, 610, 640
Polytetrafluoroethylene	$(CF_2)_n$	2.2	10^3 to 10^{11}	ESR
Polyvinylfluoride	$(C_2H_3F)_n$	1.6	10^6 to 3×10^7	315
Polyvinylpyrrolidone	$(C_6H_9NO)_n$	1.3	10^6 to 10^8	350–380
Blue cellophane (dyed cellulosics)	$(C_3H_{10}O_5)_n$	1.4	10^6 to 10^8	650
Dyed cinemoids (dyed cellulose acetate)	$(C_6H_{10}O_5)_n$	1.3	10^6 to 10^8	437, 530
Malachite green methoxide in 4-chlorostyrene	$(C_6H_7Cl)_n$	1.1	5×10^4 to 2×10^7	430, 630
Triphenylmethane cyanides in various media	C, H, N, O	0.9 to 2.0	10^4 to 10^8	430, 560, 580, 600, 625
Stilbene in polystyrene	$(C_{14}H_{12}(CH))_n$	1.1	2×10^5 to 10^8	324
Polymers with fluorescing ingredients	C, H, N, O	1 to 2	1 to 10^7	Variable

After McLaughlin et al., 1973.

REFERENCES

Aitken, M. J., Thermoluminescence Dating, Second Cairo Solid-State Conf., Cairo, Egypt, 1973.

Anon., Instrumentation for Environmental Protection-Radiation, LBL-1, Vol. 3, Lawrence Berkeley Lab., Berkeley, Calif., 1972.

Anon., Hazards Control Progress Report No. 44 — UCRL-500 01-72-3, Lawrence Livermore Lab., Livermore, Calif., 1973.

Auxier, J. A., Dissertation, Department of Nuclear Engineering, Georgia Institute of Technology, Atlanta, 1973.

Barber, D. B., LiF Response to 1.0 to 4.0 keV Electrons, Paper No. 74, 18th Ann. Meet. Health Phys. Soc., Miami, Fla., 1973.

Baxter, W. L., Exoelectron Emission from Metals, Paper, Symp. on Exoelectron Phenomena, Wayne State University, Detroit, 1973.

Becker, K. and Abdel Razek, M., Fast neutron response characteristics of TSEE dosimeters, *Nucl. Instr. Meth.*, in press.

Broerse, J. J. and Puite, K. J., Comparisons of X-Ray Dosimetry for the Coordination of Late Effects Research in Europe, in STI/PUB/338, IAEA, Vienna, 1973, 21.

Chadwick, K. H., Broeke, W. R. R., and Rintjema, D., An Intercomparison of Readout Systems for the Clear Perspex Dosimeter, IAEA-PL-479/6, in STI/PUB/338, IAEA, Vienna, 1973, 33.

Chappel, S. E. and Humphreys, J. C., Dose-rate response of a dye-polychlorostyrene film dosimeter, *Trans, Nucl. Sci.*, NS-19, 175, 1972.

Classen, I., Seelentag, W., and Waldeskog, B., Joint IAEA/WHO TLD Intercomparison Program: Evaluation of the Present Situation, IAEA-PL-479/15, in STI/PUB/338, IAEA, Vienna, 1973, 137.

Cox, F. M., Arnold, R. A., and Fabry, B. M., The Reusability of Solid TL Dosimeters and its Relation to the Maintenance of TL Standards, Paper No. 61, 18th Ann. Meet. Health Phys. Soc., Miami, Fla., 1973.

Cox, F. M. and Lucas, A. C., An Automated TLD System for Personnel Monitoring, submitted to *Health Physics,* 1973.

Distenfeld, C. H., Summary of Personnel Neutron Dosimetry Development at Brookhaven National Laboratory, Paper, 4th USAEC Workshop on Neutron Personnel Monitoring, Miami, Fla., 1973.

Distenfeld, C. H. and Klemish, J. R., Developmental Study of Personnel Neutron Dosimetry at the AGS, BNL-17452, Brookhaven Nat. Lab., Upton, N.Y., 1972.

Ehrlich, M., Dosimetry Performance Tests, IAEA-PL-479/8, in STI/PUB/338, IAEA, Vienna, 1973, 41.

Elle, D. R., Vetter, J. R., and Ziemer, P., Thermoluminescence and Microwave Induced Thermoluminescent Fading of High Dielectric Materials, Paper No. 13, 18th Ann. Meet. Health Phys. Soc., Miami, Fla., 1973.

Fitz, R. C., McNulty, P. J., and Davis, A. F., Use of Plastic Detectors to Confirm the Acceleration of Argon Nuclei to Relativistic Velocities at the Princeton Particle Accelerator, AD-749863, Air Force Cambridge Res. Lab., Hanscom Field, Mass., 1972.

Gammage, R. B. and Cheka, J. S., Toward a Practical TSEE Dosimetry System, Paper, Symp. on Exoelectron Phenomena, Wayne State University, Detroit, 1972.

Grasser, R., May, A., and Scharmann, A., Thermoluminescence of ZnS, *Z. Naturforsch.*, 27a, 228, 1972.

Griffith, R. V., Results of the 1972 Survey on Track Registration, UCRL-51362, Lawrence Livermore Lab., Livermore, Calif., 1973.

Hankins, D. E., A Small Inexpensive Albedo-neutron Dosimeter, Paper No. 106, 18th Ann. Meet. Health Phys. Soc., Miami, Fla., 1973.

Holzapfel, G. and Nink, R., Zum äußeren Photoeffekt an Elektronenhaftzentren in kristallinen Festkörpern (optisch stimulierte Exoelektronenemission — OSEE) PTB-Mitt., to be published.

IAEA, National and International Radiation Dose Intercomparisons, Proc. Panel IAEA/WHO, STI/PUB/338, Vienna, 1973.

Jacobson, A., Banks, T., and Ackerman, M., Evaluation of personnel dosimetry methods for diagnostic x-ray special procedures work, *Health Phys.*, 25, 76, 1973.

Jasiak, J. and Musialowicz, T., Investigation on track fading in Kodak personal neutron monitoring films Type A, *Nukleonika*, 17, 674, 1972.

Kelly, P., A Phenomenological Theory of Thermally Stimulated Exoelectron Emission, Paper, Symp. on Exoelectron Phenomena, Wayne State University, Detroit, 1973.

Kobalenko, L. M., Gaichenko, L. N., Lavrentovich, Y. I., and Kabakchi, A. M., Dosimetry of ionizing radiations using tinted Lavsan and Cellophane, *At. Energ.*, 31, 510, 1971.

Kohnke, E. E. and Mollenkopf, H. C., Initiation of Exoelectron Emission from MgO by Thermally Released Holes, Paper, Symp. on Exoelectron Phenomena, Wayne State University, Detroit, 1973.

Kronenberg, S., Lux, R. A., Ramm, W. J., Nilson, K. L., and Pfeffer, R. L., Anisotropic Dose and its Application in Directional Radiation Sensing, AD-750342, Army Electronics Command, Fort Monmouth, N.J., 1972.

Langmead, W. A., Personal Dosimetry Performance Tests for Radiological Protection Purposes, IAEA-PL-479/11, in STI/PUB/338, IAEA, Vienna, 1973.

LeVan, J. H., Medical Dosimetry Technology in Developing Regions of the World, Paper, IAEA-SM-160/90, in Dosimetry in Agriculture, Industry, Biology and Medicine, IAEA, Vienna, 1973, 37.

Lomanov, M. F., Shimchuk, G. G., and Yakovlev, R. M., Investigations of dose fields in a clinical proton beam by means of x-ray defectoscopic track detectors, *At. Energ.*, 34, 185, 1973.

Longworth, J. P., A Neutron Flux Spectrometer with nearly Constant Sensitivity over the Energy Range Thermal to 14 MeV, Report RD/B/N 1416, Central Electricity Generating Board, England, 1970.

Lucas, A. C., Cox, F. M., and Fabry, B. M., An Encapsulated, Reusable TL Dosimeter for Environmental Radiation Measurements, Paper No. 100, 18th Ann. Meet. Health Phys., Soc., Miami, Fla., 1973.

Matelsky, I., A Review of Ultraviolet Standards, Paper No. 48, 18th Ann. Meet. Health Phys. Soc., Miami, Fla., 1973.

McLaughlin, W. L., Hjortenberg, P. E., and Radak, B. B., Absorbed Dose Measurements with Thin Films, Paper IAEA-SM-160/32, in Dosimetry in Agriculture, Industry, Medicine and Biology, IAEA, Vienna, 1973.

Michael, H. N. and Ralph, E. K., *Dating Methods for the Archaeologist,* MIT Press, Cambridge, Mass., 1971.

Michels, J. W., *Dating Methods in Archaeology,* Seminar Press, New York, 1973.

Moos, W. S., Balamutov, V. G., and Abedin-Zadeh, R., A Multinational Survey of the Status of ^{60}Co Radiation Therapy Dosimetry, IAEA-PL-479/18, in STI/PUB/338, IAEA, Vienna, 1973, 117.

Neff, R. D., Shropshire, S. M., and Simek, J. E., A LiF TSEE Detection System for the Dosimetry Laboratory, Paper No. 70, 18th Ann. Meet. Health Phys. Soc., Miami, Fla., 1973.

Oakley, D. T., Natural Radiation Exposure in the United States, U.S. Environmental Protection Agency Report ORP/SID 72-1, 1972.

Oakley, D. T. and Moeller, D. W., Natural Radiation Exposure in the United States, Paper No. 103, 18th Ann. Meet. Health Phys. Soc., Miami, Fla., 1973.

Odor, D. L. and Ziemer, P. L., Photon Induced Fading of Lithium Borate Thermoluminescent Dosimeter, Paper No. 14, 18th Ann. Meet. Health Phys. Soc., Miami, Fla., 1973.

Paretzke, H. G., Benton, E. V., and Henke, R. P., Particle track evolution in dielectric track detectors and charge identification through track radius measurement, *Nucl. Instr. Meth.,* 108, 73, 1973a.

Paretzke, H. G., Gruhn, T. A., and Benton, E. V., The etching of polycarbonate charged particle detectors by aqueous sodium hydroxide, *Nucl. Instr. Meth.,* 107, 597, 1973b.

Patterson, H. W., Thomas, R. H., and Wallace, R., *Accelerator Health Physics,* Academic Press, New York, 1973.

Pearson, D., Wagner, J., Moran, P. R., and Cameron, J. R., Fast Neutron Dosimetry Using Activated Thermoluminescence, Paper No. 12, 18th Ann. Meet. Health Phys. Soc., Miami, Fla., 1973.

Peterson, D. D. and Regulla, D. F., The Effect of Lithium on the Yield and Energy of Thermally Stimulated Exoelectrons from Beryllium Oxide, Paper, Symp. on Exoelectron Phenomena, Wayne State University, Detroit, 1973.

Pohlit, W., TLD Probes for Intercomparison of Dosimetric Data, Paper IAEA-SM-160/7, in Dosimetry in Agriculture, Industry, Biology and Medicine, IAEA, Vienna, 1973.

Rainbolt, C., A New Concept in LiF TLD Readers, Paper No. 60, 18th Ann. Meet. Health Phys. Soc., Miami, Fla., 1973.

Ritz, V. H. and Attix, F. H., Performance of an Improved Methane-Flow Proportional Counter for TSEE. Paper, Symp. on Exoelectron Phenomena, Wayne State University, Detroit, 1973.

Shapiro, J., *Radiation Protection,* Harvard University Press, Cambridge, Mass., 1973.

Singh, M. and Sharma, A. P., Model of temperature dependence of latent image in photographic emulsions, *Indian J. Pure Appl. Phys.,* 10, 676, 1972.

Thompson, D. J., Tucker, G. E., and Armijo, B. R., Characteristics of the Harshaw TLD Cards as Used at Sandia Laboratories, Paper No. 112, 18th Ann. Meet. Health Phys. Soc., Miami, Fla., 1973.

Thorngate, J. H. and Perdue, P. T., A stable pulsed light source using low-cost light-emitting diodes, *Nucl. Instr. Meth.,* 105, 57, 1972.

United Nations, Ionizing Radiation, Levels and Effects — A Report of the U.N. Scientific Committee on the Effects of Atomic Radiation to the General Assembly, Vols. I and II, U.N., New York, 1972.

Walker, R. M., Zinner, E., and Maurette, M., Measurement of Heavy Solar Wind and Higher Energy Solar Particles During the Apollo 17 Mission, Report SPP-43, Washington University, St. Louis, 1973.

Williams, G. H., Research in TSEE Dosimetry, Contract F 41609-71-0004, Final Report, Texas Nucl. Div., Austin, 1972.

Willis, C. A., A Health Physics Library, Paper, 18th Ann. Meet. Health Phys. Soc., Miami, Fla., to be published.

Wolska-Witer, M., Koczynski, A., and Wilgocki, M., Preparation of Radiophotoluminescent Glasses for Individual Dosimetry Purposes, CLOR-91/D, Centr. Lab. Radiat. Protect., Warsaw, 1972.

Zendle, R., Schunicht, B. F., and Champagne, R. J., A Manual on Automatic and Computerized TLD System for Personnel and Environmental Radiation Exposure Monitoring, Paper No. 111, 18th Ann. Meet. Health Phys. Soc., Miami, Fla., 1973.

C

Dörschel, B., 182, 197, 203, 205, 208
Douglas, J. A., 268
Draganic, I. G., 279, 289
Dragu, A., 210
Dreis, D., 208
Dresel, H., 259, 261
Drexler, G., 20, 120
Druskina, L. S., 145
Dudley, R. A., 231, 271
Dulieu, P., 225, 229
Dunlop, J. T., 148
Duport, P., 214, 215
Durrani, S. A., 63, 181, 182, 191, 197, 209
Dutrannois, J., 5, 184, 186, 214, 268
Dutreix, A., 272
Dutreix, J., 272

E

Eastes, J. D., 93, 262
Edelmann, B. U., 34, 36, 37
Egelhaaf, H., 261
Eggermont, G., 41
Eggert, J., 1, 237, 258
Eguchi, S., 28, 35, 70, 71
Ehrlich, M., V, 5, 54, 55, 62, 231, 235, 236, 238, 242,
    256, 258, 263, 271, 300, 309
Eisen, H., 280, 286
Eisenbud, M., 270
Elle, D. R., 300
Elleman, T. S., 279
Ellis, S. C., 36, 37, 278
Elmanharawy, M. S., 72
El Naggar, S., 121
Endres, G. W. R., 46, 53, 76, 77, 93, 229
Enge, W., 184, 188, 189, 191, 193
Engelke, M. J., 78
Engelland, W., 127
Ennow, K., 135
Erdmann, U., 259
Erlenbach, H. R., 168
Espejo, D., 46
Ettinger, K. V., 63
Etzel, H. W., 141, 287
Euler, M., 120
Evans, H. J., 18
Evans, L. W., 1, 141
Evans, R. D., 269
Ewen, K., 168

F

Fabel, G. W., 289
Fabry, B. M., 302
Facey, R. A., 53
Fain, J., 178, 179
Falk, R. B., 77
Fängewisch, G. L., 211
Farmer, F. T., 14
Farrell, J. J., 280, 286

Fasso, A., 52
Faust, L. G., 261
Feagin, F. E., 259
Feher, I., 97
Feige, Y., 4, 48, 73, 87, 149
Fekula, O. M., 266
Ferraresso, G., 71
Fiedler, G., 192, 209
Fields, D. E., 46
Finston, R. A., 278
Fintelmann, D., 126
Fitch, S. H., 256
Fitz, R. C., 306
Fitzgerald, J. J., 3
Fitzgerald, K. T., 216
Fitzsimmons, C. K., 93, 262
Fitzsimmons, W. A., 288
Fix, R. C., 48
Flack, E. D., 65
Fleischer, R. L., 2, 175, 177, 178, 180, 182, 184, 187,
    190, 191, 208, 213
Fleming, S. J., 4, 27, 60, 63, 70, 94, 277, 278
Fogel, S. J., 231, 234
Forsythe, R. J., 93
Fotland, R. A., 177
Fowler, J. F., 4, 27, 43, 46, 48, 49, 53, 66, 168, 279
Fowler, P. H., 249
Francois, H., 4, 35, 70, 73, 142, 147, 148, 165, 169
Frank, A. L., 178, 211, 213
Frank, M., 4, 29, 32, 34, 42, 60, 64, 85, 125, 272, 276,
    288
Freeswick, D. C., 50
Fremlin, J. H., 191, 209
Freytag, E., 154
Fricke, H., 22, 168
Friedland, S. S., 48, 73, 87
Friedman, A. M., 189
Frieser, M., 241, 270
Frigero, N. A., 3
Fukui, K., 191
Fuller, G. E., 288, 289
Fullerton, G. D., 62
Furman, S. C., 175
Furtak, T. E., 266
Furuta, Y., 52, 53
Futtener, A. T., 12

G

Gabor, G., 28
Gaichenko, L. N., 309
Gais, P., 195, 196
Gajda, R., 116
Gal, O. S., 289
Galanter, L., 277, 278
Galkin, B. M., 51, 73, 93
Gall, A., 85
Gammage, R. B., V, 5, 112, 119, 120, 121, 122, 123,
    124, 125, 126, 127, 129, 132, 134, 135, 303
Gammertsfelder, C. C., 229
Gangadharan, P., 76, 87

Hettinger, G., 160
Heusi, K., 197, 205
Hiaoka, T., 54
Higashimura, T., 60, 73
Hillenkamp, F., 148
Hine, G. J., 255
Hiraki, H., 84
Hirasawa, Y., 245, 247
Hitomi, T., 71
Hjortenberg, P. E., 273, 287, 310
Hoegl, A., 163
Hoerlin, H., 253, 258
Hofman, E. G., 283
Hogeweg, B., 94
Holm, N. W., 22, 279
Holt, P. D., 254, 259
Holzapfel, G., 4, 31, 111, 112, 115, 116, 117, 119, 120,
  126, 128, 303
Homayooufar, M., 33, 156, 178, 265
Hood, H. P., 274
Hopkins, B. J. H., 165
Horn, P. L., 54
Horn, W., 93, 262
Horneck, G., 191
Hougs, E., 129
Houtermans, F. G., 60
Hoy, J. E., 17, 66, 77, 78
Hubeny, H., 277, 279
Hübner, K., 49, 60, 279, 281, 284, 286
Hudd, W. H. R., 77, 78
Hudis, J., 206
Huff, J. B., 281, 286
Hukkoo, R. K., 51, 55, 73
Humar, N., 72
Humpherys, K. C., 273
Humphrey, J. S., 188, 196
Humphreys, J. C., 309
Hurst, G. S., 3
Hussman, E. K., 273
Huxtable, J., 21
Huzimura, R., 71
Hvolby, J., 93

I

IAEA, V, 12, 16, 18, 22, 29, 94, 231, 260, 273, 300
Iannini, J., 263
Ibbott, G. S., 47, 51
ICRP, 2, 9, 12
ICRU, 2, 19
Ikeya, M., 67
Ilic-Papovic, J., 287
Imagawa, H., 145, 189
Ipson, S. S., 209
Irie, G., 280, 284
Isabelle, D., V, 175
Ishibashi, M., 67
Itoh, N., 67
Ivanii, G. M., 67
Iverson, G. K., 259
Iwata, S., 206, 209, 216

Izumi, Y., 245, 247

J

Jackson, J. H., 44
Jacobi, W., 213, 214, 269
Jacobs, R., 41
Jacobson, A., 302
Jacobson, J. R., 274
Jahn, W., 147, 275
Jähnert, B., 54
James, K. E., 262
James, T. H., 231, 232, 234
Jamm, J. F., 94
Janssens, A., 41
Jasiak, J., 245, 251, 268, 269, 308
Jasinska, M., 37, 38, 73
Jayachandran, C. A., 46, 47, 56
Jee, W. S. S., 209
Jeltsch, E., 272, 275, 276
Jervis, R. E., 208
Joffre, H., 6
Johns, H. E., 3, 254
Johns, T. F., 76, 91
Johnson, D. R., 5, 179, 188, 191, 198, 200, 208, 209,
  214, 215, 216
Johnson, N. M., 45
Johnson, T. L., 38, 42, 43, 44, 46, 93, 262
Jones, A. R., 4, 8, 9, 10, 11, 76, 91, 92, 93
Jones, B. E., 256, 257, 272
Jones, D. E., 22, 37, 47, 50, 65, 66, 74, 84
Jones, J. L., 48, 274
Jones, P. D., 253, 258
Jones, T. A., 3
Jowitt, D., 209
Jozefowicz, K., 206, 211
Juhasz, S., 196

K

Käämbre, H., 118
Kabakchi, A. M., 280, 284, 288, 309
Kahle, J. B., 245
Kale, L. R., 87
Kantz, A. D., 273
Kanzler, G., 122
Kapsar, B. M., 64
Karches, G. J., 259
Kartha, M., 51, 92
Kartuzhanskii, A. L., 179, 180, 181, 188
Karzmark, C. J., 46, 48, 53
Käs, H. H., 142, 145, 146, 148, 150
Kashukeev, N., 196
Kastner, J., 49, 50, 52, 54, 55, 72, 93, 95, 148, 169, 208,
  209, 216
Kaszuba, F. J., 253, 258
Katcoff, S., 206
Kathren, R. L., 46, 53, 66, 76, 155, 156, 242, 244, 261,
  268, 274
Kathuria, S. P., 40, 41, 50, 60, 62, 73

Lheureux, M., 4, 28, 60, 61
Lillicrap, S. C., 125
Lin, F. M., V, 29
Lindeken, C. L., 22, 65, 66
Lindskoug, B., 86, 93
Liniecki, J., 214
Linsley, G. S., 4, 45, 74
Lippert, J., 48, 50, 66, 68
Liss, M., 84
Littlejohn, G. J., 268
Liu, T. C., 209
Lloyd, E. A., 254
Lomanov, M. F., 306
Longworth, J. P., 302
Lorthoir, M., 28, 60
Lovett, D. B., 207, 208, 214
Lowe, C. A., 277, 278
Lu, R. H., 22, 42, 43, 66, 67, 68
Lubyanskii, G. A., 70
Lucas, A. C., 41, 63, 64, 273, 301
Luchner, K., 28, 60
Lucke, W., 50, 66
Luft, F., 1, 237, 258
Lutz, E., 162, 164
Lutz, W. J., 148
Lux, R. A., 309
Lyman, J. T., 34, 46
Lyman, T., 28, 156, 160

M

Mac Donald, J. C. F., 92
Madden, E., 241
Maffi, A., 160
Mahoney, F. J., 3
Maillie, 168
Majborn, B., 57, 68, 77
Majborn, P., 77
Makarov, Y. A., 69
Malin, K., 286
Malsky, J. S., 158, 163, 168
Manambelona, J. R., 288
Mandelberg, H. I., 259
Mandeville, C. E., 28, 58, 59
Mansfield, C. M., 93
Marato, S. P., 32, 60
Marchetti, M., 183
Marennyi, A. M., 179, 180, 181
Markov, K. P., 270
Markus, B., 6
Marrone, M. J., 38
Marsal, J., 205
Marshall, C. H., 277, 278, 282, 283
Marshall, M., 45, 47, 55, 74, 75
Marshall, T. O., 45, 74, 77, 78, 257, 272
Martensson, B. K. A., 45, 74
Martin, J. A., 48, 274
Martysh, G. G., 188
Maruyama, M., 142, 159
Mason, E. W., 4, 45, 51, 74, 85

Massey, J. B., 3
Matache, G., 169
Mateles, R. L., 279, 284
Matelsky, I., 300
Mateosian, E. D., 178
Matsumoto, K., 93, 262
Matsuo, M., 183
Mauderli, W., 252
Maul, J. E., 279
Maurette, M., 178, 201, 306
Maury, J., 210
Maushart, R., 4, 153, 162, 163, 164, 167
May, A., 146, 148, 300
Mayer, V., 28
Mayhew, M., 197, 203
Mayhugh, M. R., 4, 45, 63
Mayo, R., 281, 286
McCall, R. C., 48, 86
McCannon, D., 271
McCarthy, R., 254, 259, 271
McCray, K., 46, 54
McCullough, E. C., 62
McCurdy, D. E., 65
McDougall, D. J., 29, 70
McDougall, R. S., 69
McEachern, P., 208
McGinley, P. H., 53
McLaughlin, J. E., V, 53, 78,155
McLaughlin, W. L., 231, 235, 238, 258, 272, 273, 280, 286, 310
McLennan, J. C., 1, 111
McMillan, R. E., 22, 65, 66
McNeill, K. G., 208
McNulty, P. J., 306
McQuilling, D. W., 152
Medveczky, L., 4, 189, 210, 267
Mees, C. E., 231, 232, 234
Mehendru, P. C., 72
Mehringer, W., 277, 278
Mejdahl, V., 4, 29, 48, 50, 66, 68, 70, 77, 94, 95, 254, 259, 271
Melnikowa, G. N., 280, 284
Mengali, J., 225
Menkes, S., 155, 156, 279, 284
Menold, R., 123
Mercer, T. T., 263
Meredith, W. J., 3
Merrifield, R. E., 288
Mertens, E., 47
Meshs, T. K., 29
Metalli, P., 271
Metzner, R., 225, 228, 229
Meyer, C. T., 245
Meyer, V., 1
Michael, N. H., 300
Midkiff, A. A., 33
Mihailovic, M., 64, 71
Mihailovic, M. V., 64
Mijnheer, B. J., 205
Milavc, Z., 64
Miller, A. A., 191
Milu, C., 79

Pendurkar, H. K., 47, 87
Perdue, P. T., 302
Perelygin, V. P., 188
Perkins, D. H., 249
Perry, K. E. G., 75, 76, 91, 92
Perttala, Y., 169
Pestaner, J. F., 280, 281, 284
Petel, R., 69
Peters, I. G., 288
Peterson, D. D., 191, 303
Petralia, S., 51
Petrera, S., 189
Petrescu, A., 116, 125
Petrock, K. F., 84
Pfaff, J. A., 43, 64
Pfeffer, R. L., 309
Pfister, H., 166
Pförtner, K., 280, 284
Philbrick, C. R., 69
Philips, P. R., 200
Phillips, L. F., 268
Phykitt, H. P., 38, 75, 85, 90, 93
Piesch, E., V, 4, 9, 11, 13, 20, 60, 62, 77, 78, 142, 153,
    157, 158, 159, 160, 161, 162, 163, 164, 166, 167, 168,
    169, 195, 205, 208, 211, 262, 266, 269
Pietrzak, R., 125
Piltingsrud, H. V., 78
Pimbley, W. T., 125
Pinkerton, A. P., 54
Pipins, P. A., 118
Platzer, H., 196
Potgorsak, E. B., 289
Podgorsky, E. B., 39, 40, 41
Pohlit, W., 31, 44, 94, 300, 301
Poirier, V., 64
Poli, V., 93
Popov, V. I., 179, 180, 181
Popovic, J., 279
Poroshina, M. S., 28
Portal, G., 4, 34, 37, 38, 70, 72, 142, 147, 165, 245, 269
Potsaid, M. S., 280, 284
Powell, C. F., 249
Pradel, J., 35
Premovic, P., 289
Prepejchal, W., 148, 169
Preston, H. E., 74, 75, 76, 77, 78, 91, 92, 93
Prêtre, S., 12, 181, 195, 196, 197, 201, 204, 205, 206
Prevo, P. R., 201, 268
Price, W. E., 154, 169, 175, 178, 180, 182, 184, 190, 191
    208, 213, 214, 278
Prigent, R., 34, 38, 73
Privaova, V. E., 188
Proctor, B. E., 154, 273, 281
Prokert, K., 279, 284, 286
Prokic, M., 63
Proskina, T. I., 28, 71
Przibram, K., 141
Puite, K. J., 5, 60, 63, 94, 117, 127, 300
Pulzer, R., 49, 60
Purcell, D., 28
Pye, L. D., 274

R

Rabin, H., 154, 169, 274
Racoveanu, N., 79
Radak, B., 273, 310
Radicheva, A. A., 63
Rae, J. B., 241
Rago, P. F., 53, 54, 204, 206, 209
Rainbolt, C., 41, 63, 301
Ralarosy, J., 206
Ralph, E. K., 300
Ramm, W. J., 309
Ramsey, N. W., 288
Randall, J. T., 29, 31, 32, 114, 115
Rasp, W., 119, 134
Rassat, A., 288
Rassow, J., 54, 73, 93, 259
Rauch, J. E., 279, 286
Rebanov, V., 205
Reddy, A. R., 41, 50, 52, 53
Redpath, A. T., 47
Rees-Evans, D. B., 36, 37
Regulla, D., 4, 49, 71, 73, 93, 120, 145, 148, 153, 155,
    303
Reimer, G. M., 182
Remy, G., 206
Renard, G. J. A., 183, 190
Revzin, L. S., 66, 67
Rhyner, C. R., 38, 42, 44, 58, 59
Riborov, S., 196
Ricci, A., 93
Rich, B. L., 76, 92
Richey, J. B., 86
Richman, D., 279
Richter, B., 287
Riegert, A. L., 162
Riehl, N., 29
Rieke, J. K., 69
Rimondi, O., 47, 52, 245, 266
Rintjema, D., 309
Ritz, V. H., 5, 58, 117, 118, 121, 135, 264, 277, 279, 302
Rizzo, F. X., 277, 278
Roberts, D. R., 148, 169
Roberts, J. H., 208, 209
Robertson, M. E. A., 36, 74
Robertson, M. K., 86
Robinson, E. M., 29, 125, 214
Röhrs, H., 191, 193
Rösch, E., 230
Roesch, W. C., 3, 9, 11
Roessler, C. E., 93
Rokop, D. J., 189
Romanenko, V. P., 188
Romano, G., 189, 192
Rosenblum, B., 129
Rosenstein, M., 280, 286
Rossiter, M. J., 36, 37
Roswit, B., 168
Rothermal, H. M., 71
Rotondi, E., 5, 118, 119, 134, 135
Routti, J. T., 156
Ruault, P. A., 262

Ruden, B. I., 55, 93
Rudin, S., 69
Rutland, D., 79
Ryabov, N. V., 270
Ryba, E., 37, 38, 73
Rytilae, A., 169

S

Sachdev, M. R., 51, 73
Sagan, L. A., 2
Sakai, E., 142, 157, 158
Sakamoto, H., 5, 28, 35, 70, 71
Sakanoue, M., 175, 206
Sakurai, J., 145
Salsbery, L., 56
Samardzich, B. G., 76, 92
Samuelsson, Ch., 135
Sane, S. G., 87
Sarup, S., 28
Sasaki, M. S., 18
Sass, R. E., 245, 262
Saunders, F., 28
Saunders, J. E., 28, 85
Savignac, N. F., 94
Scarpa, G., 5, 28, 34, 52, 58, 59
Schaar, J., 195
Schäfer, H., 191, 258, 268
Schaffert, J. C., 166
Schargelina, G. A., 132
Scharmann, A., 4, 30, 41, 71, 111, 112, 115, 117, 118,
    120, 121, 122, 125, 130, 142, 145, 146, 148, 211, 278,
    300
Schayes, R., 4, 28, 60, 61, 79
Scherer, A. E., 278
Schiager, K. J., 65, 94
Schimmerling, W., 245, 262
Schlesinger, T., 48, 73, 87
Schlueter, W. A., 273
Schmidt, G. C., 1, 28
Schmidt, Th., 166
Schmitt, W., 287
Schneider, H., 94, 287
Schneider, M. F., 94
Schön, M., 29, 30
Schopper, E., 177
Schott, J. U., 177
Schreck-Köllner, H., 192, 209
Schreurs, J. W., 189
Schulman, J. H., 1, 5, 28, 34, 35, 48, 56, 63, 64, 79, 141, 142,
    142, 147, 148, 152, 153, 154, 161, 162, 164, 166, 169,
    273, 274, 278, 287
Schultz, W. W., 196, 210
Schulz, R. J., 75, 93
Schunicht, B. F., 302
Schüren, H., 183, 205, 206
Schwarz, K. K., 29
Schweitzer, G. K., 155, 254
Scott, A. G., 93, 262
Seeger, W., 272
Seelentag, W., 94, 300

Segart, O., 41
Seguin, H., 269
Seibert, G., 117, 122, 130
Seibert, J., 121, 122, 130
Seidler, K. H., 262
Seitz, M. G., 195
Selim, Y. S., 216
Sellars, J., 77
Selman, J., 3
Seren, L., 278
Sergeeva, N. A., 205
Setlow, R. B., 23
Sever, J., 149
Sgarbi, C., 189, 191
Shalek, R. J., 22, 94
Shalin, V., 205
Shambon, A., 38, 50, 65, 68, 73, 94
Shapiro, P., 53, 63, 263, 299
Sharma, S. C., 33, 156, 178, 265, 308
Shaw, K. B., 45, 74
Sherwood, H. F., 195
Shimchnk, G. G., 306
Shipley, D. B., 76
Shiragai, A., 42, 44
Shirakusa, H., 84
Shlaer, S., 20, 255, 256, 259
Shoup, R., V, 197
Shropshire, S. M., 303
Shukovsky, L. J., 22
Shur, L. L., 188
Shurcliff, W., 1, 141, 153, 161, 162, 164, 166, 169
Shuttleworth, E., 43, 56
Sidorov, T. A., 71
Siegel, V., 119, 134
Sifaki, H., 245
Simek, J. E., 303
Simmons, D. J., 216
Silverman, J., 280, 286
Singer, J., 205
Singh, D., 245, 248
Slane, D. K., 90, 93
Slattery, M. K., 28
Smith, A. R., 206, 207
Smith, D. G., 263
Smith, R. H., 169
Smolka, E. E., 40
Snyder, W. S., V, 3
Sohrabi, M., 184, 188, 200, 203, 204
Sokolov, A. D., 132
Sokolov, V. A., 70
Somasundaram, S., 51, 73, 155
Somogyi, G., 4, 179, 183, 188, 189, 210, 213, 216
Song, K. S., 203
Sonnabend, E., 93
Sootha, G. D., 72
Soudain, G., 142, 147, 165, 169, 214, 215
Sowinski, M., 177
Spanne, P., 86
Speers, R. R., 225
Spetzler, H. A. W., 288
Spiegler, G., 254, 258
Spiers, F. W., 55, 93

T

U

SUBJECT INDEX

Calcium sulfate, 1, 16, 22, 28, 33, 35, 45, 46, 49, 50, 57,
66, 67, 68, 69, 74, 75, 76, 79, 83, 95, 118, 122, 123,
130, 132, 309
Calibration, IV, 11, 19, 166, 190, 228, 241, 259, 268,
271, 275, 302
Calibration factors, see calibration
Callier coefficient, 240
Capillary, 64, 78, 81, 84, 85
Cardboard, 233
Carrier lifetime, 225, 226
Cellophane, 279, 284, 286, 309, 310
Cellulose acetate, 182, 187, 188, 189, 196, 211, 212,
232, 250, 277, 278, 283, 310
Cellulose acetobutyrate, 183, 185, 187, 188, 189, 209,
210, 278, 310
Cellulose butyroacetate, see cellulose acetobutyrate
Cellulose nitrate, 178, 179, 182, 183, 187, 188, 189, 190,
191, 195, 197, 198, 200, 203, 205, 209, 210, 211, 213,
214, 215, 216, 250, 306, 308
Cellulose propionate, 187
Cellulose triacetate, 183, 187, 214
Ceramics, 58, 119, 121
Cerenkov radiation, 159, 160
Chain reactions, 284
Characteristic curve, 236, 242
Charged particle response, 156
Chelates, 28
Chemical development, 237
Chemoluminescence, 33, 47
Chest, 14
Chromosome abberrations, 18
Cinemoids, 284, 285, 310
Circuits, 84, 96
Civil defense, 164, 166
Clay, 121
Climatic conditions, 14; see also humidity, temperature,
developing countries, and tropical countries
Clinical dosimetry, IV, 21, 22, 82, 93, 94; see also
medical dosimetry
Cobalt borosilicate glass, 274
Cold emission, 111
Color centers, 143
Commercial TLD readers, 81, 82, 83
Compensation filters, see filters
Complex-formers, 237, 238
Composition, 145, 147
Compton electrons, 47
Computer interface, 89
Conduction band, 29, 30; see also band model
Conductivity, 30, 114, 124, 288
Conductivity curves, see conductivity
Conferences on Luminescence Dosimetry, V
Contamination, 15, 166, 262
Contrast, 195, 237
Contrast amplification, see contrast
Cooling-down peaks, 113
Cormophenol, 187
Cosmic radiation, 175, 191, 193
Cost, 14, 15, 73, 74, 84, 86, 92, 167, 230
Counting gas flow, see counting gases
Counting gases, 125, 126, 303
Credit card, 77

Critical detection angle, 181
Critical organ, 3, 9, 13
Cross section, 201, 202, 206, 207, 208, 209
Crosslinking, 272, 277, 282

D

Damage constant, 225
Damage threshold, 179; see also detection threshold
Dark-current, 48, 86
Dark field illumination, 195
Dating, 70, 94, 95, 175, 300
Deadtime losses, 126; see also counting losses
Decay time, 148, 166
Decoration, 176
Degradation, 272, 277
Delta electrons, 33, 54, 135, 178, 270
Delta rays, see delta electrons
Densitometers, 241, 259, 282
Densitometry, 196, 233, 240
Dental enamel, 94
Depolarization, 46, 289
Depth-dose distribution, 6, 7, 9, 10, 13, 17, 55, 93, 160,
163, 168, 169, 208, 213, 277
Desorption, 117, 120
Detectable dose, 13, 180, 192
Detection efficiency, 201, 208
Detection threshold, see detectable dose
Developing countries, 18, 23, 94, 156
Development, photographic, 233, 235, 238, 239
Diagnostic x-rays, 2
Diamond, 288
Dicarboxylic acid, 289
Dicentrics, 18
Dielectric constant, 197
Diffuse density, 240
Diffusion length, 226
Diode base width, 220; see also silicon diodes
Dipole layer, 113
Directional response, 11, 79, 134, 153, 203, 211, 212,
230, 254, 256, 268, 269, 304, 306
Discoloration, 51, 154, 277, 287
Displacement energy, 225
Dose equivalent, 78
Dose range, 6, 13, 153, 202
Dose-rate dependence, 22, 46, 153, 235, 258
Dosimeter design, 72, 129, 164, 165
Dosimeter location, 14
Dynamic range, 167, 200, 230; see also dose range

E

Effective atomic number, 149, 278, 279
Effective energy, 20, 21, 79, 80, 162, 163, 167
Effective photon energy, see effective energy
Efficiency, 49, 181, 182, 200
Electret polarization, 46, 289
Electrochemical plating, 203
Electron affinity, 113
Electron dosimetry, see electrons

Electron emission from filters, see filters
Electron equilibrium, 19, 254
Electron microscope, 206, 234
Electron multiplier, 129
Electrons, 23, 51, 54, 55, 160, 270; see also beta emitters, beta energy
Electron spin resonance, 18, 288
Electron traps, see traps
Electro-optical devices, 195
Emission edge, 116
Emission spectrum, 49, 50, 65, 70, 71, 113
Emulsion thickness, 233
Encapsulation, 159
Energy degradation, 213
Energy dependence, 9, 20, 22, 46, 47, 53, 54, 55, 62, 65, 69, 70, 71, 78, 79, 95, 132, 133, 146, 149, 161, 162, 167, 168, 201, 202, 204, 211, 229, 230, 238, 252, 253, 254, 255, 256, 257, 258, 264, 267, 268, 269, 270, 271, 273, 274, 282, 302, 303, 304
Energy deposition, photographic films, 235
Energy distribution, 112, 113
Environmental dosimetry, 18, 66, 68, 95, 169
Environmental monitoring, see environmental dosimetry
Environmental radiation, see environmental dosimetry
EPR, see electron spin resonance
Erytherma dose, 1
Esophagus, 168
Etchants, 184, 187, 188, 189, 191
Etch pit, 175, 178, 179, 181, 191
Etch pit diameter, 192, 194
Etch products, 307
Etching, 73, 160, 179, 186, 187, 188, 190, 196; see also etchants
Etching conditions, see etching
Etching kinetics, 188
Etching speed, see etching
Ethyl cellulose, 250
Ethylenediamine tetraacetic acid, 238
Excitons, 179, 234
Exoelectron emission, 1, 2, 3, 16, 23, 24, 33, 37, 58, 70, 80, 111, 112, 303
Exoelectron photography, 129, 130
Extruded ribbons, 38, 64, 73, 76, 80, 81, 82, 87, 302
Eye lenses, 6

F

Fading, 15, 20, 21, 22, 30, 32, 33, 34, 37, 38, 42, 43, 44, 54, 57, 59, 61, 62, 64, 65, 66, 67, 68, 69, 71, 72, 85, 86, 114, 119, 124, 130, 143, 145, 150, 151, 154, 157, 167, 182, 183, 184, 211, 228, 230, 231, 235, 242, 243, 244, 245, 246, 247, 248, 249, 251, 254, 259, 265, 273, 274, 275, 279, 280, 282, 284, 286, 287, 288, 300, 301, 302, 303, 309
Fallout, 17
Film badge, see film dosimeter
Film base, 232
Film dosimeter, 1, 13, 15, 168, 231, 240, 259, 260, 261, 262, 271, 300
Film packaging, see film sealing
Film sealing, 245, 246, 247, 250, 251, 309

Filter-analytical methods, 162, 254, 258, 259
Filters, 15, 20, 21, 62, 76, 79, 80, 141, 149, 158, 161, 164, 165, 254, 255, 256, 257, 258, 259, 260, 261, 262
Filtration, see filters
Fine grain developers, 237
Finger-ring dosimeter, 165, 205, 250
Fingertip dosimetry, 14, 55, 92
First collision dose, 211, 212
Fission-foil thickness, 201, 202
Fission fragments, 181, 182, 184, 185, 186, 187, 188, 191, 192, 195, 197, 198, 199, 200, 201, 208, 209
Fission neutrons, see neutrons
Fission thresholds, 201
Fixing solutions, 233, 239
Flame polishing, 193
Fluorescence compensation, 256
Fluorescent standards, 166
Fluoride glasses, 146
Fluorimeters, 166
Fluorite, 28, 33, 49, 60, 61, 62, 64, 65, 79, 80, 117
Fluorods, 141, 148, 157, 165, 166, 168
Fog, photographic, 231, 239, 241, 242, 243, 244, 245, 246, 259
Forward current, 225, 227
Fraunhofer diffraction, 196
Free-radical photography, 176, 288
Frenkel defects, 225, 234
Frequency factor, 30, 31, 32, 44, 114, 115
Fungi, 242, 243

G

Gadolinium, 158
Gamma radiation, 13, 19, 21, 38, 39, 46, 51, 52, 54, 55, 57, 59, 62, 63, 64, 65, 68, 72, 156, 161, 235, 242
Gas discharge, 123
Gelatin, 231, 233, 247, 249, 253, 263, 265, 280, 281, 286, 287
Genetic damage, 12
Glass base composition, 141, 143, 144, 148
Glass chambers, 176
Glove boxes, 205
Glow curve, 27, 30, 31, 38, 39, 40, 50, 52, 57, 61, 64, 65, 67, 68, 69, 70, 80, 85
Glow peak, 27, 58, 64
Glucose, 69
G-M counters, 111, 125, 126
Gold thiocyanate, 232
Gonads, 6, 8, 9, 13, 14, 93, 164
Grain diameter in emulsions, see grain size
Grain size, 47, 51, 56, 62, 63, 72, 119, 231, 233, 240, 241
Graphite powder, 112, 119, 121, 132
Gut mucosa, 9, 164
G-value, 285

H

Halflife, 32
Halide acceptors, 234

Plastics, see polymer
Plutonium, 204, 205, 209
Pocket dosimeter, 168; see also ionization chambers
Point counters, 125
Poisson distribution, 270
Polyamides, 310
Polycarbonate, 175, 176, 178, 181, 182, 183, 184, 185,
    186, 187, 189, 190, 191, 194, 195, 197, 200, 203, 206,
    208, 211, 212, 213, 250, 306, 308; see also Kimfol,
    Lexan, Makrofol
Polyethylene, 27, 53, 63, 133, 159, 187, 213, 246, 251,
    264, 277, 278, 288, 304, 306, 307, 310
Polyethylene terephthalate, 250, 277, 278
Polyimide, 187
Polymer, 16, 19, 27, 121, 175, 177, 178, 190, 272, 277,
    278, 279, 280, 281
Polymer dye systems, 278, 279, 280, 281
Polymethyl methacrylate, 187, 277, 278, 310; see also
    Lucite, Perspex
Polyoxymethylene, 187
Polyphenoxide, 187
Polyphenyls, 256; see also quaterphenyl, sexiphenyl
Polypropylene, 27, 250
Polystyrene, 187, 250, 279, 282, 284, 286, 289, 310
Polytetrafluoroethylene, 310
Polytrifluorochloroethylene, 250; see also Teflon
Polyvinylacetochloride, 188
Polyvinyl alcohol, 279, 281, 282, 284, 310
Polyvinyl chloride, 188, 250, 277, 279, 309, 310
Polyvinyl fluoride, 279
Polyvinyliden chloride, 250, 278
Polyvinyl toluene, 188
Population exposure, 299
Porous glass, 286
Potassium chloride, 72, 118, 142
Potassium hydroxide, 184
Pouch paper, 250, 251, 252
Precipitation, 231
Precision, 47, 230; see also accuracy, standard deviation
Predose voltage, 142, 148, 155, 159, 168, 226, 228, 230;
    see also silicon diodes
Preheating, 43, 75, 120, 121, 122, 183
Pre-irradiation, 37, 38, 39, 65, 70, 71, 122
Presparking, 199
Processing of films, 237
Proportional counters, 125, 126, 302
Proton equilibrium, 19
Protons, 54, 63, 65, 156, 160, 207, 210, 264
Proton tracks, 182
Pulsed excitation, 148
Pushing potential, 127

Q

Quantimet, 195
Quantum efficiency, 116
Quartz, 33, 70, 86, 95, 121, 177, 181, 182, 187
Quaterphenyl, 287
Quenching, 37, 45, 64, 287
Quinine sulfate, 121

R

Radiation chemistry, 22
Radiation sterilization, 22
Radiators, 133, 209, 210, 263, 268
Radiation workers, 299, 300
Radiochemical damage mechanism, 178
Radioecology, 169
Radiography, 197, 209
Radiologists, 169
Radioluminescence, 49, 70
Radiolytic scission, 178
Radiophotoluminescence, 1, 2, 3, 22, 33, 49, 160, 306
Radiotherapy, 22
Radium, 215, 216
Radon, 94, 183, 213, 214, 215, 269, 270
Radon daughters, 270
Randall-Wilkins formula, 31, 32, 115
Rayleigh scattering, 288
Reactivation, 116
Reactors, 205, 206, 208
Reactor shielding, 94
Readout instruments, 14, 80, 86, 87, 88, 89, 90, 160
Readout temperature, 153
Reciprocity failure, 23, 235, 236, 270
Recoil nuclei, 192, 209, 210, 211, 212, 213
Recoil protons, 53, 58, 134, 176, 247, 248, 263, 265,
    302, 304, 306, 308
Recombination, 30
Recombination centers, 33, 225
Record keeping, 15
Redox potential of developers, 237
Reflectance spectrometry, 289
Reflection densitometry, 240
Registration efficiency, 159
Rem counter, 159
Replenisher solution, 239
Replica techniques, 175
Reproducibility, 62, 68, 79, 127, 132, 190, 191, 201,
    300; see also accuracy, standard deviation
Research groups, 4
Research needs, 17
Resistance heating, 27
Resolution, 23, 24, 129
Retrapping, 31, 44, 59, 61
RPL build-up, 143, 144, 148, 150, 151, 155, 165
RPL spectra, 152, 156, 157
Ruby, 69

S

Saturation, 65
Scanning electron microscopy, 175, 177, 195
Scattered radiation, 13, 19, 21
Schwarzschild effect, see reciprocity failure
Scintillators, 256, 258, 270, 287, 302
Secondary electrons, 29, 135, 255; see also delta electrons
Secondary standards, 156; see also calibration
Self-absorption, 52
Self-activation, 116
Self-dosing, 65, 78

Triboluminescence, 47
Tribo-signals, 69, 135
Triplet exciton annihilation, 288, 289
Tritium, 55, 135, 259, 261, 271
Tropical climates, 18, 244, 250, 252, 309; see also
developing countries, fading, stability

U

Ultrahigh vacuum, 129
Ultrasonic cleaning, 74, 166
Ultrasound, 84
Ultraviolet, 23, 28, 30, 49, 50, 51, 56, 58, 59, 61, 62, 66,
67, 86, 141, 153, 183, 188, 300, 307
Uranium, 192, 202, 204, 205, 206, 208, 209
Uranium miners, 94, 175, 213, 214
Uranium ore deposits, 214
Uranyl acetate, 203
Urine, 208

V

Vacuum deposition, 203

Valence band, see band model
Vibrating powder dispensers, 72
Vibration, 179
Visual track counting, 268
Volume tracks, 189
Volume traps, 113

W

Waist, 14
Washing, 155
Water, 208, 209
Weight loss, 289
Whole-body dose, 3
Work function, 122

Z

Zinc oxide, 120, 123, 300, 301
Zinc sulfide, 72, 288, 300